TRACE ATMOSPHERIC CONSTITUENTS

Volume

12

in the Wiley Series in

Advances in Environmental Science and Technology

JEROME O. NRIAGU, Series Editor

the most important recent developments in the particular field; to critically evaluate new concepts, methods, and data; and to focus attention on important unresolved or controversial questions and on probable future trends. Monographs embodying the results of unusually extensive and well-rounded investigations will also be published in the series. The net result of the new editorial policy should be more integrative and comprehensive volumes on key environmental issues and pollutants. Indeed, the development of realistic standards of environmental quality for many pollutants often entails such a holistic treatment.

JEROME O. NRIAGU, Series Editor

INTRODUCTION TO THE SERIES

The deterioration of environmental quality, which began when humankind first congregated into villages, has existed as a serious problem since the industrial revolution. In the second half of the twentieth century, under the ever-increasing impacts of exponentially growing population and of industrializing society, environmental contamination of the air, water, soil, and food has become a threat to the continued existence of many plant and animal communities of various ecosystems and may ultimately threaten the very survival of the human race. Understandably, many scientific, industrial, and governmental communities have recently committed large resources of money and humanpower to the problems of environmental pollution and pollution abatement by effective control measures.

Advances in Environmental Sciences and Technology deals with creative reviews and critical assessments of all studies pertaining to the quality of the environment and to the technology of its conservation. The volumes published in the series are expected to serve several objectives: (1) stimulate interdisciplinary cooperation and understanding among the environmental scientists; (2) provide scientists with a periodic overview of environmental developments that are of general concern or of relevance to their own work or interests; (3) provide the graduate student with a critical assessment of past accomplishment that may help stimulate interest in career opportunities in this vital area; and (4) provide the research manager and the legislative or administrative official with an assured awareness of newly developing research work on the critical pollutants and with the background information important to their responsibility.

As the skills and techniques of many scientific disciplines are brought to bear on the fundamental and applied aspects of the environmental issues, there is a heightened need to draw together the numerous threads and to present a coherent picture of the various research endeavors. This need and the recent tremendous growth in the field of environmental studies have clearly made some editorial adjustments necessary. Apart from the changes in style and format, each future volume in the series will focus on one particular theme or timely topic, starting with Volume 12. The author(s) of each pertinent section will be expected to critically review the literature and

WHITE, WARREN H., Center for Air Pollution Impact and Trend Analysis, Washington University, St. Louis, Missouri

WINCHESTER, JOHN W., Department of Oceanography, Florida State University, Tallahassee

ZAK, BERNARD D., Sandia National Laboratories, Albuquerque, New Mexico

CONTRIBUTORS

ANDERSON, LARRY G., Environmental Science Department, General Motors Research Laboratories, Warren, Michigan

BOYCE, SCOTT D., Environmental Engineering Science, W. M. Keck Laboratories, California Institute of Technology, Pasadena

CARMICHAEL, GREGORY R., Chemical and Materials Engineering Program, University of Iowa, Iowa City

CHANG, SHIH-GER, Lawrence Berkeley Laboratory, University of California, Berkeley

EATOUGH, DELBERT J., Thermochemical Institute and Department of Chemistry, Brigham Young University, Provo, Utah

HANSEN, LEE D., Thermochemical Institute and Department of Chemistry, Brigham Young University, Provo, Utah

HOFFMANN, MICHAEL R., Environmental Engineering Science, W. M. Keck Laboratories, California Institute of Technology, Pasadena

HUIE, ROBERT E., Chemical Kinetics Division, National Bureau of Standards, Washington, DC

LEVINE, STUART Z., Environmental Chemistry Division, Brookhaven National Laboratory, Upton, New York

NOVAKOV, TIHOMIR, Lawrence Berkeley Laboratory, University of California, Berkeley

PETERS, LEONARD K., Department of Chemical Engineering, University of Kentucky, Lexington

PETERSON, NORMAN C., Department of Chemistry, Polytechnic Institute of New York, Brooklyn

SCHWARTZ, STEPHEN E., Environmental Chemistry Division, Brookhaven National Laboratory, Upton, New York

SHETTER, RICHARD E., Space Physics Research Laboratory, University of Michigan, Ann Arbor

STEDMAN, DONALD H., Departments of Chemistry and Atmospheric and Oceanic Science, University of Michigan, Ann Arbor

WESELY, MARVIN L., Radiological and Environmental Research Division, Argonne National Laboratory, Argonne, Illinois

Library of Congress Cataloging in Publication Data:

Main entry under title.

Trace atmospheric constituents.

(Advances in environmental science and technology, ISSN 0065-2563 ; v. 12)
"A Wiley-Interscience publication."
Includes index.
1. Atmospheric chemistry. 2. Troposphere. 3. Acid rain. 4. Atmospheric chemistry—Research—United States. I. Schwartz, Stephen E. (Stephen Eugene) II. Series.

TD180.A38 vol. 12 [QC879.6] 628s [551.5′11] 82-16095
ISBN 0-471-87640-2

Printed in the United States of America

10 9 8 7 6 5 4 3 2 1

TRACE ATMOSPHERIC CONSTITUENTS

Properties, Transformations, and Fates

Edited by

Stephen E. Schwartz
Environmental Chemistry Division
Brookhaven National Laboratory
Upton, New York

A WILEY-INTERSCIENCE PUBLICATION

JOHN WILEY & SONS

New York • Chichester • Brisbane • Toronto • Singapore

PREFACE

The past several years have seen the development in North America, Europe, and elsewhere of an intense, heightened concern about the environmental effects of trace atmospheric constituents associated with fossil-fuel combustion for energy production. This class of substances includes not only primary pollutants such as soot and the sulfur and nitrogen oxides that are directly emitted but also the secondary pollutants such as sulfuric and nitric acids and their salts as well as ozone and other oxidants that are derived from the primary pollutants by reaction in the atmosphere. These materials may reside in the atmosphere for periods of days and be transported by the winds for distances of hundreds to thousands of kilometers. Although the presence of these materials in the atmosphere is not always directly manifest in the absence of appropriate measurement instrumentation, we have all noticed on occasion the pronounced reduction in visibility associated with such pollutants, and noticeable levels of oxidants and eye irritants have become commonplace in many locations.

Eventually these pollutant materials are removed from the atmosphere, either by incorporation into precipitation or by transport to and sorption by material surfaces. Although that may be the end of the story for the atmosphere, it is only the beginning of another story as these substances exert their effects on the biosphere. The effects of oxidants on vegetation and human health have been recognized for decades. Similarly, there is now an intense public concern in North America over the so-called acid rain phenomenon (or better "acid deposition," which phrase encompasses dry deposition as well), that is thought to be adversely affecting sensitive watersheds (and the species that inhabit them), materials (most notably steel and limestone), and perhaps vegetation (forest and agricultural productivity).

An increased public awareness of the effects of transport and deposition of air pollutants has resulted in pressures to institute control strategies to minimize emissions or otherwise to mitigate their effects. However, these strategies cannot be implemented without cost—for example, the cost of implementation of sulfur dioxide controls alone, including retrofit scrubbers, is estimated in the tens of billions of dollars. The question thus arises as to optimum control strategies: what substances, where, and how much? Whereas questions of this sort are ultimately political, a key component to developing an answer must be supplied by environmental scientists, namely, a quantitative description of the consequences that may be expected from a given hypothetical alteration of emissions. Information of this sort, as well as considerations of practicality (availability of alternative fuels, time required to bring new equipment on line, etc.) and economics (costs, benefits, and elasticity) is then used to provide an assessment of various potential strategies as input to the political process.

Narrowing our focus to atmospheric science, we find that the questions to be addressed are somewhat more limited. What changes in concentrations and deposition may be expected for a given hypothetical change in emissions? The present volume consists of timely and topical reviews of research that is directed, ultimately, to providing answers to these questions. The first three chapters examine the aqueous-phase chemistry of the nitrogen oxides (Chapter 1) and of sulfur dioxide (Chapters 2 and 3) necessary to describe incorporation of these compounds into precipitation and oxidation to the more soluble and more strongly acidic nitric and sulfuric acids. These studies bear as well on the mechanisms of toxicity of these compounds.

The next several chapters turn increasingly to field studies. Chapter 4 deals with the ubiquitous soot that consitutes the major fraction of atmospheric aerosol carbon. This chapter serves also to emphasize the interrelatedness of various pollutants by noting the catalytic role of soot for the oxidation of SO_2 in aqueous solution. Chapter 5 serves to remind us that inorganic and organic sulfur(IV) compounds are present in ambient and industrial aerosols and that these compounds may have significant health implications. Chapter 6 takes a much more global view and demonstrates, by comparison of aerosol sulfur loadings in the Northern and Southern Hemispheres, the pervasive and long-range influence of these materials. This chapter also raises important new hypotheses about the mechanism of formation of aerosol sulfate and the implications of aerosol sulfate on human health.

Chapter 7 describes one of the most powerful techniques for studying atmospheric transformations *in situ,* namely, by following a marked air parcel, and describes a number of different studies that have been carried out by various realizations of this technique. Studies described in this chapter include oxidation of SO_2 and the formation of ozone by the nitrogen oxide–hydrocarbon complex of reactions. Chapter 8 deals with another key component of the cycle of pollutants in the atmosphere, namely, the removal process. This chapter describes the eddy-correlation method for direct measurement of dry deposition rates and presents a wide variety of results that have been obtained for ozone.

The next two chapters direct our focus again to the chemistry of nitrogen compounds in the atmosphere. Chapter 9 examines transformations on the urban scale, whereas Chapter 10 places the transformations of these higher oxides in the much broader context of the global nitrogen cycle.

Ultimately, to describe the transport, transformation, and deposition of pollutant materials, we turn to computer-based models of this process, two aspects of which are treated in the final two chapters of this volume. Chapter 11 examines the complex of photochemical and thermal gas-phase reactions occurring in the nitrogen oxide–sulfur oxide–ozone–hydrocarbon system and examines means of decreasing the number of species that must be actively modeled while still preserving the chemical integrity of the system. Such a decrease is necessary, given the size limitations even of modern large-scale computers, if this complex of reactions is to be incorporated into a grid-type model that will describe both the spatial and temporal evolution of this chemistry. Finally Chapter 12 describes developments in constructing models of this sort and how the various elements of the entire process—emissions, transport, chemical reaction, and deposition—may be combined to yield a quantitative description of this process.

At a recent meeting that dealt with formulating a program of research and assessment for the national acid rain problem, one of my colleagues in the assessment business leveled the charge that those of us in the research business seemed less interested in solving the acid rain problem than in carrying out our research efforts, which he characterized as "hobbies." Although initially I felt defensive at my colleague's accusation, in retrospect I feel less so. Taking a dictionary definition of hobby as "an engrossing topic plan, etc., to which one constantly reverts; also an occupation or interest to which one gives his spare time," I would confess that such a definition rather closely fits the research process. Narrowing our perspective to the contributions of this volume, we see that without exception these represent the authors' involvement in engrossing topics or plans. I can attest as well that the preparation of these articles has resulted from each of the authors giving generously of his spare time. So if we sometimes appear to slight the application of our research to questions of social concern, I would suggest that this may reflect not so much a lack of concern as an intensity of involvement arising from a conviction that correct answers to these questions will come only from rigorous scientific research.

STEPHEN E. SCHWARTZ

Upton, New York
September 1982

CONTENTS

TRACE ATMOSPHERIC CONSTITUENTS

1

KINETICS OF REACTIVE DISSOLUTION OF NITROGEN OXIDES INTO AQUEOUS SOLUTION

Stephen E. Schwartz

Environmental Chemistry Division
Department of Energy and Environment
Brookhaven National Laboratory
Upton, New York 11973

Warren H. White

Center for Air Pollution Impact and Trend Analysis
Washington University
St. Louis, Missouri 63130

1. INTRODUCTION

It would be difficult to cite a process so common to the industrial chemical scene and yet posing so severe a challenge to our comprehension of its fundamental mechanisms as that of the absorption of nigrogen oxides in water to produce nitric acid.

Carberry (1959)

The reactive dissolution of nitrogen oxides into aqueous solution, i.e., physical dissolution of NO_2 and/or NO followed by aqueous-phase reaction to form nitrous and/or nitric acids, has been postulated to be potentially important as a mechanism of removal of these oxides from the atmosphere, as a means of acidification of precipitation, and as an influence on the composition of atmospheric aerosols (Durham et al., 1981; Lee and Schwartz, 1981a; Liljestrand and Morgan, 1981; Liss, 1976; Orel and Seinfeld, 1977; Söderlund and Svennson, 1976) (see also Chapter 9 in this volume by Anderson and Chapter 10 by Stedman and Shetter). Aqueous-phase reactions of nitrogen oxides are also of interest in environmental science in connection with the mechanism of toxicity of these gases to animals (Goldstein et al., 1977; Mustafa et al., 1979; Pryor and Lightsey, 1981) and to vegetation (Hill, 1971; National Research Council, 1977). These reactions are of interest as well in the search for possible water-based scrubbers for removal of nitrogen oxides from flue gases (Counce, 1980; Faucett et al., 1978) and in development and interpretation of wet analytical methods for analyses of these gases in flue gases (Driscoll, 1974) or ambient air (National Research Council, 1977).

To quantitatively describe aqueous-phase reactions of the nitrogen oxides, it is necessary to develop an understanding of the fundamental equilibrium and kinetic properties of this system. In a previous article (Schwartz and White, 1981) we reviewed the chemical equilibria in this system and presented a fairly consistent picture of these equilibria, including the Henry's law (solubility) equilibria of the reactive oxides NO_2, N_2O_3, and N_2O_4. (The Henry's law constant of NO, which is reversibly and nonreactively soluble in aqueous solution in the absence of other reactive substances, is well established.) It was shown that the nitrogen oxides exhibit only limited solubility according to Henry's law but substantially greater reactive solubility, i.e., by undergoing oxidation or reduction to the much more soluble nitrite or nitrate anions. At low partial pressures of these gases, characteristic of atmospheric conditions, such reactive dissolution will be controlled largely if not entirely by the chemical kinetic rate. Here we examine the kinetics of these reactions.

A parallel route for incorporation of the nitrogen oxides into aqueous solution is gas-phase reaction to form the oxyacid (nitrous or nitric acid) followed by uptake of the gaseous acid by liquid water. Gas-phase formation of HNO_2 and HNO_3 takes place in the atmosphere largely by free-radical reactions the kinetics of which are reasonably well described (Baulch et al., 1980; Crutzen, 1979); direct formation of HNO_2 or HNO_3 by gas-phase molecular reaction, if it occurs at all, is quite slow (Carberry, 1959; England and Corcoran, 1974; Kaiser and Wu, 1977a,b). Aqueous-phase reactions of HNO_2 and HNO_3 (i.e., establishing the respective ionization equilibria) would be expected to be quite fast (Eigen et al., 1964), and consequently the rate of dissolution of gaseous HNO_2 and HNO_3 will be limited entirely by mass transport. Such mass-transport rates are rather well described theoretically for conditions pertinent to either cloud droplets (Freiberg and Schwartz, 1981) or falling raindrops (Hales, 1972), and the rates of in-cloud and below-cloud scavenging of nitric acid vapor have recently been evaluated on this basis (Levine and Schwartz, 1982). Consequently, we do not address these processes further here.

To facilitate the discussion of studies of the aqueous-solution kinetics of the nitrogen oxides, we first present a brief, general review in Section 2 of the kinetics of gas–liquid reactions. In general, the reactive dissolution of a gas into a liquid is a complex process that involves mass transport in both phases and solution-phase reaction of the dissolved gas and whose rate is governed by the combined rates of these several processes. Consequently, to present an interpretation of various laboratory studies, it is necessary to consider these mass-transport processes in some detail. In Section 3 we introduce a nomenclature that facilitates the discussion of the nitrogen oxide reactions in particular by taking into account the rapid equilibria between the so-called simple oxides (NO_2 and NO) and the compound oxides (N_2O_4 and N_2O_3).

The review of the kinetic studies is organized about techniques that have been employed in these studies. These are as follows:

Section 4. Indirect inference from kinetic studies of the decomposition of nitrous acid, kinetics of isotope exchange, diazotization reaction, and so on.

Section 5. Direct measurement of aqueous-phase kinetics.

Section 6. Direct measurement of the rate of uptake of nitrogen oxides by liquid water.

The results of these experiments are expressed as products of various powers of the Henry's law coefficients and kinetic rate constants as appropriate to the various types of experiment. In Section 7 these results are examined in relation to each other to obtain recommended values for the elementary thermochemical and kinetic constants that describe these reactions. Finally, studies of the interaction of nitrogen oxides with dissolved materials are reviewed in Section 8.

As becomes evident below, there is a large body of work pertinent to the reactive dissolution of the nitrogen oxides into aqueous solution. The great majority of this work has been motivated by the role of these reactions in the industrially important manufacture of nitric acid; for a historical account of this technology, the interested reader is referred to the monograph by Chilton (1968). The present chapter relies heavily on this literature. The question thus naturally arises as to whether studies conducted under conditions widely differing from those of the intended application (specifically, partial pressures of the nitrogen oxides orders of magnitude greater than those characterizing the ambient atmosphere) are of utility in the present review. In reply, we would assert that, provided the elementary reactions that comprise the mechanisms are correctly elucidated, the rates inferred from these mechanisms would be applicable under any conditions of interest, specifically including atmospheric conditions of low partial pressures. Nevertheless, we must remember that under laboratory conditions of high partial pressures of the nitrogen oxides, second- and higher-order reactions

Table 1. Reactions of Nitrogen Oxides and Oxyacids[a]

[1]	$2NO_2 \overset{W}{=} 2H^+ + NO_2^- + NO_3^-$
[2]	$NO + NO_2 \overset{W}{=} 2H^+ \ 2NO_2^-$
[3]	$3NO_2 \overset{W}{=} 2H^+ + 2NO_3^- + NO$
[4]	$N_2O_4 \overset{W}{=} 2H^+ + NO_2^- + NO_3^-$
[5]	$N_2O_3 \overset{W}{=} 2H^+ + 2NO_2^-$
[6]	$N_2O_5 \overset{W}{=} 2H^+ + 2NO_3^-$
[7]	$2NO_2 = N_2O_4$
[8]	$NO + NO_2 = N_2O_3$
[9]	$3NO_2 = N_2O_5 + NO$
[10]	$3H^+ + 3NO_2^- \overset{W}{=} H^+ + NO_3^- + 2NO$
[11]	$NO_2 + NO_2^- = NO_3^- + NO$
[I1]	$HNO_3(aq) = H^+ + NO_3^-$
[I2]	$HNO_2(aq) = H^+ + NO_2^-$

[a] Reaction numbers correspond to those introduced by Schwartz and White (1981). The prefix G, A, or M to the reaction number is used to denote a gas-, aqueous-, or mixed- (gaseous oxide, aqueous acid) phase reaction. Aqueous acids are expressed as ionized (dissociated) forms. The addition of a prime to a reaction number, e.g., [A11′], denotes nitrous acid expressed as the undissociated form. Equilibria [I1] and [I2] represent the acid dissociation equilibria for nitric and nitrous acid. The symbols $\overset{W}{=}$ and $\overset{W}{\rightarrow}$ refer to equilibria or reactions involving water as a reagent in which H_2O is not explicitly denoted as a reagent. Water activity is taken as unity unless otherwise specified.

Table 2. Equilibria Involving Gaseous and Aqueous Nitrogen Oxides and Oxyacids[a]

Equilibrium Number	Equilibrium	Equilibrium Constant	Value	Units	ΔH° (kcal mol^{-1})
[G7]	$2NO_2(g) = N_2O_4(g)$	$p_{N_2O_4}/p^2_{NO_2}$	6.86	atm^{-1}	−13.7
[G8]	$NO(g) + NO_2(g) = N_2O_3(g)$	$p_{N_2O_3}/p_{NO}p_{NO_2}$	0.535	atm^{-1}	−9.5
[H_{NO}]	$NO(g) = NO(aq)$	$[NO(aq)]/p_{NO}$	1.93(−3)[b]	M atm^{-1}	−2.94
[I2]	$HNO_2(aq) = H^+ + NO_2^-$	$[H^+][NO_2^-]/[HNO_2]$	5.1 (−4)	M	2.5[c]
[M1′]	$2NO_2(g) \overset{w}{=} HNO_2(aq) + H^+ + NO_3^-$	$[HNO_2][H^+][NO_3^-]/p^2_{NO_2}$	4.78(5)	M^3 atm^{-2}	−25.6
[M2′]	$NO(g) + NO_2(g) \overset{w}{=} 2HNO_2(aq)$	$[HNO_2]^2/p_{NO}p_{NO_2}$	1.26(2)	M^2 atm^{-2}	−18.2
[M11′]	$NO_2(g) + HNO_2(aq) = H^+ + NO_3^- + NO(g)$	$[H^+][NO_3^-]p_{NO}/[HNO_2]p_{NO_2}$	3.79(3)	M	−7.4

[a] From Schwartz and White (1981). Equilibrium constants and reaction enthalpies are given for T = 25°C = 298 K; equilibrium constants for other temperatures may be evaluated by means of ΔH°.

[b] The notation 1.93(−3) represents 1.93×10^{-3}.

[c] The values for ΔH° [I2] and ΔH° [V2] given by Schwartz and White (1981, Table VII) should be corrected to read 2.5 and −7.0 kcal mol^{-1}, respectively.

will be favored and might mask any first-order reaction that was occurring in parallel, whereas under atmospheric conditions such a first-order mechanism might be dominant. Consequently, we must keep open the possibility of such first-order mechanisms, either with water itself or with other dissolved materials. This possibility can be addressed only by direct experimentation at low partial pressures. As is noted in Section 8, there exist numerous qualitative indications of the existence of such additional pathways for the reactive dissolution of the nitrogen oxides. In this context the present review will serve as well as a guide to the formulation of laboratory experiments to elucidate these pathways.

As noted above, this chapter represents an extension of, and relies heavily on, the authors' previous review (Schwartz and White, 1981) of the chemical and phase (Henry's law) equilibria in this system. In particular, we retain the system of labeling of chemical reactions adopted there; this nomenclature is summarized in Table 1. For convenience, in Table 2 we also present a summary of pertinent equilibrium constant expressions and numerical values for equilibrium constants and for reaction enthalpies. It should be pointed out that we employ as standard states 1 atm (for gases) and 1 M (for solutes); for convenience, units are retained with numerical values of the equilibrium constants (Daniels and Alberty, 1975). The Henry's law equilibrium between a gaseous substance X(g) and the dissolved substance X(aq),

$$\mathrm{X(g)} = \mathrm{X(aq)} \qquad [\mathrm{H_X}]$$

represents a special case of the standard state convention adopted. The equilibrium constant expression for this reaction is

$$H_X = \frac{[\mathrm{X(aq)}]}{p_X}, \tag{1}$$

and it is seen that the Henry's law coefficient (H_X) has units M atm^{-1}.

2. KINETICS OF GAS–LIQUID REACTIONS

In this section the formalism used to describe the rates of gas–liquid reactions is briefly outlined. Some of this discussion is merely definitional, i.e., expressing the rates and reagent concentrations in terms of gas- or aqueous-phase species, and the interconversion between the corresponding expressions. However, much of the discussion addresses the relationships between reaction rates and rate constants as a function of the degree of mixing maintained between the reagent concentrations in the two phases. This discussion outlines criteria for ascertaining the mixing condition that describes a given experimental situation and presents expressions for evaluating rates of reactive uptake of gases for these various situations. These expressions are utilized in the analysis presented in the subsequent sections.

2.1. Aqueous-Phase Kinetics

We first consider a general aqueous-phase reaction of stoichiometry

$$m_1X_1(aq) + m_2X_2(aq) + \cdots \rightarrow n_1Y_1(aq) + n_2Y_2(aq) + \cdots . \quad [A12]$$

The rate of this reaction is defined conventionally and unambiguously (e.g., Johnston, 1966) as

$$\begin{aligned} R_{A12} &= -\frac{1}{m_1}\frac{d_{12}[X_1(aq)]}{dt} = -\frac{1}{m_2}\frac{d_{12}[X_2(aq)]}{dt} = \cdots \\ &= \frac{1}{n_1}\frac{d_{12}[Y_1(aq)]}{dt} = \frac{1}{n_2}\frac{d_{12}[Y_2(aq)]}{dt} = \cdots . \end{aligned} \quad (2)$$

Here the subscripted differential is employed to specify the rate of the indicated process, which need not (as in the case of consecutive or parallel reactions) be equal to the net rate of change in the concentration of the indicated species. The rate of aqueous-phase reaction is conventionally expressed in units moles (of reaction, as written) liter^{-1} s^{-1}, i.e., M s^{-1}. Frequently, and in the case of elementary reactions necessarily (Johnston, 1966), the reaction rate expression is given by the molecularity of the reaction, i.e.,

$$R_{A12} = k_{A12}[X_1]^{m_1}[X_2]^{m_2}, \quad (3)$$

where the subscript A to the rate constant k denotes that this rate expression refers to the aqueous-phase reaction [A12]. Specializing to a reaction involving a single reagent,

$$mX(aq) \rightarrow \text{products}, \quad [A13]$$

the rate expression is

$$R_{A13} = -\frac{1}{m}\frac{d_{13}[X]}{dt} = k_{A13}[X]^m. \quad (4)$$

An important relationship that, when applicable, relates the forward and reverse rate constants of a reaction may be developed as follows. Provided the forward rate of reaction may be expressed as a product of species concentrations (the exponents need not necessarily be equal to the respective stoichiometric coefficients)

$$R_f = k_f[X_1]^{p_1}[X_2]^{p_2} \cdots$$

applicable over the entire extent of reaction, and similarly for the reverse reaction

$$R_r = k_r[Y_1]^{q_1}[Y_2]^{q_2} \cdots ,$$

then at equilibrium (where the net rate of reaction is zero) $R_f - R_r = 0$ and hence the ratio of the products of species concentrations is equal to the ratio of rate constants, itself a constant:

$$\frac{[Y_1]^{q_1}[Y_2]^{q_2}\cdots}{[X_1]^{p_1}[X_2]^{p_2}\cdots} = \frac{k_f}{k_r} = K. \tag{5}$$

The constant K is equal to the equilibrium constant of the reaction (or to some power of the equilibrium constant). Relationship (5) necessarily holds for elementary reactions but is of much greater generality, holding, for example, in the cases of pre- or postequilibria and/or steady-state intermediates. In fact, Eq. (5) may obtain even in cases in which the rate law does not coincide with the molecularity of the reaction. Equation (5) permits evaluation of the reverse-rate constant from a known forward-rate constant and serves as an important check on the validity of an assumed mechanism and/or on the accuracy of measured rate constants.

2.2. Mixed-Phase Kinetics

We now consider a mixed-phase reaction, in which a gaseous reagent X(g) dissolves in water and reacts by aqueous-phase reaction, as described by the mechanism

$$\mathrm{X(g)} \rightleftharpoons \mathrm{X(aq)}$$

$$m\mathrm{X(aq)} \rightarrow \text{products}. \qquad [\mathrm{A13}]$$

Such a situation necessarily involves competition between the rate of reaction of the physically dissolved gas and the replenishment of the concentration of the dissolved gas by mass transport in each of the two phases and dissolution of the reagent gas at the interface. It is beyond the scope of the present review to survey this literature; an excellent treatment is given by Danckwerts (1970). We restrict the present discussion to a brief outline of situations pertinent to studies that are examined in this review. In this outline it is useful to consider the characteristic rate of aqueous-phase reaction as an "independent variable" that may be compared to the rate of aqueous-phase mixing as governed by convection and/or diffusion. At one extreme, that denoted by Danckwerts as the case of a "very slow reaction," we have a situation in which the rate of reaction is sufficiently slow, compared to the rate at which the phase equilibrium is restored, that the concentration of the aqueous-phase reagent is uniform throughout the aqueous phase and equal to the value given by Henry's law evaluated with respect to the bulk gas-phase concentration. (Here and in the discussion that follows we consider gas-phase mass transport to be sufficiently rapid that the gas-phase profile of the reagent concentration is essentially uniform; if this is not the case, the rate of gas-phase mass transport must be ascertained and appropriate corrections applied.)

A second limiting situation of interest is that of convective-controlled uptake. In this situation reaction is sufficiently slow to allow saturation of a region of the liquid near the surface on a time scale that is rapid compared to that of convective mixing. This saturated volume is then convected into the bulk, where reaction proceeds rapidly compared to the time scale of return of this parcel of material to the surface. In this limiting situation the rate of uptake is controlled entirely by the physical (Henry's law) solubility of the gas and the rate of convective mixing. Finally, we consider the extreme of a reaction sufficiently fast that the uptake occurs entirely within a thin film near the interface in which there is a steep concentration gradient of the reagent governed by the competition between reaction and diffusion of the material from the interface; this concentration profile is established rapidly in comparison to the rate of convective mixing.

As may be anticipated, the above situations are idealizations. However, these idealizations may often be closely approached in laboratory studies, permitting interpretation of measured rates of uptake in terms of the fundamental quantities (Henry's law coefficients and kinetic rate constants) of interest. We should note also that studies may be profitably conducted as well under phase-mixing conditions intermediate to these idealized situations, for which semiempirical models are employed to describe the mass transport and kinetics. We proceed now to outline the several mass-transport situations and to develop expressions relating the rate of uptake as measured in laboratory studies to the fundamental quantities of interest.

2.3. Phase-Mixed Limit (Very Slow Reaction)

We first consider reactions in the phase-mixed limit. For Henry's law satisfied, i.e.,

$$[\mathrm{X(aq)}] = \mathrm{H_X} p_\mathrm{X}, \tag{6}$$

it is seen that the rate of aqueous-phase reaction of a single reagent X reacting according to [A13] may be expressed in terms of the gas-phase partial pressure p_X as

$$R_\mathrm{A13} = \mathrm{H_X^{\mathit{m}}} k_\mathrm{A13} p_\mathrm{X}^{m}. \tag{7}$$

Consequently, it is convenient to define a rate constant k_M13 for the mixed-phase reaction

$$m\mathrm{X(g)} \rightarrow \text{products} \qquad \text{[M13]}$$

such that the rate is expressed as

$$R_\mathrm{A13} = k_\mathrm{M13} p_\mathrm{X}^{m}. \tag{8}$$

It should be noted that the rate of the mixed-phase reaction remains expressed in aqueous-phase concentration units, i.e., in units of moles per liter (of solution) per second. Comparison of (7) and (8) yields the relationship

$$k_\mathrm{M13} = \mathrm{H_X^{\mathit{m}}} k_\mathrm{A13}. \tag{9}$$

Thus the Henry's law coefficient provides the link between the aqueous- and mixed-phase rate coefficients. Since, as is generally the case in experimental studies of mixed-phase reactions in the phase-mixed limit, it is the partial pressure of the reagent and not its aqueous-phase concentration that is known, it is the mixed-phase k_M that is determined. Thus for a reactive gas, e.g., NO_2, the product $H_X^m k_A$ might be known, but not the individual factors. However, in order to establish that the system under examination satisfies the phase-mixed conditions, it is imperative that these factors be determined separately. Specifically, knowledge of k_A permits evaluation of the time quantity τ_r characteristic of chemical reaction in the aqueous phase. This characteristic time is given by

$$\tau_r = \begin{cases} k_A^{-1}, & m = 1 \\ (H_X p_X k_A)^{-1}, & m = 2 \\ (H_X p_X)^{(m-1)} k_A^{-1}, & m = m \end{cases}$$

where m is the reaction order. Comparison of τ_r with the characteristic time τ_{replen} of replenishment of the aqueous phase reagent is necessary for addressing the assumption of phase mixedness in a system under examination; specifically, the phase-mixed assumption is valid only if $\tau_r \gg \tau_{replen}$ (Lee and Schwartz, 1981b; Schwartz and Freiberg, 1981).

2.4. Convective-Controlled Uptake

In this limiting situation it is assumed that the rate of uptake of the gas is controlled by the rate of convection of the material present at the surface into the bulk aqueous phase. If the concentration at the surface is denoted [X(s)], the rate of uptake (in molar per second) is described as

$$R_{abs} = k_{s \to a}[X(s)],$$

where $k_{s \to a}$ is a stochastic first-order "rate coefficient" that characterizes the frequency with which material at the surface is convected into the bulk solution. It is further assumed that Henry's law is satisfied at the surface, i.e., that the surface concentration of the reacting gas is in equilibrium with the gas phase

$$[X(s)] = H_X p_X,$$

and hence

$$R_{abs} = k_{s \to a} H_X p_X. \tag{10}$$

The rate coefficient $k_{s \to a}$ is not an intrinsic property of the reaction system under consideration but depends on the nature and intensity of convective mixing that pertains to a given physical situation (i.e., is "apparatus dependent"). Under favorable circumstances $k_{s \to a}$ may be determined by appropriate characterization measurements. Under such circumstances the

Henry's law coefficient of the reactive gas may be determined directly from the measured rate of uptake since this rate of uptake depends only on the physical solubility of the gas and is independent of the details of the aqueous-phase reaction.

2.5. Slow Reactions

We now consider the situation intermediate to the two limiting cases just discussed. The model describing this situation again posits a surface layer saturated in the reagent gas. Additionally, a uniform reagent concentration is assumed in the bulk aqueous phase having a steady-state value governed by the rates of convective mass-transport coupling the bulk and surface regions and chemical reaction. Again denoting the surface aqueous-phase concentration [X(s)], we may describe the rates of the aqueous-phase mass-transport process governing the bulk concentration [X(aq)] as

$$\mathrm{X(s)} \xrightarrow{k_{\mathrm{s}\to\mathrm{a}}} \mathrm{X(aq)}$$

and

$$\mathrm{X(aq)} \xrightarrow{k_{\mathrm{a}\to\mathrm{s}}} \mathrm{X(s)}.$$

For a first-order reaction

$$\mathrm{X(aq)} \xrightarrow{k_{\mathrm{A}}} \text{products},$$

the steady-state concentration [X(a)] is given by

$$\frac{d[\mathrm{X(a)}]}{dt} = 0 = k_{\mathrm{s}\to\mathrm{a}}[\mathrm{X(s)}] - k_{\mathrm{a}\to\mathrm{s}}[\mathrm{X(aq)}] - k_{\mathrm{A}}[\mathrm{X(aq)}],$$

whence

$$[\mathrm{X(aq)}] = \frac{k_{\mathrm{s}\to\mathrm{a}}[\mathrm{X(s)}]}{k_{\mathrm{a}\to\mathrm{s}} + k_{\mathrm{A}}} = \frac{\mathrm{H_X} k_{\mathrm{s}\to\mathrm{a}} p_{\mathrm{X}}}{k_{\mathrm{a}\to\mathrm{s}} + k_{\mathrm{A}}}, \tag{11}$$

where Henry's law equilibrium has been assumed at the surface. Finally, we observe that $k_{\mathrm{s}\to\mathrm{a}} = k_{\mathrm{a}\to\mathrm{s}}$, since, in the absence of reaction ($k_{\mathrm{A}} = 0$), X(aq) is in Henry's law equilibrium with the gas,

$$[\mathrm{X(aq)}] = \mathrm{H_X} p_{\mathrm{X}}, \tag{6}$$

whereas by (11), for $k_{\mathrm{A}} = 0$,

$$[\mathrm{X(aq)}] = \mathrm{H_X} \frac{k_{\mathrm{s}\to\mathrm{a}}}{k_{\mathrm{a}\to\mathrm{s}}} p_{\mathrm{X}}.$$

In subsequent discussion we may thus refer to the aqueous-phase mixing coefficient without indication of direction, i.e., $k_{\mathrm{as}} = k_{\mathrm{a}\to\mathrm{s}} = k_{\mathrm{s}\to\mathrm{a}}$.

Returning to the reactive system, we observe that the overall rate of absorption is given by

$$R_{abs} = k_A[X(aq)] = \frac{k_A H_X k_{as}}{k_{as} + k_A} p_X. \qquad (12)$$

Measurement of R_{abs} as a function of k_{as} permits both H_X and k_A to be determined. For high values of k_{as}, which promote phase mixing, one obtains the phase-mixed condition as a limit to (12)

$$R_{abs} = H_X k_A p_X, \qquad k_{as} \gg k_A \qquad (13)$$

equivalent to (7) for $m = 1$. On the other hand, for lower values of k_{as} such that $k_{as} \ll k_A$, one obtains the situation described above in which the rate of absorption is controlled entirely by the rate of convective mass transport,

$$R_{abs} = H_X k_{as} p_X, \qquad k_{as} \ll k_A. \qquad (14)$$

Generalization of the slow reaction treatment to reaction of arbitrary order has been given, among others, by Astarita (1967).

2.6. Diffusion-Controlled Uptake (Very Fast Reaction)

Laboratory experiments may also be profitably conducted under conditions in which the characteristic time of reaction is shorter than that of mass transport (by either convection or molecular diffusion). Under such conditions the reaction takes place entirely in a thin region near the surface in which there is a nonuniform reagent concentration profile that is governed by the interworking of reaction and molecular diffusion. The rate of reaction, which is proportional to the surface area of contact between the two phases, has been treated for a variety of chemical mechanisms according to this so-called penetration theory (Danckwerts, 1970). For a reaction that is mth order in a single reagent X, the steady-state rate of uptake of the reagent species is given (Danckwerts, 1970, p. 49) by

$$J_X = \left(\frac{2m}{m + 1}\right)^{1/2} D_X^{1/2} k_A^{1/2} H_X^{(m+1)/2} p_X^{(m+1)/2} \qquad (15)$$

where J_X is the flux of gas X into the liquid (kmol m^{-2} s^{-1}), D_X is the aqueous-phase diffusion coefficient of the dissolved reagent X (m^2 s^{-1}), and k_A is the mth-order aqueous-phase rate coefficient ($M^{(1-m)}$ s^{-1}).

A further word about units may be pertinent here. We have chosen to employ MKS units, as indicated. Consistent with these units, the appropriate unit for concentration would be kmol m^{-3}. However, since this unit is numerically equal to mol liter^{-1} or molar (M), we may continue to employ molar units for concentrations and for compound quantities, e.g., H and k_A. Also, we should point out that the rate expression (15) differs by a factor of $m^{1/2}$ from that frequently employed (e.g., Danckwerts, 1970), as a consequence of differing convention in the definition of the rate coefficient. We continue to employ the convention (4) for an mth order reaction, $d[\mathrm{X}]/dt = -mk[\mathrm{X}]^m$, rather than the expression $d[\mathrm{X}]/dt = -k[\mathrm{X}]^m$ commonly employed in kinetic studies by penetration theory.

If the steady-state rate J_{X} is measured for known p_{X}, and if D_{X} is known or, as is generally the case, can be accurately estimated (Wilke and Chang, 1955), such measurements yield the product

$$k_{\mathrm{A}}^{1/2}\mathrm{H_X}, \qquad m = 1$$

$$k_{\mathrm{A}}^{1/2}\mathrm{H_X}^{3/2}, \qquad m = 2$$

$$k_{\mathrm{A}}^{1/2}\mathrm{H_X}^{(m+1)/2}, \qquad m = m.$$

Thus kinetic studies in this diffusion-controlled limit in conjunction, for example, with kinetic studies in the phase-mixed limit can yield both $\mathrm{H_X}$ and k_{A}, as is desired.

In the case of a first-order reaction, the expression for the time dependence of Q, the amount of reagent taken up per unit area in a time of contact t, has been given (Danckwerts, 1950) as

$$Q_{\mathrm{X}}(t) = D_{\mathrm{X}}^{1/2}k_{\mathrm{A}}^{1/2}\mathrm{H_X}p_{\mathrm{X}}\left(t + \frac{1}{2k_{\mathrm{A}}}\right), \qquad k_{\mathrm{A}}t > 1. \tag{16}$$

It may be seen that measurement of Q_{X} as a function of contact time permits evaluation of both k_{A} and $\mathrm{H_X}$, provided D_{X} is known or can be accurately estimated. Also, k_{A} may be determined without knowledge of D_{X} (or for that matter even p_{X}) from the slope:intercept ratio.

2.7. Fast Reactions

We now treat the situation intermediate between that of convective- and diffusive-controlled uptake. This situation, in which the transport of reagent occurs by mechanically induced convection in addition to diffusion, continues to exhibit a rate of uptake of the reactive gas proportional to the area of the gas–liquid interface, but this rate is enhanced by the convective mass-transport process. An expression for this rate, developed for the so-called film model of the absorption into agitated liquids, is given by Danckwerts (1970, p. 107) for reaction by a first-order or pseudo-first-order mechanism as

$$J_{\mathrm{X}} = \mathrm{H_X}p_{\mathrm{X}}(D_{\mathrm{X}}k_{\mathrm{A}})^{1/2}\coth\left[\frac{(D_{\mathrm{X}}k_{\mathrm{A}})^{1/2}}{k_{\mathrm{L}}}\right] \tag{17}$$

where k_{L}, having units m s^{-1}, is the liquid-phase mass-transfer coefficient. For a given system, k_{L} may be determined from measurement of the rate of uptake of a gas of a known Henry's law coefficient. For low values of k_{L}, the rate of uptake given by Eq. (17) approaches that given by Eq. (15)

for mass transport by diffusion only:

$$J_X \approx H_X p_X (D_X k_A)^{1/2}, \qquad k_L \leq 0.7\ (D_X k_A)^{1/2}. \tag{18}$$

At high values of k_L the rate of uptake becomes controlled entirely by the rate of convective mass-transport and independent of the chemical–kinetic rate constant k_A:

$$J_X \approx H_X p_X k_L, \qquad k_L \geq 1.8\ (D_X k_A)^{1/2}. \tag{19}$$

Equation (19) is seen to exhibit identical dependence on H_X and p_X as that given by (14), differing only insofar as the rate of uptake is expressed per unit area of gas–liquid interface [Eq. (19)] or per unit liquid volume [Eq. (14)]. Introducing the ratio of the interface area to liquid volume a (units: length^{-1}) and observing that $R_{abs} = aJ$, we see that equations (14) and (19) are identical if we identify k_{as} with $k_L a$; the latter is the nomenclature that is conventionally employed in the chemical engineering literature. For further discussion, the reader is referred to Chapter 6 of the monograph of Danckwerts (1970).

2.8. Gas-Phase Mass Transport

Before concluding this discussion, we wish to emphasize that the preceding review has assumed that the interfacial partial pressure of the reactive gas is equal to the partial pressure in the bulk gas. If this condition is not met, as might be the case for a fast aqueous-phase reaction, the interfacial partial pressure of the reactive gas must be employed in the above expressions. The decrease in partial pressure from the bulk to the interface may be inferred from J for known gas-phase mass-transport coefficient k_G. In this regard it should be noted also that for a situation in which the rate of uptake is controlled by gas-phase mass transport, the rate of uptake again is proportional to the first power of p_X, irrespective of the rate and mechanism of aqueous-phase reaction.

2.9. Apparent Reaction Order

One means of experimentally identifying the situation that governs a particular system under investigation is examination of the power law dependence of the rate of absorption on the partial pressure of the reagent gas, i.e., the apparent reaction order referred to the reagent gas concentration. This dependence is summarized in Table 3 for the several limiting conditions for various aqueous-phase reaction orders. As we have noted, the apparent gas-phase order is equal to the aqueous-phase order in the phase-mixed limit, but the relationship between the two orders becomes weaker as the situation becomes increasingly mass-transport controlled. Also, we would

Table 3. Apparent Reaction Order in Gas-Phase Reagent

Aqueous-Phase Reaction Order	Phase-Mixed Limit	Convective Mass-Transport Controlled	Diffusive Mass-Transport Controlled
0	0	1	$\frac{1}{2}$
$\frac{1}{2}$	$\frac{1}{2}$	1	$\frac{3}{4}$
1	1	1	1
2	2	1	$\frac{3}{2}$
3	3	1	2
m	m	1	$(m + 1)/2$

point out that the several situations identified are idealizations and that intermediate situations would be expected. The criteria that must be satisfied for a gas–liquid reaction to be described by one of the three limiting situations outlined above are summarized in Table 4.

3. RATE EXPRESSIONS FOR NITROGEN OXIDE REACTIONS

As noted in Section 1, the study of the reactive dissolution of nitrogen oxides is complicated by the fact that these oxides undergo rapid interconversion (in both gas and aqueous phases) between the so-called simple oxides (NO and NO_2) and the compound oxides (N_2O_3 and N_2O_4). As is shown in Sections 5.3 and 5.4, the aqueous-phase interconversion equilibria are established rapidly in comparison to the rates of the subsequent hydrolysis reactions (to form nitrous and/or nitric acids). Consequently, it is not possible by studies of the kinetics of the latter reactions to ascertain whether the immediate precursor of reaction is a single molecule of the compound oxide or two molecules of the simple oxides. Therefore, the choice of whether to express the reaction kinetics in terms of the concentration of the compound

Table 4. Mixing Conditions for Limiting Regimes of Gas–Liquid Reactions[a]

Regime	Conditions
Phase mixed	$\tau_r \gtrsim 10\tau_m$ or $\tau_r \gtrsim 10\tau_d$
Convective mass-transport controlled	$\tau_r \lesssim 0.1\tau_m$ and $\tau_r \gtrsim 3\tau_m^2/\tau_d$
Diffusive mass-transport controlled	$\tau_r \lesssim 0.1\tau_d$ and $\tau_r \lesssim 0.38\tau_m^2/\tau_d$

[a] Times τ_r, τ_m, and τ_d are the characteristic time or reaction, convective mixing, and molecular diffusion, as defined by $\tau_r = \{(-d_r[A]/dt)/[A]\}^{-1}$, $\tau_m = (k_L a)^{-1}$, and $\tau_d = (Da^2)^{-1}$, respectively. In these equations k_L is the liquid-side mass-transfer coefficient, a is the interfacial area per unit liquid volume, and D is the aqueous-phase diffusion coefficient of the reagent species. Adapted from Lee and Schwartz (1981b).

oxide or the simple oxides is arbitrary. However, as we have seen in Section 2, under non-phase-mixed conditions the description of the reaction kinetics depends strongly on the reaction order—actually the order in transported species. Thus it is convenient to be able to express the reaction rate in terms of the concentration of either the simple or compound oxides, depending on which form is predominately present. Although such treatment is convenient, one must not lose sight of the relationships between the different kinetic expressions, since they ultimately pertain to the same reactions. These relationships are developed here.

3.1. Reactions [1] and [4]

We first consider the aqueous-phase reaction

$$2NO_2(aq) \xrightarrow{w} HNO_2 + H^+ + NO_3^- . \qquad [A1']$$

This reaction might occur as an elementary reaction as written, in which case the rate of formation of nitrate ion product would be

$$\frac{d}{dt}[NO_3^-] = \frac{d_{A1'}}{dt}[NO_3^-] \equiv R_{A1'} = k_{A1'}[NO_2(aq)]^2;$$

alternatively, the reaction might occur by way of an intermediate N_2O_4 according to the mechanism

$$2NO_2(aq) \rightarrow N_2O_4(aq) \qquad [A7]$$

$$N_2O_4(aq) \xrightarrow{w} HNO_2 + H^+ + NO_3^- , \qquad [A4']$$

in which case the rate of formation of nitrate ion would be given by

$$\frac{d}{dt}[NO_3^-] = \frac{d_{A4'}}{dt}[NO_3^-] \equiv R_{A4'} = k_{A4'}[N_2O_4(aq)].$$

As we have noted, the aqueous-phase equilibrium [A7] appears to be established rapidly on a time scale relative to that of the overall reaction. Under the assumption that equilibrium [A7] obtains, then

$$[N_2O_4(aq)] = K_{A7}[NO_2(aq)]^2,$$

and hence the rate of reaction [A4′] can be equally well expressed as a function of the concentration of $NO_2(aq)$, viz.,

$$R_{A4'} \equiv \frac{d_{A4'}}{dt}[NO_3^-] = k_{A4'}K_{A7}[NO_2(aq)]^2.$$

Because of the identical dependence on $[NO_2(aq)]^2$ of the rate of nitrate formation by each of these two mechanisms, it is impossible to distinguish the contributions of the two paths to the total rate under conditions in which equilibrium [A7] obtains (Bray, 1932). Consequently, it becomes notationally convenient to consider the two mechanisms collectively as

$$2NO_2(aq)(N_2O_4(aq)) \overset{w}{\rightarrow} HNO_2 + H^+ + NO_3^- . \qquad [A1'/4']$$

The rate of this "hybrid" reaction is the sum of the rates of the two separate mechanisms,

$$\begin{aligned} R_{A1'/4'} &\equiv R_{A1'} + R_{A4'} \\ &= \frac{d_{A1'}}{dt}[NO_3^-] + \frac{d_{A4'}}{dt}[NO_3^-] \\ &= k_{A1'}[NO_2(aq)]^2 + k_{A4'}K_{A7}[NO_2(aq)]^2 \\ &= k_{A1'/4'}[NO_2(aq)]^2, \end{aligned} \qquad (20)$$

where we have introduced the hybrid rate constant,

$$k_{A1'/4'} \equiv k_{A1'} + k_{A4'}K_{A7}. \qquad (21)$$

For purposes of comparing kinetic measurements by various techniques, it is sometimes convenient to express the rate of reaction [A1′/4′] in terms of N_2O_4 as the reagent. We express this hybrid reaction as

$$N_2O_4(2NO_2) \overset{w}{\rightarrow} HNO_2 + H^+ + NO_3^- \qquad [A4'/1']$$

having rate

$$R_{A4'/1'} = k_{A4'/1'}[N_2O_4(aq)] \qquad (22)$$

where

$$k_{A4'/1'} = \frac{k_{A1'/4'}}{K_{A7}} = \frac{k_{A1'}}{K_{A7}} + k_{A4'}. \qquad (23)$$

3.2. Reverse Reaction

It is useful to extend the treatment of indistinguishable reactions to the reverse reactions

$$HNO_2 + H^+ + NO_3^- \overset{w}{\rightarrow} 2NO_2(aq) \qquad [-A1']$$

and

$$HNO_2 + H^+ + NO_3^- \overset{w}{\rightarrow} N_2O_4(aq). \qquad [-A4']$$

The rates of these reactions are given respectively by

$$R_{-A1'} = k_{-A1'}[HNO_2][H^+][NO_3^-]$$

and

$$R_{-A4'} = k_{-A4'}[HNO_2][H^+][NO_3^-].$$

Hence the total rate of the reverse reaction may be expressed as

$$R_{-A1'/4'} \equiv R_{-A1'} + R_{-A4'} = (k_{-A1'} + k_{-A4'})[HNO_2][H^+][NO_3^-]$$

where $R_{-A1'/4'}$ is the rate of the hybrid reaction

$$HNO_2 + H^+ + NO_3^- \rightarrow 2NO_2(N_2O_4). \qquad [-A1'/4']$$

The rate coefficient for the hybrid reaction is seen to be equal to the sum of the rate coefficients for the two individual reactions

$$k_{-A1'/4'} = k_{-A1'} + k_{-A4'}. \tag{24}$$

At equilibrium the forward and reverse rates for each reaction are equal (principle of detailed balance):

$$k_{A1'}[NO_2(aq)]^2 = k_{-A1'}[HNO_2][H^+][NO_3^-]$$

$$k_{A4'}[N_2O_4(aq)]^2 = k_{A4'}K_{A7}[NO_2(aq)]^2 = k_{-A4'}[HNO_2][H^+][NO_3^-]$$

whence

$$k_{A1'/4'}[NO_2(aq)]^2 = k_{-A1'/4'}[HNO_2][H^+][NO_3^-]$$

whence

$$\frac{k_{A1'/4'}}{k_{-A1'/4'}} = \frac{[HNO_2][H^+][NO_3^-]}{[NO_2(aq)]^2} = K_{A1'}. \tag{25}$$

Thus it is seen that the forward and reverse hybrid reactions [A1′/4′] and [−A1′/4′], although not elementary reactions, exhibit the properties of elementary reactions that the rates are expressed in terms of the molecularity of the reaction, that the forward:reverse rate constant ratio is equal to the equilibrium constant, and that the rate constant is independent of composition.

3.3. Phase-Mixed System

The above considerations may be directly extended to the mixed-phase reaction in the phase-mixed limit. By (7) we obtain for the rate of the mixed-phase reaction [M1′]

$$R_{M1'} = k_{A1'}H^2_{NO_2}p^2_{NO_2}$$

and similarly for [M4′],

$$R_{M4'} = k_{A4'}H_{N_2O_4}p_{N_2O_4}.$$

However, under the assumption of gas-phase equilibrium [G7], it follows that

$$p_{N_2O_4} = K_{G7}p^2_{NO_2},$$

and thus the rate of reaction [M4′] can be expressed as a function of the partial pressure of NO_2,

$$R_{M4'} = K_{G7}k_{A4'}H_{N_2O_4}p^2_{NO_2}.$$

Hence we may define the hybrid mixed-phase reaction

$$2NO_2(g)(N_2O_4(g)) \rightarrow HNO_2 + H^+ + NO_3^- \qquad [M1'/4']$$

having rate

$$\begin{aligned} R_{M1'/4'} &\equiv R_{M1'} + R_{M4'} \\ &= k_{A1'}H^2_{NO_2}p^2_{NO_2} + K_{G7}k_{A4'}H_{N_2O_4}p^2_{NO_2} \\ &= k_{M1'/4'}p^2_{NO_2}, \end{aligned}$$

where we introduce the hybrid mixed-phase rate constant

$$k_{M1'/4'} \equiv k_{A1'}H^2_{NO_2} + K_{G7}k_{A4'}H_{N_2O_4}. \qquad (26)$$

From examination of thermochemical cycles (Schwartz and White, 1981) it is readily established that the Henry's law coefficients of NO_2 and N_2O_4 are related by

$$H_{N_2O_4} = \frac{K_{A7}}{K_{G7}}H^2_{NO_2}; \qquad (27)$$

hence one obtains the relationship between the mixed- and aqueous-phase rate constants,

$$k_{M1'/4'} = H^2_{NO_2}(k_{A1'} + K_{A7}k_{A4'}) = H^2_{NO_2}k_{A1'/4'}. \qquad (28)$$

In consideration of reactions under phase-mixed conditions, $k_{A1'/4'}$ may thus be treated as if it were the rate constant of an elementary reaction second-order in NO_2(aq) [see Eq. (9)].

Analogously to the previous discussion, the rate constant for the hybrid reverse mixed-phase reaction

$$HNO_2 + H^+ + NO_3^- \xrightarrow{w} 2NO_2(g)(N_2O_4(g)) \qquad [-M1'/4']$$

is given by

$$k_{-M1'/4'} = k_{-M1'} + k_{-M4'}, \qquad (29)$$

and the relationship to the equilibrium constant is

$$K_{M1'} = \frac{k_{M1'/4'}}{k_{-M1'/4'}}.$$

3.4. Reactions [2] and [5]

The concept of indistinguishable reactions applies also to formation of nitrous acid from NO and NO_2 by the reaction

$$NO(aq) + NO_2(aq) \xrightarrow{w} 2HNO_2 \qquad [A2']$$

or by the sequence

$$NO(aq) + NO_2(aq) = N_2O_3(aq) \tag{[A8]}$$

$$N_2O_3(aq) \xrightarrow{w} 2HNO_2. \tag{[A5′]}$$

This leads to definition of the hybrid aqueous-phase reactions

$$NO(aq) + NO_2(aq)(N_2O_3(aq)) \xrightarrow{w} 2HNO_2 \tag{[A2′/5′]}$$

and

$$2HNO_2 \xrightarrow{w} NO(aq) + NO_2(aq)(N_2O_3(aq)) \tag{[−A2′/5′]}$$

having rate constants

$$k_{A2'/5'} = k_{A2'} + k_{A5'}K_{A8} \tag{30}$$

and

$$k_{-A2'/5'} = k_{-A2'} + k_{-A5'}, \tag{31}$$

such that the rates of the hybrid reactions may be evaluated as

$$R_{A2'/5'} = k_{A2'/5'}[NO(aq)][NO_2(aq)] \tag{32}$$

and

$$R_{-A1'/5'} = k_{-A2'/5'}[HNO_2]^2. \tag{33}$$

Similarly, we define the hybrid mixed-phase reactions

$$NO(g) + NO_2(g)(N_2O_3(g)) \xrightarrow{w} 2HNO_2(aq) \tag{[M2′/5′]}$$

and

$$2HNO_2(aq) \xrightarrow{w} NO(g) + NO_2(g)(N_2O_3(g)) \tag{[−M2′/5′]}$$

having rate constants

$$k_{M2'/5'} = k_{A2'}H_{NO}H_{NO_2} + K_{G8}k_{A5'}H_{N_2O_3} \tag{34}$$

and

$$k_{-M2'/5'} = k_{-M2'} + k_{-M5'}. \tag{35}$$

3.5. Reactions Involving Nitrous Acid and/or Nitrite

One further notational consideration addresses the relationship between the kinetics of reactions involving nitrous acid and the corresponding reactions involving nitrite (as distinguished here by the presence or absence, respectively, of primes in the reaction numbers). Consider the reactions

$$2NO_2(aq) \xrightarrow{w} H^+ + NO_3^- + HNO_2 \tag{[A1′]}$$

and

$$2NO_2(aq) \xrightarrow{w} 2H^+ + NO_3^- + NO_2^-. \tag{[A1]}$$

Kinetically the nitrous acid dissociation equilibrium (as any protonation equilibrium) may be expected to be established quite fast (Eigen et al., 1964), and thus under ordinary circumstances we are unable to distinguish, say, between direct reaction by [A1] or reaction [A1′] followed by ionization of the HNO_2 product. The rate of reaction by the sum of both paths is

$$R_{A1} + R_{A1'} = -\frac{1}{2}\frac{d[NO_2(aq)]}{dt} = (k_{A1} + k_{A1'})[NO_2(aq)]^2. \quad (36)$$

For simplicity, and without confusion, we choose not to retain the distinction between the two paths and simply denote the sum as

$$R_{A1*} \equiv R_{A1} + R_{A1'} = k_{A1*}[NO_2(aq)]^2, \quad (37)$$

where we introduce

$$k_{A1*} \equiv k_{A1} + k_{A1'}.$$

Similar considerations apply regarding the reverse reaction, which may occur as

$$H^+ + NO_3^- + HNO_2 \overset{w}{\rightarrow} 2NO_2(aq) \quad [-A1']$$

and/or

$$2H^+ + NO_3^- + NO_2^- \overset{w}{\rightarrow} 2NO_2(aq) \quad [-A1]$$

with rates

$$R_{-A1'} = k_{-A1'}[H^+][NO_3^-][HNO_2] \quad (38)$$

and

$$R_{-A1} = k_{-A1}[H^+]^2[NO_3^-][NO_2^-]. \quad (39)$$

Assuming that the nitrous acid dissociation equilibrium [I2] obtains, i.e., $K_{12} = [H^+][NO_2^-]/[HNO_2]$, we may write R_{-A1} in terms of the associated species as

$$R_{-A1} = k_{-A1}K_{12}[H^+][NO_3^+][HNO_2] \quad (40)$$

and in turn may write the total reaction rate by both paths as

$$\begin{aligned} R_{-A1} + R_{-A1'} &= (k_{-A1'} + k_{-A1}K_{12})[H^+][NO_3^-][HNO_2] \\ &= k_{-A1*}[H^+][NO_3^-][HNO_2] \end{aligned} \quad (41)$$

where we have introduced $k_{-A1*} \equiv k_{-A1'} + K_{-A1}K_{12}$. From detailed balance considerations it is seen that k_{A1*} and k_{-A1*} are related as

$$\frac{k_{A1*}}{k_{-A1*}} = K_{A1'},$$

and thus that the hybrid reaction [A1*]

$$2NO_2(aq) \overset{w}{\rightleftharpoons} H^+ + NO_3^- + HNO_2(H^+ + NO_2^-) \quad [A1*]$$

may be treated as an elementary reaction. We utilize this hybrid notation as well to avoid being burdened with retaining the distinction between the two separate paths. Furthermore, the preceding discussion concerning the hybrid reactions [A1′/4′], [M1′/4′], etc. may be extended to encompass this further hybridization by replacing the primes with asterisks.

These hybrid reactions and the corresponding rate expression will be useful in the discussion of laboratory studies of the rates of the aqueous-phase and mixed-phase reactions of the nitrogen oxides and oxyacids. In particular, for the discussion of experimental studies of reactions [1/4] and [2/5], we establish the convention of expressing the kinetics of these reactions in the hybrid notation regardless of the formulation employed by the original investigators. The validity of this treatment, which depends on equilibria [A7] and [A8] being rapidly established relative to the time scale of reactions [A1*/4*] and [A2*/5*], respectively, is addressed in Sections 5.3 and 5.4.

4. INDIRECT MEASUREMENTS OF PHASE-MIXED KINETICS

Although there is a large body of work directly bearing on the aqueous and mixed-phase kinetics of the reactions of the nitrogen oxides with water (Sections 5 and 6), there is also considerable literature that pertains indirectly to the kinetics of these reactions. Specifically, there are a number of studies of aqueous-phase reactions of the nitrogen oxides and oxyacids that can be interpreted in terms of a sequence of elementary reactions whose rate-determining step is one of the elementary reactions of interest here. From the measured kinetics of the overall reactions, it is possible to infer the kinetics of the reactions of interest. These studies are reviewed in this section.

4.1. Reaction [M1*/4′]

The kinetics of the mixed-phase reaction

$$2NO_2(g)(N_2O_4(g) \xrightarrow{w} H^+ + NO_3^- + HNO_2(aq)(H^+ + NO_2^-) \qquad [M1^*/4^*]$$

may be inferred from studies of the kinetics of nitrous acid decomposition and formation, isotope exchange, and electrochemical reduction. These studies are reviewed here.

4.1.1. Nitrous Acid Decomposition

The earliest reliable measurement of the kinetics of reaction [M1*/4*] resulted from a remarkable study of the kinetics of nitrous acid decomposition

$$3HNO_2(aq) \xrightarrow{w} H^+ + NO_3^- + 2NO(g) \qquad [M10^*]$$

and formation

$$H^+ + NO_3^- + 2NO(g) \xrightarrow{w} 3HNO_2(aq) \qquad [-M10^*]$$

by E. Abel and H. Schmid (1928a–f). A major advance in the understanding of these reactions was achieved by the recognition of the need for a well-defined NO partial pressure and for sufficient interphase mixing to assure that the Henry's law equilibrium for NO was maintained. Under these conditions the stoichiometry of nitrous acid decomposition was established as that of reaction [M10*], and the initial rate law was found to be

$$R_{M10^*} \equiv \frac{1}{3}\frac{d_{M10^*}}{dt}[HNO_2] = k_{fwd}\frac{[HNO_2]^4}{p_{NO}^2}, \tag{42}$$

where k_{fwd} denotes an empirical rate coefficient for reaction [M10*] in the forward direction. (We point out that we have expressed R_{M10^*} consistent with the modern convention [Eq. (2)]. This convention differs by a factor of 3 from that employed by the original investigators, who reported R_{M10^*} as $-d_{M10^*}[HNO_2]/dt$). The observed rate law, which excited considerable theoretical interest because of the unusual order in reagents (differing markedly from the reaction stoichiometry), was interpreted in terms of the following mechanism:

$$4HNO_2(aq) \overset{w}{=} 2NO_2(aq)(N_2O_4(aq)) + 2NO(aq) \qquad [-2A2']$$

$$2NO_2(aq)(N_2O_4(aq)) \overset{w}{\rightarrow} H^+ + NO_3^- + HNO_2(H^+ + NO_2^-) \qquad [A1^*/4^*]$$

$$2NO(aq) = 2NO(g) \qquad [-2H_{NO}]$$

where reactions [A2′] and [H_{NO}] are assumed to be in equilibrium and [A1*/4*] is the rate-determining step. Assumption of equilibrium [A7] permits the rate of reaction [A1*/4*] (and hence of the overall reaction) to be expressed in terms of the unknown concentration of the intermediate $NO_2(aq)$ [see Eq. (20)] as

$$R_{M10^*} = R_{A1^*/4^*} = k_{A1^*/4^*}[NO_2(aq)]^2$$

Assumption of equilibrium [A2′] permits $[NO_2(aq)]$ to be expressed in terms of $[NO(aq)]$,

$$[NO_2(aq)] = K_{A2'}^{-1}\frac{[HNO_2]^2}{[NO(aq)]}, \tag{44}$$

and the further assumption of the Henry's law equilibrium for NO (phase-mixed limit) allows $[NO_2(aq)]$ to be expressed in terms of p_{NO},

$$[NO_2(aq)] = H_{NO}^{-1}K_{A2'}^{-1}\frac{[HNO_2]^2}{p_{NO}}.$$

Hence for the rate of the overall reaction [M10*], one obtains

$$R_{M10^*} = k_{A1^*/4^*}H_{NO}^{-2}K_{A2'}^{-2}\frac{[HNO_2]^4}{p_{NO}^2}. \tag{45}$$

Comparison of equations (42) and (45) shows that the hybrid rate constant

$k_{A1^*/4^*}$ is related to the measured rate coefficient k_{fwd} as

$$k_{A1^*/4^*} = H_{NO}^2 K_{A2'}^2 k_{fwd}. \tag{46}$$

Equation (46) is not directly applicable since the aqueous-phase equilibrium constant $K_{A2'}$ is not firmly established (Schwartz and White, 1981). However, we observe [see Eq. (28)] that

$$k_{M1^*/4^*} = H_{NO_2}^2 k_{A1^*/4^*} \tag{47}$$

and hence

$$k_{M1^*/4^*} = H_{NO_2}^2 H_{NO}^2 K_{A2'}^2 k_{fwd}.$$

Observing (see Tables 1 and 2) that the expressions for the mixed- and aqueous-phase equilibria [M2′] and [A2′] are related by the Henry's law coefficients for NO and NO_2 as

$$K_{A2'} = \frac{K_{M2'}}{H_{NO} H_{NO_2}}, \tag{48}$$

one obtains as the rate coefficient for the hybrid mixed-phase reaction

$$k_{M1^*/4^*} = K_{M2'}^2 k_{fwd}. \tag{49}$$

Abel and H. Schmid (1928c) obtained the value $k_{fwd} = 2.6 \times 10^{-1}$ atm^2 M^{-3} s^{-1} at 25°C and zero ionic strength; following those authors (see also H. Schmid, 1940), but employing the value $K_{M2'} = 1.26 \times 10^2$ M^2 atm^{-2} from Table 2, we obtain the value of $k_{M1^*/4^*}$ given in Table 5, 4.06×10^3 M atm^{-2} s^{-1}.

4.1.2. Temperature Dependence

The temperature dependence of the empirical rate coefficient k_{fwd} was studied by Abel et al. (1930), who found an activation energy of 28.6 kcal mol^{-1}. This may be combined with $\Delta H°[M2'] = -18.2$ kcal mol^{-1} (Table 2) to yield an activation energy associated with $k_{M1^*/4^*}$ of -7.8 kcal mol^{-1}. This negative activation energy, which indicates a decrease in the reaction rate with increasing temperature, undoubtedly reflects a negative enthalpy of solution of NO_2 (decrease in the Henry's law coefficient for NO_2 with increasing temperature) that outweighs the presumably positive activation energy associated with $k_{A1^*/4^*}$.

4.1.3. Reverse Reaction

In their study of the reverse of [M10′], i.e., the formation of nitrous acid from NO(g) and nitric acid, Abel and H. Schmid (1928d) obtained the rate law

$$R_{-M10^*} \equiv \frac{1}{3} \frac{d_{-M10^*}}{dt} [HNO_2] = k_{rev}[H^+][NO_3^-][HNO_2]. \tag{50}$$

Table 5. Experimental Determinations of $k_{-A1^*/4^*}$ and $k_{M1^*/4^*}$ at 25°C

Reference	$k_{-A1^*/4^*}$ ($M^{-2}\ s^{-1}$)	$k_{M1^*/4^*}$ (M atm^{-2} s^{-1})	$k_{M4^*/1^*}$ (M atm^{-1} s^{-1})
Abel and H. Schmid (1928c)		4.06(3)	5.92(2)
Abel and H. Schmid (1928d)	8.9(−3)	4.25(3)[a]	6.20(2)
G. Schmid and Bähr (1964)	9.3(−3)[b]	4.42(3)[a,b]	6.44(2)
Komiyama and Inoue (1978)		8.63(3)	12.6 (2)
Komiyama and Inoue (1978)		11.4 (3)[c]	7.45(2)[c]
Jordan and Bonner (1973)	7.2(−3)	3.4 (3)[d]	5.0 (2)
Heckner (1973)		12.9 (3)	18.8 (2)

[a] Computed from $k_{-A1^*/4^*}$ by (52).

[b] Measurements at 20°C corrected to 25°C.

[c] Measurements at 15°C, not temperature corrected.

(Again we note that the rate convention employed here differs by a factor of 3 from that employed by the original investigators.) The rate law (50) exhibits the interesting feature that it is autocatalytic, i.e., that the rate of formation of the product HNO_2 is proportional to the HNO_2 concentration. This rate law also differs from the reaction stoichiometry, but it is seen that the equilibrium constant expression, obtained by equating R_{M10^*} and R_{-M10^*}, nonetheless exhibits the appropriate dependence on reagent concentrations [see Eq. (5)]. This rate law was interpreted in terms of the reverse of the decomposition mechanism:

$$HNO_2 + H^+ + NO_3^- \overset{w}{\rightarrow} 2NO_2(aq)(N_2O_4(aq)) \qquad [-A1^*/4^*]$$

$$2NO_2(aq)(N_2O_4(aq)) + 2NO(aq) \overset{w}{=} 4HNO_2(aq) \qquad [2A2^*]$$

$$2NO(g) = 2NO(aq). \qquad [2H_{NO}]$$

For reaction [−A1*/4*] rate limiting,

$$R_{-M10^*} = R_{-A1^*/4^*} = k_{-A1^*/4^*}[HNO_2][H^+][NO_3^-]; \qquad (51)$$

i.e., $k_{-A1^*/4^*}$ is equal to the empirically observed rate coefficient k_{rev}. We have noted [Eq. (25)] that the ratio of the forward and reverse hybrid rate coefficients is equal to the equilibrium constant $K_{A1'}$ and hence

$$k_{A1^*/4^*} = K_{A1'}k_{-A1^*/4^*} = K_{A1'}k_{rev},$$

and in turn [Eq. (28)] we obtain for the mixed-phase rate coefficient

$$k_{M1^*/4^*} = H_{NO_2}^2 K_{A1'}k_{rev},$$

We further observe (see Tables 1 and 2) that $K_{A1'} = K_{M1'}/H_{NO_2}^2$, and hence

$$k_{M1^*/4^*} = k_{M1'}k_{rev}. \qquad (52)$$

Using the value of k_{rev} ($8.9 \times 10^{-3}\ M^{-2}\ s^{-1}$) determined by Abel and

H. Schmid (1928d) at 25°C and zero ionic strength and the value $K_{M1'} = 4.78 \times 10^5$ M^3 atm^{-2} given in Table 2, we obtain $k_{M1*/4*} = 4.25 \times 10^3$ M atm^{-2} s^{-1}, in substantial agreement with that obtained from measurement of the forward reaction rate.

A similar study, but at higher ionic strength, was carried out by G. Schmid and Bähr (1964). Those authors obtained the value for k_{rev} at 20°C, extrapolated to zero ionic strength, 5.56×10^{-3} M^{-2} s^{-1}. Employing the value $K_{M1'} = 9.91 \times 10^5$ M^3 atm^{-2} obtained from Table 2 for 20°C, we obtain $k_{M1*/4*} = 5.51 \times 10^3$ M atm^{-2} s^{-1} at 20°C. Correcting this value to 25°C by means of the activation energy obtained by Abel et al. (1930), we obtain $k_{M1*/4*} = 4.42 \times 10^3$ M atm^{-2} s^{-1}, in close agreement with the earlier determinations.

4.1.4. *Non-Phase-Mixed Study*

A study of nitrous acid decomposition similar to that of Abel and H. Schmid was carried out by Komiyama and Inoue (1978). In this study, however, the aqueous-phase concentration of NO was not maintained equal to a value in Henry's law equilibrium with the gas phase but was maintained, rather, in steady state, as controlled by the formation reaction

$$3HNO_2(aq) \rightarrow H^+ + NO_3^- + 2NO(aq) \qquad [A10^*]$$

and mass transfer of the NO from the solution into a flowing sparge gas,

$$NO(aq) \xrightarrow{k_{as}} NO(g),$$

[see Eq. (11) describing steady-state reagent concentrations in the slow-reaction regime]. The mass-transfer rate, which was controlled by the physical characteristics of the reactor and the flow rate of helium used as a sparge gas, was determined in separate experiments employing CO_2. Under these conditions the rate law observed for reaction [A10*] was

$$R_{A10*} = k'_{fwd} k_{as}^{2/3} [HNO_2]^{4/3}. \qquad (53)$$

This rate law was shown to be consistent with the Abel–Schmid mechanism as given above, except that NO(aq) is interpreted not as being in Henry's law equilibrium, but in steady state, as controlled by the net rate of [−A2*] and the mass-transfer rate. Under this assumption the steady-state equations for NO(aq) and NO_2(aq) are

$$\frac{d[NO(aq)]}{dt} = 0 = k_{-A2*}[HNO_2]^2 - k_{A2*}[NO(aq)][NO_2(aq)] - k_{as}[NO(aq)] \qquad (54)$$

and

$$\frac{d[NO_2(aq)]}{dt} = 0 = k_{-A2*}[HNO_2]^2 - k_{A2*}[NO(aq)][NO_2(aq)] - 2k_{A1*/4*}[NO_2(aq)]^2. \qquad (55)$$

From (54) and (55) we obtain

$$[NO(aq)] = \frac{2k_{A1*/4*}}{k_{as}}[NO_2(aq)]^2.$$

Assumption of equilibrium [A2′] permits $[NO_2(aq)]$ to be evaluated by (44) as

$$[NO_2(aq)] = 2^{-1/3}K_{A2'}^{-1/3}k_{A1*/4*}^{-1/3}k_{as}^{1/3}[HNO_2]^{2/3}.$$

Hence, by (43) we obtain

$$R_{A10*} = 2^{-2/3}K_{A2'}^{-2/3}k_{A1*/4*}^{1/3}k_{as}^{2/3}[HNO_2]^{4/3}. \quad (56)$$

Comparison of (56) with (53) gives

$$k_{A1*/4*} = 4K_{A2'}^2 k_{fwd}'^3.$$

Finally, from (45) and (49) we obtain

$$k_{M1*/4*} = 4k_{fwd}'^3 \frac{K_{M2'}^2}{H_{NO}^2}. \quad (57)$$

From the value $k'_{fwd} = 7.97 \times 10^{-3}\ M^{-1/3}\ s^{-1/3}$ determined by Komiyama and Inoue for 25°C and $K_{M2'}$ and H_{NO} as given in Table 2, we obtain $k_{M1*/4*} = 8.63 \times 10^3\ M\ atm^{-2}\ s^{-1}$, a factor of roughly 2 greater than the values obtained in the studies (Abel and H. Schmid, 1928d; G. Schmid and Bähr, 1964) in which NO was maintained at its Henry's law concentration. The authors suggest that in those studies the Henry's law equilibrium may not have been maintained; alternatively, the question may be raised as to whether Komiyama and Inoue's model, which assumes a uniform NO concentration and in turn a uniform NO_2 concentration, correctly describes their experimental situation, in which [NO(aq)] is maintained by competition between formation by chemical reaction and removal by mass transport from the aqueous phase into the gas phase. The existence of a nonuniform concentration profile for $NO_2(aq)$ over a significant portion of the liquid volume would result in an average rate different from that computed by using the mean $NO_2(aq)$ concentration, since this rate is quadratic in $[NO_2(aq)$ [Eq. (43)]. In turn, the resulting value of $k_{A1*/4*}$ would exhibit a corresponding departure from the true value. Because of the uncertainty raised by the possibility of nonuniform concentration profiles, we question the numerical accuracy of the value of $k_{M1*/4*}$. Nonetheless, the observed reaction-rate law, in expectation with that based on the Abel–Schmid mechanism, and the proximity of the rate constant with that obtained in the earlier studies, lends support to the mechanistic interpretation by Abel and H. Schmid.

4.1.5. Isotope Exchange Study

The rate of reaction [−A10*] may be determined also from measurement of the rate of nitrogen exchange between NO and nitric acid in aqueous solution, which has been shown to be limited by NO_3^-–HNO_2 exchange (Jordan and Bonner, 1973). In that study isotopically labeled NO was brought into contact with aqueous HNO_3 solution. Initially the exchange rate was quite slow. However, after an induction period that was attributed to formation of an equilibrium concentration of HNO_2, a faster linear exchange rate was observed. (In separate experiments the buildup of HNO_2, monitored spectrophotometrically, was shown to take place on a time scale corresponding to the induction period of the exchange reaction.) The rate law for nitrogen exchange, derived from the measured rate of isotopic equilibration, was $R_N = k_{ex}[H^+][NO_3^-][HNO_2]$, where $[HNO_2]$ was evaluated as the equilibrium concentration as determined by the NO and HNO_3 concentrations. This rate law is identical to that observed by Abel and H. Schmid (1928d) for reaction [−A10*] [Eq. (51)]. Although it was not appreciated by the original authors, their data are represented by a rate constant that also exhibits close quantitative agreement with the rate constant of Abel and H. Schmid. The concentration and rate data for the several runs reported by Jordan and Bonner as well as the activity data for nitric acid at these concentrations as obtained by interpolation from the data of Davis and DeBruin (1964) are presented in Table 6. The data in the last column represent the rate constant for reaction [−A10*] under the assumption that this reaction is the rate-limiting step in the nitrogen exchange. The mean rate constant $(7.15 \pm 6\%) \times 10^{-3}$ $M^{-2}\ s^{-1}$ lies within some 20% of the values determined by Abel and H. Schmid (1928d) and G. Schmid and Bähr (1964) as shown in Table 5.

4.1.6. Electrochemical Kinetic Studies

The rate of reaction [−A10*] has been measured indirectly also in studies of the time dependence of the electrochemical reduction of nitric acid to

Table 6. Determination of $k_{-A1^*/4^*}$ from Isotopic Exchange Data of Jordan and Bonner (1973)[a]

$[HNO_3]$ (M)	$[HNO_2]$ (M)	p_{NO} (atm)	R_N ($M\ s^{-1}$)	$y_\pm$[b] (—)	α[b] (—)	$a_\pm$ (M)	$\frac{R_N}{[HNO_2]a_\pm^2}$ ($M^{-2}\ s^{-1}$)
0.914	8.2(−2)	0.218	2.8(−4)	0.76	0.98	0.68	7.5(−3)
0.609	7.1(−2)	0.189	1.1(−4)	0.74	0.98	0.45	7.8(−3)
0.499	6.4(−2)	0.259	6.7(−5)	0.74	0.98	0.37	7.8(−3)
0.450	6.1(−2)	0.228	3.7(−5)	0.74	0.99	0.33	5.5(−3)
0.440	6.0(−2)	0.203	4.3(−5)	0.74	0.99	0.32	6.9(−3)
0.308	5.2(−2)	0.228	2.3(−5)	0.74	0.99	0.23	8.8(−3)
0.301	5.0(−2)	0.226	1.4(−5)	0.74	0.99	0.22	5.7(−3)
							(7.2 ± 6%[c])(−3)

[a] 25°C.

[b] Activity coefficient $y_\pm$ and fractional dissociation α obtained by interpolation from data of Davis and DeBruin (1964).

[c] Standard deviation of the mean.

nitrous acid following sudden application of a constant electrode potential (Heckner, 1973; G. Schmid and Krichel, 1964). The reaction mixture consisted of a dilute solution of nitrous acid in rather concentrated nitric or mixed nitric and perchloric acids (≥3 M total acid concentration). After application of the constant electrode potential, the current was found to exhibit a well-defined maximum (20–400 ms depending on conditions) followed by a slow decrease. G. Schmid (1961) has interpreted such current–time observations in terms of concurrent diffusion and homogeneous reaction of a species (HNO_2) undergoing autocatalytic decomposition. The differential equation describing the species concentration C is

$$\frac{\partial C}{\partial t} = D\frac{\partial^2 C}{\partial x^2} + k_{\text{auto}}C, \tag{58}$$

where k_{auto} is the rate coefficient for the autocatalytic reaction. The solution of (58) yields the time dependence of the current i,

$$i = FC_0\left(\frac{D}{\pi t}\right)^{1/2}\exp(k_{\text{auto}}t),$$

where F is the Faraday constant, D is the diffusion coefficient of the species, and C_0 is the initial (uniform) concentration of the species. It is seen that the current exhibits a maximum at time $t_m = (2k_{\text{auto}})^{-1}$, whence

$$k_{\text{auto}} = (2t_m)^{-1}. \tag{59}$$

The observations were interpreted in terms of the following mechanism for the cathode half-reaction:

$$2(HNO_2 + H^+ + e^- \xrightarrow{\text{w}} NO) \qquad [\text{A14}]$$

$$H^+ + NO_3^- + 2NO \xrightarrow{\text{w}} 3HNO_2 \qquad [-\text{A10*}]$$

$$3H^+ + NO_3^- + 2e^- \xrightarrow{\text{w}} HNO_2. \qquad [\text{A15}]$$

Here [A14] represents the heterogeneous electrode reaction, which is posited to be fast, and the homogeneous reaction [−A10*] is assumed to be the rate-determining step. We have seen [Eq. (51)] that reaction [−A10*] has the rate

$$R_{-\text{A10*}} = k_{-\text{A1*/4*}}a_{HNO_2}a_{H^+}a_{NO_3^-}, \tag{60}$$

which is taken to be the rate of the overall reaction [15]. Here we have utilized activities in place of concentrations because of the rather high acid concentrations employed. Introducing activity coefficients into (60), we obtain

$$\frac{d[HNO_2]}{dt} = R_{\text{A15}}$$

$$= k_{-\text{A1*/4*}}y_{H^+}y_{NO_3^-}y_{HNO_2}[H^+][NO_3^-][HNO_2].$$

For fixed $[H^+]$ and $[NO_3^-]$, we see that reaction [A15] is autocatalytic in $[HNO_2]$; i.e., that

$$\frac{d[HNO_2]}{dt} = k_{auto}[HNO_2]$$

where

$$k_{auto} = k_{-A1^*/4^*}[H^+][NO_3^-]y_{H^+}y_{NO_3^-}y_{HNO_2},$$

thus permitting k_{auto} and, in turn $k_{-A1^*/4^*}$ to be determined [Eq. (59)] from the measured time to maximum current t_m.

Measurements of t_m and, in turn, $k_{-A1^*/4^*}$ were presented as a function of nitric acid concentration by G. Schmid and Krichel (1964); values of $k_{-A1^*/4^*}$ ranged from $4.2 \times 10^{-2}\ M^{-2}\ s^{-1}$ upward, increasing with increasing acid concentration. These values were indicated by the authors as being entirely consistent with the values of the empirical rate coefficient k_{rev} [Eq. (50)] for reaction [−M10*], as determined by G. Schmid and Bähr (1964). Similarly, in the work of Heckner (1973) measurements of t_m extrapolated to zero ionic strength led to a value of $k_{-A1^*/4^*}$ of $(2.7 \pm 20\%) \times 10^{-2}\ M^{-2}\ s^{-1}$. In this case also quantitative agreement was indicated between the value of the rate constant $k_{-A1^*/4^*}$ inferred from electrochemical measurements and the empirical coefficient k_{rev} (Abel and H. Schmid, 1928d; G. Schmid and Bähr, 1964).

Unfortunately, the close quantitative agreement between the chemical–kinetic and electrochemical experiments appears to be fortuitous. We have noted in conjunction with Eq. (50) that the expression for R_{-M10^*} adopted by Abel and H. Schmid (as well as by G. Schmid and Bähr) differs by a factor of 3 from modern convention. The value of $k_{-A1^*/4^*}$ obtained from the electrochemical measurements must be identified not with the rate constant given by Abel and H. Schmid (denoted k_2 by those authors), but rather with their rate constant *divided by 3*, i.e., the quantity k_{rev} as defined here. In view of the apparently close agreement found between $k_{-A1^*/4^*}$ determined by chemical kinetics and in the electrochemical studies, we are somewhat disappointed to have discovered this discrepancy. On the other hand, we take some solace in the qualitative agreement between the predicted dependence of the electrochemical transients on species concentrations that supports the mechanism of reaction [A10*] initially proposed by Abel and H. Schmid.

4.1.7. Summary

Rate constants $k_{-A1^*/4^*}$ and $k_{M1^*/4^*}$ determined by the several studies that were conducted at low ionic strength or that can be extrapolated to zero ionic strength are given in Table 5. Comparison of these results shows quite close agreement (8% spread) for the studies by Abel and H. Schmid (1928c,d) and G. Schmid and Bähr (1964). Since these studies represent independent investigations by two different methods (decomposition and formation of nitrous acid) characterized by a large number of individual determinations, including examination of dependence on ionic strength and temperature, we place the greatest confidence in these results.

To facilitate subsequent comparison of these measurements to measurements by other techniques, it is convenient to express the rate constant in terms of the compound oxide N_2O_4 as the reagent. In this notation the forward reaction is written as

$$N_2O_4(g)(2NO_2(g)) \xrightarrow{w} H^+ + NO_3^- + HNO_2(H^+ + NO_2^-) \qquad [M4^*/1^*]$$

having rate

$$R_{M4^*/1^*} = k_{M4^*/1^*} p_{N_2O_4},$$

where

$$k_{M4^*/1^*} = \frac{k_{M1^*/4^*}}{K_{G7}}.$$

Values of $k_{M4^*/1^*}$, obtained with $K_{G7} = 6.86 \text{ atm}^{-1}$ at 25°C (Table 2), are also given in Table 5.

The temperature dependence of $k_{M4^*/1^*}$ may be expressed as an activation energy evaluated as

$$E_a[M4^*/1^*] = E_a[M1^*/4^*] - \Delta H°[G7].$$

From the value $E_a[M1^*/4^*] = -7.8 \text{ kcal mol}^{-1}$ given above and $\Delta H°[G7]$ given in Table 2, we obtain $E_a[M4^*/1^*] = +5.9 \text{ kcal mol}^{-1}$.

4.1.8. *Examination of Assumptions*

Before concluding the discussion of reaction [A1*/4*], it is of interest to examine more closely the mechanistic interpretation of the nitrous acid decomposition kinetics as initially set forth by Abel and H. Schmid, to address the assumptions of that mechanism. In particular, it was assumed that the reaction

$$2HNO_2(aq) \overset{w}{=} NO_2(aq) + NO(aq) \qquad [-A2']$$

could be considered to be at equilibrium. A necessary condition that must be satisfied for this equilibrium to obtain is that the rate of formation of $NO_2(aq)$ by reaction [−A2′] greatly exceed the rate of its depletion by reaction [A1*/4*],

$$2NO_2(aq)(N_2O_4(aq) \xrightarrow{w} HNO_2 + H^+ + NO_3^-, \qquad [A1^*/4^*]$$

i.e., that

$$k_{-A2'}[HNO_2]^2 \gg 2k_{A1^*/4^*}[NO_2(aq)]^2. \qquad (61)$$

For evaluation of these rates, it is necessary to anticipate values of the two rate constants; based on the discussion presented below, we take $k_{-A2'} =$

$5.6\ M^{-1}\ s^{-1}$ and $k_{A1*/4*}$ as $6 \times 10^7\ M^{-1}\ s^{-1}$. Also, to evaluate $[NO_2(aq)]$, we take $K_{A2'} = 8 \times 10^6$, consistent with $H_{NO_2} = 1 \times 10^{-2}$ M atm^{-1}. With these quantities we evaluate the rates of reactions [−A2′] and [A1*/4*] for the concentrations employed by Abel and H. Schmid (1928c, p. 293), as shown in Table 7. It is seen that the rate of reaction [A2′] greatly exceeds (three to four orders of magnitude) that of reaction [A1*/4*] for the concentrations employed, thus satisfying condition (61). Also shown is the rate of the overall reaction [A10′] evaluated by Eq. (42); this rate is seen to coincide with that of the rate-determining step [A1*/4*]. It is thus established that the Abel–Schmid mechanism is entirely consistent with values of the several constants of the magnitude assumed.

Similar examination has been made of the data of Komiyama and Inoue (1978) shown in Table 8 for representative conditions of that study. In most cases $R_{-A2'}$ sufficiently exceeds $2R_{A1*/4*}$ that the assumption of equilibrium [A2′] appears justified. However, at low $[HNO_2]$ and high k_{as}, $2R_{A1*/4*}$ becomes appreciable in comparison to $R_{-A2'}$ (as great as 40%), suggesting that this assumption may not be appropriate under these conditions.

4.2. Reaction [M2*/5*]

The rate coefficient $k_{M2*/5*}$ of the mixed-phase reaction

$$NO(g) + NO_2(g)(N_2O_3(g)) \xrightarrow{w} 2HNO_2(2H^+ + 2NO_2^-) \qquad [M2*/5*]$$

may be inferred indirectly from measurement of the kinetics of the reverse reaction, i.e., from measurement of the kinetics of formation of $N_2O_3(aq)$ from $HNO_2(aq)$,

$$2HNO_2(aq) \rightarrow N_2O_3(aq), \qquad [-A5']$$

Table 7. Rates of Elementary Reactions in HNO_2 Decomposition Study by Abel and H. Schmid

Run Number	$[HNO_2]^a$ (M)	$p_{NO}{}^a$ (atm)	[NO(aq)] (M)	$[NO_2(aq)]^b$ (M)	$R_{-A2'}{}^b$ (M s^{-1})	$R_{A1*/4*}{}^b$ (M s^{-1})	$R_{A10', meas}{}^a$ (M s^{-1})
17	0.025	0.47	9.1(−4)	8.6(−8)	3.5(−3)	4.5(−7)	4.6(−7)
18	0.040	0.47	9.1(−4)	2.2(−7)	9.0(−3)	2.9(−6)	3.0(−6)
19	0.025	0.286	5.5(−4)	1.4(−7)	3.5(−3)	1.2(−6)	1.2(−6)
2	0.075	0.98	1.9(−3)	3.7(−7)	3.2(−2)	8.3(−6)	8.6(−6)
7	0.050	0.965	1.9(−3)	1.7(−7)	1.4(−2)	1.7(−6)	1.7(−6)
20	0.040	0.495	9.6(−4)	2.1(−7)	9.0(−3)	2.5(−6)	2.8(−6)

[a] Concentrations and rates given by Abel and H. Schmid (1928c, p. 293).

[b] Constants employed: $K_{A2'} = 8 \times 10^6$; $k_{-A2'} = 5.6\ M^{-1}\ s^{-1}$; $k_{A1*/4*} = 6 \times 10^7\ M^{-1}\ s^{-1}$; k_{fwd} [Eq. (42)] $= 2.6 \times 10^{-1}$ atm^2 M^{-3} s^{-1}.

Table 8. Rates of Elementary Reactions in HNO_2 Decomposition Study by Komiyama and Inoue

k_{as}[a] (s^{-1})	$[HNO_2]$[a] (M)	[NO(aq)][b] (M)	$[NO_2(aq)]$[b] (M)	$R_{-A2'}$[b] ($M\ s^{-1}$)	$R_{A1*/4*}$[b] ($M\ s^{-1}$)	$R_{A10',meas}$[a] ($M\ s^{-1}$)
2.4(−1)	5(−2)	3.7(−4)	8.5(−7)	1.4(−2)	4.3(−5)	7.2(−5)
2.4(−1)	1(−2)	4.3(−5)	2.9(−7)	5.6(−4)	5.1(−6)	
2.4(−1)	1(−3)	2.0(−6)	6.3(−8)	5.6(−6)	2.4(−7)	
2.4(−1)	1(−4)	9.2(−8)	1.4(−8)	5.6(−8)	1.1(−8)	
1.2(−1)	5(−2)	4.6(−4)	6.7(−7)	1.4(−2)	2.7(−5)	3.9(−5)
1.2(−1)	1(−2)	5.4(−5)	2.3(−7)	5.6(−4)	3.2(−6)	
1.2(−1)	1(−3)	2.5(−6)	5.0(−8)	5.6(−6)	1.5(−7)	
1.2(−1)	5(−4)	1.0(−6)	3.1(−8)	1.4(−6)	5.9(−8)	
4.5(−2)	5(−2)	6.4(−4)	4.9(−7)	1.4(−2)	1.4(−5)	2.1(−5)
4.5(−2)	1(−2)	7.5(−5)	1.7(−7)	5.6(−4)	1.7(−6)	
4.5(−2)	1(−3)	3.5(−6)	3.6(−8)	5.6(−6)	7.8(−8)	
4.5(−2)	5(−4)	1.4(−6)	2.3(−8)	1.4(−6)	3.1(−8)	
7.7(−3)	5(−2)	1.2(−3)	2.7(−7)	1.4(−2)	4.4(−6)	5.8(−6)
7.7(−3)	1(−2)	1.3(−4)	9.3(−8)	5.6(−4)	5.2(−7)	
7.7(−3)	1(−3)	6.3(−6)	2.0(−8)	5.6(−6)	2.4(−8)	
7.7(−3)	5(−4)	2.5(−6)	1.3(−8)	1.4(−6)	9.5(−9)	

[a] Concentrations and rates from Komiyama and Inoue (1978), Figures 4 and 5.
[b] Evaluated by using $K_{A2'} = 8 \times 10^6$, $k_{-A2'} = 5.6\ M^{-1}\ s^{-1}$, $k_{A1*/4*} = 6 \times 10^7\ M^{-1}\ s^{-1}$.

as

$$k_{M2*/5*} = k_{-A5'} K_{M2'}. \tag{62}$$

The rate coefficient $k_{-A5'}$ has not been measured directly but has been inferred from the rate of diazotization, nitrosation, and isotope exchange reactions that are interpreted as proceeding through N_2O_3 intermediate. If the rate-limiting step of those reactions is [−A5′], the rate constant $k_{-A5'}$ may be determined from the rate of the overall reaction.

4.2.1. Diazotization Studies

The rate and mechanism of diazotization reactions by N_2O_3 in aqueous solution has been of interest for some time (Ridd, 1961). Of importance in the present context is the observation (Bunton et al., 1961; Hughes and Ridd, 1958; Kalatzis and Ridd, 1966) that the diazotization or nitrosation of amines under certain conditions exhibits a rate law of the form

$$R_d = k_{obs}[HNO_2]^2 \tag{63}$$

independent of the concentration of the amine. Moreover, over a limited range of basicity, the observed rate constant is independent of the nature of the amine. This rate law has been interpreted in terms of the mechanism

$$2HNO_2 \xrightarrow{k_{-A5'}} N_2O_3$$

$$N_2O_3 \xrightarrow{k_{A5'}} 2HNO_2$$

$$N_2O_3 + ArNH_2 \xrightarrow{k_d} ArN_2^+ + NO_2^- .$$

Making the steady-state assumption for $[N_2O_3]$, one obtains

$$R_d = k_d[N_2O_3][ArNH_2]$$

$$= \frac{k_{-A5'}k_d[ArNH_2][HNO_2]^2}{k_{A5'} + k_d[ArNH_2]} .$$

For $k_d[ArNH_2] \gg k_{A5'}$,

$$R_d = k_{-A5'}[HNO_2]^2,$$

establishing identification of the observed rate constant k_{obs} [Eq. (63)] with $k_{-A5'}$.

Values of k_{obs} have been reported by Hughes and Ridd (1958) and Kalatzis and Ridd (1966), respectively, for diazotization of aniline and for nitrosation of *N*-methyl aniline at 0°C. Evaluation of k_{obs} from the experimental measurements is somewhat sensitive to the value of the acid dissociation constant of nitrous acid, K_{12}, employed. For $K_{12} = 3.5 \times 10^{-4}$ M at 0°C (as calculated from Table 2), we obtain $k_{obs} = 0.78$ M^{-1} s^{-1} at 0°C for both reactions. By the identification above, this value may be taken for $k_{-A5'}$ at this temperature.

Recently a value of 12.4 kcal mol^{-1} has been reported for the activation energy of $k_{-A5'}$ based on the temperature dependence of the rate of aniline diazotization (Ford, 1980). This value, in conjunction with the above value of $k_{-A5'}$ at 0°C, yields the value 5.3 M^{-1} s^{-1} at 25°C.

4.2.2. *Isotope Exchange Studies*

The rate and mechanism of oxygen–isotope exchange between nitrous acid and water has been examined by Bunton et al. (1959; 1961). The isotopic equilibration was found to exhibit the rate law

$$R_{ex} = k_{ex}[HNO_2]^2.$$

The exchange reaction also was interpreted as proceeding by means of N_2O_3 as an intermediate:

$$HONO + HONO^* \xrightarrow{\text{slow}} N_2O_3 + H_2O^*$$

$$H_2O + N_2O_3 \xrightarrow{\text{fast}} 2HONO,$$

leading to identification of k_{ex} with $k_{-A5'}$.

For k_{ex} at 0°C, Bunton et al. (1959) give 0.51 M^{-1} s^{-1}, in reasonably good agreement with the value given for $k_{A5'}$ based on diazotization kinetics, 0.78 $M^{-1} s^{-1}$. Subsequently (Bunton et al., 1961), the diazotization and exchange reactions were studied under "identical conditions," and "almost identical" reaction rates were reported, although, unfortunately, no details of that study have been given.

Bunton et al. (1959) also examined the isotope exchange kinetics at 25°C, from which a value of $k_{ex} = 5.8 \pm 15\%$ $M^{-1} s^{-1}$ may be inferred, based on two measurements, in close agreement with the 25°C value for $k_{-A5'}$ obtained from diazotization kinetics. We thus take the value $k_{-A5'} = (5.6 \pm 15\%)$ $M^{-1} s^{-1}$ representing the mean of these two determinations. This value of $k_{-A5'}$, in conjunction with $K_{M2'} = 126$ M^2 atm^{-2} at 25°C (Table 2), permits $k_{M2^*/5^*}$ to be evaluated by Eq. (63). The resulting value, 7.0×10^2 M atm^{-2} s^{-1}, is given in Table 9. Also given is $k_{M5^*/2^*}$, evaluated as $K_{M2^*/5^*}/K_{G8}$.

4.2.3. *Temperature Dependence*

As noted above, the activation energy of $k_{-A5'}$ inferred from diazotization kinetics (Ford, 1980) was 12.4 kcal mol^{-1}. The value inferred from the isotope exchange kinetics (Bunton et al., 1959) is 15.8 kcal mol^{-1}, yielding an average value and range of 14.1 ± 1.7 kcal mol^{-1}. From (63) it is seen that the activation energy associated with $k_{M2^*/5^*}$ may be evaluated as

$$E_a[\text{M2*/5*}] = E_a[-\text{A5}'] + \Delta H°[\text{M2}'].$$

From the value $\Delta H°[\text{M2}'] = -18.2$ kcal mol^{-1} (Table 2) we obtain $E_a[\text{M2*/5*}] = -(4.1 \pm 1.7)$ kcal mol^{-1}. The overall negative activation energy (increasing rate with decreasing temperature) would appear to reflect an increased solubility at lower temperatures outweighing a presumably decreasing rate of reaction [A2*/5*]. We note (Table 2) the negative enthalpy of solution of NO, $\Delta H°(H_{NO}) = -2.9$ kcal mol^{-1}.

The activation energy associated with $k_{M5^*/2^*}$ evaluated as

$$E_a[\text{M5*/2*}] = E_a[\text{M2*/5*}] - \Delta H°[\text{G8}]$$

Table 9. Experimental Determination of $k_{-A5'}$ and $k_{M2^*/5^*}$ at 25°C[a]

Rate Constant	Diazotization	Isotope Exchange	Mean	Units
$k_{-A5'}$	5.3[a]	5.8[b]	5.6 ± 15%	$M^{-1} s^{-1}$
$k_{M2^*/5^*}$[c]	6.7(2)	7.3(2)	7.0(2)	M atm^{-2} s^{-1}
$k_{M5^*/2^*}$[d]	1.25(3)	1.37(3)	1.3(3)	M atm^{-1} s^{-1}

[a] From measurements of Hughes and Ridd (1958) and Kalatzis and Ridd (1966) at 0°C, as corrected to 25°C by employing the activation energy reported by Ford (1980).

[b] Measurements by Bunton et al. (1959).

[c] Evaluated as $k_{M2^*/5^*} = k_{-A5'}K_{M2'}$ for $K_{M2'} = 126$ M^2 atm^{-2}.

[d] Evaluated as $k_{M5^*/2^*} = k_{M2^*/5^*}/K_{G8}$ for $K_{G8} = 0.535$ atm^{-1}.

is $+5.4 \pm 1.7$ kcal mol^{-1}, where we have employed the value of $\Delta H°$[G8] in Table 2.

4.3. Reaction [M11′]

In a recent study Epstein et al. (1980) have examined the kinetics of the oxidation of ferrous ion by nitric acid,

$$3Fe^{2+} + 4H^+ + NO_3^- \overset{W}{=} 3Fe^{3+} + NO.$$

The kinetics of this system are complex and, as the authors point out, have yet to be entirely elucidated. The approach taken by Epstein et al. was to describe the evolution of the concentrations of the several species in terms of a set of coupled ordinary differential equations in these concentrations. The computed time profiles of various observables (extinction, redox potential) were compared to observation as a function of parameters (rate constants, H_{NO_2}, initial HNO_2 concentration), and values of these parameters were inferred from the best fit to observation. By this approach the authors deduced inter alia the rate of the reaction

$$H^+ + NO_3^- + NO(aq) = HNO_2 + NO_2(aq). \qquad [-A11']$$

The rate expression given was

$$R_{-A11'} = k_{(64)}[H^+]^2[NO_3^-][NO(aq)]; \qquad (64)$$

we choose to write this as

$$R_{-A11'} = k_{-A11'}[H^+][NO_3^-][NO(aq)], \qquad (65)$$

where $k_{-A11'}$, the effective rate constant for reaction [$-$A11′] as written, incorporates the additional hydrogen-ion dependence, i.e., $k_{-A11'} = k_{(64)}[H^+]$. From the numerical value given by Epstein et al. for $k_{(64)}$, we obtain $k_{-A11'} = 7.5\ [H^+]\ M^{-2}\ s^{-1}$ at 23°C. The uncertainty in the value of the rate constant given by Epstein et al. was claimed to be "factor-of-two"; however, ionic strength corrections, which were estimated only by means of the Debye–Hückel formula, which would break down at concentrations well below the ionic strength of 2.1 M employed, led the authors to state that the values of the rate constants obtained were "probably accurate to no better than an order of magnitude."

Knowledge of the value of the effective rate constant $k_{-A11'}$ permits evaluation of the rate constant of the reverse reaction

$$NO_2(g) + HNO_2(aq) \rightarrow H^+ + NO_3^- + NO \qquad [M11']$$

by means of the equilibrium constant of the reaction

$$NO_2(g) + HNO_2 = NO(aq) + H^+ + NO_3^- \qquad [M11\dagger]$$

as

$$k_{M11'} = K_{M11\dagger}k_{-A11'}.$$

The equilibrium constant $K_{M11\dagger}$ may be evaluated as $K_{M11'}H_{NO}$ and has the value 7.58 M^2 atm^{-1} at 25°C (data from Table 2). The resulting value of $k_{M11'}$ is $5.7 \times 10^1[H^+]$ atm^{-1} s^{-1}.

In view of the indirect means by which Epstein et al. inferred the rate of reaction [−A11′] in their system, it is of interest to address the consistency of this rate with other experimental studies. We first consider studies of reaction

$$2NO(aq) + H^+ + NO_3^- \overset{w}{\rightarrow} 3HNO_2(aq). \qquad [-A10']$$

Assuming that reaction (−A11′) occurs (as claimed by Epstein et al.), the overall reaction [−M10′] may proceed by way of not only the sequence discussed in Section 4.1, but also the sequence

$$2NO(g) = 2NO(aq) \qquad [2H_{NO}]$$

$$NO(aq) + H^+ + NO_3^- \rightarrow HNO_2 + NO_2(aq) \qquad [-A11']$$

$$NO(aq) + NO_2(aq) \overset{w}{=} 2HNO_2 \qquad [A2']$$

$$2NO(g) + H^+ + NO_3^- \overset{w}{\rightarrow} 3HNO_2 \qquad [-M10']$$

For reaction [A2′] fast, reaction [−A11′] is rate determining, and the rate of [−A10′] may be evaluated as

$$R_{-A10'} = k_{-A11'}[NO(aq)][H^+][NO_3^-]. \qquad (66)$$

For NO at Henry's law equilibrium (phase-mixed limit), this rate is

$$R_{-M10'} = k_{-A11'}H_{NO}p_{NO}[H^+][NO_3^-]. \qquad (67)$$

This predicted rate law stands at variance with the observed rate law for this reaction discussed above (Section 4.1),

$$R_{-M10'} = k_{rev}[H^+][NO_3^-][HNO_2]. \qquad (50)$$

This inconsistency may be addressed further by comparing the rates given by (67) and (50) with experimental data. This is done in Table 10, without correction for ionic strength. It is seen that the rate of reaction [−M10′] predicted by the mechanism incorporating the reaction proposed by Epstein et al. exceeds the measured rate by one to two orders of magnitude. Furthermore, this rate law predicts neither the observed dependence on $[HNO_2]$ (cf. runs 5 and 6) nor the observed independence on p_{NO} (cf. runs 19 and Table 9). It is thus established that this rate law is inconsistent with the measured rates.

It is of interest also to explore the consequences of the rate of reaction [A11′] proposed by Epstein et al. (1980) on the kinetics of nitrous acid

Table 10. Comparison of Measured[a] and Calculated Rates of Reaction [−M10′]

Run Number	$[H^+]$ (M)	$[NO_3^-]$ (M)	$[HNO_2]$ 10^{-3} M	p_{NO} (atm)	$R_{-M10'}$ $(M\ s^{-1})$ Measured	Eq. (50)	Eq. (67)
19	0.050	0.060	10.0	1	1.9(−7)	2.7(−7)	2.2(−6)
20	0.050	0.120	10.0	1	3.9(−7)	5.3(−7)	4.3(−6)
13–15	0.197	0.206	8.8	1	1.8(−6)	3.2(−6)	1.2(−4)
5	0.203	0.162	9.8	1	1.5(−6)	2.9(−6)	9.8(−5)
6	0.203	0.162	2.4	1	3.8(−7)	7.0(−7)	9.7(−5)
Table 9	0.050	0.060	10.0	0.514	2.1(−7)	2.7(−7)	1.1(−6)

[a] Data of Abel and H. Schmid (1928d, pp. 141–142).

decomposition. If, as was assumed in the Abel–Schmid mechanism treated above, it is assumed that reaction [−A2′] is in equilibrium, the subsequent occurrence of reaction [A11′] would yield the following mechanism:

$$2HNO_2 \overset{w}{=} NO_2(aq) + NO(aq) \qquad [-A2']$$

$$NO_2(aq) + HNO_2 \rightarrow H^+ + NO_3^- + NO(aq) \qquad [A11']$$

$$2NO(aq) = 2NO(g) \qquad [-2H_{NO}]$$

$$3HNO_2 \overset{w}{\rightarrow} H^+ + NO_3^- + 2NO(g) \qquad [M10']$$

i.e., the overall stoichiometry that is observed. This mechanism yields the kinetic rate law

$$R_{M10'} = k_{A11'}K_{-A2'}\frac{[HNO_2]^3}{[NO(aq)]}$$

$$= k_{M11'}K_{M2'}^{-1}\frac{[HNO_2]^3}{p_{NO}} \qquad (68)$$

The form of (68) is seen to be inconsistent with the observed rate law (Abel and H. Schmid, 1928c),

$$R_{M10'} = k_{fwd}\frac{[HNO_2]^4}{p_{NO}^2}, \qquad (42)$$

raising further concern about the validity of the rate law (64) for reaction [−A11′] proposed by Epstein et al. This inconsistency is addressed quantitatively in Table 11, in which the rate of reaction [M10′] evaluated by (68) is compared to the measured rates. Again the rate law inferred from (64) is entirely at variance with the measurements. In particular, those runs (5, 6, 14–16) for which the calculated rate exceeds the measured rate (by a factor of as much as 4) would appear to exclude a rate law of the form proposed by Epstein et al.

Table 11. Comparison of Measured[a] and Calculated Rates of HNO_2 Decomposition

Run Number	$[HNO_2]$ 10^{-3} M	$[H^+]$ 10^{-3} M	$R_{-M10'}$ $(M\ s^{-1})$ Measured	Eq. (68)
5	75	99	8.6(−6)	1.9(−5)
6	100	105	3.0(−5)	4.8(−5)
10	25	0.6	9.1(−8)	4.2(−9)
11	50	1.2	1.9(−6)	6.8(−8)
12	50	0.6	1.6(−6)	3.4(−8)
13	50	0.6	1.8(−6)	3.4(−8)
14	50	60.0	2.1(−6)	3.4(−6)
15	50	52.5	1.7(−6)	6.6(−6)
16	50	150	1.8(−6)	8.5(−6)

[a] Concentrations and measurements from Abel and H. Schmid (1928c, pp. 288, 292); $p_{NO} \approx 1$ atm.

Further evidence against the rate law proposed by Epstein et al. may be adduced from the observation of an induction period in the autocatalytic reaction

$$H^+ + NO_3^- + 2NO(g) \xrightarrow{w} 3HNO_2. \qquad [-M10']$$

An example is given by Jordan and Bonner (1973). For 0.46 M HNO_3 and $p_{NO} = 0.20$ atm, the time taken to reach equilibrium HNO_2 concentration (0.06 M) was 20–60 min. This may be compared to that calculated for the mechanism [$2H_{NO}$]–[−A11′]–[A2′] with rate law (67). The time required to reach equilibrium by this rate may be approximated as

$$\tau_e \approx \frac{[HNO_2]_e}{d[HNO_2]/dt}$$

$$\approx \frac{[HNO_2]_e}{3R_{-M10'}}.$$

Evaluation of this quantity for the concentrations given yields $\tau_e \approx 1$ min, much shorter than that observed.

In conclusion of this section it would appear that little case can be made for the rate expression (64) advanced by Epstein et al. (1980). The evidence in favor of this expression is indirect, and as the authors themselves point out, one must be extremely cautious in drawing quantitative inferences about rate constants from simulations of such complex systems. There is, moreover, strong evidence against the rate expression (64) from several studies. We thus conclude that the rate of the elementary reaction [−A11′] is substantially less than that postulated by Epstein et al. and hence that evidence in support of this reaction must be discounted.

5. DIRECT MEASUREMENT OF AQUEOUS-PHASE KINETICS

In view of the difficulties of interpreting indirect studies and in view also of the difficulty of direct measurement of mixed-phase reaction rates in the nitrogen oxide–water system, an attractive alternative approach to these quantities is by means of direct measurement of the aqueous-phase rate constants. Measurement of these rate constants is based on the concept of rapidly preparing an aqueous solution of the nitrogen oxide(s) of interest and monitoring the subsequent evolution of the dissolved species. If this can be done on a time scale short compared to coupling to the gas phase (e.g., "outgassing"), the possibility exists of determining the aqueous-phase rate constant of the reaction under examination. We address in this section studies of this sort where the aqueous nitrogen oxide(s) are produced by pulse radiolysis or flash photolysis or are directly injected into water under high pressure.

5.1. Reaction [A7]

In the study by Grätzel et al. (1969), $NO_2(aq)$ was produced by pulse radiolysis of an aqueous solution of nitrite in the presence of N_2O [the experiment is outlined more completely in the discussion of equilibrium [A7] given by Schwartz and White (1981)]. The concentration of NO_2 initially formed was determined absolutely by pulse dosimetry. The decay of $[NO_2]$, which was monitored by optical absorption at 400 nm, took place initially on a microsecond time scale and subsequently on a millisecond time scale.

The initial rapid decrease in $[NO_2(aq)]$ is ascribed to the reaction

$$2NO_2(aq) \xrightarrow{k_{A7}} N_2O_4(aq).$$

A second-order initial rate law was observed

$$\frac{d[NO_2]}{dt} = -2k_{A7}[NO_2]^2 \tag{69}$$

for initial $[NO_2]$ ranging from 8×10^{-6} to 30×10^{-6} M. The authors derived k_{A7} from the initial rate data only (i.e., neglecting the back reaction). From the integrated form of (69), one obtains for $\tau_{3/4}$, the time for the absorbance of NO_2 to decrease to three-fourths of its initial value,

$$\tau_{3/4} = \frac{1}{3[NO_2]_0}\frac{1}{2k_{A7}}.$$

From a graph of $\tau_{3/4}$ versus $[NO_2]_0^{-1}$ (as determined from the measured absorbance at 400 nm, making use of the extinction coefficient also derived in that study), Grätzel et al. (1969) obtained the value $k_{A7} = (4.5 \pm 1.0) \times 10^8\ M^{-1}\ s^{-1}$.

5.2. Reaction [A8]

In the presence of excess NO (2×10^{-5} to 1×10^{-4} M), NO_2 formed in the pulse radiolysis was found to react according to

$$NO(aq) + NO_2(aq) \underset{k_{-A8}}{\overset{k_{A8}}{\rightleftharpoons}} N_2O_3(aq);$$

the kinetics of approach to equilibrium [A8] was monitored (Grätzel et al., 1970) by the rate of increase of absorption at 260 nm (N_2O_3) or by the rate of decrease of absorption at 400 nm (NO_2). The N_2O_3 concentration was observed to exhibit the time dependence

$$[N_2O_3] = [N_2O_3]_f(1 - e^{-t/\tau}),$$

where $[N_2O_3]_f$ represents the final value of $[N_2O_3]$. From consideration of reversible pseudo-first-order kinetics ([NO(aq)] is in excess and effectively constant), the time constant τ is given (Eigen and DeMaeyer, 1963) as

$$\tau^{-1} = k_{A8}[NO] + k_{-A8'},$$

and hence, by (5),

$$\tau^{-1} = k_{A8}([NO] + K_{A8}^{-1}).$$

From a graph of τ^{-1} versus [NO], k_{A8} was determined as 1.1×10^9 M^{-1} s^{-1} (20°C).

5.3. Reaction [A1*/4*]

As discussed in Section 3.1, $NO_2(aq)$ and/or $N_2O_4(aq)$ react with water according to the stoichiometry

$$2NO_2(aq)(N_2O_4(aq)) \xrightarrow{W} H^+ + NO_3^- + HNO_2(H^+ + NO_2^-). \quad [A1*/4*]$$

If, as is the case even in the pulse radiolysis experiments, equilibrium [A7] is established more rapidly than is [A1*/4*], one cannot distinguish between the two paths. We have chosen [see Eq. (20)] to express the rate of reaction referred to two NO_2 molecules as reagents, i.e.,

$$R_{A1*/4*} = k_{A1*/4*}[NO_2]^2. \quad (70)$$

5.3.1. *Pulse Radiolysis and Flash Photolysis Studies*

In the pulse radiolysis experiment by Grätzel et al. (1969), which was conducted in slightly basic solution, the rate of [A1*/4*] was monitored as a decrease in $[OH^-]$, itself monitored by electrical conductivity. Under these conditions

$$R_{A1*/4*} = -\frac{1}{2}\frac{d[OH^-]}{dt} = -\frac{1}{2}\frac{d[N(IV)]}{dt}, \quad (71)$$

where $[N(IV)] \equiv [NO_2] + 2[N_2O_4]$. Assuming equilibrium [A7] to be established, i.e., $[N_2O_4] = K_{A7}[NO_2]^2$, one may express $[NO_2]$ in terms of $[N(IV)]$ as

$$[NO_2] = \frac{(8K_{A7}[N(IV)] + 1)^{1/2} - 1}{4K_{A7}}. \tag{72}$$

Observing that $-d[N(IV)]/dt = 2k_{A1^*/4^*}[NO_2]^2$, one obtains a differential equation in $[N(IV)]$,

$$-\frac{d[N(IV)]}{dt} = k_{A1^*/4^*}K_{A7}\left\{[N(IV)] - \frac{K_{A7}}{4}\left[\left(\frac{8[N(IV)]}{K_{A7}} + 1\right)^{1/2} - 1\right]\right\}. \tag{73}$$

Equation (73) may exhibit apparent first- or second-order kinetics in the limit as N(IV) is present predominantly as N_2O_4 or NO_2, respectively, as governed by equilibrium [A7]. The distribution of N(IV) between NO_2 and N_2O_4 is shown in Figure 1 as a function of total N(IV) concentration. From inspection of this diagram we may place the bounds on the N(IV) concen-

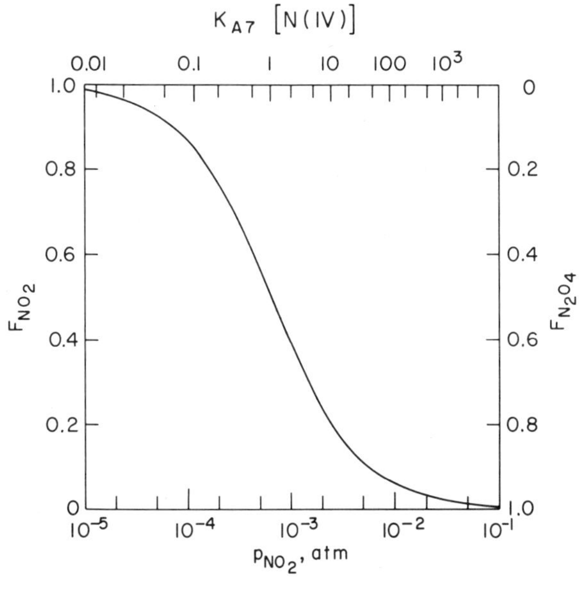

Figure 1. Equilibrium distribution of aqueous N(IV) between monomeric and dimeric forms as a function of K_{A7} [N(IV)] (top scale) and of p_{NO_2} (bottom scale); F_{NO_2} and $F_{N_2O_4}$ represent fraction of N(IV) present as NO_2 and N_2O_4, respectively. In evaluation as function of p_{NO_2}, H_{NO_2} was taken as 1.2 M atm^{-1} and K_{A7} as 6.5×10^4 M^{-1}.

tration regime for first- or second-order kinetics as

$$-\frac{d[\mathrm{N(IV)}]}{dt} = \begin{cases} \dfrac{k_{\mathrm{A1^*/4^*}}}{K_{\mathrm{A7}}}[\mathrm{N(IV)}], & [\mathrm{N(IV)}] \gtrsim 50K_{\mathrm{A7}}^{-1} \quad (74) \\ \\ 2k_{\mathrm{A1^*/4^*}}[\mathrm{N(IV)}]^2, & [\mathrm{N(IV)}] \lesssim 0.06K_{\mathrm{A7}}^{-1} \quad (75) \end{cases}$$

Experimentally it was found that the half-life for reaction [A1*/4*] increased from 3 to 14 ms with decreasing $[\mathrm{N(IV)}]_0$ in the range $(0.46\text{–}6) \times 10^{-6}$ M (cf. $K_{\mathrm{A7}}^{-1} = 15.3 \times 10^{-6}$ M). Grätzel et al. interpreted these data according to mixed-order kinetics [Eq. (73)], making use of their value of K_{A7}. The value of the rate constant determined in this way, expressed as $k_{\mathrm{A1^*/4^*}}$, is $k_{\mathrm{A1^*/4^*}} = (6.5 \pm 10\%) \times 10^7\ \mathrm{M^{-1}\ s^{-1}}$.

Reaction [A1*/4*] was studied also (by pulse radiolysis and flash photolysis) by Treinin and Hayon (1970). In this study NO_2(aq) in the absence of NO(aq) was produced by pulse radiolysis of nitrite solution in the presence of N_2O [the same method as employed by Grätzel et al. (1969)] or by flash photolysis of nitrate in the presence of nitrite:

$$NO_3^- + h\nu \rightarrow NO_2 + O^-$$

followed by

$$O^- + NO_2^- \xrightarrow{\mathrm{w}} NO_2 + 2OH^-. \qquad [A16]$$

The extent of reaction was followed by monitoring NO_2 or N_2O_4 spectrophotometrically. Treinin and Hayon obtained essentially the same value of K_{A7} as was obtained by Grätzel et al. However, in the kinetic study, in contrast to the findings of Grätzel et al., Treinin and Hayon observed an apparent first-order decay of $[NO_2]$ that was independent of $[NO_2]$ throughout the range studied [$(0.25\text{–}22) \times 10^{-6}$ M, corresponding to [N(IV)] in the range $(0.26\text{–}100) \times 10^{-6}$ M]. The value of the observed first-order rate constant was $k_{\mathrm{obs}} \approx 300\ \mathrm{s^{-1}}$ at 25°C, with the reaction half-life in no case greater than 3.6 ms (cf. 3–14 ms observed by Grätzel et al.). These authors also addressed the possible dependence of k_{obs} on nitrite and nitrate concentrations (≤ 0.1 M) and pH (5.0–11.4) and found k_{obs} to be independent of these variables.

The observations by Treinin and Hayon may be interpreted as follows, bearing in mind that in this study the observed reactant was NO_2. We note that

$$2[NO_2]\frac{d[NO_2]}{dt} = K_{\mathrm{A7}}^{-1}\frac{d[N_2O_4]}{dt},$$

and hence

$$\frac{d[\mathrm{N(IV)}]}{dt} \equiv \frac{d[NO_2]}{dt} + 2\frac{d[N_2O_4]}{dt} = (1 + 4K_{\mathrm{A7}}[NO_2])\frac{d[NO_2]}{dt}.$$

From Eq. (70) we thus obtain

$$-\frac{d[NO_2]}{dt} = \frac{2k_{A1^*/4^*}[NO_2]^2}{1 + 4K_{A7}[NO_2]}. \tag{76}$$

At high $[NO_2]$ this reduces to

$$-\frac{d[NO_2]}{dt} = \frac{k_{A1^*/4^*}}{2K_{A7}}[NO_2], \qquad [NO_2] \gtrsim 2.5\, K_{A7}^{-1}. \tag{77}$$

Setting (77) equal to the observed first-order rate constant, we obtain $k_{A1^*/4^*} = (4.7 \pm 22\%) \times 10^7\ M^{-1}\ s^{-1}$, which is in reasonable agreement with the value determined by Grätzel et al. Unfortunately for this interpretation, it is seen that the condition (77) is scarcely fulfilled at the highest initial NO_2 concentrations employed by Treinin and Hayon ($4K_{A7}[NO_2]_{max} = 7$). Thus, to account for the observed first-order kinetics, Treinin and Hayon proposed an additional reaction sequence

$$NO_2(aq) + X \rightarrow NO_2 \cdot X \qquad [A17]$$

$$NO_2 \cdot X + NO_2(aq) \xrightarrow{W} 2H^+ + NO_2^- + NO_3^-(+X) \qquad [A18]$$

where X (which may be water itself) is present at constant composition. For [A17] fast, the sequence [A17]–[A18] has rate

$$R_{A17-A18} = k_{A17}[NO_2],$$

and hence (70) becomes

$$-\frac{d[N(IV)]}{dt} = 2k_{A1^*/4^*}[NO_2]^2 + 2k_{A17}[NO_2]$$

leading to

$$-\frac{d[NO_2]}{dt} = 2k_{A17}[NO_2]\left\{\frac{1 + (k_{A1^*/4^*}/k_{A17})[NO_2]}{1 + 4K_{A7}[NO_2]}\right\}. \tag{78}$$

Equation (78) is seen to explain the first-order decay of $[NO_2]$ with $k_{obs} = 2k_{A17}$, provided the quantity in braces remains approximately constant over the range of $[NO_2]$ investigated. The latter condition is satisfied for $k_{A1^*/4^*} = 4K_{A7}k_{A17} = 2K_{A7}k_{obs}$ and is seen to be equivalent to the high-$[NO_2]$ limit to (76); i.e., $k_{A1^*/4^*} = 4.7 \times 10^7\ M^{-1}\ s^{-1}$, which agrees with $k_{A1^*/4^*}$ determined by Grätzel et al. The apparent first-order rate constant was determined as $k_{A17} = k_{obs}/2 = (1.5 \pm 22\%) \times 10^2\ s^{-1}$ at 25°C. The temperature dependence of the hydrolysis rate was examined as well. For temperatures of up to 76°C, NO_2 continued to react with apparent first-order kinetics, with an apparent activation energy of 4.7 kcal mol^{-1}.

The experimental results of Treinin and Hayon disagree with those of Grätzel et al., which showed an increase in the characteristic lifetime of aqueous N(IV) well beyond the value given by the first-order mechanism proposed by Treinin and Hayon. We are unable to explain this inconsistency.

However, we are somewhat uneasy with Treinin and Hayon's explanation for the constancy of k_{obs} in terms of the coincidence of $k_{A1*/4*}$ and $4K_{A7}k_{A17}$, especially as this coincidence is required to obtain over the temperature range 25–76°C.

Reaction [A1*/4*] was investigated by Ottolenghi and Rabani (1968) in the ferrocyanide-sensitized flash photolysis of nitrate. On ultraviolet excitation, ferrocyanide efficiently ejects an electron, which, the authors argued, reacts with nitrate to form NO_2:

$$e(aq)^- + NO_3^- \xrightarrow{w} NO_2 + 2OH^-.$$

The NO_2 thus formed reacted either with ferrocyanide

$$NO_2 + Fe(CN)_6^{4-} \rightarrow Fe(CN)_6^{3-} + NO_2^- \qquad [A19]$$

or by reaction [A1*/4*]; the competition between these reactions was varied by varying $[NO_2]_0$ and/or $[Fe(CN)_6^{4-}]$. The time-dependent concentrations (over some tens of milliseconds) of NO_2, $Fe(CN)_6^{4-}$, and $Fe(CN)_6^{3-}$ were inferred from the optical extinction measured at 420 nm. At low $[NO_2]_0$, NO_2 reacted exclusively by reaction [A19], permitting determination of k_{A19}. With increasing $[NO_2]_0$, 1.6–7.3 × 10^{-6} M, an increasing proportion of NO_2 reacted by [A1*/4*]; $k_{A1*/4*}$ was determined by fit to the time-dependent absorption measurements as (3.8 ± 1.4) × 10^7 M^{-1} s^{-1} (temperature unspecified but presumably 20–25°C).

In our judgment this experiment should be taken as yielding only semiquantitative information on the rate and mechanism of reaction [A1*/4*]. First, we note that Ottolenghi and Rabani's estimated error bars on the extinction coefficient of NO_2 [obtained by difference in the presence of the much more intensely absorbing $Fe(CN)_6^{4-}$] are ±50%. Second, reaction [A1*/4*] was apparently interpreted entirely as a second-order reaction in NO_2,

$$-\frac{d[NO_2]}{dt} = 2k_{A1*/4*}[NO_2]^2$$

rather than by mixed-order kinetics [Eq. (76)], although mixed-order kinetic analysis is mandatory at the apparent NO_2 concentrations employed ($4K_{A7}[NO_2]_0$ = 0.4–1.9). The failure to use a mixed-order analysis would result in a value of $k_{A1*/4*}$ that is too low; compare this with the value of 6.5 × 10^7 M^{-1} s^{-1} determined by Grätzel et al. More important, however, is the fact that there was no indication of a process (e.g., reaction [17] giving rise to a first-order rate law for reaction [A1*/4*], although this process would dominate at low $[NO_2]$ if the rate law advanced by Treinin and Hayon were correct. Thus this experiment may be taken as evidence against the first-order rate law advanced by Treinin and Hayon.

5.3.2. Flow-Tube Study

Reaction [A1*/4*] was studied also by Moll (1966) by a rapid-mixing flow technique. Liquid N_2O_4 was injected under pressure into a rapidly flowing turbulent water stream. The rate of reaction as a function of distance downstream was determined by monitoring the longitudinal temperature profile, under the assumption that the heat evolution and hence the local temperature rise was proportional to the local rate of reaction. Rather high initial N_2O_4 concentrations were employed (8×10^{-2} M), corresponding to $[\mathrm{N(IV)}]_0 K_{A7} \approx 5000$, justifying the interpretation of the data in terms of first-order kinetics as treated by Moll. For the studies at 20°C Moll obtained a $1/e$ lifetime $\tau_{obs} = 3.77 \times 10^{-3}$ s, corresponding to $k_{A1^*/4^*} = 1.73 \times 10^7\ \mathrm{M^{-1}\,s^{-1}}$, a factor of roughly 4 and 2.3 lower than the values obtained by Grätzel et al. (1969) and Treinin and Hayon (1970), respectively.

A possible source of systematic error in Moll's experiment is that of a transition from first- to second-order kinetics as N(IV) is no longer predominately present as N_2O_4. Reexamination of Moll's data suggests the possibility of a somewhat greater initial rate, consistent with this interpretation. Utilizing only the initial data points from Moll's run ($x < 7$ cm), we obtain an upward revision of the rate coefficient ($\tau_{obs} = 2.56 \times 10^{-3}$ s; $k_{A1/4^*} = 2.6 \times 10^7\ \mathrm{M^{-1}\,s^{-1}}$), which nonetheless remains a factor of 2.5 and 1.6 lower than the values given by Grätzel et al. and Treinin and Hayon, respectively.

5.3.3. *Comparison of Rates of Reactions* [A7] *and* [A1*/4*]

Before concluding the discussion of reaction [A1*/4*], we observe that it is possible to address the relative rates of reactions [A7] and [A1*/4*] and thereby to address the assumption inherent throughout this discussion that equilibrium [A7] is achieved rapidly compared to the time scale of reaction [A1*/4*]. A quantitative test of this assumption is achieved by comparison of the relaxation time of reaction [A7], $\tau^{(7)}$, with the characteristic time of reaction [A1*/4*], $\tau^{(1/4)}$; a necessary and sufficient condition for equilibrium [A7] to be maintained is that $\tau^{(1/4)} \gg \tau^{(7)}$. The characteristic time $\tau^{(1/4)}$ may be evaluated as

$$\tau^{(1/4)} = (k_{A1^*/4^*}[\mathrm{NO_2(aq)}])^{-1}.$$

The relaxation time $\tau^{(7)}$ may be evaluated (Eigen and De Maeyer, 1963) as

$$\tau^{(7)} = (4k_{A7}[\mathrm{NO_2(aq)}] + k_{-A7})^{-1}.$$

By (5) $k_{-A7} = k_{A7}/K_{A7}$, and hence

$$\frac{\tau^{(1/4)}}{\tau^{(7)}} = \frac{k_{A7}}{k_{A1^*/4^*}}\left(4 + \frac{K_{A7}^{-1}}{[\mathrm{NO_2(aq)}]}\right).$$

From the data of Grätzel et al. (1969) $k_{A7} = 4.5 \times 10^8\ \mathrm{M^{-1}\,s^{-1}}$, $k_{A1^*/4^*} = 6 \times 10^7\ \mathrm{M^{-1}\,s^{-1}}$, and $K_{A7}^{-1} = 1.4 \times 10^{-5}$ M (see also Section 7.1). Thus it is seen that $\tau^{(1/4)}/\tau^{(7)}$ is at least as great as 30 and correspondingly greater for $[\mathrm{NO_2(aq)}]$ less than 10^{-5} M. This agreement thus establishes the appropriateness of the hybrid reaction treatment that has been adopted.

5.4. Reaction [A2*/5*]

In the study by Grätzel et al. (1970) the rate of hydrolysis of the equilibrium mixture of $NO{-}NO_2{-}N_2O_3$,

$$NO(aq) + NO_2(aq)(N_2O_3(aq)) \xrightarrow{w} 2HNO_2(2H^+ + 2NO_2^-), \qquad [A2^*/5^*]$$

was monitored as the rate of decrease in $[N_2O_3(aq)]$, as measured by absorption spectrophotometry at 260 nm. Under conditions of excess [NO(aq)], $[N_2O_3(aq)]$ was observed to decrease according to first-order kinetics, i.e.,

$$-\frac{1}{[N_2O_3(aq)]}\frac{d[N_2O_3(aq)]}{dt} = \tau_{obs}^{-1}. \tag{79}$$

The observed time constant τ_{obs} may be related to the hybrid rate constant $k_{A2^*/5^*}$ as follows. First, we note that at fixed [NO(aq)]—as is the case for $[NO(aq)] \gg [NO_2(aq)] + [N_2O_3(aq)]$—the ratio of $[N_2O_3]$ to the sum of $[N_2O_3] + [NO_2]$ is given by

$$\frac{[N_2O_3(aq)]}{[N_2O_3(aq)] + [NO_2(aq)]} = \frac{K_{A8}[NO(aq)]}{K_{A8}[NO(aq)] + 1} \tag{80}$$

a quantity that remains constant, independent of the extent of hydrolysis. Consequently, $[N_2O_3(aq)]$ will decrease in proportion to the sum of $[NO_2(aq)] + [N_2O_3(aq)]$,

$$\begin{aligned} -\frac{d[N_2O_3(aq)]}{dt} &= R_{A2^*/5^*}\frac{[N_2O_3(aq)]}{[N_2O_3(aq)] + [NO_2(aq)]} \\ &= R_{A2^*/5^*}\frac{K_{A8}[NO(aq)]}{K_{A8}[NO(aq)] + 1}. \end{aligned} \tag{81}$$

Since [see Eq. (32)]

$$R_{A2^*/5^*} = k_{A2^*/5^*}[NO(aq)][NO_2(aq)] \tag{82}$$

we obtain

$$\tau_{obs} = k_{A2^*/5^*}^{-1}([NO(aq)]^{-1} + K_{A8}). \tag{83}$$

Grätzel et al. (1970) obtained $\tau_{obs} = 3.2 \times 10^{-3}$ s for $[NO(aq)] = 1.04 \times 10^{-4}$ M at 20°C. Unfortunately, the dependence of τ_{obs} on [NO(aq)] was not examined. Such measurements would enhance our confidence in the interpretation of the measurements as well as provide an independent measurement of K_{A8}. Using the measurement of τ_{obs} at the single NO(aq) concentration employed and Grätzel et al.'s value of K_{A8} ($1.37 \times 10^4\ M^{-1}$), we obtain $k_{A2^*/5^*} = 7.4 \times 10^6\ M^{-1}\ s^{-1}$.

The possible interference of $NO_2(N_2O_4)$ hydrolysis (reaction [A1*/4*]) must be considered, since if this reaction were occurring in parallel to [A2*/5*], part of the decrease in $[NO_2]$ (and, in turn, of the observed $[N_2O_3]$) would be misascribed to [A2*/5*]. The ratio of rates of NO_2 loss by these two reactions is

$$\frac{R_{A1^*/4^*}}{R_{A2^*/5^*}} = \frac{k_{A1^*/4^*}[NO_2]}{k_{A2^*/5^*}[NO]} . \tag{84}$$

For Grätzel et al.'s conditions of [NO] = 1.04×10^{-4} M and initial [NO_2] = 2×10^{-6} M, this quantity is initially 0.17, decreasing as [NO_2] decreases during the course of the reaction. (Here we have used the values of $k_{A1^*/4^*}$ and $k_{A2^*/5^*}$ given above.) Thus the value of $k_{A2^*/5^*}$ may be somewhat too great, reflecting a slight contribution of reaction [A1*/4*]. For this reason also it would seem advisable to study reaction [A2*/5*] as a function of [NO(aq)] since the relative contribution of reaction [A1*/4*] would be lessened at higher [NO(aq)].

Reaction [A2*/5*] was studied also by Treinin and Hayon. Under the condition of high excess NO(aq) ([NO(aq)] = 1.9×10^{-3} M), N_2O_3(aq) (in equilibrium with NO_2(aq), prepared by flash photolysis of nitrite solution and monitored by ultraviolet (UV) absorption spectrophotometry, presumably at 280 nm) decayed by pseudo-first-order kinetics [Eq. (79)]. The value of τ_{obs} was observed to decrease with increasing [OH^-] in the pH range 7–10, as given by

$$\tau_{obs}^{-1} = (2 \times 10^3 + 10^8[OH^-])\ s^{-1},$$

indicating a hydroxide-ion-dependent mechanism in addition to reaction directly by [A2*/5*]. For pH $\lesssim$ 8, reaction [A2*/5*] is seen to dominate the hydrolysis mechanism with $\tau_{obs} = 0.5 \times 10^{-3}$ s. This value of τ_{obs} may be combined with K_{A8} [Eq. (83)] to yield $k_{A2^*/5^*}$. Again, unfortunately, the dependence of τ_{obs} on [NO(aq)] was not examined. Using Treinin and Hayon's value K_{A8} ($5 \times 10^4\ M^{-1}$), we obtain $k_{A2^*/5^*} = 1.0 \times 10^8\ M^{-1}\ s^{-1}$, more than an order of magnitude greater than that derived from the data of Grätzel et al. Alternatively, if we use Grätzel et al.'s value of K_{A8}, we obtain $k_{A2^*/5^*} = 2.8 \times 10^7\ M^{-1}\ s^{-1}$, still a factor of 4 greater than the value derived from the measurements by Grätzel et al.

The rate of decrease of [N_2O_3(aq)] was monitored also in the flash photolysis of nitrite solutions in the absence of added NO. Under these conditions, equal initial concentrations are expected for NO(aq) and NO_2(aq) since, according to the interpretation of this system advanced by Treinin and Hayon, the initial photolysis

$$NO_2^- + h\nu \rightarrow NO + O^- \tag{A20}$$

is followed by

$$O^- + NO_2^- \xrightarrow{w} NO_2 + 2\,OH^- . \tag{A16}$$

For reaction of NO and NO_2 entirely by [A2*/5*], it is seen that there is no net reaction, whereas for reaction of $2NO_2$ by [A1*/4*], one has the overall reaction stoichiometry

$$3NO_2^- \xrightarrow{w} 2NO + NO_3^- + 2OH^- .$$

Under conditions of low intensity, steady illumination of 2.7×10^{-4} M NO_2^- solution at 229 nm, Treinin and Hayon found a quantum yield for NO_2^- depletion of less than 10^{-3}. In the flash photolysis reactions also there was "no" net reaction of NO_2^-, as measured in the final NO_2^- absorbance, despite a "high" yield for the primary photolytic reaction [A20].

In the studies of the flash photolysis of nitrite in the absence of added NO, N_2O_3 (as monitored at 280 nm) was found to exhibit apparent first-order kinetics with $k_{obs} = (510 \pm 60)\ s^{-1}$; NO_2, monitored at 400 nm, also exhibited apparent first-order kinetics with $k_{obs} = (293 \pm 47)\ s^{-1}$. The measurements were interpreted as if the hydrolysis took place entirely by reaction [A2*/5*], in conformity with the observation of no net reaction. For hydrolysis by reaction [A2*/5*] only, the concentrations of NO and NO_2 would decrease at the same rate and thus remain equal throughout the course of the hydrolysis. Thus $[N_2O_3]$ would be proportional to $[NO_2]^2$; the approximate factor of 2 between the observed rate constants for N_2O_3 and NO_2 was taken to reflect this proportionality. However, for $[N_2O_3] \propto [NO_2]^2$, N_2O_3 would be expected to exhibit mixed-order kinetics, as was anticipated also for NO_2 in the study of reaction [A1*/4*] [Eq. (76)]. The failure to observe the transition from first- to second-order kinetics raises uncertainty with this study and indicates the need for further systematic investigation of this reaction system.

An additional serious concern with the photolysis mechanism proposed by Treinin and Hayon and/or with our understanding of both reactions [A1*/4*] and [A2*/5*] is raised by the absence of net reaction in the photolysis of nitrite solution. If, as Treinin and Hayon propose, NO(aq) and NO_2(aq) are formed in equal concentrations and hence by (84), the ratio of $R_{A1^*/4^*}$ (which leads to net reaction) to $R_{A2^*/5^*}$ (which does not) should be equal to the ratio of the rate constants $k_{A1^*/4^*}/k_{A2^*/5^*}$. The latter ratio is equal to 8.8 (as evaluated from the data due to Grätzel et al.) or 0.5–1.7 (as evaluated from the data of Treinin and Hayon). Such values are entirely inconsistent with the observation of a net yield for loss of NO_2^- of 10^{-3}, despite a high primary photolytic yield.

The validity of treating reaction [A2*/5*] as a hybrid reaction may be examined (as was done above for reaction [A1*/4*]) by comparing the characteristic time for reaction [A2*/5*], $\tau^{(2/5)}$, with the relaxation time for equilibrium [A8], $\tau^{(8)}$. The characteristic time $\tau^{(2/5)}$ is given by

$$\tau^{(2/5)} = \{k_{A2^*/5^*}([NO(aq)] + [NO_2(aq)])\}^{-1},$$

and the relaxation time $\tau^{(8)}$ is given by

$$\begin{aligned}\tau^{(8)} &= \{k_{A8}([NO(aq)] + [NO_2(aq)]) + k_{-A8}\}^{-1}\\ &= \{k_{A8}([NO(aq)] + [NO_2(aq)]) + K_{A8}^{-1})\}^{-1}.\end{aligned}$$

The ratio

$$\frac{\tau^{(2/5)}}{\tau^{(8)}} = \frac{k_{A8}}{k_{A2^*/5^*}}\left(1 + \frac{K_{A8}}{[NO(aq)] + [NO_2(aq)]}\right)$$

may be evaluated for known k_{A8}, $k_{A2^*/5^*}$, and K_{A8}. Although there is some spread in the values of these constants as determined by Grätzel et al. (1970) and Treinin and Hayon (1970), it appears that the ratio $\tau^{(2/5)}/\tau^{(8)}$ exceeds an order of magnitude, supporting the hybrid reaction treatment.

5.5. Summary

The directly measured rate constants given above for the several reactions studied are summarized in Table 12. For reactions [A7] and [A8], only the measurements by Grätzel et al. (1969, 1970) are available. These, however, derive from measurements over a rather wide range of reagent concentration that conform well with mechanistic expectation, so we feel rather confident in the mechanism and the numerical values of the rate constants.

With respect to reactions [A1*/4*] and [A2*/5*], there is substantial divergence in the observations and the mechanistic interpretation as well as in the numerical values of the rate constants. With respect to reaction [A1*/4*] based on our review we place highest confidence in the work of Grätzel et al. (1969), although we are somewhat unsettled by Treinin and Hayon's failure to observe a transition from first- to second-order kinetics in their study. With respect to reaction [A2*/5*], the situation is worse still. Neither Grätzel et al.'s nor Treinin and Hayon's measurements were taken over an appropriately wide range of reagent concentrations. Still more unsettling are the quantum yield studies by Treinin and Hayon, which if their photolytic mechanism is correct, reopen the entire question of mechanism of these two reactions.

To facilitate comparison of these measurements of the rate constants for reaction [A1*/4*] to measurements by other techniques, it is convenient to express the rate constant in terms of the compound oxide N_2O_4 as the reagent. In this notation the reaction is expressed as

$$N_2O_4(aq)(2NO_2(aq)) \xrightarrow{W} H^+ + NO_3^- + HNO_2(H^+ + NO_2^-) \qquad [A4^*/1^*]$$

having rate

$$R_{A4^*/1^*} = k_{A4^*/1^*}[N_2O_4(aq)]$$

where

$$k_{A4^*/1^*} = k_{A1^*/4^*}/K_{A7}.$$

Similarly, for reaction [A2*/5*], we express the reaction alternatively as

$$N_2O_3(aq)(NO(aq) + NO_2(aq)) \xrightarrow{W} 2HNO_2(2H^+ + 2NO_2^-) \qquad [A5^*/2^*]$$

with rate

$$R_{A5^*/2^*} = k_{A5^*/2^*}[N_2O_3(aq)]$$

where

$$k_{A5^*/2^*} = k_{A2^*/5^*}/K_{A8}.$$

Values of $k_{A4^*/1^*}$ and $k_{A5^*/2^*}$ are given in Table 13.

Table 12. Directly Measured Aqueous-Phase Rate Constants

	Reaction	k ($M^{-1}\ s^{-1}$)	T (°C)	Reference
[A7]	$2NO_2(aq) \rightarrow N_2O_4(aq)$	$4.5 \pm 1.0(8)$	20	Grätzel et al. (1969)
[A8]	$NO(aq) + NO_2(aq) \rightarrow N_2O_3(aq)$	1.1 (9)	20	Grätzel et al. (1970)
[A1*/4*]	$2NO_2(aq)(N_2O_4(aq)) \xrightarrow{w} H^+ + NO_3^- + HNO_2(H^+ + NO_2^-)$	$6.5 \pm 0.7(7)$	20	Grätzel et al. (1969)
		$4.7 \pm 1.0(7)$[a]	25	Treinin and Hayon (1970)
		2.6 (7)[b]	20	Moll (1966)
[A2*/5*]	$NO(aq) + NO_2(aq)(N_2O_3(aq)) \xrightarrow{w} 2HNO_2(2H^+ + 2NO_2^-)$	7.4 (6)	20	Grätzel et al. (1970)
		1.0 (8)	25	Treinin and Hayon (1970)
		2.8 (7)[c]	25	Treinin and Hayon (1970)

[a] Plus an additional pseudo-first-order component: $R_{A1^*/4^*} = k_{A17}[NO_2(aq)]$, with $k_{A17} = (1.5 \pm 0.3) \times 10^2\ s^{-1}$.

[b] As recalculated; see text.

[c] As recalculated by using K_{A8} from Grätzel et al. (1970); see text.

Table 13. Directly Measured Aqueous-Phase Rate Constants, Referred to Compound Oxides

	Reaction	k (s^{-1})	T (°C)	Reference
[A4*/1*]	$N_2O_4(aq)(2NO_2(aq)) \xrightarrow{w} H^+ + NO_3^- + HNO_2(H^+ + NO_2^-)$	1.0(3)	20	Grätzel et al. (1969)
		6.0(2)	25	Treinin and Hayon (1970)
		4.0(2)	20	Moll (1966)
[A5*/2*]	$N_2O_3(aq)(NO(aq) + NO_2(aq)) \xrightarrow{w} 2HNO_2(2H^+ + 2NO_2^-)$	5.3(2)	20	Grätzel et al. (1970)
		2.0(3)	25	Treinin and Hayon (1970)

6. MIXED-PHASE STUDIES OF REACTIVE DISSOLUTION OF NITROGEN OXIDES BY LIQUID WATER

As has been indicated in the preceding discussion, the reactive uptake of the nitrogen oxides into aqueous solution under non-phase-mixed conditions is a complex process that is fraught with interpretational uncertainty. Nevertheless, a number of studies have been carried out under conditions that are sufficiently well characterized to permit interpretation in terms of the elementary processes (equilibrium constants and rate constants) of the present concern. We review these studies here insofar as they lead, in our judgment, to such interpretations. If the results of these studies are correctly interpreted, they should, of course, be consistent with studies by other methods described.

A major difficulty in interpretation of non-phase-mixed studies of the absorption of the nitrogen oxides has been that of inferring the reaction mechanism. The difficulty of these studies is due not only to the usual ambiguities of such studies [i.e., the mechanism of mass transport and the phase(s) in which mass transport is controlled and in which reaction takes place] but also, in the case of the nitrogen oxides, to the multiplicity of possible reaction mechanisms. Here the peculiar property of the nitrogen oxides to form compound oxides (N_2O_3, N_2O_4) comes into play, since interpretation of the competition between mass transport and reaction depends on knowledge of the order of reaction in diffusing species. Thus in the case of reaction [A1*/4*],

$$2NO_2(aq)(N_2O_4(aq)) \overset{w}{\rightarrow} H^+ + NO_3^- + HNO_2(H^+ + NO_2^-), \qquad [A1*/4*]$$

the order in diffusing species will be 2 or 1, respectively, as NO_2 or N_2O_4 is the predominant aqueous N(IV) species as dictated by equilibrium [A7]. For N_2O_4 predominant, the reaction is first order in the diffusing species and consequently exhibits a first-power dependence on $p_{N_2O_4}$, irrespective of the mechanism and extent of aqueous-phase mass transport, as indicated in Table 3. This first-power dependence on $p_{N_2O_4}$, in turn, through equilibrium [G7], results in a second-power dependence on p_{NO_2}. In contrast, for NO_2 the predominant aqueous phase species, the rate of reaction will exhibit a dependence on p_{NO_2} to the second, first, or three-halves power as the phase mixing in the system is in the phase-mixed, convective-mass-transport, or molecular-diffusion regimes, respectively. The apparent reaction orders with respect to gas-phase NO_2 are summarized in Table 14 for the several regimes and for various aqueous-phase reaction orders [referred to NO_2(aq)] for the dimer N_2O_4 being the predominant N(IV) species present in and diffusing in the aqueous solution. These apparent reaction orders may be compared with those given in Table 3 for the monomer being the transported species. This comparison is facilitated by examination of Figure 2, which summarizes the apparent reaction orders for the various regimes.

One point that must again be emphasized is that the power dependences

Table 14. Apparent Reaction Order in Gas-Phase NO_2 for Principal Aqueous-Phase Species Being N_2O_4

Reaction Order in NO_2(aq)	Phase-Mixed Limit	Convective Mass-Transport Controlled	Diffusive Mass-Transport Controlled
0	0	2	1
$\frac{1}{2}$	$\frac{1}{2}$	2	$\frac{5}{4}$
1	1	2	$\frac{3}{2}$
2	2	2	2
3	3	2	$\frac{5}{2}$
m	m	2	$m/2 + 1$

on p_{NO_2} must be considered to be limiting cases and that intermediate cases may be expected as a consequence both of the transition between mass-transport regimes and of the transition between N_2O_4 and NO_2 as the predominant aqueous N(IV) species. In the case of reaction [A1*/4*] both of these "dimensions" may be expected to be affected as the partial pressure of the reagent gas is varied. Consequently, one must be highly cautious in accepting any determination of Henry's law coefficient or of kinetic rate constant from direct measurement of the rate of uptake of NO_2 in a mixed-phase system. In particular, one test that may be readily applied is that the

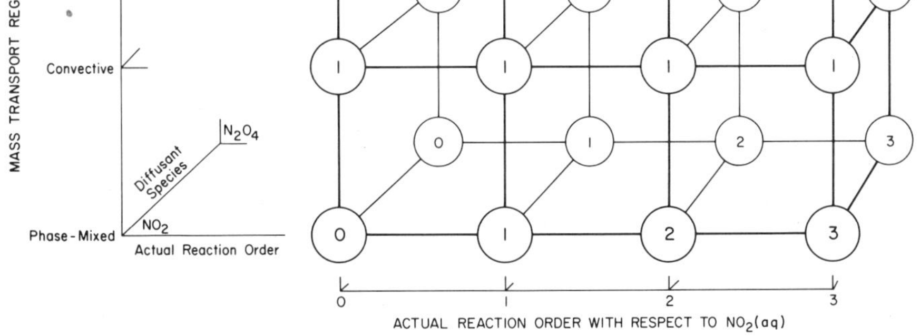

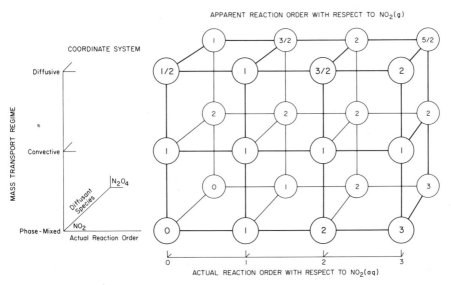

Figure 2. Apparent order of mixed-phase reactive uptake of NO_2 (N_2O_4) referred to the partial pressure of NO_2, displayed for limiting cases of reaction order, phase-mixing regime, and diffusing species (NO_2 or N_2O_4). Apparent reaction order is indicated by the numerical value displayed at the appropriate coordinate location.

predominant N(IV) species present be consistent with the model employed in the interpretation. Unfortunately, this test has not previously been applied, even in studies subsequent to the determination of K_{A7} by Grätzel et al. (1969).

This review of direct measurements of the rate of uptake of nitrogen oxides into aqueous solution is organized about the dependence of this rate on the partial pressures of the oxides (apparent reaction order referred to the gaseous reagents). We first consider studies of the uptake of NO_2 (N_2O_4) and then consider studies of the simultaneous uptake of NO and NO_2 (N_2O_3).

6.1. Second-Order Uptake of NO_2

As noted in Section 1, there has been a deep and continuing interest in the mechanism and rate of reaction of $NO_2(g)$ [and/or $N_2O_4(g)$] at high partial pressures (10^{-3}–10^{-1} atm) with liquid water because of the role of this process in the industrially important manufacture of nitric acid. Much of the early work (e.g., Chambers and Sherwood, 1937) suggested that the rate of uptake was controlled by gas-phase mass transport. Subsequent investigators (e.g., Denbigh and Prince, 1947) carried out experiments such that gas-phase mass transport was not controlling, permitting investigation of chemical–kinetic influences on the rate of reactive dissolution.

A second persistent question has been the phase (gas or aqueous) in which the reaction takes place. Peters and colleagues (Peters and Holman, 1955; Peters et al., 1955) have argued that at least some of the reaction taking place under laboratory conditions must be occurring in the gas phase, as evidence by mist formation when gaseous NO_2 and N_2O_4 are mixed with water vapor. Carberry (1959), on the other hand, argues against significant gas-phase reaction, pointing out that mist formation is found only when a condensed phase is present. Subsequent work (England and Corcoran, 1974; Kaiser and Wu, 1977a) has provided a rate constant for the gas-phase reaction [G1]—in fact an upper limit, because of the possibility of heterogeneous contribution to the reaction rate—that establishes that the gas-phase reaction is slow in comparison to measured aqueous-phase rates. It now seems generally accepted (e.g., Sherwood et al., 1975) that the mixed-phase reactions under examination can be accounted for entirely by reactions taking place in the aqueous phase.

A significant advance in the understanding of this system was made by Wendel and Pigford (1958), who quantitatively addressed both the mass transport of NO_2 (N_2O_4) in the gas phase (which leads to a reduction in partial pressure at the gas-liquid interface, compared to that in the bulk gas) and the rate of aqueous-phase reaction. In that study it was found that the rate of uptake expressed as amount of N_2O_4 taken up in unit time per unit area of interface was given by

$$J_{N_2O_4} = q p_{N_2O_4(i)} \tag{85}$$

where $p_{N_2O_4(i)}$ refers to the interfacial partial pressure, and q is a constant. (This proportionality to $p_{N_2O_4}$ is, of course, equivalent to a proportionality to $p^2_{NO_2}$, i.e., an apparent reaction order of 2 referred to $NO_2(g)$.) This first-order dependence on $p_{N_2O_4}$ (or second-order dependence on p_{NO_2}) has been confirmed in subsequent investigations, which have somewhat refined the value of q from that given by Wendel and Pigford, as listed in Table 15.

The studies of Wendel and Pigford and subsequent investigators have been interpreted by means of "penetration theory" (see Section 2) for a pseudo-first-order aqueous-phase reaction of N_2O_4, i.e., dissolution and diffusion of N_2O_4 and reaction by

$$N_2O_4 \xrightarrow{w} H^+ + NO_3^- + HNO_2(H^+ + NO_2^-); \qquad [A4^*]$$

such a pseudo-first-order reaction ($m = 1$) would give rise to the observed dependence of $J_{N_2O_4}$ on the first power of the interfacial partial pressure of N_2O_4 [see Eq. (15)]. If this interpretation is correct, which we believe it to be, we may identify the observed q [Eq. (85)] with the product $H_{N_2O_4}(k_{A4^*}D_{N_2O_4})^{1/2}$, where $D_{N_2O_4}$ is the diffusion coefficient of N_2O_4 in aqueous solution. Since this diffusion coefficient may be accurately estimated [e.g., by the semiempirical correlation of Wilke and Chang (1955)], it is seen that measurement of q leads to the product $H_{N_2O_4}k^{1/2}_{A4^*}$. As has been recognized by numerous investigators, the product $H_{N_2O_4}k^{1/2}_{A4^*}$ may be resolved as is desired into its component factors using additional information from other experiments; we defer that discussion for the present.

An alternative interpretation of the observation of a first-power dependence on $p_{N_2O_4}$ might be formulated as follows. As noted above, this pressure dependence is equivalently expressed as a second-order dependence on p_{NO_2}. If NO_2 rather than N_2O_4 were the principal form of the dissolved reagent, such a power-law dependence would be predicted [Eq. (15) or Table 3] for a third-order reaction in $NO_2(aq)$,

$$3NO_2(aq) \rightarrow \text{products}.$$

The choice between these two interpretations may be addressed by considering the distribution of dissolved N(IV) between the two species NO_2 and N_2O_4, since it is the order of the reaction *in diffusing species* that determines the exponent m in Eq. (15) (e.g., Brian, 1964). Expressed as a function of NO_2 partial pressure, the fraction of dissolved N(IV) that is present as NO_2 is

$$F_{NO_2} = \frac{[NO_2(aq)]}{[NO_2(aq)] + 2[N_2O_4(aq)]} = \frac{1}{1 + 2H_{NO_2}K_{A7}p_{NO_2}} \qquad (86)$$

The partial pressure corresponding to a given value of F_{NO_2} is

$$p_{NO_2} = (2H_{NO_2}K_{A7})^{-1}\frac{1 - F_{NO_2}}{F_{NO_2}} \qquad (87)$$

and may be evaluated for assumed equilibrium constants and specified

Table 15. Measurements of the Area Rate Constant for Uptake of N_2O_4 by Aqueous Solution in Diffusion-Controlled Regime

Study	Method[a]	p_{NO_2} (10^{-2} atm)	q^b (10^{-3} kmol m^{-2} s^{-1} atm^{-1})	$H_{N_2O_4}k^{1/2}_{A4^*/1^*}$ [c] (M atm^{-1} s$^{-1/2}$)
Denbigh and Prince (1947)	WWC	3–19	0.37 ± 16%[d,e]	
Caudle and Denbigh (1953)	WWC	0.6–5	0.4–1.6[d]	
Wendel and Pigford (1958)	WWC	3–9	0.26–0.34[d]; 0.58	16
Dekker et al. (1959)	WWC	1.6–7	1.1 ± 13%	29
Kramers et al. (1961)	LWJ	4–14	0.83 ± 3%[f]	22
Gerstacker (1961)	LWJ	9–21	1.05 ± 5%	28
Bartholomé and Gerstacker (1961)	LWJ	10	0.82 ± 5%	22
Corriveau and Pigford (1971)	WS	1.0–2.1	0.47	13
Chilton and Knell (1972)	WWC	1.4–3.6	1.4	37
Hoftyzer and Kwanten (1972)	LWJ	4–40	0.93	25
Kameoka and Pigford (1977)	WS	0.7–1.3	0.69	18
Komiyama and Inoue (1980)[g]	B	0.06–0.08	1.5(1.7)[h]	45(45)[h]
"	F	0.04–0.10	2.4 (2.7)[h]	72(72)[h]

[a] Methods: WWC, wetted wall column; LWJ, laminar water jet; WS, wetted sphere; B, bubbler; F, flat-surface contactor.

[b] Defining equation $J_{N_2O_4} = qp_{N_2O_4}$; all values referred to $p_{N_2O_4}$ at gas–aqueous interace and are for 25°C unless otherwise indicated.

[c] For $D_{N_2O_4} = 1.40 \times 10^{-9}$ m^{-2} s^{-1} at 25°C and 1.10×10^{-9} m^{-2} s^{-1} at 15°C.

[d] Referred to bulk pressure.

[e] NO present, 0.06–0.2 atm.

[f] Derived by interpolation from measurements at 20 and 30°C.

[g] Measurements at 15°C; 0.01 M NaOH.

[h] Corrected to 25°C by using temperature data of Kramers et al. (1961).

F_{NO_2}. Here we have used the values of $H_{NO_2} = 1.2 \times 10^{-2}$ M atm^{-1} and $K_{A7} = 6.5 \times 10^4$ M^{-1} recommended by Schwartz and White (1981) from analysis of thermochemical cycles. However, it should be emphasized that the subsequent discussion does not rely quantitatively on these values—they are introduced here only for the purpose of ascertaining the predominant aqueous-phase N(IV) species. The quantities F_{NO_2} and $F_{N_2O_4}(= 1 - F_{NO_2})$ are shown as function of p_{NO_2} in Figure 1. It is seen that for partial pressures of NO_2 in excess of 6×10^{-3} atm the equilibrium fraction of dissolved N(IV) present as N_2O_4 exceeds 90%. Since this condition is satisfied by essentially all the experiments summarized in Table 15, we may rule out the possibility that the observed pressure dependence of $J_{N_2O_4}$ (as $p_{N_2O_4}$ or as $p^2_{NO_2}$) results from a third-order reaction in NO_2(aq) and thus may confirm the interpretation of a pseudo-first-order reaction in N_2O_4(aq).

The preceding discussion serves to illustrate another point. Because of the distinction in order between reactions [A1*] and [A4*], examination of this reaction in penetration theory might be considered a potential means of distinguishing between these two mechanisms. However, as we have noted, the penetration theory distinguishes the order in diffusing species, not the order of the reaction. Thus, provided any departure from the equilibrium [A7] relaxes fast in comparison to the characteristic time of reaction [A1*/4*], the two mechanisms cannot be distinguished by this type of study. Consequently, the observed quantity q must be identified with $H_{N_2O_4}(k_{A4*/1*}D_{N_2O_4})^{1/2}$ rather than with $H_{N_2O_4}(k_{A4*}D_{N_2O_4})^{1/2}$.

The interpretation of the several experiments presented in Table 15 by penetration theory also requires demonstration, other than by the first-order dependence on $p_{N_2O_4}$, that the reaction occurs in the diffusion-controlled regime since, as may be seen by reference to Table 3, such a first-order dependence is expected for the several mass-transport regimes. The penetration theory is applicable provided the liquid-phase mass-transfer coefficient k_L is sufficiently small [Eq. (18)] and also provided the characteristic depth of reagent penetration is much less than the physical extent of the liquid. First, with respect to the depth of reagent penetration, this may be estimated as $\delta = (D/kA)^{1/2}$. For $D \sim 1 \times 10^{-9}$ m^2 s^{-1} and $k_A \sim 10^3$ s^{-1} (see Table 13 as well as Section 7.1), the penetration depth is estimated as $\sim$1 μm, well less than the thickness of the films, jets, etc. employed. Turning to convective mixing, we note that this represents a potential source of error that must be controlled by experimental design or otherwise excluded. In the case of experiments by wetted-wall or laminar-jet absorbers, this condition would appear to be at least closely approached. In the case of the experiments employing bubbler and flat surface absorbers (Komiyama and Inoue, 1980) this condition may be addressed by means of criterion (18); if the apparatus parameters given by the authors and the estimate $k_A \sim 10^3$ s^{-1} are used, the condition of a diffusive-controlled reaction would again appear to be satisfied.

Examination of the data in Table 15 shows a reasonable consistency in the values of q obtained in the several studies Wendel and Pigford (1958) through Kameoka and Pigford (1977). These studies lead to a recommended value of q of $0.9 \pm 0.3 \times 10^{-3}$ kmol m^{-2} s^{-1} atm^{-1}, where the uncertainty is based on the spread in the values reported. The values of Komiyama and Inoue (1980), which were obtained at 15°C, differ significantly (even when corrected for temperature) from the remaining determinations and even from each other. An error with the bubbler contactor might arise from error in the interfacial area, which was determined from photographs; it is difficult to account for a potential source of error with the flat-surface contactor. It is unfortunate that these authors did not address the discrepancy between their results obtained with the two different contactors. Evidently the authors used the data obtained with the bubbler contactor in their evaluation of elementary quantities ($H_{N_2O_4}$, $k_{A4^*/1^*}$, etc.).

As noted in Section 2, the aqueous-phase diffusion coefficient of a dissolved gas may be estimated fairly confidently by using a semiempirical correlation such as that given by Wilke and Chang (1955). This procedure gives $D_{N_2O_4} = 1.4 \times 10^{-9}$ m^2 s^{-1} at 25°C. Consequently, we obtain $H_{N_2O_4}k_{A4^*/1^*}^{1/2} = (2.3 \pm 0.8) \times 10^1$ M atm^{-1} s$^{-1/2}$ at 25°C. This value is compared below to other experimental measures of $H_{N_2O_4}$ and $k_{A4^*/1^*}$.

As outlined in Section 2, measurement of the amount of gaseous reagent $Q_X(t)$ taken up by the liquid as a function of contact time t under conditions of diffusive mass transport provides further information concerning the elementary constants of the system. As may be seen from Eq. (16), a plot of the quantity $Q_X(t)/t$ versus t^{-1} is expected to be linear with slope $q'p_X/2$, where $q' = H_X D_X^{1/2} k_X^{-1/2}$. Such studies as a function of contact time have been carried out by varying the length of jet exposed to the gaseous reagent; the results are shown in Table 16. On the basis of the spread in the data, we present a recommended value of $q' = (2.6 \pm 1.2) \times 10^{-6}$ kmol m^{-2} atm^{-1} at 25°C. Again using $D_{N_2O_4} = 1.4 \times 10^{-9}$ m^2 s^{-1}, we obtain $H_{N_2O_4}k_{A4^*/1^*}^{-1/2} = (6.9 \pm 3.2) \times 10^{-2}$ M atm^{-1} s$^{1/2}$ at 25°C.

The values q and q' may be combined to yield $k_{A4^*/1^*}$ and $H_{N_2O_4}$ as

$$k_{A4^*/1^*} = \frac{q}{q'}$$

$$H_{N_2O_4} = \left(\frac{qq'}{D_{N_2O_4}}\right)^{1/2}.$$

The values of these quantities obtained from the several studies in which q' was determined are also given in Table 16. We defer further discussion to Section 7.1.

The temperature dependence of the rate of uptake of N_2O_4 has been examined by several investigators, as noted in Table 17. The temperature dependence of q may be interpreted as yielding an overall activation energy associated with the quantity $H_{N_2O_4}D_{N_2O_4}^{1/2}k_{A4^*/1^*}^{1/2}$, which is equal to the sum

Table 16. Measurements of the Time Dependence of the Amount of Uptake of N_2O_4 by Aqueous Solution

Study	q'^{a} (10^{-6} kmol m^{-2} atm^{-1})	$H_{N_2O_4}k_{A4*/1*}^{-1/2}$ (M atm^{-1} s$^{1/2}$)	$k_{A4*/1*}$ (s^{-1})	$H_{N_2O_4}{}^{d}$ (M atm^{-1})
Kramers et al. (1961)	2.88^{b}	7.7(−2)	2.9(2)	1.31
Gerstacker (1961)	≤1.19			
Bartholomé and Gerstacker (1961)	3.58	9.6(−2)	2.3(2)	1.45
Hoftyzer and Kwanten (1972)	1.39^{c}	3.7(−2)	6.7(2)	0.96

[a] Defining equation: $d[Q_{N_2O_4}(t)/t]/dt = q'p_{N_2O_4}$; all values are referred to $p_{N_2O_4}$ at the gas–aqueous interface and are for 25°C.

[b] Derived by interpolation from measurements at 20 and 30°C.

[c] Original data not presented; deduced from reported derived quantities.

[d] For $D_{N_2O_4} = 1.40 \times 10^{-9}$ m^2 s^{-1}.

of the activation energies associated with the several factors, i.e.,

$$E_a(H_{N_2O_4}D_{N_2O_4}^{1/2}k_{A4*/1*}^{1/2}) = \Delta H°(H_{N_2O_4}) + \tfrac{1}{2}E_a(D_{N_2O_4}) + \tfrac{1}{2}E_a[A4*/1*].$$

The value of $E_a(D_{N_2O_4})$, which represents the activation energy associated with diffusion in aqueous solution, may be inferred from Wilke and Chang (1955) to be 4.45 kcal mol^{-1}, as governed principally by the temperature dependence of the solvent viscosity. The term $\Delta H°(H_{N_2O_4})$ represents the heat of (nonreactive) solution of N_2O_4 and would be expected to be negative, of magnitude several kilocalories per mole, corresponding to a decreasing solubility with increasing temperature. The term $E_a[A4*/1*]$, representing the activation energy of reaction, would be expected to be positive. Thus the overall activation enthalpy might be either positive or negative, depending on the relative magnitude of the several terms.

Examination of the data in Table 17 indicates that $E_a(H_{N_2O_4}k_{A4*/1*}^{1/2})$ is in the range −4 to +1.3 kcal mol^{-1}. Looking more closely at these studies, we would hazard the guess that this quantity lies in the range 0.5 ± 0.7 kcal mol^{-1}, since the value given by Dekker et al. (1959) is superceded by the later, higher value from the same group (Kramers et al., 1961). The value reported by Hoftyzer and Kwanten (1972) represents by far the widest temperature range and hence would be expected to have the greatest accuracy. Unfortunately, this quantity is given only in a review article, which neither describes the experiment nor reports the original data, which do not appear to have been published.

With respect to $E_a(H_{N_2O_4}k_{A4*/1*}^{-1/2})$, there are substantially fewer data available and little consistency in those. In particular, there appears to be an

Table 17. Temperature Dependence of $H_{N_2O_4} k_{A4^*/1^*}^{1/2}$ and $H_{N_2O_4} k_{A4^*/1^*}^{-1/2}$ [a]

Study	Temperature Range (°C)	E_a (kcal mol^{-1}) $H_{N_2O_4} D_{N_2O_4}^{1/2} k_{A4^*/1^*}^{1/2}$	$H_{N_2O_4} k_{A4^*/1^*}^{1/2}$		$H_{N_2O_4} D_{N_2O_4}^{1/2} k_{A4^*/1^*}^{-1/2}$	$H_{N_2O_4} k_{A4^*/1^*}^{-1/2}$
Denbigh and Prince (1947)	25–40	1.61	−0.62			
Wendel and Pigford (1958)	25–40	−0.88	−3.11			
Dekker et al. (1959)	25–35	−1.74	−3.96			
Kramers et al. (1961)	20–30	2.56	0.33		−2.34	−4.57
Bartholomé and Gerstacker (1961)	5–25	2.13	−0.09	(5–15°C)	−1.80	−4.08
				(15–25°C)	−17.43	−19.67
Hoftyzer and Kwanten (1972)	3–75	3.48	1.25			

[a] The values of $E_a(H_{N_2O_4} k_{A4^*/1^*}^{1/2})$ and $E_a(H_{N_2O_4} k_{A4^*/1^*}^{-1/2})$ were evaluated by using $E_a(D_{N_2O_4}) = 4.45$ kcal mol^{-1}.

anomaly in the measurements by Bartholomé and Gerstacker (1961). Consequently, little inference can be drawn about this quantity.

These studies of the temperature dependence of reaction [4*/1*] is related to other temperature-dependence studies in Section 7.3.

6.2. $\frac{3}{2}$-Order Uptake of NO_2

Several investigators (Komiyama and Inoue, 1980; Sada et al., 1979; Takeuchi et al., 1977) have reported, at NO_2 partial pressures lower than those for which a second-order rate law is observed, a rate of uptake of NO_2 proportional to the $\frac{3}{2}$ power of p_{NO_2},

$$J_{NO_2} = q p_{NO_2}^{3/2}.$$

The results of these studies are summarized in Table 18. The interpretation that has been advanced by all of these investigators to account for this observed rate law is that of a second-order reaction of dissolved NO_2 in the fast-reaction limit, NO_2 being the diffusant species, for which J_{NO_2} is given by Eq. (15) with $m = 2$, viz.,

$$J_{NO_2} = (\tfrac{4}{3})^{1/2} D_{NO_2}^{1/2} k_{A1^*/4^*}^{1/2} H_{NO_2}^{3/2} p_{NO_2}^{3/2}. \tag{88}$$

According to this interpretation, the following identification can be made:

$$q = (\tfrac{4}{3})^{1/2} D_{NO_2}^{1/2} k_{A1^*/4^*}^{1/2} H_{NO_2}^{3/2}. \tag{89}$$

Again the diffusion coefficient may be estimated rather closely, based on the correlation by Wilke and Chang (1955). In the case of NO_2 there is some uncertainty in application of the Wilke–Chang formula because of the unknown molar volume V of the liquid at its normal boiling point. Taking $V = 35.1$ cm^3 mol^{-1} [cf. 36.4 and 34.0 cm^3 mol^{-1} for N_2O and CO_2, respectively (Himmelblau, 1964)] we obtain the values of D_{NO_2} presented in Table 18. The resulting values of $H_{NO_2}^{3/2} k_{A1^*/4^*}^{1/2}$, evaluated by Eq. (89), are also shown. However, on the basis of the discussion below, we consider the interpretation of these data by means of equations (88) and (89) not to be entirely justified and thus would expect some inaccuracy in the values of $H_{NO_2}^{3/2} k_{A1^*/4^*}^{1/2}$ thus evaluated.

Fundamentally, the argument against interpretation of the $\frac{3}{2}$-order data by Eq. (88) is that NO_2 is not the sole aqueous phase N(IV) species as is assumed in the derivation of that equation. That this is the case may be seen by inspection of Figure 1, which indicates a substantial proportion of N(IV) present as N_2O_4 at the partial pressures employed in all of the studies listed in Table 18. Thus, unless the values of H_{NO_2} and/or K_{A7} utilized in preparing Figure 1 are greatly in error, the condition that NO_2 be the predominant diffusing species is not satisfied, and the interpretation of the rate of uptake by means of equation (88) is not quantitatively correct. The value of

Table 18. $\frac{3}{2}$**-Order Uptake of** NO_2

Study	Method[a]	T (°C)	p_{NO_2} (10^{-4} atm)	q[b] (10^{-3} kmol m^{-2} s^{-1} $atm^{-3/2}$)	D_{NO_2} (10^{-9} m^2 s^{-1})	$H_{NO_2}^{3/2} k_{A1*/4*}^{1/2}$[c] (M $atm^{-3/2}$ $s^{-1/2}$)	$F_{N_2O_4}$[d] (%)
Takeuchi et al. (1977)	F	25	1.0–4.7	2.6	2.00	50	13–42
		15	0.9–3.8	3.8	1.57	83	
		10	0.4–3.7	5.7	1.30	137	
Sada et al. (1979)	F	25	2.9–8.1	1.74	2.00	34	29–56
Komiyama and Inoue (1980)[e]	F	15	1.0–2.9	1.16	1.57	25	13–29
"	B	15	0.34–1.8	0.79	1.57	17	5–22

[a] Methods: F, flat-surface contactor; B, bubbler.

[b] Defining equation: $J_{NO_2} = q p_{NO_2}^{3/2}$.

[c] Defining equation: $H_{NO_2}^{3/2} k_{A1*/4*} = q/(4/3)^{1/2} D_{NO_2}^{1/2}$.

[d] Evaluated from Figure 1 for $H_{NO_2} = 1.2 \times 10^{-2}$ M atm^{-1} and $K_{A7} = 6.5 \times 10^4$ M atm^{-1}.

[e] 0.01 M NaOH.

$H_{NO_2}^{3/2}k_{A1^*/4^*}^{1/2}$ determined from the data of the several studies is seen to decrease with decreasing p_{NO_2} and resulting decrease of $F_{N_2O_4}$. Of the several studies, those by Komiyama and Inoue (1980) would appear to exhibit the least influence from diffusion and reaction of N_2O_4 occurring in parallel with diffusion and reaction of NO_2.

6.3. First-Order Uptake of NO_2

A rate of uptake of NO_2 into aqueous solution proportional to the first power of p_{NO_2} has been reported by Andrew and Hanson (1961) and Komiyama and Inoue (1980). The former study was interpreted as convective mass-transport-controlled reactive uptake of NO_2 according to the overall reaction

$$4NO_2(g) + 2H_2O(l) \rightarrow 2H^+ + 2NO_3^- + 2HNO_2(g).$$

The rate of uptake of NO_2 per unit interfacial area J_{NO_2} was determined from the measured efficiency of removal of NO_2 from a flowing gas stream by a sieve-plate reactor of known mass-transfer characteristics. At the lowest NO_2 partial pressures employed, $(1.8\text{–}6.1) \times 10^{-4}$ atm, the authors reported a constant efficiency of removal of NO_2 from the gas stream flowing through their reactor, i.e., a removal rate proportional to the first power of the partial pressure of the gas.

For convective mass-transport-controlled uptake, the rate of uptake would be related to the Henry's law coefficient of the gas as

$$J_{NO_2} = k_L H_{NO_2} p_{NO_2} \tag{90}$$

[see Eqs. (14) and (19)]. However, since by the analytical method employed, the evolved nitrous acid was equivalent to nitrogen dioxide by the equilibrium

$$2HNO_2(g) = NO(g) + NO_2(g) + H_2O(g), \qquad [-G2]$$

the authors interpreted the measured net rate of uptake of NO_2 as three-fourths the rate of uptake of physically dissolved NO_2; i.e., $J_{meas} = \frac{3}{4}k_L H_{NO_2} p_{NO_2}$. The value of H_{NO_2} thus obtained was 4.1×10^{-2} M atm^{-1}.

The problem of evolution of HNO_2 was obviated by Komiyama and Inoue (1980), who employed alkaline solution (0.01 M NaOH). These authors also extended the range of measurement to a partial pressure lower than that employed by Andrew and Hanson (1.2×10^{-5} atm). In contrast to those authors, Komiyama and Inoue found that the ratio of the rate of uptake of NO_2 to the partial pressure of the gas continued to decrease with decreasing partial pressure of NO_2 to partial pressure as low as 2×10^{-5} atm. Using the limiting low-pressure rate of uptake and interpreting their measured rate of uptake according to Eq. (90), Komiyama and Inoue reported a value of $H_{NO_2} = 2.35 \times 10^{-2}$ M atm^{-1} at 15°C, somewhat lower than the value reported by Andrew and Hanson. The results of these experiments are summarized in Table 19.

For comparison of the values of H_{NO_2} obtained by Komiyama and Inoue (1980) to other values at higher temperature, it is desirable to estimate the temperature dependence of this quantity on the basis of an estimated enthalpy of solution. Such an estimate was given by Schwartz and White (1981) by use of the second-law heat of solution of O_3 to estimate that for NO_2; the resulting corrected value for H_{NO_2} at 25°C was given as 1.9×10^{-2} M atm^{-1}.

It is necessary also to address the assumption, inherent in interpretation of these studies, that the dissolved N(IV) species is present as NO_2, not N_2O_4. The equilibrium fraction of dissolved N(IV) present as N_2O_4, $F_{N_2O_4}$ may be evaluated [Eq. (86)] for specified values of K_{A7}, H_{NO_2} and p_{NO_2}. Using the value of K_{A7} given by Grätzel et al. (1969) and H_{NO_2} given by Andrew and Hanson, we find $F_{N_2O_4}$ to range from 49 to 76% for the range partial pressures that appeared to indicate a constant efficiency of uptake. This high a fraction of N_2O_4 would appear to invalidate Andrew and Hanson's interpretation of their study (unless the value of K_{A7} determined by Grätzel et al. is greatly in error) and indeed raises questions regarding the accuracy of their measurement method. We are thus inclined to reject Andrew and Hanson's value for H_{NO_2}.

Similar examination applied to the results of Komiyama and Inoue (1980) indicates that the fraction of N(IV) present as N_2O_4 was 4–7% at the partial pressures employed in that study, and thus it would appear that N_2O_4 would be making only a slight contribution to the rate of uptake attributed to NO_2. On this basis we are inclined to decrease the reported value of H_{NO_2} slightly (6%) to account for this contribution. The resulting value of H_{NO_2} is 2.2×10^{-2} M atm^{-1} at 15°C or, using the temperature coefficient estimate given above, 1.8×10^{-2} M atm^{-1} at 25°C.

The values of H_{NO_2} obtained by Komiyama and Inoue (1980) are compared below to the results of other studies. However, it is useful at this point to note that these authors' value of H_{NO_2} may be combined with their value of $H_{NO_2}^{3/2} k_{A1^*/4^*}^{1/2}$ obtained with the same apparatus (bubbler contactor) to yield $k_{A1^*/4^*} = 2.12 \times 10^7$ M^{-1} s^{-1}. These quantities may also be combined with

Table 19. First-Order Uptake of NO_2

Study	Method[a]	T (°C)	p_{NO_2} (10^{-5} atm)	H_{NO_2} (M atm^{-1})
Andrew and Hanson (1961)	SP	25	18–61	4.1 (−2)[b]
Komiyama and Inoue (1980)	B	15	1.2–2.4	2.35(−2)[c]

[a] Methods: SP, sieve plate; B, bubbler.

[b] Defining equation: $J_{NO_2} = \frac{3}{4} H_{NO_2} k_L p_{NO_2}$.

[c] Defining equation: $J_{NO_2} = H_{NO_2} k_L p_{NO_2}$.

Table 20. Evaluation of Equilibrium and Rate Constants from Komiyama and Inoue (1980)

Quantity	Value	Units
H_{NO_2}[a]	2.35(−2)	$M\ atm^{-1}$
$H_{NO_2}^{3/2}k_{A1*/4*}^{1/2}$[a]	17	$M\ atm^{-3/2}\ s^{-1/2}$
$H_{N_2O_4}k_{A4*/1*}^{1/2}$[a]	46	$M\ atm^{-1}\ s^{-1/2}$
$k_{A1*/4*}$	2.1 (7)	$M^{-1}\ s^{-1}$
$H_{N_2O_4}$	2.7	$M\ atm^{-1}$
K_{A7}	7.5 (4)	M^{-1}
$k_{A4*/1*}$	2.8 (2)	s^{-1}

[a] Data from experiments with bubbler contactor at 15°C. Other data employed: $D_{NO_2} = 1.57 \times 10^{-9}\ m^2\ s^{-1}$; $D_{N_2O_4} = 1.08 \times 10^{-9}\ m^2\ s^{-1}$; $K_{G7} = 15.3\ atm^{-1}$.

their value of $H_{N_2O_4}k_{A4*/1*}^{1/2}$ to yield $H_{N_2O_4}$ and K_{A7} since the latter quantities are related to the measured quantities by

$$H_{N_2O_4} = H_{NO_2}^2 K_{A7} K_{G7}^{-1} \tag{91}$$

and

$$k_{A1*/4*} = k_{A4*/1*} K_{A7}. \tag{92}$$

The values of $H_{N_2O_4}$ and K_{A7} may thus be evaluated in terms of the primary measured quantities as

$$H_{N_2O_4} = \frac{(H_{N_2O_4}k_{A4*/1*}^{1/2})^2 H_{NO_2} K_{G7}}{(H_{NO_2}^{3/2}k_{A1*/4*}^{1/2})^2} \tag{93}$$

and

$$K_{A7} = \frac{(H_{N_2O_4}k_{A4*/1*}^{1/2})^2 K_{G7}^2}{(H_{NO_2}^{3/2}k_{A1*/4*}^{1/2})^2 H_{NO_2}} \tag{94}$$

Finally, $k_{A4*/1*}$ may be evaluated by use of (92). The quantities employed in the calculation and the resulting derived quantities are summarized in Table 20. These results permit further examination of the identity of the transported aqueous-phase N(IV) species. In particular, for the pressures leading to the observed $\frac{3}{2}$-order uptake, $(0.34\text{–}1.8) \times 10^{-4}$ atm, $F_{N_2O_4}$ is calculated to range from 11 to 39%, again indicating the possibility of inaccuracy in the value of $H_{NO_2}^{3/2}k_{A1*/4*}$ derived in this pressure range.

The experimental parameters furnished by Komiyama and Inoue also permit the evaluation of the characteristic times of their mixed-phase reaction system to allow determination of whether the criteria of the several limiting regimes are fulfilled, as listed in Table 4. These characteristic times are given in Table 21. The criteria for diffusive mass-transport-controlled

Table 21. Characteristic Times in Study of Reaction [A1*/4*] by Komiyama and Inoue (1980)

Process	Characteristic Time	Value (s)
Convective mixing[a]	τ_m	4.3(1)
Diffusion (N_2O_4)[b]	$\tau_d(N_2O_4)$	5.5(4)
Diffusion (NO_2)[b]	$\tau_d(NO_2)$	3.8(4)
	$\tau_m^2/\tau_d(N_2O_4)$	3.3(−2)
	$\tau_m^2/\tau_d(NO_2)$	4.8(−2)
Reaction (N_2O_4)[c]	$\tau_r(N_2O_4)$	3.6(−3)
Reaction (NO_2)[d]	$\tau_r(NO_2)$	
p_{NO_2} (atm)		
1.8(−4)		5.6(−3)
5.3(−5)		1.9(−2)
3.4(−5)		3.0(−2)
2.4(−5)		4.2(−2)
1.2(−5)		8.4(−2)

[a] Evaluated as $\tau_m = (k_L a)^{-1}$ for $k_L = 1.8 \times 10^{-4}$ m s^{-1} and $a = 1.3 \times 10^2$ m^{-1} for the bubbler contactor.

[b] Evaluated as $\tau_d = (Da^2)^{-1}$ for $D_{NO_2} = 1.57 \times 10^{-9}$ m^2 s^{-1} and $D_{N_2O_4} = 1.08 \times 10^{-9}$ m^2 s^{-1}.

[c] Evaluated as $\tau_r(N_2O_4) = k_{A4^*/1^*}^{-1}$, for value of $k_{A4^*/1^*}$ given in Table 19.

[d] Evaluated as $\tau_r(NO_2) = (2H_{NO_2}k_{A1^*/4^*}p_{NO_2})^{-1}$, for values of H_{NO_2} and $k_{A1^*/4^*}$ given in Table 20.

reaction are

$$\tau_r \lesssim 0.1\tau_d \tag{95}$$

and

$$\tau_r \lesssim 0.38\,\frac{\tau_m^2}{\tau_d}\,. \tag{96}$$

Comparison of $\tau_r(N_2O_4)$ with $\tau_d(N_2O_4)$ and with $\tau_m^2/\tau_d(N_2O_4)$ shows that these criteria are fulfilled, i.e., that the interpretation of the second-order uptake of NO_2 (i.e., first order in N_2O_4) as diffusive mass-transport controlled leads to no inconsistency.

A similar examination may be made for the $\frac{3}{2}$-order uptake of NO_2 (1.8×10^{-4} atm $\geq p_{NO_2} \geq 3.4 \times 10^{-5}$ atm), but for a second-order reaction it must be noted that τ_r is a function of p_{NO_2}. Condition (95) is readily met for both extremes of p_{NO_2}. However, at the lower end of the partial pressure range for which the $\frac{3}{2}$-power law was observed, τ_r considerably exceeds the bound $0.38\ \tau_m^2/\tau_d$ (1.8×10^{-2} s) that represents the onset of significant (>10%) enhancement to the rate of uptake of NO_2 above that given by diffusion alone, as a consequence of convective mass transport. Alternatively, condition (96) may be used to set the lower bound to NO_2 partial pressures for which one may analyze the rate of uptake according to diffusive mass-transport-controlled reaction (5.3×10^{-5} atm). However, although this has the effect of satisfying the mass-transport condition, the range of contribution of N_2O_4 to dissolved N(IV) over the $\frac{3}{2}$-power regime increases to 16–39%.

The criteria for convective mass-transport-controlled uptake (Table 4) are

$$\tau_r \lesssim 0.1\tau_m \tag{97}$$

and

$$\tau_r \gtrsim 3\frac{\tau_m^2}{\tau_d}. \tag{98}$$

Condition (97) is readily met for both extremes of partial pressure [(1.2–2.4) $\times 10^{-5}$ atm] that gave rise to a first-order power law for the dependence of the rate of uptake on p_{NO_2}. However, condition (98) is seen not to be met (or even closely approached) for either extreme of p_{NO_2} ($3\tau_m^2/\tau_d = 1.5 \times 10^{-1}$ s). This finding contradicts the assumption that the rate of uptake of NO_2 in this pressure range was controlled by convective mass transport. This inconsistency forces us to conclude that the interpretation of the data of Komiyama and Inoue (1980) leading to the values of the equilibrium and rate constants given in Table 21 cannot be correct. In particular, the value of H_{NO_2} and/or $H_{NO_2}^{3/2}k_{A1^*/4^*}^{1/2}$ must be significantly in error. These considerations are addressed further in Section 7.1.

6.4. Mixed-Order Uptake of NO_2 at Very Low Partial Pressures

The reactive dissolution of NO_2 by water at low partial pressures was recently reëxamined by Lee and Schwartz (1981b). The gas was brought into contact with the liquid as finely dispersed bubbles produced by flowing through a disk-frit. The concentration of dissolved product species was monitored by electrical conductivity, permitting high sensitivity. The mass-transfer time constant of the reactor, which was varied, was determined from the rate of approach to saturation following introduction of CO_2 into the flowing gas, also monitored by electrical conductivity. Nitrogen dioxide partial pressures ranged from 1×10^{-7} to 8×10^{-4} atm. The stoichiometry of the reaction was established by analysis of the aqueous-phase products to be that of reaction [M1], i.e., $2NO_2(g) \xrightarrow{W} 2H^+ + NO_2^- + NO_3^-$. The power dependence of the reaction rate on p_{NO_2} was approximately $\frac{3}{2}$; however, systematic variations about this value were observed. The rate of uptake at the low end of the partial pressure range (1×10^{-7}–1×10^{-5} atm) was interpreted in terms of a model of a second-order reaction in the slow-reaction limit [see Eq. (12) for a first-order reaction]. This model is briefly summarized here (cf. Astarita, 1967). Henry's law equilibrium is assumed for the surface concentration of dissolved NO_2,

$$[NO_2(s)] = H_{NO_2} p_{NO_2}.$$

The aqueous-phase NO_2 concentration is assumed to be governed by convective mass transport and second-order reaction:

$$NO_2(s) \underset{k_{as}}{\overset{k_{as}}{\rightleftharpoons}} NO_2(aq)$$

$$2NO_2(aq) \xrightarrow{k_{A1^*/4^*}} 2H^+ + NO_2^- + NO_3^- .$$

Assumption of a steady-state concentration for $NO_2(aq)$ yields

$$[NO_2(aq)] = \frac{(1 + 8\tau_m k_{A1^*/4^*} H_{NO_2} p_{NO_2})^{1/2} - 1}{4\tau_m k_{A1^*/4^*}} \tag{99}$$

and, in turn,

$$R_{1^*/4^*} = \frac{1}{2\tau_m}\left\{H_{NO_2} p_{NO_2} + \frac{1 - (1 + 8\tau_m k_{A1^*/4^*} H_{NO_2} p_{NO_2})^{1/2}}{4\tau_m k_{A1^*/4^*}}\right\}. \tag{100}$$

Here τ_m is the convective mixing time constant (k_{as}^{-1}). This model led to a treatment, in terms of the two parameters H_{NO_2} and $k_{A1^*/4^*}$, that accounted for the dependence of the rate of uptake on p_{NO_2} and τ_m to within 10% over the indicated range in p_{NO_2} and over a threefold variation in τ_m. The values of H_{NO_2} and $k_{A1^*/4^*}$ obtained from this treatment were $(7.0 \pm 0.5) \times 10^{-3}$ M atm^{-1} and $(1.0 \pm 0.1) \times 10^8$ m^{-1} s^{-1}, respectively, at 22°C, where the error bars were empirically estimated. At higher NO_2 partial pressures ($p_{NO_2} \gtrsim 3 \times 10^{-5}$ atm) the rate of reactive dissolution was systematically greater than that accounted for by the slow-reaction treatment. This was attributed to the presence of $N_2O_4(aq)$ at these high partial pressures.

It is of interest to examine the relative values of the characteristic mixing time in the study of Lee and Schwartz to address self-consistency of their treatment. In particular, that treatment requires that the characteristic time of reaction τ_r exceed $3\tau_m^2/\tau_d$, consistent with the assumed absence of reaction in the surface film. The comparison of these quantities is given in Table 22; $3\tau_m^2/\tau_d$ is evaluated as $3\tau_m^2 D_{NO_2} a^2$, where for the surface : volume ratio a, the value 8 cm^{-1} has been employed, representing an upper-limit estimate to this quantity for an aqueous bubble column (Calderbank, 1967). It is seen that the required condition is fulfilled (with the possible exception of a single point) and thus that there is no inconsistency in the authors' treatment.

A concern that must be raised regarding the interpretation of the kinetic data of Lee and Schwartz (1981b) is that of appropriateness of the slow-reaction model to a second-order reaction. Although the model posits that the bulk solution in which reaction takes place is characterized by a uniform concentration (as determined by the competition of convective mass transport and reaction), it is clear that this concentration must exhibit a distribution reflecting the stochastic nature of the convective mixing process; it

Table 22. Examination of Surface Reaction in the Study by Lee and Schwartz (1981b)

p_{NO_2} (atm)	τ_r (s)[a]
9.71(−6)	7.4(−2)
1.66(−6)	4.3(−1)
4.90(−7)	1.5
1.04(−7)	6.9
τ_m (s)	$3\tau_m^2/\tau_d$ (s)[b]
1.69	1.1(−2)
2.14	1.8(−2)
3.17	3.9(−2)
4.27	7.0(−2)
5.26	10.6(−2)

[a] Evaluated as $(2H_{NO_2}k_{A1^*/4^*}p_{NO_2})^{-1}$, for $H_{NO_2} = 7 \times 10^{-3}$ M atm^{-1} and $k_{A1^*/4^*} = 1.0 \times 10^8$ M^{-1} s^{-1} as given by Lee and Schwartz.

[b] Upper-limit estimate, evaluated as $3\tau_m^2 D_{NO_2} a^2$ for a taken as 8 cm^{-1} (Calderbank, 1967).

is recalled that k_{as} represents only an average rate coefficient for convective mixing. In the case of a first-order reaction, this treatment would appear to be appropriate even for a nonuniform bulk concentration, since the local rate of reaction R is proportional to the first power of the local reagent concentration [X], and, in turn, the average reaction rate $\overline{R}$ is proportional to the average concentration $\overline{[X]}$ as inferred from k_{as}. However, in the case of a second-order reaction, this approach must be questioned since the average reaction rate $\overline{R}$ is proportional to $\overline{[X]^2}$, which cannot be taken as $\overline{[X]}^2$.

In the absence of a more detailed treatment, it does not appear possible to estimate the magnitude or even the direction of the errors in $k_{A1^*/4^*}$ and H_{NO_2} resulting from application of the uniform concentration model. One would note, however, that at the two extremes of the slow reaction regime, the second-order reaction approaches the phase-mixed and convectively controlled situations, just as is the case with first-order reactions. At both extremes the model would appear to apply exactly, irrespective of reaction order, leading to unequivocal determination of H_{NO_2} (convective-controlled limit, $\tau_m/\tau_r \gg 1$) and $H_{NO_2}^2 k_{A1^*/4^*}$ (phase-mixed, $\tau_m/\tau_r \ll 1$). One finds from Table 22 that τ_m/τ_r in the experiment of Lee and Schwartz ranged from 0.24 to 71, i.e., fairly closely approaching both extremes. The close adherence of the measurements to the predictions of the model [Eq. (100)] over the entire range of τ_m/τ_r (cf. Figure 6 of Lee and Schwartz, 1981*b*) lends empirical support to application of this model to the case of a second-order reaction.

The values of H_{NO_2} and $k_{A1^*/4^*}$ obtained by Lee and Schwartz are compared to other determinations in Section 7.1.

6.5. Uptake of NO and $NO_2(N_2O_3)$

The presence of NO in addition to NO_2 admits the possibility of reaction

$$NO(aq) + NO_2(aq)(N_2O_3(aq)) \xrightarrow{w} 2HNO_2(2H^+ + 2NO_2^-) \qquad [A2^*/5^*]$$

in addition to reaction [A1*/4*]. Studies of the uptake of NO + $NO_2(N_2O_3)$ interpreted according to penetration theory have been conducted by Hofmeister and Kohlhaas (1965), Corriveau and Pigford (1971), and Komiyama and Inoue (1980).

The study by Hofmeister and Kohlhaas employed a laminar water jet of variable path length (2–6 cm). The water jet was exposed to a NO–NO_2 gas mixture in which the composition was varied from 10 to 80% N(IV) (N(IV) $\equiv NO_2 + 2N_2O_4$); the total pressure was not specified, nor was it indicated whether an inert gas such as N_2 was present. The rates of reactions [A1*/4*] and [A2*/5*] were determined from measurements of the amounts of uptake of total acid (Q_{H^+}) and nitrite ($Q_{NO_2^-}$) as

$$\begin{aligned} Q_{N_2O_4} &= Q_{H^+} - Q_{NO_2^-} \\ Q_{N_2O_3} &= Q_{NO_2^-} - \tfrac{1}{2}Q_{H^+} \end{aligned} \qquad (101)$$

From the ratio of the slope to intercept of plots of the amount of uptake of N_2O_3 as a function of contact time [see Eq. (16)] the authors reported $k_{A5^*/2^*} = 68\ s^{-1}$ at 25°C; we have replotted the data for N(IV) = 40, 50, 60, 70, and 80% and obtained $k_{A5^*/2^*}$ = 117, 99, 53, 68, and 111 s^{-1}, yielding an average $k_{A5^*/2^*} = (90 \pm 12)\ s^{-1}$. The authors present a single plot of $Q_{N_2O_3}$ versus $p_{N_2O_3}$ (for a specified contact time). From the slope of this plot (and from the slope-intercept ratio of the plots of $Q_{N_2O_3}$ vs. contact time) it is possible to infer the flux of N_2O_3 and, in turn, obtain the value $q_{N_2O_3} = 3.3 \times 10^{-3}$ kmol m^{-2} atm^{-1}, although this is done with considerable trepidation since the data employed in evaluating $p_{N_2O_3}$ were not specified.

A further concern that must be raised regarding this study is that of application of penetration theory to this system as if the two reactions [A1*/4*] and [A2*/5*] take place in parallel without interaction. We noted above that the quantitative application of penetration theory requires knowledge of the identity of the diffusing species. In an aqueous system containing both N_2O_3 and N_2O_4 it is not clear *a priori* that the two reactions will occur without interaction, e.g., by reaction such as $2N_2O_3(aq) = N_2O_4(aq) + 2NO(aq)$. A test for possible interaction can be made by examining $Q_{N_2O_4}$ as a function of contact time, as also determined by Hofmeister and Kohlhaas. Values of $k_{A4^*/1^*}$ obtained in this way for N(IV) = 50, 60, 70, and 80% were 64, 580, 53, and 58 s^{-1}, respectively—i.e., mutually inconsistent and substantially different from values determined from studies with N_2O_4 only, as listed in Table 16. Consequently, the determination of $k_{A5^*/2^*}$ must be viewed with caution.

In a similar study (Corriveau and Pigford, 1971) a wetted-sphere absorber was employed to determine the rate of uptake of N_2O_3 from the mixed N_2O_3–N_2O_4 system. Again the separate rates of uptake were determined from measured rates of uptake of acidity and nitrite. The apparatus did not afford time resolution; thus only $q_{N_2O_3} = H_{N_2O_3}(k_{5^*/2^*}D_{N_2O_3})^{1/2}$ was determined. The authors present a graph of their data plotted as $J_{N_2O_3}$ versus interfacial N_2O_3 partial pressure; from the slope of this graph $q_{N_2O_3}$ is found to be 1.59×10^{-3} kmol m^{-2} s^{-1} atm^{-1} at 25°C. Unfortunately, examination of a plot of $J_{N_2O_4}$ versus $p_{N_2O_4}$ which should be linear with slope of $q_{N_2O_4}$, exhibits no such regularity and very little resemblance to the authors' plot of such data for N_2O_4 only. Thus we are again left with a result that must be viewed with caution.

The results of these two studies are summarized in Table 23.

In the study by Komiyama and Inoue (1980) the simultaneous uptake of NO and NO_2 was studied at much lower pressures of the reagent gases ($p_{NO} \sim 10^{-3}$ atm; $p_{NO_2} \sim 10^{-4}$ atm). The rate law describing this uptake was reported as

$$J_{NO} = J_{NO_2} = q(p_{NO}p_{NO_2})^{3/4}. \tag{102}$$

Under the conditions of these studies the rate of nitrate production was reported to be "negligibly small," and, indeed, for a given p_{NO_2} the reported rate of uptake in the presence of NO was substantially greater than in the absence of NO. Thus for $p_{NO_2} = 4.1 \times 10^{-5}$ atm and $p_{NO} = 4.2 \times 10^{-4}$ atm the rate of uptake of NO_2 was eightfold greater than in the absence of NO. This finding implies a rapid aqueous-phase reaction of NO and NO_2.

To address the reaction rate quantitatively, Komiyama and Inoue carried out a series of runs at a fixed ratio of $p_{NO}/p_{NO_2} = 10.2$. This ratio of partial pressures was selected with the objective of maintaining equal interfacial

Table 23. Direct Contact Measurements of Rate of Uptake of N_2O_3 (25°C)

Study	q^a (10^{-3} kmol m^{-2} s^{-1} atm^{-1})	$H_{N_2O_3}k_{A5^*/2^*}^{1/2}$ [b] (M atm^{-1} $s^{-1/2}$)
Hofmeister and Kohlhaas (1965)	3.3	8.8
Corriveau and Pigford (1971)	1.6	4.3

[a] Defining equation: $J_{N_2O_3} = qp_{N_2O_3}$.

[b] Evaluated as $J_{N_2O_3}/D_{N_2O_3}^{1/2}$ for $D_{N_2O_3}$ taken as 1.4×10^{-9} m^2 s^{-1}.

concentrations of the dissolved gases, i.e.,

$$H_{NO}p_{NO} = H_{NO_2}p_{NO_2} \tag{103}$$

for the value of H_{NO_2} that had been inferred from the authors' study of the uptake of NO_2 outlined above. This condition was desired so that the diffusion and reaction of NO and NO_2 could be treated identically to a second-order reaction, i.e.,

$$J_{NO} = J_{NO_2} = \tfrac{1}{2}(\tfrac{4}{3}\overline{D}k_{A2^*/5^*})^{1/2}(H_{NO}p_{NO}H_{NO_2}p_{NO_2})^{3/4} \tag{104}$$

[see Eq. (88)]. Such interpretation would apply provided the following conditions are satisfied:

1. $H_{NO}p_{NO} = H_{NO_2}p_{NO_2}$.
2. The rate of reaction [A2*/5*] exceeds the rate of [A1*/4*] to such an extent that depletion of N(IV) by the latter reaction is negligible; i.e., $k_{A2^*/5^*}H_{NO}H_{NO_2}p_{NO}p_{NO_2} \gg k_{A1^*/4^*}H_{NO_2}^2p_{NO_2}^2$.
3. The principal aqueous-phase diffusing species are NO(aq) and NO_2(aq) (and not N_2O_3(aq) or N_2O_4(aq)).
4. The diffusion coefficients D_{NO} and D_{NO_2} are approximately equal and may be represented by an average value $\overline{D}$.
5. The reaction is sufficiently rapid that it may be treated as diffusive mass-transport controlled.

Under the assumption that the several conditions are satisfied, $k_{A2^*/5^*}$ may be evaluated from the measured value of q as

$$k_{A2^*/5^*} = \frac{3q^2}{\overline{D}(H_{NO}H_{NO_2})^{3/2}}\,. \tag{105}$$

Because of the assumption inherent in the experimental design that [NO(s)] = [NO_2(s)], it is necessary in evaluation $k_{A2^*/5^*}$ by (105) to employ Komiyama and Inoue's value of H_{NO_2}, i.e., 2.35×10^{-2} M atm^{-1} at 15°C; $H_{NO} = 2.30 \times 10^{-3}$ M atm^{-2} and $\overline{D}$ is evaluated as $\tfrac{1}{2}(D_{NO} + D_{NO_2}) = \tfrac{1}{2}(1.95 + 1.57) \times 10^{-9}$ m^2 s^{-1} = 1.74×10^{-9} m^2 s^{-1}. From the measured value of q (1.1×10^{-3} kmol m^{-2} s^{-1} atm$^{-3/2}$) one obtains $k_{A2^*/5^*} = 5.3 \times 10^9$ M^{-1} s^{-1}. This rate constant is essentially equal to the diffusion-controlled encounter rate constant of the two species (Benson, 1960), thus implying, if this interpretation is correct, an extremely rapid reaction between NO and NO_2.

In an additional series of experiments the ratio p_{NO}/p_{NO_2} was varied over the range 3–20; expression (102 describing the rate of uptake continued to obtain, with the value of q unchanged from that in the series of runs in which [NO(s)] and [NO_2(s)] were maintained putatively equal. This behavior is inconsistent with the model leading to (105). However, departure from (105) might not be expected to be too great, as can be seen by the following argument, which rests on an alternative interpretation of these experiments.

Assume $H_{NO}p_{NO} \gg H_{NO_2}p_{NO_2}$. Then the aqueous-phase NO concentration may be approximated as uniform at the value $H_{NO}p_{NO}$ since depletion of NO(aq) by reaction would be much less than that of NO_2(aq). In this limit the rate of uptake may be approximated as a pseudo-first-order reaction of NO_2 and

$$J_{NO_2} = H_{NO_2}[k^{(1)}D_{NO_2}]^{1/2}; \tag{106}$$

here $k^{(1)} = k_{A2^*/5^*}[NO(aq)] = k_{A2^*/5^*}H_{NO}p_{NO}$. Hence

$$J_{NO_2} = (D_{NO_2}k_{A2^*/5^*})^{1/2}(H_{NO}p_{NO})^{1/2}(H_{NO_2}p_{NO_2}). \tag{107}$$

More generally, J_{NO_2} may be written as

$$J_{NO_2} = \left(\frac{\overline{D}}{3}\right)^{1/2} k_{A2^*/5^*}^{1/2}(H_{NO}H_{NO_2}p_{NO}p_{NO_2})^{3/4}f(\alpha) \tag{108}$$

where α is the dimensionless ratio

$$\alpha \equiv \frac{H_{NO}p_{NO}}{H_{NO_2}p_{NO_2}}$$

and $f(\alpha)$ is also dimensionless. For $\alpha = 1$, $f = 1$; i.e., the situation is that leading to (104). For $\alpha \gg 1$, comparison of (104) and (107) shows that

$$f \sim 3^{1/2}\left(\frac{D_{NO_2}}{\overline{D}}\right)^{1/2}\alpha^{-1/4} \tag{109}$$

A sketch of the function f is shown in Figure 3. (Here and subsequently we approximate $\overline{D} = D_{NO_2}$.) This function is approximately equal to 1 for α within an order of magnitude of unity and for values of α outside this range decreases only slowly. Because of this slow variation it is seen that irrespective of the value of α, departure from Eq. (104) would not be expected to be too great for a limited range of α. Thus for a spread in α of a factor of 6.7 (corresponding to the range of p_{NO}/p_{NO_2} employed by Komiyama and Inoue), one would expect a "scatter" of at most $\pm 25\%$, depending on the value of α, from the dependence given by (104). This "scatter" is consistent with that observed by Komiyama and Inoue. Thus the variation of the rate of uptake as $(p_{NO}p_{NO_2})^{3/4}$ need not be taken as evidence for the validity of Komiyama and Inoue's value of H_{NO_2}.

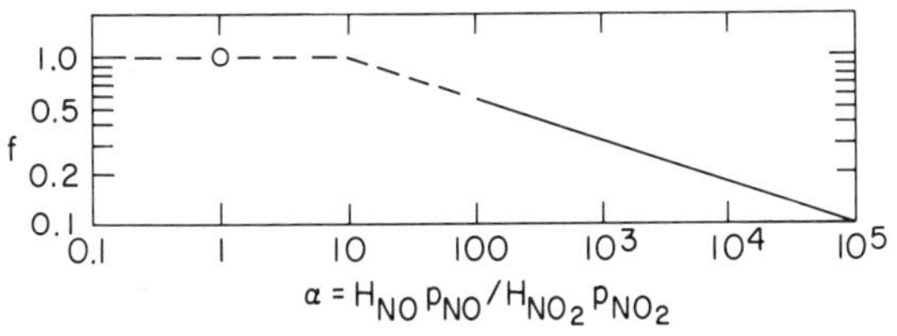

Figure 3. Ratio of rate of uptake of NO_2 under diffusive mass-transport-controlled conditions for $H_{NO}p_{NO}/H_{NO_2}p_{NO_2} = \alpha \neq 1$ to that for $\alpha = 1$.

The alternative interpretation developed here leads to

$$H_{NO_2}^{3/2} k_{A2*/5*} = \frac{3q^2}{H_{NO}^{3/2} D_{NO_2} \bar{f}^2} \tag{110}$$

[see Eq. (108)]; here $\bar{f}$ is the value of f at the midrange of p_{NO}/p_{NO_2}. For application of (110), it is necessary to know $\bar{f}$, which, however, depends on H_{NO_2}. Fortunately, this dependence is weak. In particular for $p_{NO}/p_{NO_2} = 10.2$, values of H_{NO_2} within an order of magnitude of Komiyama and Inoue's value of 2.35×10^{-2} M atm^{-1} would lead to $\bar{f} \approx 1$. Hence for H_{NO_2} in this range, (110) together with Komiyama and Inoue's measured value of q yields $H_{NO_2}^{3/2} k_{A2*/5*} = 2.1 \times 10^7$ m$^{1/2}$atm$^{-3/2}$ s^{-1} for values of H_{NO_2} within the range 2×10^{-3} M atm$^{-1} \leq H_{NO_2} \leq 2 \times 10^{-1}$ M atm^{-1}. It should be noted that although this discussion satisfactorily accounts for the lack of sensitivity of the rate law to the ratio p_{NO}/p_{NO_2}, it does little to allay the concern raised above regarding the large value of $k_{A2*/5*}$. Indeed, for H_{NO_2} less than Komiyama and Inoue's value of 2.35×10^{-2} M atm^{-1}, a value of $k_{A2*/5*}$ is required that is even greater than that given above. We return to further consideration of this value of $H_{NO_2}^{3/2} k_{A2*/5*}$ in Section 7.2.

The two interpretations of the simultaneous $\frac{3}{4}$-order uptake of NO and NO_2 are summarized in Table 24.

Before concluding this discussion, we briefly consider the several other assumptions that were made permitting the uptake of NO and NO_2 to be treated as a diffusion mass-transport-controlled reaction of these two oxides. Again we note that a significant fraction of dissolved N(IV) is present as N_2O_4; this fraction ranges from 6 to 39% for the NO_2 partial pressures employed, where for this evaluation we have employed $H_{NO_2} = 1.2 \times 10^{-2}$ M atm^{-1} and $K_{A7} = 6.5 \times 10^4$ M^{-1}. (If Komiyama and Inoue's values of these constants are employed, $F_{N_2O_4}$ ranges from 13 to 59%.) Thus concern must be raised again regarding the accuracy of the constants determined. Similar examination may be made of the concentration of N_2O_3 relative to that of NO or NO_2; in this case the equilibrium concentration of the compound oxide appears to be less than about 10%.

Table 24. Alternative Interpretations of Simultaneous $\frac{3}{4}$-Order Uptake of NO and NO_2[a]

Quantity	Value	Units
q^b	1.1(−3)	kmol m^{-2} s^{-1} atm$^{-3/2}$
$k_{A2*/5*} = 3q^2/\bar{D}(H_{NO}H_{NO_2})^{3/2}$ [c]	5.3(9)	M^{-1} s^{-1}
$H_{NO_2}^{3/2} k_{A2*/5*} = 3q^2/H_{NO}^{3/2} D_{NO_2} \bar{f}^2$ [d]	2.1(7)	M$^{1/2}$ s^{-1} atm$^{-3/2}$

[a] Data of Komiyama and Inoue (1980), 15°C.

[b] Defining equation: $J_{NO} = J_{NO_2} = q(p_{NO}p_{NO_2})^{3/4}$.

[c] For assumed $H_{NO}p_{NO} = H_{NO_2}p_{NO_2}$; $\bar{D} = 1.74 \times 10^{-9}$ m^2 s^{-1}.

[d] For assumed $H_{NO}p_{NO} \gtrsim 2H_{NO_2}p_{NO_2}$; $D_{NO_2} = 1.57 \times 10^{-9}$ m^{-2} s^{-1}; $\bar{f} = 1$.

With respect to mass transport we would note that for the reaction to be in the diffusive regime, the criterion $\tau_r < 0.38\ \tau_m^2/\tau_d$ must be satisfied. For interpretation as diffusion-controlled reaction of NO and NO_2 [Eq. (104)], this criterion must be satisfied for both NO and NO_2, where

$$\tau_r(NO_2) = (H_{NO}p_{NO}k_{A2^*/5^*})^{-1}$$

and

$$\tau_r(NO) = (H_{NO_2}p_{NO_2}k_{A2^*/5^*})^{-1}.$$

Evaluation of these quantities for values of H_{NO_2} and $k_{A2^*/5^*}$ proposed by Komiyama and Inoue (1980) leads to no inconsistency. For interpretation as diffusion-controlled uptake of NO_2 (with excess [NO]), the requirement must be met only for NO_2; again, no inconsistency is indicated.

7. SYNTHESIS

The preceding sections have reviewed measurements, obtained by a variety of techniques, of the Henry's law solubility and/or kinetics of aqueous-phase reaction of the nitrogen oxides in liquid water. These experiments lead to results that may be expressed in terms of products of various powers of the Henry's law coefficients of the appropriate species and/or rate constants of the appropriate reactions. For the most part, however, we have refrained from combining the results of these studies to yield the values of the Henry's law coefficients and rate constants, reserving that component of this review until the various studies themselves have been reviewed, so that all relevant studies may be taken into account. Having completed the review of the various individual experiments, we proceed in this section to consider the results of these experiments in their entirety with the objective of obtaining recommended values of the Henry's law coefficients and aqueous-phase rate constants that exhibit the greatest consistency with this body of experimental data. As is seen later, it is not possible to ascribe unique values to these quantities because of unrecognized determinate errors in the various studies and/or inappropriate application of the model by which the experimental measurement is interpreted. Nevertheless, as is seen, it is possible to recommend values of these quantities that lie within rather narrow uncertainty bounds.

7.1. Reaction [1*/4*]

In the case of reaction [1*/4*] the quantities for which there are experimental measurements are listed in Table 25. Examination of this set of quantities shows that these experiments may be expressed in terms of the five elementary constants, H_{NO_2}, $H_{N_2O_4}$, K_{A7}, $k_{A1^*/4^*}$, and $k_{A4^*/1^*}$. However, we

Table 25. Experimentally Measured Quantities Pertinent to Mixed- and Aqueous-Phase Reactions [1] and [4]

Measured Quantity	Method	Data Source
1. $H_{N_2O_4}k_{A4^*/1^*}(\equiv k_{M4^*/1^*})$	Indirect—nitrous acid decomposition, etc.	Table 5
2. $k_{A4^*/1^*}$	Aqueous pulse radiolysis, etc.	Table 13
3. $H_{N_2O_4}k_{A4^*/1^*}^{1/2}$	Direct contact, penetration theory	Table 15
4. $H_{N_2O_4}k_{A4^*/1^*}^{-1/2}$	Direct contact, penetration theory	Table 16
5. $H_{NO_2}^2k_{A1^*/4^*}(\equiv k_{M1^*/4^*})$	Indirect—nitrous acid decomposition, etc.	Table 5
6a. H_{NO_2}	Thermochemical cycles (paths II and III)	Schwartz and White (1981)
6b. H_{NO_2}	Convective mass-transport uptake of NO_2	Table 19
6c. H_{NO_2}	Slow-reaction study of NO_2 uptake	Section 6.4
7. $k_{A1^*/4^*}$	Slow-reaction study of NO_2 uptake	Section 6.4
8. $H_{NO_2}^{3/2}k_{A1^*/4^*}^{1/2}$	Direct contact, penetration theory	Table 18
9. K_{A7}	Pulse radiolysis	Schwartz and White (1981)

note that the relationships

$$H_{N_2O_4} = H_{NO_2}^2 K_{A7} K_{G7}^{-1} \tag{111}$$

and

$$k_{A4^*/1^*} = k_{A1^*/4^*} K_{A7}^{-1} \tag{112}$$

introduce two constraints on this set of constants and thus that the number of independent elementary constants (degrees of freedom) necessary for description of this reaction system is reduced to three. (Here and subsequently we consider the gas-phase equilibrium constant K_{G7} well established and not subject to further refinement in consideration of these experiments.) For the moment, however, it is useful to retain a redundant set of constants, since, depending on the predominant aqueous-phase species (NO_2 or N_2O_4), various experimental results are expressible in terms of either H_{NO_2} and $k_{A1^*/4^*}$ or $H_{N_2O_4}$ and $k_{A4^*/1^*}$.

In the discussion that follows we employ a graphical method that appears to be quite useful for the interpretation of these experiments. As is seen later, this method allows the data from all pertinent experiments to be compared and used in selecting a set of recommended elementary constants and in estimating the uncertainty of these constants. This approach thus contrasts with algebraic approaches that have been employed previously [e.g., Sherwood et al. (1975) and Stedman (1979), as well as the primary literature already cited]. We first consider experiments the results of which are expressible in terms of only $H_{N_2O_4}$ and/or $k_{A4^*/1^*}$ (Table 25, items 1–4). For example, consider a given measurement of $k_{M4^*/1^*}$ ($= H_{N_2O_4}k_{A4^*/1^*}$), as listed in Table 5, e.g., the value 5.92×10^2 M atm^{-1} s^{-1} derived from Abel and H. Schmid (1928c). This result can be considered a single equation in the two unknowns $H_{N_2O_4}$ and $k_{A4^*/1^*}$, which can be satisfied by any pair of values of $H_{N_2O_4}$ and $k_{A4^*/1^*}$ such that their product is equal to the measured value. This set of solutions constitutes a curve in the plane whose axes are $k_{A4^*/1^*}$ and $H_{N_2O_4}$, representing the locus of points (in $k_{A4^*/1^*}$–$H_{N_2O_4}$ space) that satisfy this particular experimental result. For the coordinate axes taken as log $k_{A4^*/1^*}$ and log $H_{N_2O_4}$, this locus of points is represented by a straight line, as given in Figure 4, line *e*. The slope of this line is -1, corresponding to the equation

$$\log H_{N_2O_4} + \log k_{A4^*/1^*} = \log(592).$$

The other experimental determinations of $H_{N_2O_4}k_{A4^*/1^*}$ in Table 5 lead to the

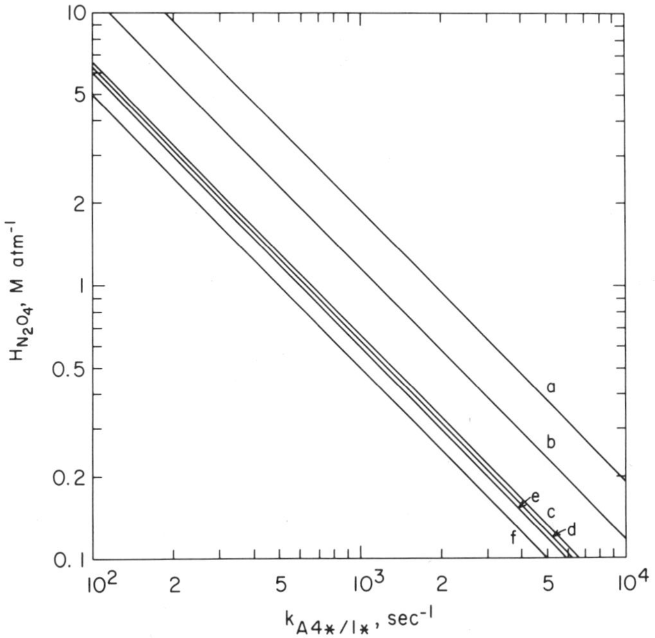

Figure 4. Measurements of $H_{N_2O_4}k_{A4^*/1^*}$ displayed in log $H_{N_2O_4}$–log $k_{A4^*/1^*}$ plane: (*a*) Heckner (1973); (*b*) Komiyama and Inoue (1978); (*c*) G. Schmid and Bähr (1964); (*d*) Abel and H. Schmid (1928d); (*e*) Abel and H. Schmid (1928c); (*f*) Jordan and Bonner (1973). All measurements at or corrected to 25°C.

remaining lines shown in Figure 4, i.e., to a family of parallel lines all with slope of −1. In principle, for all these measurements carried out at (or corrected to) the same temperature, all these lines should be coincident. The "spread" in these lines represents experimental error.

We now consider other experiments. In Figure 5 are shown the results of experiments measuring $H_{N_2O_4}k_{A4^*/1^*}^{1/2}$ and $H_{N_2O_4}k_{A4^*/1^*}^{-1/2}$ (Tables 15 and 16). These experiments lead to lines (in the log $k_{A4^*/1^*}$–log $H_{N_2O_4}$ plane) of slope −1/2 and +1/2, respectively. Again considerable spread in the data is evidenced, as seen in the breadth of the two sets of lines. Also shown in Figure 5, as indicated by the filled circles, are the points of intersection as inferred from each of the three experiments in which both $H_{N_2O_4}k_{A4^*/1^*}^{1/2}$ and $H_{N_2O_4}k_{A4^*/1^*}^{-1/2}$ were measured. These points represent presumptively unique solutions of these sets of data for $H_{N_2O_4}$ and $k_{A4^*/1^*}$ (two equations in two unknowns), and indeed, assuming highly accurate and precise measurements, such a procedure would lead to a correct evaluation of these quantities. However, as may be seen in Figure 5, the uncertainty in the measured data admits a considerable spread in the values of $H_{N_2O_4}$ and $k_{A4^*/1^*}$ that may be inferred from the region of intersection of these "bands."

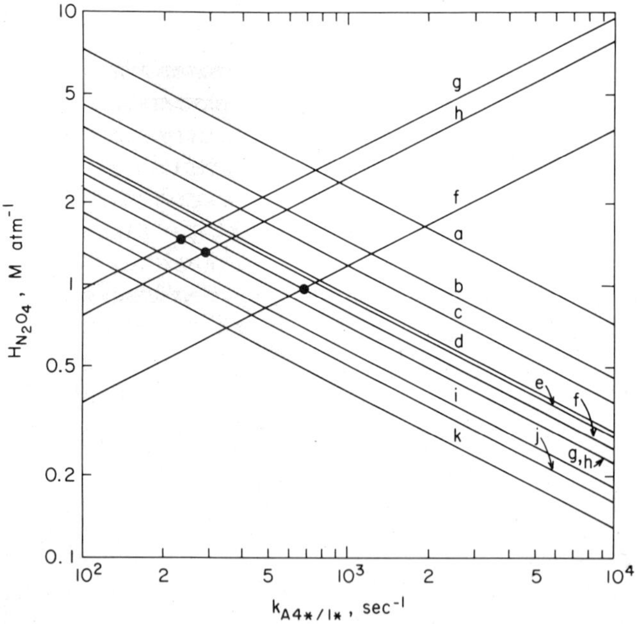

Figure 5. Measurements of $H_{N_2O_4}K_{A4^*/1^*}^{1/2}$ and $H_{N_2O_4}k_{A4^*/1^*}^{-1/2}$ displayed in log $H_{N_2O_4}$–log $k_{A4^*/1^*}$ plane (filled circles indicate intersections of individual experimental deteminations): (*a*) Komiyama and Inoue (1980), flat-surface contactor; (*b*) ibid., bubbler contactor; (*c*) Chilton and Knell (1972); (*d*) Dekker et al. (1959); (*e*) Gerstacker (1961); (*f*) Hoftyzer and Kwanten (1972); (*g*) Kramers et al. (1961); (*h*) Bartholomé and Gerstacker (1961); (*i*) Kameoka and Pigford (1977); (*j*) Wendel and Pigford (1958); (*k*) Corriveau and Pigford (1971). All measurements at or corrected to 25°C.

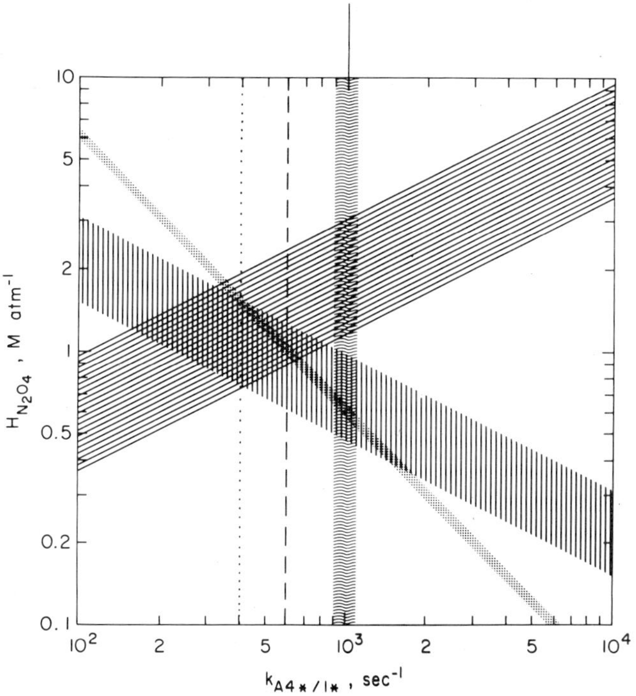

Figure 6. Summary plot of data on reactive dissolution of N_2O_4, displayed in log $H_{N_2O_4}$ –log $k_{A4*/1*}$ plane. ░░░░░, $H_{N_2O_4}k_{A4*/1*}$; ||||||||||||, $H_{N_2O_4}k_{A4*/1*}^{1/2}$; //////////////, $H_{N_2O_4}k_{A4*/1*}^{-1/2}$; ≋≋≋≋, -·-·-·, . . ., measurements of $k_{A4*/1*}$ by Grätzel et al. (1969), Treinin and Hayon (1970), and Moll (1966), respectively. All measurements at or corrected to 25°C except Grätzel et al., 20°C.

Comparison of Figures 4 and 5 suggests the utility of plotting these data on the same graph. This is done in Figure 6, where the spread of the data of the various types of experiment is now indicated by bands rather than lines representing the individual studies. In selecting the width of these bands we have somewhat arbitrarily excluded outlying results. In particular, in the case of $H_{N_2O_4}k_{A4*/1*}$ we rely heavily on the data of Abel and H. Schmid (1928c and d) and G. Schmid and Bähr (1964), for the reasons given in Section 4.1. Similarly in the case of determinations of $H_{N_2O_4}k_{A4*/1*}^{1/2}$ the band shown in Figure 6 reflects the recommended uncertainty in this quantity given in Section 6.1.

Also shown in Figure 6 are the results of the three studies leading to determination of $k_{A4*/1*}$, shown as vertical lines (Moll, 1966; Treinin and Hayon, 1970) or as a vertical band, representing the uncertainty estimate of the original authors (Grätzel et al., 1969). As previously noted, of the several studies we place highest confidence in the work of Grätzel et al. (1969), in view of the direct nature of the measurement, high reproducibility, and large number experimental runs.

Examination of Figure 6 shows rather satisfying agreement among the several types of experiment. In the absence of additional information the region of intersection of the bands resulting from the several types of experiment might serve as the basis for recommending values of $H_{N_2O_4}$ and $k_{A4*/1*}$. However, we defer such discussion to follow consideration of experiments yielding information on H_{NO_2} and $k_{A4*/1*}$.

Experimental results expressible in terms of H_{NO_2} and/or $k_{A1*/4*}$ only (Table 25, items 5–8) are displayed in Figure 7, as lines or bands in the log $k_{A1*/4*}$–log H_{NO_2} plane. (It should be pointed out that, to facilitate subsequent comparison with the N_2O_4 data, the scale of the log H_{NO_2} axis in this plot is twice that of the log $k_{A1*/4*}$ axis.) Values of H_{NO_2} obtained thermochemically are represented by the two overlapping horizontal bands, where the width of the band represents the propagated uncertainty estimate in the original equilibrium measurement(s). The value of H_{NO_2} obtained by Komiyama and Inoue (1980) as corrected here for temperature dependence (estimated) and N_2O_4 solubility is shown as a horizontal line, H_{NO_2} = 1.8

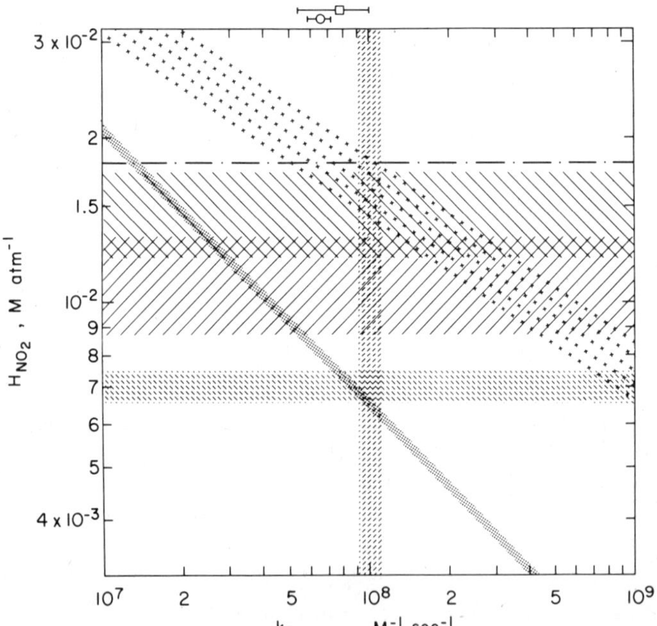

Figure 7. Summary plot of data on reactive dissolution of NO_2 displayed in log H_{NO_2}–log $k_{A1*/4*}$ plane. [pattern], $H^2_{NO_2}k_{A1*/4*}$; [pattern], $H^{3/2}_{NO_2}k^{1/2}_{A1*/4*}$, Komiyama and Inoue (1980), 15°C; [pattern] and [pattern], H_{NO_2} and $k_{A1*/4*}$, Lee and Schwartz (1981a), 22°C; [pattern] and [pattern], H_{NO_2} evaluated thermochemically, Schwartz and White (1981); -·-·-·-, H_{NO_2}, Komiyama and Inoue (1980), corrected, as indicated in text, to 25°C and for N_2O_4 solubility. Points and error bars at top of figure indicate measurements of K_{A7} when this figure is used in conjunction with Figure 6; see text. ○, Grätzel et al. (1969), 20°C; □, Treinin and Hayon (1970). All data for 25°C except as indicated.

$\times 10^{-2}$ M atm^{-1}. [The uncorrected value, not shown, is 2.35×10^{-2} M atm^{-1}. Also not shown is the value given by Andrew and Hanson (1961), 4×10^{-2} M atm^{-1}.] The value of H_{NO_2} (and estimated uncertainty) obtained by Lee and Schwartz (1981b) in their study of reaction [A1*/4*] in the slow-reaction regime is represented by a horizontal band; those authors' value and uncertainty of $k_{A1^*/4^*}$ are represented by a vertical band. Also represented in Figure 7 are the values of $k_{M1^*/4^*}$ ($= H^2_{NO_2}k_{A1^*/4^*}$) shown as a band of slope $-1/2$ (one decade per two decades); these data, of course, represent the same experimental measurements that led to $H_{N_2O_4}k_{A4^*/1^*}$ shown in Figure 6. Finally, we show as a band of slope $-1/3$ measurements of $H^{3/2}_{NO_2}k^{1/2}_{A1^*/4^*}$ as derived from the study by Komiyama and Inoue (1980); the limits of this band represent determinations with the two different contactors employed.

Examination of the NO_2 data as displayed in Figure 7 also reveals a reasonably well defined region of intersection, although there is some evident discrepancy. Such discrepancy would be lessened if the values of H_{NO_2} and $H^{1/2}_{NO_2}k^{3/2}_{A1^*/4^*}$ indicated in Figure 7 were shown to be too great; we proceed to present arguments that suggest that this may be the case.

With respect to the values of H_{NO_2} as determined by Komiyama and Inoue (1980), we have already addressed (Section 6.3 and Table 22) the question of whether the conditions for uptake solely by convective mass transport, as assumed by the authors, were in fact fulfilled. In that discussion it was shown that this condition was not fulfilled for the values of $k_{A1^*/4^*}$ that they obtained. This discussion can be extended to arbitrary, unknown values of $k_{A1^*/4^*}$ as follows. For a second-order reaction in which the characteristic reaction time τ_r is given by

$$\tau_r = (2H_{NO_2}k_{A1^*/4^*}p_{NO_2})^{-1}, \tag{113}$$

the condition for convective mass-transport-controlled uptake

$$\tau_r \gtrsim \frac{3\tau_m^2}{\tau_d} \tag{98}$$

is equivalent to

$$H_{NO_2}k_{A1^*/4^*} \le \left(2p_{NO_2}\frac{3\tau_m^2}{\tau_d}\right)^{-1}. \tag{114}$$

This condition is readily displayed as a graph in the H_{NO_2}–$k_{A1^*/4^*}$ plane. In Figure 8 condition (114) is indicated for the extremes of p_{NO_2} that were employed by Komiyama and Inoue (1980) leading to their determination of H_{NO_2} (Table 19); here the mass-transport parameters given in Table 21 have been employed. Condition (114) is satisfied only for values of H_{NO_2} and $k_{A1^*/4^*}$ such that the point (H_{NO_2}, $k_{A1^*/4^*}$) lies below and to the left of the line indicated for the partial pressure of the measurement. The line $H_{NO_2}k_{A1^*/4^*} = (2p_{NO_2} \cdot 3\tau_m^2/\tau_d)^{-1}$ evaluated at the lowest partial pressure employed (1.2×10^{-5} atm) for the stated mixing conditions intersects the line $H_{NO_2} = 2.35 \times 10^{-2}$ M atm^{-1} (as obtained from the measured rate of uptake under the assumption that the solubility is dominated by the Henry's law solubility) at the value of $k_{A1^*/4^*} = 1.2 \times 10^7$ M^{-1} s^{-1}. Thus the assumption that the reactive contribution to the uptake of NO_2 under these conditions is small ($\lesssim$10%) compared to that due to the physical solubility is valid only for $k_{A1^*/4^*} \lesssim 1.2 \times 10^7$ M^{-1} s^{-1}. If this condition is not satisfied, the value of H_{NO_2} inferred from the measurement will be too high because of the increased solubility due to reaction. This would certainly be the case, both for the value of $k_{A1^*/4^*}$ indicated by Komiyama and Inoue's own data (2.1×10^7 M^{-1} s^{-1}) and *a fortiori* for a value of $k_{A1^*/4^*}$ as great as that indicated by Lee and Schwartz (1981b) (1×10^8 M^{-1} s^{-1}). This discussion thus reinforces the previous inference that the value of H_{NO_2} given by Komiyama and Inoue (1980) is somewhat too great.

We now address measurements of $H_{NO_2}^{3/2} k_{A1^*/4^*}^{1/2}$. We first consider the appropriateness of treating these measurements according to diffusive mass-transport-controlled reaction. As noted in Section 6.3, the condition to be

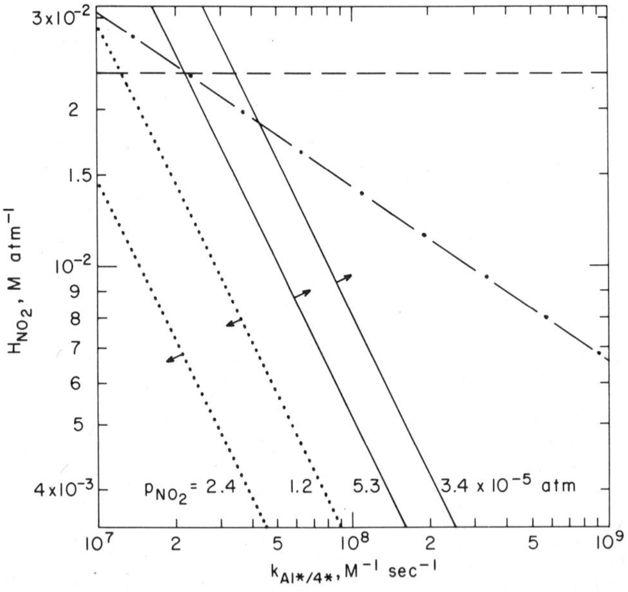

Figure 8. Examination of mass-transport regimes in study by Komiyama and Inoue (1980). ----- and -·-·-·-·- represent reported values of H_{NO_2} and $H_{NO_2}^{3/2} k_{A4^*/1^*}^{1/2}$, respectively (bubbler contactor, 15°C). Dotted lines represent bound to convective mass-transport regime for specified mass-transport conditions and for indicated p_{NO_2}; determination of H_{NO_2} at a given p_{NO_2} is valid only if point $k_{A1^*/4^*}, H_{NO_2}$ lies below and to the left of this bound. Solid lines represent bound to diffusive mass-transport regime; determination of $H_{NO_2}^{1/2} k_{A1^*/4^*}^{1/2}$ is valid only if point $k_{A1^*/4^*}, H_{NO_2}$ lies above and to the right of the bound.

satisfied is that

$$\tau_r \lesssim \frac{0.38\tau_m^2}{\tau_d}, \tag{96}$$

which is equivalent to

$$H_{NO_2}k_{A1^*/4^*} \geq \left(2p_{NO_2}\frac{0.38\tau_m^2}{\tau_d}\right)^{-1}. \tag{115}$$

Condition (115) is shown in Figure 8 for $p_{NO_2} = 5.3$ and 3.4×10^{-5} atm; the condition is satisfied for values of H_{NO_2} and $k_{A1^*/4^*}$ lying above and to the right of the indicated bounds at the respective partial pressures. It is seen that condition (115) would be satisfied for the range of partial pressures employed by Komiyama and Inoue [$(0.34–1.8) \times 10^{-4}$ atm] for, say, $H_{NO_2} = 1 \times 10^{-2}$ atm and $k_{A1^*/4^*} = 1 \times 10^8$ M^{-1}. As indicated in the discussion in Section 6.2, only for the very low end of the range of partial pressures employed [$(3.4–5.3) \times 10^{-5}$ atm] would there appear to be a concern that condition (115) is not satisfied for the values of H_{NO_2} and $k_{A1^*/4^*}$ that come under consideration from examination of Figures 7 and 8. Thus we cannot account for a positive error in $H_{NO_2}^{3/2}k_{A1^*/4^*}^{1/2}$ on this basis.

We now address other potential sources of inaccuracy in measurements of $H_{NO_2}^{3/2}k_{A1^*/4^*}^{1/2}$. Measurements of this quantity from all available sources, as listed in Table 18, are shown in Figure 9. It may be seen that considerable spread is exhibited by these data. This spread would appear to be due both to the fact that the several studies were not conducted at the same temperature and to an increased fraction of dissolved N_2O_4 contributing to the measured rate of uptake at the higher NO_2 partial pressures. Thus the two determinations given by Takeuchi et al. (1977) at 15 and 25°C suggest a factor of ~1.65 decrease in $H_{NO_2}^{3/2}k_{A1^*/4^*}^{1/2}$ between these temperatures, although this inference must be regarded as somewhat tentative since an increase in temperature would most likely lead also to an increase in K_{A7} and hence in $F_{N_2O_4}$, the fraction of N(IV) present as N_2O_4. We have also noted that in the study by Komiyama and Inoue (1980) $F_{N_2O_4}$ would appear to be on the order of 20%. These considerations would appear to suggest that the band of slope $-1/3$ in Figure 7 may be too high by a factor of roughly $(1.2 \times 1.65)^{2/3} = 1.6$. Lowering of this band by such a factor would bring it substantially closer to the region of intersection defined by the remaining methods indicated in Figure 7.

We now turn to the simultaneous consideration of both sets of measurements, i.e., the "N_2O_4 data" shown in Figure 6 and the "NO_2 data" shown in Figure 7. This simultaneous consideration can be effected by superimposing the two graphs. (This can be facilitated by making a transparent overlay of Figure 7 to the scale of Figure 6.) In superimposing Figures 6 and 7 it is required that the bands representing the $H_{NO_2}^2k_{A1^*/4^*}$ and $H_{N_2O_4}k_{A4^*/1^*}$ data be superimposed. This constraint allows relative trans-

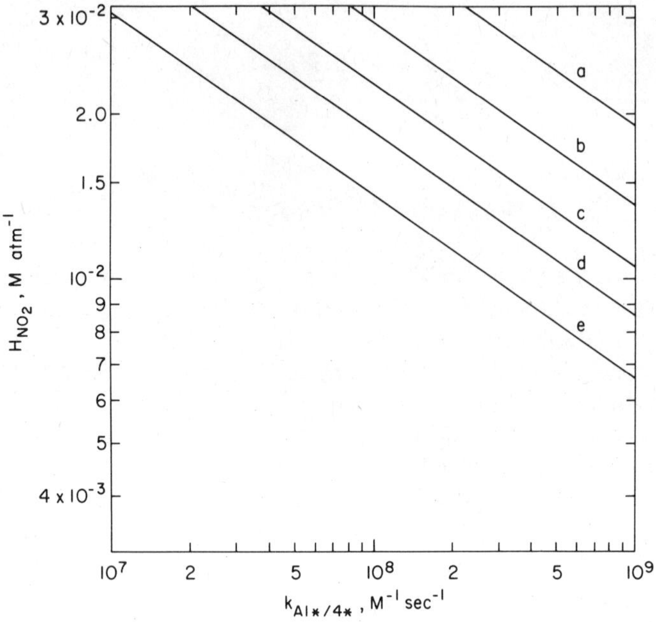

Figure 9. Measurements of $H_{NO_2}^{3/2}k_{A1^*/4^*}^{1/2}$ displayed in the log H_{NO_2}–log $k_{A1^*/4^*}$ plane: (*a*) Takeuchi et al. (1977), 15°C; (*b*) ibid., 25°C; (*c*) Sada et al. (1979), 25°C; (*d*) Komiyama and Inoue (1980), flat contactor, 15°C; (*e*) ibid., bubbler contactor, 15°C.

lation of the two figures in only one dimension, i.e., parallel to a line of slope − 1 in the log $H_{N_2O_4}$–log $k_{A4^*/1^*}$ plane. This single degree of freedom represents the variable K_{A7}. The constraint that the two bands representing $H_{NO_2}^2 k_{A1^*/4^*}$ and $H_{N_2O_4}k_{A4^*/1^*}$ be superimposed is mathematically equivalent to the constraint

$$H_{N_2O_4}k_{A4^*/1^*} = H_{NO_2}^2 k_{A1^*/4^*}K_{G7}^{-1} \tag{116}$$

obtained from (111) and (112), for specified values of the gas-phase equilibrium constant K_{G7}, which we have taken as well established. For any given relative position of Figures 6 and 7 [subject to (116)] the value of K_{A7} may be read off [by Eq. (112)] as the ratio of the abscissa scales of the two graphs. Once the relative position of the two graphs is fixed (i.e., K_{A7} is specified), it is seen that specification of the vertical coordinate specifies both H_{NO_2} and $H_{N_2O_4}$ and similarly that specification of a horizontal coordinate axis specifies both $k_{A1^*/4^*}$ and $k_{A4^*/1^*}$. Thus the three degrees of freedom in this system may be specified by relative translation of the two graphs (parallel to the "northwest–southeast" axis) and by specification of the two coordinates of a point in the plane.

At this juncture we should again point out [as discussed by Schwartz and

White (1981)] that the equilibrium constant K_{A7} has also been investigated experimentally. In two studies this quantity has been reported as (6.54 ± 10%) × 10^4 M^{-1} [Grätzel et al. (1969); 20°C] and (7.7 ± 30%) × 10^4 M^{-1} [Treinin and Hayon (1970); 25°C]; the error bars represent estimates given by Schwartz and White. These values of the equilibrium constant K_{A7} may be displayed on the superposition of Figures 6 and 7 by the device at the top of the figures. Shown at the top of Figure 7 are points at $k_{A1^*/4^*} = 6.54 \times 10^7$ and 7.7×10^7 M^{-1} s^{-1} and the corresponding error bars. For the two figures superimposed such that the value $k_{A4^*/1^*} = 10^3$ s^{-1} indicated by the vertical line at the top of Figure 6 overlays, say, the point $k_{A1^*/4^*} = 6.54 \times 10^7$ M^{-1} s^{-1} on Figure 7, then $K_{A7} = k_{A1^*/4^*}/k_{A4^*/1^*} = 6.54 \times 10^4$ M^{-1}.

Armed with Figures 6 and 7, we are now prepared to select a recommended set of elementary constants (and associated uncertainties) that describe the physical solubility and reaction kinetics in this system. The criterion that we have used in this selection is that the recommended range for the several constants should be such that it encompass at least partially the bands representing the several experimental methods, unless (as is the case with the measurement of $H_{NO_2}^{3/2} k_{A1^*/4^*}^{1/2}$) there is reason to suspect systematic error in the measurement. With respect to direct measurements of $k_{A4^*/1^*}$, we have included the determinations by Grätzel et al. (1969) and Treinin and Hayon (1970) but do not extend the range to encompass the measurement of Moll (1966) in view of the indirect nature of those measurements. By this admittedly arbitrary procedure we have selected the recommended range for the several constants given in Table 26. Within that range we have selected, again arbitrarily, a self-consistent [by Eqs. (111) and (112)] set of recommended values for the several quantities. These recommended values and uncertainty ranges should be useful for many purposes (e.g., estimation of rates in the ambient atmosphere) and should, additionally, serve as a point of departure for subsequent experiments that would more precisely define these quantities.

Table 26. Recommended Values and Uncertainties for Equilibrium and Rate Constants Describing Reactive Dissolution of NO_2 and N_2O_4 in Dilute Aqueous Solution, 20–25°C

Quantity	Minimum	Recommended	Maximum	Units
H_{NO_2}	0.7	1.0	1.3	10^{-2} M atm^{-1}
$H_{N_2O_4}$	0.45	1.0	2	M atm^{-1}
K_{A7}	6	7	9	10^4 M^{-1}
$k_{A1^*/4^*}$	0.35	0.7	1.0	10^8 M^{-1} s^{-1}
$k_{A4^*/1^*}$	0.6	1.0	1.6	10^3 s^{-1}

7.2. Reaction [2*/5*]

The quantities pertinent to the kinetics of reaction [2*/5*] for which there are experimental measurements are listed in Table 27. The Henry's law coefficient of NO is well established. The remaining elementary constants in which the measurements may be expressed are H_{NO_2}, $H_{N_2O_3}$, K_{A8}, $k_{A2^*/5^*}$, and $k_{A5^*/2^*}$. As was the case in the discussion of reaction [1*/4*], these quantities are not independent but are related in this case by

$$H_{N_2O_3} = H_{NO_2}K_{A8}(H_{NO}K_{G8}^{-1}) \tag{117}$$

and

$$k_{A5^*/2^*} = k_{A2^*/5^*}K_{A8}^{-1}, \tag{118}$$

where the gas-phase equilibrium constant K_{G8} is also taken as well established. Again for the moment we consider the N_2O_3 and NO + NO_2 experiments separately, returning subsequently to consideration of the entire set of measurements.

Measurements that can be expressed in terms of $H_{N_2O_3}$ and/or $k_{A5^*/2^*}$, displayed in the log $H_{N_2O_3}$–log $k_{A2^*/5^*}$ plane, are shown in Figure 10. Again, assuming accurate data, the lines representing the various measurements should intersect at a point. The departure from that situation indicated in Figure 10 represents inaccuracy in the measurements and/or interpretation. Inspection of Figure 10 suggests that the value of ($H_{N_2O_3}$, $k_{A5^*/2^*}$) obtained from the measurements by Hofmeister and Kohlhaas (1965) is considerably at variance with the remaining data. Earlier we noted potential concern with both this study, as well as that of Corriveau and Pigford (1971) in view of possible interaction with transport and reaction of NO_2 or N_2O_4; the prox-

Table 27. Experimentally Measured Quantities Pertinent to Mixed- and Aqueous-Phase Reactions [2] and [5]

Measured Quantity	Method	Data Source
1. $H_{N_2O_3}$	Thermochemical cycle ($K_{A5'}$)	Schwartz and White (1981)
2. $H_{N_2O_3}k_{A5^*/2^*}(\equiv k_{M5^*/2^*})$	Isotope exchange, diazotization	Table 9
3. $H_{N_2O_3}k_{A5^*/2^*}^{1/2}$	Direct contact, penetration theory	Table 23
4. $k_{A5^*/2^*}$	Aqueous pulse radiolysis	Table 13
5. H_{NO_2}	Review	Table 26
6. $H_{NO_2}k_{A2^*/5^*}(\equiv k_{M2^*/5^*}/H_{NO})$	Isotope exchange, diazotization	Table 9
7. $H_{NO_2}^{3/2}k_{A2^*/5^*}$	Direct contact, penetration theory	Table 24

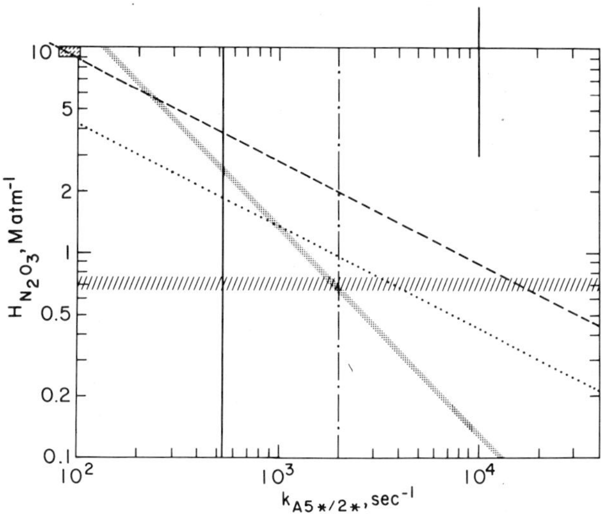

Figure 10. Summary plot of data on reactive dissolution of N_2O_3, displayed in log $H_{N_2O_3}$ –log $k_{A5*/2*}$ plane. ////////, $H_{N_2O_3}$ evaluated from data of Markovits et al. (1981); ░░░░, $H_{N_2O_3}k_{A5*/2*}$ inferred from diazotization and isotope exchange kinetics; ———— , and -·-·-·, $k_{A5*/2*}$ measured in pulse radiolysis (Grätzel et al., 1970) and flash photolysis (Treinin and Hayon, 1970), respectively; .... and ----, $H_{N_2O_3}k^{1/2}_{A5*/2*}$, measured by direct contact (Corriveau and Pigford, 1971; Hofmeister and Kohlhaas, 1965, respectively); ░░░░, region of ($H_{N_2O_3}$,$k_{A5*/2*}$) inferred from Hofmeister and Kohlhaas (1965). All measurements at or corrected to 25°C except for Grätzel el al. (1970), 20°C.

imity of the line representing the data of Corriveau and Pigford to the region of intersection of the remaining data may well be fortuitous. We note as well the factor of 4 discrepancy between the values of $k_{A5*/2*}$ reported by Grätzel et al. (1970) and Treinin and Hayon (1970); as noted in Section 5.4, neither measurement was taken over a sufficiently wide range of conditions as to command great confidence. Thus the region of intersection indicated in Figure 10 cannot be considered as setting very tight bounds on $k_{A5*/2*}$ or $H_{N_2O_3}$. A range of (0.4–3) $\times$ 10^3 s^{-1} for $k_{A5*/2*}$ and 0.6–3 M atm^{-1} for $H_{N_2O_3}$ would appear to be suggested.

We now consider measurements expressible in terms of $k_{A2*/5*}$ and/or H_{NO_2} (Figure 11). Here the band for H_{NO_2} represents the recommended range for that quantity presented in Table 26. The paucity of data precludes any extended discussion. However, it may be noted that the measurements of $H^{3/2}_{NO_2}k_{A2*/5*}$ [derived from the reinterpretation of the data of Komiyama and Inoue (1980)] suggest a value of $k_{A2*/5*}$ in excess of 10^{10} M^{-1} s^{-1}, i.e., unrealistically high.

We now consider both the N_2O_3 and the NO + NO_2 data simultaneously, as may be effected by superimposing Figures 10 and 11. Here the constraint

$$H_{N_2O_3}k_{A5*/2*} = H_{NO_2}k_{A2*/5*}(H_{NO}K^{-1}_{G8}) \quad (119)$$

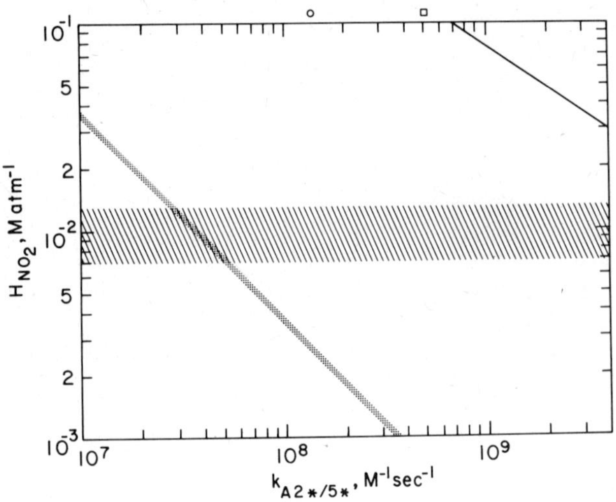

Figure 11. Summary plot of data on simultaneous reactive dissolution of NO and NO_2, displayed in log H_{NO_2}–log $k_{A2^*/5^*}$ plane. \\\\\\\\\\\\ , H_{NO_2}, recommended value and uncertainty (Table 26); ▒▒▒▒, $H_{NO_2}k_{A2^*/5^*}$ inferred from diazotization and isotope exchange kinetics; ———, $H_{NO_2}^{3/2}k_{A2^*/5^*}$ as inferred from measurements by Komiyama and Inoue (1980), 15°C. Points at top of diagram indicate measurements of K_{A8} when this diagram is used in conjunction with Figure 10; see text. ○, Grätzel et al. (1970), 20°C; □, Treinin and Hayon (1970). All data for 25°C except as indicated.

obtained from (117) and (118) requires superposition of the bands representing $H_{NO_2}k_{A2^*/5^*}$ and $H_{N_2O_3}k_{A5^*/2^*}$, with relative translation permitted only parallel to a line of slope -1. For any given relative position of Figures 10 and 11, the value of K_{A8} may be read as the ratio of the two abscissa scales. Also indicated by the device at the top of Figures 10 and 11 are the values of K_{A8} determined by Grätzel et al. (1970) and by Treinin and Hayon (1970).

From the range of values of the several elementary constants obtained in the superposition of Figures 10 and 11 we have selected a recommended set of these quantities and estimated uncertainties, where the uncertainties encompass the measurements shown in Figures 10 and 11 (with the exception of the value for $H_{NO_2}^{3/2}k_{A2^*/5^*}$). In this process we have employed the value and uncertainties for H_{NO_2} from Table 26 without additional refinement. The recommended values and uncertainties are given in Table 28.

7.3. Temperature Dependence

As noted above, the temperature dependence of the several experimentally measured quantities of describing the rate of uptake or reaction of the nitrogen oxides in water may be expressed as an activation energy associated

Table 28. Recommended Values and Uncertainties for Equilibrium and Rate Constants Describing Reactive Dissolution of NO + $NO_2(N_2O_3)$ in Dilute Aqueous Solution (20–25°C)

Quantity	Minimum	Recommended	Maximum	Units
H_{NO_2}	0.7	1.0	1.3	10^{-2} M atm^{-1}
$H_{N_2O_3}$	0.6	1.0	2.3	M atm^{-1}
K_{A8}	1.4	3	5	10^4 M^{-1}
$k_{A2^*/5^*}$	0.1	0.3	1	10^8 M^{-1} s^{-1}
$k_{A5^*/2^*}$	0.4	1.0	2	10^3 s^{-1}

with the process, evaluated as

$$E_a(q) = -R \frac{d \ln q}{dT^{-1}} .$$

In principle, the activation energies associated with the several experiments can be resolved into the activation energies and enthalpies of reactions of the several elementary processes, entirely analogously to the resolution of the rate quantities themselves into the rate coefficients and equilibrium constants of the elementary reactions. In practice, however, the available data for the temperature dependence of both reaction systems [1*/4*] and [2*/5*] are quite limited, and consequently we are able to carry out this resolution only to a very limited extent.

The available data describing the temperature dependence of reactions [1*/4*] and [2*/5*] are summarized in Table 29. From this table it is seen that only in the case of data pertaining to the temperature dependence of $H_{N_2O_4}$ and $k_{A4^*/1^*}$ is there sufficient information to allow resolution into the separate terms. This resolution is shown graphically in Figure 12, which may be interpreted entirely analogously to Figure 6, except that the axes are now E_a[A4*/1*] and $\Delta H°(H_{N_2O_4})$. The region of intersection of the data represented in this figure is encompassed by the recommended values

Table 29. Temperature Dependence of Experimentally Determined Rate Quantities in Reactions [1*/4*] and [2*/5*]

Quantity	E_a (kcal mol^{-1})	Reference to Text
$k_{M4^*/1^*} \equiv H_{N_2O_4} k_{A4^*/1^*}$	+5.9	Section 4.1
$H_{N_2O_4} k_{A4^*/1^*}^{1/2}$	+0.5 ± 0.7	Section 6.1
$H_{N_2O_4} k_{A4^*/1^*}^{-1/2}$	−4.3 ± 0.3	Section 6.1
$k_{M1^*/4^*} \equiv H_{NO_2}^2 k_{A1^*/4^*}$	−7.8	Section 4.1
$k_{M5^*/2^*} \equiv H_{N_2O_3} k_{A5^*/2^*}$	+5.4 ± 1.7	Section 4.2
$k_{M2^*/5^*} \equiv H_{NO} H_{NO_2} k_{A2^*/5^*}$	−4.1 ± 1.7	Section 4.2

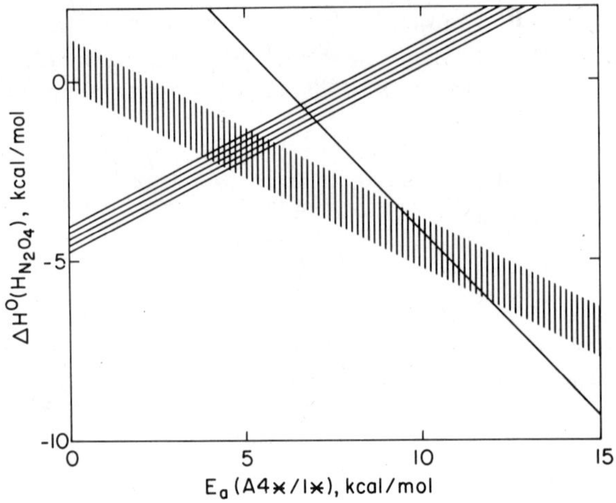

Figure 12. Summary plot of data on temperature dependence of reactive dissolution of N_2O_4 displayed in the $\Delta H°[H_{N_2O_4}]$, $E_a[A1^*/4^*]$ plane. ———, $E_a(H_{N_2O_4}k_{A4^*/1^*})$; ||||||||||||||||||||, $E_a(H_{N_2O_4}k^{1/2}_{A1^*/4^*})$; ////////////////, $E_a(H_{N_2O_4}k^{-1/2}_{A4^*/1^*})$.

$E_a[A^*/1^*] = 7.5 \pm 2.5$ kcal mol^{-1} and $\Delta H°(H_{N_2O_4}) = -(2.5 \pm 1.5)$ kcal mol^{-1}. Insufficient information is available to permit resolution of the remaining temperature dependence data summarized in Table 29 into the corresponding enthalpies of solution and activation energies for the indicated elementary reactions.

7.4. Sources of Error

It is of interest to speculate on the causes for the lack of closer agreement between the several experiments that have been reviewed. Much of the above discussion has focused on appropriateness of the mixed-phase reaction model employed to describe the system under examination and on correct identification of the transported species, NO_2 or N_2O_4, and to the extent possible corrections have been made or the studies excluded from the final comparisons. Less easily addressed are possible systematic errors, such as in calibration of mass-transport properties and extinction coefficients. Even more difficult to address are considerations of reagent purity, but it must be emphasized that both neat NO_2 (N_2O_4) and dilute NO_2 in N_2 or air as supplied by manufacturers are frequently contaminated with lower oxides and/or with oxyacids and further that NO_2 may be inadvertently contaminated by conventional gas-handling apparatus. The effects of contaminants, including contaminants in water, would be most pronounced in studies at low NO_2 partial pressures, and as noted in Section 8, there have been a number of studies in which these effects are indicated. We would thus urge extreme attention to purity of reagents in any future studies.

8. OTHER STUDIES AND SOLUTE EFFECTS

In this section we consider other studies examining the rate of uptake of nitrogen oxides by aqueous solution, including studies of the effects of dissolved solutes. In some of these studies, as reviewed in Section 8.1, the phase-mixing conditions were not sufficiently well characterized to allow quantitative interpretation in terms of Henry's law coefficients and kinetic rate constants. Nonetheless, it remains possible to gain insight into the processes taking place in these experiments, and, as we shall see, it would appear that reaction of NO_2 with dissolved impurities may influence the observed rate of uptake.

The interaction of NO and NO_2 with dissolved solutes in aqueous solution has received comparatively little study, much of that qualitative, although in some cases it is possible to derive quantitative kinetic information making use of solubility coefficients obtained above. Such interactions are of potential importance not only in interpreting laboratory studies, but also in describing reactions of NO and NO_2 with dissolved materials present in atmospheric and surface liquid water. A further application has included development of efficient means of sampling nitrogen oxides for chemical analysis, although interest in such methods has somewhat waned with the advent of chemiluminescent methods for the determination of these gases. An additional application is the search for possible wet scrubbers for removal of nitrogen oxides from flue gases (Counce, 1980; Faucett et al., 1978).

8.1. NO_2 + Liquid Water

Measurements of the fraction of NO_2 taken up by bubbling NO_2 in air through water (NO_2 partial pressures 3×10^{-7} to 0.12 atm) were reported by Borok (1960). At higher NO_2 partial pressures the fractional absorption of NO_2 decreased with decreasing partial pressures, as expected for uptake by the second-order reaction [A1*/4*]. However, for partial pressures below 2.5×10^{-6} atm, the fraction absorbed was found to increase with decreasing partial pressure. These results, if correct, would imply a mechanism for removal of $NO_2(g)$ other than reaction [A1*/4*], although insufficient information is available for identification of this mechanism. Subsequent work, such as that by Lee and Schwartz (1981b), has given no indication of any departure from a second-order rate law for the uptake of NO_2 by water, thus suggesting the possibility that the increased efficiency for removal of NO_2 at low partial pressures found by Borok may have been due to contamination by reactive impurities.

The experimental data of Borok (1960) were subsequently interpreted to yield a "Henry's law" coefficient for NO_2 (Hermann and Matteson, 1980; Matteson, 1979) evaluated as the ratio of the dissolved nitrite concentration in the bubbler solution to the partial pressure of NO_2 in the gas that was passed through the bubbler. The value obtained, $H_{NO_2} = 0.55$ M atm^{-1} (Matteson, 1981), is some fiftyfold greater than that which has been inferred in the present analysis. This high value for H_{NO_2} undoubtedly reflects the reactive solubility of NO_2 (by reaction [A1*/4*] and/or with impurities) rather than the purely physical solubility that is represented by Henry's law.

The experimental studies by Herrmann and Matteson (1980), in which suspended water droplets (3-mm diameter) were exposed to a flowing mixture of NO_2 in nitrogen [$p_{NO_2} = (1\text{–}3) \times 10^{-4}$ atm] might also be mentioned here. The water drops were removed from the flowing gas stream after exposures of 5–30 s and subsequently analyzed for NO_2^-. Experimental results were presented as the average aqueous-phase nitrite concentration (in the collected drops) as a function of the duration of exposure of the drop to the flowing gas mixture for various values of the water vapor supersaturation ratio (ratio of the water vapor pressure to the equilibrium water vapor pressure of the suspended drop at the temperature of the liquid). At longer exposure times ($\gtrsim$30 s) the aqueous-phase nitrite concentration approached a constant "saturation" value. The authors attribute this "saturation" to a solubility equilibrium between NO_2(g) and nitrite ion NO_2^-, which they proceed to address in terms of their value (0.55 M atm^{-1}) for the "Henry's law" coefficient for NO_2. It must be emphasized that the equilibrium in these studies, if any, between NO_2(g) and NO_2^- is *not* the Henry's law equilibrium, which strictly denotes the equilibrium between gaseous NO_2 and the physically dissolved material, since NO_2^- can be formed from NO_2 only by chemical reaction. It is further readily established that the material analyzed as nitrite cannot have been present as physically dissolved NO_2; for $p_{NO_2} = 10^{-4}$ atm, the characteristic time describing the removal of NO_2(aq) by reaction [A1*/4*], evaluated as $(2H_{NO_2}p_{NO_2}k_{A1^*/4^*})^{-1}$, would be approximately 5×10^{-3} s, i.e., orders of magnitude less than the duration of exposure of the drops to NO_2(g). Hence the time dependence of the approach to "saturation" cannot have been due simply to the kinetics of mass transport and physical dissolution (as assumed by Herrmann and Matteson) but must reflect the reactive dissolution as well. It is possible also to address whether the "saturation" reported by Herrmann and Matteson represents achievement of chemical equilibrium in this system. An approach to this is to consider the characteristic time of reaching equilibrium for reaction [M1′]:

$$2NO_2(g) \overset{w}{=} HNO_2(aq) + H^+ + NO_3^- \qquad [M1']$$

evaluated as

$$\tau^{(1')} = \frac{[NO_3^-]_{eq}}{H_{NO_2}^2 k_{A1^*/4^*} p_{NO_2}^2}. \qquad (120)$$

Here it is assumed that the only reaction taking place is [M1′] and further that the reaction takes place under phase-mixed conditions, i.e., aqueous phase saturated in NO_2. In fact, neither assumption would be expected to hold for the NO_2 partial pressures and drop sizes employed, and consequently (120) would underestimate the actual characteristic time of achieving equilibrium. Even so, the values of $\tau^{(1')}$ obtained by (120), (3.9–17) × 10^2 s, substantially exceed the contact times characteristic of the approach to "saturation." Consequently, it would appear that the "saturation" reported by Herrmann and Matteson (1980) is also not due to chemical equilibrium. In principle, it would appear possible to interpret these measurements in terms of the pertinent mass-transport rates, reaction kinetics, and chemical equilibria. Unfortunately, however, since the concentration measurements are presented only as the ratio to an unspecified "saturation" concentration, such interpretation cannot be given.

An interesting experimental study that deserves mention is that by Beilke (1970), who addressed the "washout" of NO_2 (as well as SO_2 and CO_2) by recirculating artificial "rain" falling through a chamber containing the gas (initial concentration 3×10^{-8} atm). Exposure times extended up to 30 min for rainfall rates ranging from 1 to 50 mm h^{-1}. Experimental results were expressed in terms of a washout coefficient λ (units: s^{-1}) describing the exponential decrease (with time) of the concentration of the gas remaining in the chamber. Values of λ varied with rainfall rate and mean drop diameter; the latter dependence is indicative of mass-transport restriction to the rate of uptake of the gas by the falling droplets. In the case of NO_2 there was some indication of a nonexponential decay, i.e., values of λ decreased during the experiment. A typical value of λ was 1.6×10^{-4} s^{-1}, for rainfall rate $I = 2.8 \times 10^{-7}$ m s^{-1} (1 mm h^{-1}), mean drop diameter of 0.85 mm, and temperature 20°C.

It is instructive to compare the observed value of the washout coefficient to values that might be expected from equilibrium or kinetic considerations. If it is assumed that the decrease in gas-phase concentration is due entirely to loss as dissolved material in the artificial rain, it is readily established that the washout coefficient may be evaluated in terms of the concentration of the dissolved material as

$$\lambda = \frac{\overline{C}I\mathbf{R}T}{pZ}$$

where λ = washout coefficient (s^{-1})
$\overline{C}$ = average aqueous-phase concentration (M)
I = rainfall rate (m s^{-1})
$\mathbf{R}$ = universal gas constant (liter atm mol^{-1} K^{-1})
T = absolute temperature (K)
p = gas-phase partial pressure (atm)
Z = chamber height (m)

For dissolution of NO_2 by Henry's law only, $\overline{C} = H_{NO_2}p_{NO_2}$, and $\lambda = H_{NO_2}I\mathbf{R}T/Z$. For the conditions given above λ is evaluated to be 7×10^{-8} s^{-1}, where we have taken the chamber height as 1 m (cf. chamber volume

1 m^3). Since this value is orders of magnitude lower than the observed washout coefficient at this partial pressure (1.6×10^{-4} s^{-1}), substantial reactive dissolution of NO_2 is indicated. It is possible also to evaluate λ under the assumption that reaction [A1] proceeds to equilibrium rapidly on the time scale of the exposure of the drops to the gas. In this case $\bar{C}$ = $[NO_3^-] + [NO_2^-] + [HNO_2] \approx 1 \times 10^{-3}$ M and $\lambda \approx 0.2$ s^{-1}, orders of magnitude greater than the observed rate. This comparison both indicates the high potential washout rate if this equilibrium were to be reached under the conditions of the experiment and at the same time establishes that the system is far from equilibrium.

Finally, it is possible to evaluate the washout rate that might be expected if the uptake rate were controlled by reaction kinetics. Here we need to estimate the time t that the drops are exposed to the gas in the chamber. This quantity may be evaluated assuming (as suggested by Beilke) ballistic trajectories for the drops; again for assumed $Z = 1$ m, $t \approx 1$ s. The washout rate λ is then evaluated as $\lambda = 2R_{1^*/4^*}$ $\mathbf{I}RT$ t/pZ; for $R_{1^*/4^*}$ evaluated as $2H^2_{NO_2}k_{A1^*/4^*}p^2_{NO_2}$ (i.e., no mass-transport limitation) $\lambda \approx 4 \times 10^{-9}$ s^{-1}, orders of magnitude less than the observed value. The fact that the observed value so greatly exceeds the value calculated for uptake only by reaction [A1*/4*] establishes that there is a mechanism of uptake other than this reaction. Such a mechanism might be the reaction of NO_2 with dissolved impurities, which would be consistent with the observation, noted above, that λ decreased with time during an experimental run, perhaps reflecting depletion of the impurity reagent species.

Recently Ohta et al. (1981) reported results of measurements of NO_2 partial pressures in effluent gas (initially containing NO_2 in the range 6×10^{-7} to 10^{-4} atm; carrier gas unspecified) bubbled through distilled, deionized water. Within 20 min the partial pressure of NO_2 in the effluent gas was found to reach a constant value; the relationship between the effluent partial pressure and the final composition of the solution was reported as

$$[H^+][NO_3^-][HNO_2] = K'_{obs}p^4_{NO_2}, \tag{121}$$

where K'_{obs} has the value 2.4×10^3 M^3 atm^{-4} and the expression is valid for 6×10^{-7} atm $\leq p_{NO_2} \leq 4 \times 10^{-5}$ atm for 16–17°C. Final concentrations of aqueous species were unspecified but evidently were in the range 1–40 $\times 10^{-6}$ M over the indicated partial pressure range. We note that by the acid dissociation equilibrium for nitrous acid, Eq. (121) may be written in terms of the dissociated species as

$$[H^+]^2[NO_3^-][NO_2^-] = K_{obs}p^4_{NO_2} \tag{122}$$

where $K_{obs} = K'_{obs}K_{12} = 1.2$ and where we have made use of the value $K_{12} = 5.1 \times 10^{-4}$ M employed by Ohta et al. The observed relationship (121) between solution composition and NO_2 partial pressure was attributed by the authors to establishment of equilibrium [M1′] in their system. However, the apparent value for $K_{M1'}$ was much less than that given by other investigators (Pick, 1920; Schwartz and White, 1981) and depended as well on the second power of NO_2 partial pressure.

We have severe misgivings about this interpretation of measurements of Ohta et al. First, we note that there is no theoretical justification for a dependence of an equilibrium constant on the second (or any) power of one of the reagent partial pressures; alternatively, an observed relationship such as (121) should be reflective of the reaction stoichiometry. Moreover, the authors fail to demonstrate that chemical equilibrium is in fact achieved; certainly the evidence adduced, namely, that the effluent NO_2 partial pressure has become invariant, can in no way be taken as indicative of achievement of chemical equilibrium. Rather, this condition would be expected generally with such an experimental arrangement for kinetic and/or mass-transport-controlled reactive uptake, with the effluent partial pressure governed by the amount of reagent gas taken up by the solution during the contact time.

It is of interest to speculate on possible reasons that might account for Ohta et al.'s observations. Denoting the concentration $[H^+] = C$, we find that from the stoichiometry of reaction [M1], Eq. (122) gives for the amount of material taken up by solution

$$C = (4K_{obs})^{1/4} p_{NO_2}, \tag{123}$$

i.e., proportional to the first power of the partial pressure of NO_2. We note that for convective-controlled uptake

$$\frac{dC}{dt} = k_L a H_{NO_2} p_{NO_2},$$

or for a contact time t,

$$C = k_L a H_{NO_2} p_{NO_2} t. \tag{124}$$

Assuming that in all cases a similar contact time was employed (the authors indicate that "equilibrium" was attained within 20 min), we may identify (123) and (124), obtaining

$$k_L a = \frac{(4K_{obs})^{1/4}}{H_{NO_2} t}.$$

For $t = 20$ min and $H_{NO_2} = 1 \times 10^{-2}$ M atm^{-1}, we obtain an estimate $k_L a \approx 0.1$ s^{-1}; such a value is not unreasonable for a laboratory bubbler and supports the suggestion that uptake in this system was convectively controlled. Further examination of this suggestion is obtained by comparison of τ_r evaluated for the partial pressures indicated, with estimates of τ_d and of τ_m^2/τ_d (see Table 21). These comparisons indicate that both criteria for convective-controlled uptake [Eqs. (97) and (98)] are fulfilled, thus lending further support to this interpretation of these results. Unfortunately, further analysis cannot be undertaken because of the lack of data in the original paper. However, it seems indisputable that the data do not support the authors' claim of chemical equilibrium.

8.2. NO_2 + Aqueous Solution

Oxidation of dissolved NO_2 by O_2 is a potentially important atmospheric reaction in view of the high partial pressure of O_2 and the large negative free-energy change for this reaction. Although reaction of gaseous NO_2 and O_2 is not reported, aqueous-phase reaction must be considered because of possible solvent stabilization of intermediate species. Lee and Schwartz (1981a) have examined, by means of a bubbler reactor (cf. Lee and Schwartz, 1981b), the rate of formation of ionic products when NO_2 is reacted with water alternately by use of N_2 and air as carrier gas. Within the sensitivity of the measurements (~8%), no change in reaction rate was observed for NO_2 partial pressures of as low as 5×10^{-8} atm. At that value of p_{NO_2} for the phase-mixing conditions employed, the measured value of $R_{A1^*/4^*}$ was 1.1×10^{-11} M s^{-1} (22°C). Hence an upper limit to the rate of aqueous-phase reaction of NO_2 with dissolved O_2 (in equilibrium with air) expressed as a first-order mixed-phase rate constant may be given as $0.08 \times R_{A1^*/4^*}/p_{NO_2} \approx 2 \times 10^{-5}$ M atm^{-1} s^{-1}.

Several studies have addressed the influence of acid, base, or other dissolved species on the rate of uptake of N_2O_4 by water as measured in the fast-reaction regime. Generally these studies have indicated that such interaction is slight. Thus Kameoka and Pigford (1977) reported that the rate of uptake of N_2O_4, $q \equiv (J_{N_2O_4}/p_{N_2O_4})$ is unchanged for absorption into water or 0.05 M H_2SO_4; for absorption into 0.2 M NaOH, an increase in q of 7% was reported. Hoftyzer and Kwanten (1972) reported a slight decrease of q with increasing ionic strength for absorption into solutions (concentration range unspecified) of NaOH, Na_2SO_4, NaCl, HNO_3, K_2SO_4, and $Al_2(SO_4)_3$ but again no specific interaction with any of these substances. It may be noted that at higher concentrations of NaOH (1–5 M) the rate of uptake has been found to decrease (Chilton and Knell, 1972), apparently because of the increased viscosity and consequent diminished rate of diffusion in the absorbing solution (Kameoka and Pigford, 1977).

An enhancement of the rate of N_2O_4 uptake was reported (Chilton and Knell, 1972) for a solution of potassium iodide (1 M). This enhanced rate of uptake presumably reflects the role of $NO_2(N_2O_4)$ in oxidizing the iodide ion (Szabó et al., 1956).

$$NO_2 + I^- \rightarrow NO_2^- + \tfrac{1}{2}I_2$$

The influence of dissolved ionic species has also been examined at lower NO_2 partial pressures, at which NO_2 rather than N_2O_4 would be the predominant transported species. The rate of uptake of NO_2 by NaOH solution (0.063 M) was reported to be the same as that into water (Sada et al., 1979) for $p_{NO_2} = (3\text{–}8) \times 10^{-4}$ atm. We note also that the study by Komiyama and Inoue (1980) (p_{NO_2} as low as 1×10^{-5} atm) employed 0.01 M NaOH. Similarly, Takeuchi et al. (1977) found the rate of uptake into urea solutions (0.1 and 1 M) unchanged from that into water. However, substantial enhancement of the rate of uptake was reported (Takeuchi et al., 1977) for solutions of oxidizing (MnO_4^-) or reducing species (I^-, SO_3^{2-}, HSO_3^-).

Although reaction of $NO_2(N_2O_4)$ with dissolved S(IV) has been examined in several studies (Nash, 1979; Sato et al., 1979; Takeuchi, 1977) these reactions are not yet well understood. Evidently NO_2 may either act as an oxidizing agent (Nash, 1979) or interact more specifically with the dissolved S(IV) (Sato et al., 1979). The latter authors report, in addition to formation of sulfate and dithionate $S_2O_6^{2-}$, the formation of hydroxylamine disulfonate $HON(SO_3)_2^{2-}$. The sequence of reactions leading to these products is not understood. In particular, possible reactions of nitrite (or nitrous acid) may be implicated; in this connection we note that hydroxylamine disulfonate is the principal product of the reaction of nitrite with sulfite (Oblath et al., 1981; Seel and Knorre, 1961).

Quantitative information on the rate of reaction of NO_2 with dissolved S(IV) is available from the study by Takeuchi et al. (1977). In that study the rate of NO_2 uptake was measured with a bubbler reactor as a function of p_{NO_2} (2×10^{-5} to 2×10^{-4} atm) and S(IV) concentration. The apparent reaction order with respect to p_{NO_2} was found to decrease with increasing concentration of SO_3^{2-} or HSO_3^-, approaching unity at concentrations of 0.01 and 0.2 M, respectively, The rates of these reactions were interpreted according to the reaction

$$NO_2(aq) + S(IV)(aq) \rightarrow \text{products} \qquad [A21]$$

taking place in the diffusion-controlled regime. (The products of reaction were not determined.) For the rate of [A21] much greater than that of [A1*/4*] the uptake of NO_2 is pseudo-first-order. For NO_2, the predominant aqueous-phase N(IV) species, the rate of uptake of NO_2 per unit interfacial area is given by (15) by setting $m = 1$ and $k_A = k_{A21}[S(IV)]$, i.e.,

$$J_{NO_2} = H_{NO_2}(k_{A21}D_{NO_2})^{1/2}[S(IV)]^{1/2}p_{NO_2}.$$

The results of these studies are summarized in Table 30. These studies would appear to establish a fairly fast reaction of dissolved NO_2 with sulfite and bisulfite ions: for $H_{NO_2} \approx 1 \times 10^{-2}$ M atm^{-1} (Table 26), $k_{A21} \approx 1 \times 10^7$ and 3×10^5 M^{-1} s^{-1}, respectively. Again it should be noted that there may be some inaccuracy in the rate constant values given because of the fraction of dissolved N(IV) present as N_2O_4 in equilibrium with NO_2 in the partial pressure range investigated (4×10^{-5} to 2×10^{-4} atm), viz., 5–30%.

Quantitative information on the kinetics of the reaction of NO_2 with ferrous ion (Fe^{2+}) may be obtained, albeit indirectly, from the kinetics of the oxidation of Fe^{2+} by nitrous acid (Abel et al., 1936). In that study the

Table 30. Measurements of Rate of Reaction of NO_2 with SO_3^{2-} and HSO_3^-[a]

Reaction	$H_{NO_2}(k_{A21}D_{NO_2})^{1/2}$ (kmol m^{-2} s^{-1} atm^{-1} $M^{-1/2}$)	$H_{NO_2}k_{A21}^{1/2}$[b] ($M^{1/2}$ atm^{-1} $s^{-1/2}$)	k_{A21}[c] M^{-1} s^{-1}
$NO_2(aq) + SO_3^{2-} \rightarrow$ products	1.55(−3)	34.6	1(7)
$NO_2(aq) + HSO_3^- \rightarrow$ products	2.33(−4)	5.22	3(5)

[a] Data of Takeuchi et al. (1977) for 25°C.

[b] Evaluated for $D_{NO_2} = 2.00 \times 10^{-9}$ m^2 s^{-1}.

[c] Evaluated for $H_{NO_2} = 1 \times 10^{-2}$ M atm^{-1}.

observed rate law was

$$\frac{d[Fe^{3+}]}{dt} = k_{obs} \frac{[Fe^{2+}][HNO_2]^2}{p_{NO}} . \qquad (125)$$

This rate law was interpreted according to the following mechanism, which is similar to that inferred for nitrous acid decomposition (Abel and Schmid, 1928c):

$2HNO_2 \overset{w}{=} NO(aq) + NO_2(aq)$	(rapid equilibrium)	[−A2′]
$NO_2(aq) + Fe^{2+} \rightarrow NO_2^- + Fe^{3+}$	(slow, rate determining)	[A22]
$NO_2^- + H^+ = HNO_2$	(rapid equilibrium)	[−I2]
$NO(aq) = NO(g)$	(rapid equilibrium)	[$-H_{NO}$]
$Fe^{2+} + HNO_2 + H^+ = Fe^{3+} + NO(g) + H_2O$		[A23]

Equilibria [A2′] and [H_{NO}] maintain an aqueous-phase NO_2 concentration

$$[NO_2(aq)] = \frac{1}{K_{A2'}H_{NO}} \frac{[HNO_2]^2}{p_{NO}} . \qquad (126)$$

Hence

$$\begin{aligned} R_{A23} = \frac{d[Fe^{3+}]}{dt} &= k_{A22}[Fe^{2+}][NO_2(aq)] \\ &= \frac{k_{A22}}{K_{A2'}H_{NO}} \frac{[Fe^{2+}][HNO_2]^2}{p_{NO}} . \end{aligned} \qquad (127)$$

Comparison of (127) with (125) gives

$$\begin{aligned} k_{A22} &= k_{obs}K_{A2'}H_{NO}, \\ &= k_{obs}\frac{K_{M2'}}{H_{NO_2}} , \end{aligned} \qquad (128)$$

where we have used Eq. (48). From the value $k_{obs} = 4.0\ M^{-2}\ atm^{-1}\ s^{-1}$ at 25°C obtained by Abel et al. (1936) we obtain $k_{A22} = 5.0 \times 10^4\ M^{-1}\ s^{-1}$, where we have employed the values $K_{M2'} = 1.26 \times 10^2\ M^2\ atm^{-2}$ (Table 2) and $H_{NO_2} = 1 \times 10^{-2}\ M\ atm^{-1}$ (Table 26).

In addition to the studies noted above, a number of studies have qualitatively addressed the interaction of NO_2 with dissolved aqueous species by examining the fraction of NO_2 absorbed by passing a dilute gaseous mixture of NO_2 through a bubbler containing the solution of interest. The uptake of NO_2 is particularly enhanced by reducing species I^-, HSO_3^-, SO_3^{2-}, S^{2-}, $S_2O_3{}^{2-}$, $S_2O_4{}^{2-}$, Sn^{2+}, Cu^+, Fe^{2+}, $Fe(CN)_6^{4-}$ and other ferrous complexes, and AsO_2^- (Christie et al., 1970; Kobayashi et al., 1977, 1980; Nash, 1970a). Addition of arsenite ion (AsO_2^-) to 0.1 *M* NaOH has been shown to substantially enhance NO_2 absorption efficiency (Christie et al., 1970; Margeson et al., 1977; Merryman et al., 1973; Nash, 1970a), with the absorbed NO_2 reduced with high yield to nitrite ion (Christie et al., 1970). In the case of the other species listed above, analysis of nitrogen products does not appear to have been conducted, although evolution of NO was noted for a number of these species (Kobayashi et al., 1977). Kobayashi et al. have pointed out a rough correlation between the efficiency of NO_2 removal by a solution of a given species and the redox potential of that species: the better the reducing agent, the more efficient the removal.

Of the several studies cited, that of Nash (1970a) merits further discussion. First, NO_2 (ca. 3 ppm) was bubbled through the solution under examination. In the absence of a reactive solute the fraction of NO_2 absorbed under the conditions employed was low (water and 0.05 M sulfuric acid, 7%; 0.1 M alkali, 14%). The concentration of the reactive solute was then adjusted until the fraction of NO_2 removed was 50 or 90%, and the corresponding solute concentration was reported. Although the details of mass transport and reaction are not known, this procedure would appear to permit evaluation of relative rate constants (rate constant ratios) since for a fixed fraction of NO_2 absorbed, similar spatial profiles of NO_2 concentration would be expected for different solutes. Nash's results are shown in Table 31. Also shown are the rate constants for reaction of $NO_2(aq)$ with the various solutes X, evaluated relative to the rate constant for reaction of NO_2 with sulfite ion ($1 \times 10^7\ M^{-1}\ s^{-1}$) inferred above from the work by Takeuchi et al. (1977), as

$$k_{NO_2\text{-}X} = k_{A21} \frac{C_{SO_3^{2-}}}{C_X}. \qquad (129)$$

Several of the rate constants evaluated in this way are seen to be quite large, approaching the diffusion controlled encounter rate constant (ca. $5 \times 10^9\ M^{-1}\ s^{-1}$) (Benson, 1960). Reaction products were not identified, although Nash noted the development of a yellow color on reaction of NO_2 with phenoxide ion and aromatic bases, indicating formation of a nitro compound. Reaction of NO_2 with dissolved organics and organic amines in aqueous solution has been reported also by Kobayashi et al. (1977), Levaggi et al. (1972), and Challis and Kyrtopoulos (1976, 1977).

Interaction of NO_2 with dissolved oxidants has received relatively little study. Reaction is reported with MnO_4^- (Takeuchi et al., 1977), XeO_3, and OsO_4 (Rigdon and Crawford, 1971). No analysis of products has been reported. Aqueous-phase oxidation of NO_2 to NO_3^- by hydrogen peroxide has been employed as the basis of a method for determination of gas-phase NO_2. Martens et al. (1973) report a high conversion of NO_2 (ca. 2×10^{-4} atm) to NO_3^- in 2 h by a solution of 1 M H_2O_2. The mechanism and kinetics apparently have not been studied. The overall reaction may well proceed by [A1*/4*] followed by oxidation of nitrite ion (Anbar and Taube, 1954). In view of the potential atmospheric importance of direct reaction of NO_2

Table 31. Relative Reactivities of Various Solutes with $NO_2(aq)$, Based on Data of Nash (1970a)

Solute, X	C_{50}[a] (mM)	C_{90}[a] (mM)	k_{NO_2-X} ($M^{-1}\,s^{-1}$)
Neutral solutions			
Azide, N_3^-	1800	—[c]	7.2(4)
Thiosulfate, $S_2O_3^{2-}$	1250	—	1.0(5)
Sulfanilate	1000	—	1.3(5)
Iodide, I^-	125	1200	1.1(6)
Aniline	45	—	2.9(6)
Sulfite, SO_3^{2-}	13	150	1(7)
Ferrocyanide, $Fe(CN)_6^{4-}$	13	150	1.0(7)
p-Anisidine	3	—	4.3(7)
Alkaline solution, pH 13			
Barbituric acid	30	—	4.3(6)
Phenol	5	100	[1.5–2.6](7)
Arsenite	3	100	[1.5–4.3](7)
Acetylacetone	4	40	3.5(7)
p-Cresol	0.7	8	1.9(8)
Thymol	0.1	1	1.4(9)
o-Methoxyphenol	0.07	0.7	2.0(9)

[a] Concentrations at which fractional removal of NO_2 was 50% or 90%, respectively.

[b] Evaluated relative to value for sulfite, $k_{NO_2-SO_3^{2-}} = 1 \times 10^7\ M^{-1}\,s^{-1}$, derived from data of Takeuchi (1977).

[c] Blanks indicate that 90% removal of NO_2 was not attained for saturated solution.

with H_2O_2 and also of the large negative free-energy change of reaction, the kinetics of this process merit further investigation.

8.3. Aqueous Methods for Chemical Analysis of NO_2

A large number of studies have addressed the removal of NO_2 from a gas mixture by bubbling through aqueous solution as a means of sampling NO_2 from ambient air or from flue gases (Driscoll, 1974) for the purpose of chemical analysis. By and large, these methods have been plagued with low and variable efficiency of sampling and variable and nonstoichiometric product yield. The present discussion briefly examines these methods, with particular attention to aqueous-phase reactions of NO_2.

A variety of aqueous solutions have been employed as absorber solutions. Methods initially proposed included absorption directly into an acetic acid solution containing diazotization and coupling reagents (Saltzman, 1954) or into sodium hydroxide solution (Jacobs and Hochheiser, 1958). It was subsequently shown that the sodium hydroxide abosrber suffered from low and variable collection efficiency (Christie, 1970; Huygen and Steerman, 1971; Meadows and Stalker, 1966; Merryman et al., 1973; Mulik et al., 1974; Nash, 1970b), and consequently a variety of additives have been employed to enhance this efficiency: arsenite (Christie, 1970; Merryman et al., 1973); guaiacol (*o*-methoxy phenol) (Nash, 1970b); and various amines and/or sulfonates (Huygen and Steerman, 1971; Mulik et al., 1974). With respect to absorption directly into solution of the diazotization and coupling reagents, there is also evidence that the presence of these reagents may affect absorption efficiency. Saltzman (1954) proposed the possibility of direct reaction of NO_2 with sulfanilic acid in addition to reaction [1*/4*]; we note in this context that reaction of NO_2 with sulfanilate anion was shown by Nash (1970a). On the other hand, Huygen (1970) indicates that the reaction occurs with the amine coupling reagent. In this context we also note that evolution of NO has been observed when NO_2 is bubbled through the Saltzman absorber solution (Ellis, 1964; Levaggi et al., 1972).

In addition to considerations of absorber efficiency, there has been extended discussion in the literature addressing the stoichiometry of the reactive dissolution process. If dissolution of NO_2 into aqueous solution were to proceed entirely by reaction [1*/4*],

$$2NO_2(g) + H_2O(l) \rightarrow 2H^+ + NO_2^- + NO_3^-, \qquad [1^*/4^*]$$

then the stoichiometric coefficient (defined here as apparent moles nitrite produced in solution per mole of $NO_2(g)$ absorbed) would be 0.5. In some studies a calibration factor has been reported as moles nitrite formed per mole of $NO_2(g)$ introduced into the bubbler; in this case, if absorption of NO_2 were not complete, the calibration factor so defined would be correspondingly diminished. The choice of nitrite as the referenced anion is a consequence of the analytical method generally employed, which is nominally specific to nitrite ion, viz., acid diazotization of sulfanilic acid followed by coupling reaction to form a diazo dye species having a high extinction coefficient.

The literature regarding the stoichiometric coefficient appropriate for the various analytical methods (and modifications) is also quite extensive and cannot be reviewed here. Suffice it to say that the stoichiometric coefficient is generally higher than the value 0.5 that would be expected for uptake of NO_2 by reaction [1*/4*] only; values reported have ranged from 0.7 to 1 (Buck and Stratmann, 1967; Crecelius and Forwerg, 1970; Huygen and Steerman, 1971; Saltzman, 1954; Scaringelli et al., 1970). Evidently as well, the stoichiometric coefficient may depend on the NO_2 partial pressure, increasing with decreasing NO_2 partial pressure (Buck and Stratmann, 1967; Crecelius and Forwerg, 1970), although the lack of such a dependence has been emphasized by others (e.g., Scaringelli et al., 1970).

It is beyond the scope of the present review to account for all the above observations; nonetheless, it is possible to apply the results of the review of NO_2 aqueous kinetics to gain insight into the processes taking place when NO_2 is sampled by means of an aqueous bubbler. Specifically, we address conditions that must be satisfied for NO_2 to be effectively removed from a gas mixture by aqueous-phase reaction on bubbling the gas through aqueous solution. An upper limit to the amount of NO_2 removed may be evaluated by assuming the liquid volume to be saturated in NO_2 at the incident partial pressure; whether this condition is satisfied depends on the rate of convective mixing compared to that of depletion of aqueous NO_2 by reaction. Under the assumption that the liquid is saturated in NO_2, the rate of reaction of NO_2 is

$$M_{\text{react}} = V_l \mathrm{H}_{NO_2} p_{NO_2} k^{(1)} \tag{130}$$

where M_{react} is the amount of NO_2 reacted per unit time (mol s^{-1}), V_l is the liquid volume (liters), and $k^{(1)}$ is the pseudo-first-order rate constant for reaction of dissolved NO_2; for a second-order reaction with rate constant $k^{(2)}$, $k^{(1)}$ would be evaluated as $k^{(2)}$ times the concentration of the second reagent.

The rate of depletion of NO_2 evaluated by (130) is to be compared to the rate at which NO_2 enters the bubbler in the incoming gas mixture, evaluated as

$$M_{\text{enter}} = \frac{p_{NO_2} F_g}{\mathbf{R}T} \tag{131}$$

where M_{enter} also has units of moles per second, and F_g is the gas flow rate in liters per second.

A necessary condition, which must be satisfied for the scrubber to remove the incident NO_2 with high efficiency, is that M_{react} evaluated by (130) greatly exceed M_{enter}, i.e., that

$$k^{(1)} \gg \frac{F_g}{V_l \mathrm{H}_{NO_2} \mathbf{R}T}\,. \tag{132}$$

The right-hand side of (132) may be evaluated for conditions typically employed (Saltzman, 1954), i.e., $F_g = 6.7 \times 10^{-3}$ liter s^{-1} (0.4 liter/min) and $V_l = 1.5 \times 10^{-2}$ liter; the resulting value (ca. 2 s^{-1}) may be compared with $k^{(1)}$ evaluated for various mechanisms. For reaction of NO_2 by reaction [1*/4*], $k^{(1)} = k_{A1*/4*} \mathrm{H}_{NO_2} p_{NO_2} = 1.0 \times 10^{-2}\ s^{-1}$ for $p_{NO_2} = 10^{-8}$ atm. Clearly, reaction [1*/4*] cannot be an effective means for removal of NO_2 in the bubbler. This calculation thus supports the empirical finding, noted above, of a low efficiency for absorption of NO_2 by water or by 0.1 M OH^- solution as employed in the Jacobs–Hochheiser method. We conclude further that for NO_2 to be efficiently removed from the incident gas mixture, as is shown to be the case with the Saltzman reagent (e.g., Levaggi et al., 1972) or with 0.1 M OH^- solution containing guaiacol (Nash, 1970b) or arsenite (Christie et al., 1970), there must be direct reaction of dissolved NO_2 with the dissolved reagent(s) present; i.e., subsequent reaction of nitrite would not account for the enhanced absorption. This conclusion may be substantiated further by evaluating $k^{(1)}$ for NO_2 reaction with these reagents. For example, for NO_2 reacting with sulfanilic acid, as specified in the Saltzman (1954) method (3×10^{-2} M), we employ the rate constant 1.3×10^5 M^{-1} s^{-1} evaluated (Table 31) from the data of Nash (1970a), under the assumption that the rate constant for sulfanilate anion as studied by Nash is not too dissimilar from that for sulfanilic acid. The resulting value, $k^{(1)} \approx 4 \times 10^3$ s^{-1}, easily satisfies condition (132), strongly supporting the suggestion (Saltzman, 1954) of a direct reaction of NO_2 with sulfanilic acid. Such a direct reaction occurring in parallel with reaction [1*/4*] could account for the observed stoichiometric factor greater than 0.5 expected for uptake by reaction [1*/4*] only. Furthermore, the proportion of NO_2 uptake by direct reaction with dissolved reagent rather than by reaction [1*/4*] would be expected to increase with decreased NO_2 partial pressure as the concentration of physically dissolved NO_2 decreased, consistent with the observations by Buck and Stratmann (1967) and Crecelius and Forwerg (1970). Similar evaluations may be made, again using the rate constants evaluated in Table 31, for NO_2 reacting with guaiacol or arsenite, again establishing that the direct reaction of dissolved NO_2 with these reagents is responsible for the high absorption efficiencies obtained.

8.4. NO + Aqueous Solution

As noted above, NO is reversibly soluble and nonreactive in liquid water in the absence of other reagents. The aqueous-phase reactions of NO in the presence of NO_2 have also been discussed. We thus restrict this discussion to other aqueous-phase reactions of NO that may be pertinent to the incorporation of this gas into ambient liquid water.

The aqueous-phase reaction of NO with O_2,

$$2NO(aq) + O_2(aq) \rightarrow 2NO_2(aq),$$

or perhaps the overall reaction

$$2NO(aq) + O_2(aq) \xrightarrow{w} 2H^+ + NO_2^- + NO_3^-$$

may be significant in view of the known gas-phase reaction

$$2NO(g) + O_2(g) \rightarrow 2NO_2(g).$$

The rate of the aqueous-phase reaction does not appear to have been determined. Although Seddon and Sutton (1963) report that for $[NO] = 1 \times 10^{-4}$ M and $[O_2] = 1 \times 10^{-3}$ M, NO is "rapidly and quantitatively" oxidized to HNO_2, presumably the reaction is second order in NO and first order in O_2, as is the gas-phase reaction (Baulch et al., 1980). Hence, in view of the low solubilities of both NO and O_2 ($H \approx 2 \times 10^{-3}$ M atm^{-1}, corresponding to $C(aq)/C(g) = HRT \approx 0.05$), one would anticipate that the rate of aqueous-phase reaction might be less than that of the gas-phase reaction at the equilibrium partial pressures of the gases and thus negligibly slow at characteristic ambient partial pressures.

Nitric oxide is nonreactive with acid, base, and a variety of anions and cations: Cl^-; ClO_4^-; PO_4^{-3}; SO_4^{2-}; BO_3^{3-}; acetate; Na^+; and Li^+ (Armor, 1974). It reacts as a complexing agent with transition metal ions such as Fe^{2+} and Cr^{2+} (Armor and Buchbinder, 1973; Kustin et al., 1966) and is capable of oxidizing mild reducing agents such as Sn(II) (Nunes and Powell, 1970a). Nitric oxide is oxidized by strong oxidants: MnO_4^-; ClO_2^- [for both of which evolution of NO_2 was noted (Kobayashi et al., 1977)]; and H_2O_2. The species MnO_4^- (e.g., Groth and Calabro, 1969) and H_2O_2 (e.g., Martens et al., 1973) are commonly employed as oxidants in solutions for absorbing nitrogen oxides for analytical determination. The mechanism and kinetics of oxidation of NO by H_2O_2 are not well understood. Bajeva et al. (1979) report a study of the rate uptake of NO (0.07–0.24 atm) by aqueous H_2O_2 (0.13–0.5 M). The rate of uptake $J_{NO} = qp_{NO}[H_2O_2]^{1/2}$ was interpreted according to mixed-phase reaction in the fast reaction (diffusion-controlled) regime. For $[H_2O_2] \gg H_{NO}p_{NO}$, this expression for the rate of uptake indicates a rate law for aqueous-phase reaction

$$R = k^{(2)}[H_2O_2][NO]. \tag{133}$$

At 25°C the value $k^{(2)} = 5.7 \times 10^2$ M^{-1} s^{-1} was obtained. Seddon and Sutton (1963), however, report a much slower reaction rate. For $[H_2O_2] = 10^{-4}$ M and $[NO] = 2 \times 10^{-3}$ M (i.e., $p_{NO} = 1$ atm), the reaction rate is reported as less than 2×10^{-10} M s^{-1}; i.e., assuming rate law (133) $k^{(2)} \leq 1 \times 10^{-3}$ M^{-1} s^{-1}. The reason for the discrepancy is not known.

Aqueous-phase reaction of NO with dissolved S(IV) has been studied by Nunes and Powell (1970b), Takeuchi et al. (1977), and Martin et al. (1981). The latter authors, working in acid solution (pH = 2), report no reaction and are able to place an upper limit to the reaction rate. For a rate law taken as first order in S(IV) and NO, Martin et al. report

$$R < 1 \times 10^{-2}[S(IV)][NO]. \tag{134}$$

In contrast, in basic solution NO evidently undergoes rapid reaction with S(IV). Nunes and Powell (1970b), working with solutions 0.1–1 M in $[OH^-]$, report a reaction rate of the form

$$R = k_1[NO][S(IV)] + k_2[NO]; \tag{135}$$

at 25°C $k_1 = 0.45$ M^{-1} s^{-1} and $k_2 = 0.13$ s^{-1}. The product *N*-nitroso-hydroxylamine-*N*-sulfonate, $O_3S(NO)NO^{2-}$, is stable in alkaline solution but liberates N_2O as the solution is made acidic. Rapid reaction of NO with sulfite ion is reported also by Takeuchi et al. (1977). In that study a stirred bubbler was employed for introduction of NO ($1.5–10 \times 10^{-4}$ atm in N_2)

into sodium sulfite solution (0.02–0.8 M). Uptake of NO was in the fast reaction (diffusion controlled regime), i.e., independent of the convective mass-transport coefficient k_L. The rate of uptake was reported as

$$J_{NO} = q p_{NO}^{3/2}, \tag{136}$$

where $q = 1.2 \times 10^{-4}$ kmol m^{-2} s^{-1} atm$^{-3/2}$ was independent of $[SO_3^{2-}]$ over the concentration range studied. This finding suggests a rate law

$$\frac{d[NO]}{dt} = -2k^{(2)}[NO]^2, \tag{137}$$

where $k^{(2)}$ is a pseudo-second-order rate constant. The value obtained for $k^{(2)}$, 6.0×10^8 M^{-1} s^{-1} at 25°C, is quite high; compare this with the diffusion-controlled encounter rate coefficient (ca. 5×10^9 M^{-1} s^{-1}) (Benson, 1960). The range of applicability of (134)–(137) in terms of both $[SO_3^{2-}]$ and pH is not known. Further investigation of this reaction system appears warranted.

9. SUMMARY

The reactive dissolution of nitrogen oxides NO_2 and NO into aqueous solution is a potentially important process relative to the fate of these substances in the atmosphere. The quantitative description of this process requires knowledge of the pertinent equilibrium constants (including Henry's law coefficients) and kinetic rate constants (Anderson, 1982; Durham et al., 1981; Lee and Schwartz, 1981a; Peters and Carmichael, 1982). These quantities are directly relevant as well to development of analytical methods for these compounds, development of flue-gas scrubbers, and interpretation of the mechanism of toxicity of these compounds to plant and animal systems.

There is a wide and diverse literature pertaining to the mechanism and kinetics of the reactive dissolution of nitrogen oxides into aqueous solution, including indirect measurements of phase-mixed kinetics and measurements of aqueous-phase kinetics, as well as measurements of the kinetics of reactive uptake by means of gas–aqueous reactors. This literature is critically reviewed, and recommended values of the several equilibrium and kinetic constants are given, along with uncertainty estimates based somewhat subjectively on the spread in the data. Reactions of NO_2 and NO with dissolved solutes are also reviewed, and, where possible, estimates of kinetic rate constants are presented. However, there remain large uncertainties in the stoichiometry, mechanism, and kinetics of some of these reactions.

ACKNOWLEDGMENTS

This work was supported in part by the High Altitude Pollution Program of the Office of Environment and Energy, Federal Aviation Administration, and was performed under the auspices of the United States Department of

Energy under Contract No. DE-AC02-76CH00016. The authors wish to acknowledge the generous cooperation of their colleagues in preparation of this chapter: Dr. Leonard Newman, Head, Environmental Chemistry Division; and Drs. Yin-Nan Lee and George Y. Markovits, who have been engaged in related laboratory studies.

NOMENCLATURE

p_X	Partial pressure of species X (atm)
[X]	Aqueous-phase concentration of species X (M)
a_X	Activity of species X (m)
y_X	Activity of coefficient of species X
H_X	Henry's law coefficient of species X (M atm^{-1})
K	Equilibrium constant. May refer to gas-, aqueous-, or mixed-phase equilibrium as indicated by subscript G, A, or M
R	Aqueous-phase rate (M s^{-1})
k	Rate coefficient. May refer to aqueous- or mixed-phase reaction as indicated by subscript A or M
k_{as}	Stochastic rate coefficient for convective mass transport between surface and bulk aqueous solution (s^{-1})
k_L	Liquid-side mass-transfer coefficient (m s^{-1})
m	Order of aqueous-phase reaction
a	Specific interfacial area or surface area per unit volume of solution (m^{-1})
J_X	Flux of species X into solution (kmol m^{-2} s^{-1})
$Q_X(t)$	Amount of species X taken up by solution in time t (kmol m^{-2})
D_X	Aqueous-phase diffusion coefficient of species X (m^2 s^{-1})
q	Empirical coefficient relating flux and partial pressure
τ	Characteristic time of indicated process (s)
τ_r	Characteristic time of reaction
τ_m	Characteristic time of convective mixing
τ_d	Characteristic time of diffusion
F_{NO_2}, $F_{N_2O_4}$	Fraction of N(IV) present as NO_2 or N_2O_4, respectively
E_a	Empirical energy of activation of indicated process (kcal mol^{-1})

REFERENCES

Abel, E. and Schmid, H. (1928a and b). Kinetics of nitrous acid. I. Introduction and overview. II. Orientational experiments. *Z. Phys. Chem.* **132,** 55–77 (in German).

Abel, E. and Schmid, H. (1928c). Kinetics of nitrous acid. III. Kinetics of nitrous acid decomposition. *Z. Phys. Chem.* **134,** 279–300 (in German).

Abel, E. and Schmid, H. (with Babad, S.) (1928d). Kinetics of nitrous acid. IV. Kinetics of the formation of nitrous acid from nitric acid and nitric oxide. *Z. Phys. Chem.* **136,** 135–145 (in German).

Abel, E. and Schmid, H. (with Babad, S.) (1928e). Kinetics of nitrous acid. V. Kinetics the nitrous acid–nitric acid–nitric oxide reaction. *Z. Phys. Chem.* **136,** 419–429 (in German).

Abel, E. and Schmid, H. (1928f). Kinetics of nitrous acid. VI. The equilibrium of the nitrous acid–nitric acid–nitric oxide reaction in connection with its kinetics. *Z. Phys. Chem.* **136,** 430–436 (in German).

Abel, E., Schmid, H., and Römer, E. (1930). Kinetics of nitrous acid. VII. Rate and temperature. *Z. Phys. Chem.* **148,** 337–349 (in German).

Abel, E., Schmid, H., and Pollak, F. (1936). Kinetics of the oxidation of ferrous ion by nitrous acid. *Sitzber. Akad. Wiss. Wien IIb* **145,** 731–749 (in German).

Anbar, M. and Taube, H. (1954). Interaction of nitrous acid with hydrogen peroxide. *J. Am. Chem. Soc.* **76,** 6243–6247.

Anderson, L. G. (1982). Fate of nitrogen oxides in urban atmospheres (Chapter 9, present volume).

Andrew, S. P. S. and Hanson, D. (1961). The dynamics of nitrous gas absorption. *Chem. Eng. Sci.* **14,** 105–113.

Armor, J. N. (1974). Influence of pH and ionic strength upon solubility of NO in aqueous solution. *J. Chem. Eng. Data* **19,** 82–84.

Armor, J. N. and Buchbinder, M. (1973). Reduction of coordinated nitrosyls. Preparation, characterization, and reduction of nitrosyl-pentaaquochromium (2^+). *Inorg. Chem.* **12,** 1086–1090.

Astarita, G. (1967). *Mass Transfer with Chemical Reaction,* Elsevier, Amsterdam, pp. 31–32.

Bajeva, K. K., Rao, D. S., and Sarkar, M. K. (1979). Kinetics of absorption of nitric oxide in hydrogen peroxide solution. *J. Chem. Eng. Jap.* **12,** 322–325.

Bartholomé, E. and Gerstacker, H. (1961). The penetration theory and its application to kinetic investigations of physical and chemical gas absorption into liquids. *Alta Tech. Chim., Accad. Naz. dei Lincei, Rome* 177–198 (in German). Also available in *Scuola Azione,* Part I (7), 61–117 (1963–1964) (in Italian).

Baulch, D. L., Cox, R. A., Hampson, R. F., Jr., Kerr, J. A., Troe, J., and Watson, R. T. (1980). Evaluated kinetic and photochemical data for atmospheric chemistry. *J. Phys. Chem. Ref. Data* **9,** 295–469.

Beilke, S. Laboratory investigations on washout of trace gases, in *Precipitation Scavenging (1970),* USAEC Symposium Series 22, December 1970, pp. 261–269.

Benson, S. W. (1960). *The Foundations of Chemical Kinetics,* McGraw-Hill, New York, p. 498.

Borok, M. T. (1960). Dependence of the degree of absorption of nitrogen dioxide in water on its concentration in a gaseous mixture. *J. Appl. Chem. USSR* **33,** 1761–1766.

Bray, W. C. (1932). The mechanism of reactions in aqueous solution. Examples involving equilibria and steady states. *Chem. Rev.* **10,** 161–177.

Brian, P. L. T. (1964). Gas absorption accompanied by an irreversible reaction of general order. *Am. Inst. Chem. Eng. J.* **10,** 5–10.

Buck, M. and Stratmann, H. (1967). The joint and separate determination of nitrogen monoxide and nitrogen dioxide in the atmosphere. *Staub* (in English), **27** (6), 11–15.

Bunton, C. A., Llewellyn, D. R., and Stedman, G. (1959). Oxygen exchange between nitrous acid and water. *J. Chem. Soc.* **1959,** 568–573.

Bunton, C. A., Burch, J. E., Challis, B. C., and Ridd, J. H. (1961). Unpublished work cited by Ridd (1961).

Calderbank, P. H. (1967). Mass transfer in fermentation equipment, in *Biochemical and Biological Engineering Science,* N. Blakebrough, Ed., Academic, New York, pp. 101–180.

Carberry, J. J. (1959). Some remarks on chemical equilibrium and kinetics in the nitrogen oxides–water system. *Chem. Eng. Sci.* **9,** 189–194.

Caudle, P. G. and Denbigh, K. G. (1953). Kinetics of the absorption of nitrogen peroxide into water and aqueous solutions. *Transact. Faraday Soc.* **49,** 39–52.

Challis, B. C. and Kyrtopoulos, S. A. (1976). Nitrosation under alkaline conditions. *J. Chem. Soc. Chem. Commun.* **1976,** 877–878.

Challis, B. C. and Kyrtopoulos, S. A. (1977). Rapid formation of carcinogenic *N*-nitrosamines in aqueous alkaline solutions. *Br. J. Cancer* **35,** 693–696.

Chambers, F. S. and Sherwood, J. K. (1937). The equilibrium between nitric oxide, nitrogen peroxide and aqueous solutions of nitric acid. *J. Am. Chem. Soc.* **59,** 316–319.

Chilton, T. H. (1968). *Strong Water,* MIT Press, Cambridge, MA.

Chilton, T. H. and Knell, E. W. (1972). The absorption of nitrogen oxides, Part 2: Absorption of nitrogen peroxide in alkaline. American Institute of Chemical Engineers, Pacific Chemical Engineering Conference (PACHEC), pp. 75–81.

Christie, A. A., Lidzey, R. G., and Radford, D. W. F. (1970). Field methods for the determination of nitrogen dioxide in air. *Analyst* (*Lond.*), **95,** 519–524.

Corriveau, C. E., Jr. and Pigford, R. L. (1971). Absorption of N_2O_3 into water. University of California, Lawrence Radiation Laboratory Report, UCRL-20479.

Counce, R. M. (1980). The scrubbing of gaseous nitrogen oxides in packed towers. Oak Ridge National Laboratory, Oak Ridge, TN, Report ORNL-5676.

Crecelius, H.-J. and Forweg, W. (1970). Investigations of the "Saltzman factor," *Staub* (in English), **30** (7), 23–25.

Crutzen, P. J. (1979). The role of NO and NO_2 in the chemistry of the troposphere and stratosphere. *Ann. Rev. Earth Plant Sci.* **7,** 443–472.

Danckwerts, P. V. (1950). Absorption by simultaneous diffusion and chemical reaction. *Transact. Faraday Soc.* **46,** 300–304.

Danckwerts, P. V. (1970). *Gas–Liquid Reactions,* McGraw-Hill, New York.

Daniels, F., and Alberty, R. A. (1975). *Physical Chemistry,* 4th ed., Wiley, New York, p. 157.

Davis, W., Jr. and DeBruin, H. J. (1964). New activity coefficients of 0–100 percent aqueous nitric acid. *J. Inorg. Nucl. Chem.* **26,** 1069–1083.

Dekker, W. A., Snoeck, E., and Kramers, H. (1959). The rate of absorption of NO_2 in water. *Chem. Eng. Sci.* **11,** 61–71.

Denbigh, K. G. and Prince, A. J. (1947). Kinetics of nitrous gas absorption in aqueous nitric acid. *J. Chem. Soc.* **1947,** 790–801.

Driscoll, J. N. (1974). Flue gas monitoring techniques, in *Nitrogen Oxides,* Ann Arbor Science, Ann Arbor, MI, Chapter 7, pp. 219–263.

Durham, J. C., Overton, J. H., and Aneja, V. P. (1981). Influence of gaseous nitric acid on sulfate production and acidity in rain. *Atmosph. Environ.* **15,** 1059–1068.

Eigen, M. and De Maeyer, L. (1963). Relaxation Methods, in *Technique of Organic Chemistry,* A. Weissberger, Ed., Vol. 8, Part 2, 2nd ed., Interscience, New York, pp. 895–1054.

Eigen, M., Kruse, W., Maass, G., and DeMaeyer, L. (1964). Rate constants of protolytic reactions in aqueous solution. *Progress in Reaction Kinetics, Vol. 2,* G. Porter, Ed., Pergamon, Oxford, pp. 286–318.

Ellis, C. F. (1964). A suggested procedure for converting NO in low concentration to NO_2. *Internatl. J. Air Water Pollut.* **8,** 297–299.

England, C. and Corcoran, W. H. (1974). Kinetics and mechanisms of the gas-phase reaction of water vapor and nitrogen dioxide. *Ind. Eng. Chem. Fundam.*, **13,** 373–384.

Epstein, I. R., Kustin, K., and Warshaw, L. J. (1980). A kinetics study of the oxidation of iron(II) by nitric acid. *J. Am. Chem. Soc.* **102,** 3571–3578.

Faucett, H. L., Maxwell, J. D., and Burnett, T. A. (1978). Technical assessment of NO_x removal processes for utility application. Electric Power Research Institute, Palo Alto, CA, Report EPRI-AF-568.

Ford, S. (1980). Unpublished study. Private communication from G. Stedman, University College of Swansea, Wales.

Freiberg, J. and Schwartz, S. E. (1981). Oxidation of SO_2 in aqueous droplets: Mass-transport limitation in laboratory studies and the ambient atmosphere. *Atmosph. Environ.* **15,** 1145–1154.

Gerstacker, H. (1961). Discussion of "Absorption of nitrogen tetroxide by water jets." *Chem. Eng. Sci.* **14,** 124–125.

Goldstein, E., Peek, N. F., Parks, N. J., Hines, H. H., Steffey, E. P., and Tarkington, B. (1977). Fate and distribution of inhaled nitrogen dioxide in rhesus monkeys. *Am. Rev. Resp. Disease* **115,** 403–412.

Grätzel, M., Henglein, A., Little, J., and Beck, G. (1969). Pulse radiolytic investigation of some elementary processes of oxidation and reduction of nitrite ion. *Ber. Bunsenges.* **73,** 646–653 (in German).

Grätzel, M., Taniguchi, S., and Henglein, A. (1970). Pulse radiolytic investigation of NO oxidation and the equilibrium $N_2O_3 = NO + NO_2$ in aqueous solution. *Ber. Bunsenges.* **74,** 488–492 (in German).

Groth, R. H. and Calabro, D. S. (1969). Evaluation of Saltzman and phenoldisulfonic acid methods for determining NO_x in engine exhaust gases. *J. Air Pollut. Control Assoc.* **19,** 884–887.

Hales, J. M. (1972). Fundamentals of the theory of gas scavenging by rain. *Atmosph. Environ.* **6,** 635–659.

Heckner, H. N. (1973). Potentiostatic switching experiments for the cathodic reduction of nitrous acid in perchloric acid with the addition of nitric acid. *Electroanalyt. Chem. Interfacial Electrochem.* **44,** 9–20.

Herrmann, J. P. and Matteson, M. J. (1980). Nitrogen dioxide absorption in evaporating and condensing water droplets, in *Advances in Environmental Science and Engineering,* Vol. 3, J. R. Pfafflin and E. N. Ziegler, Eds., Gordon and Breach, New York, pp. 92–99.

Hill, A. C. (1971). Vegetation: A sink for atmospheric pollutants. *J. Air Pollut. Control Assoc.* **21,** 341–347.

Himmelblau, P. M. (1964). Diffusion of dissolved gases in liquids. *Chem. Rev.* **64,** 527–550.

Hofmeister, H.-K. and Kohlhaas, R. (1965). The absorption of NO–NO_2 mixtures via a laminar water column. *Ber. Bunsenges.* **69,** 232–238 (in German).

Hoftyzer, P. J. and Kwanten, F. J. G. (1972). Absorption of nitrous gases, in *Gas Purification Processes,* G. Nonhebel, Ed., Newnes–Butterworths, London, pp. 164–187.

Hughes, E. P. and Ridd, J. H. (1958). Nitrosation, diazotisation, and deamination. Part III. Zeroth-order diazotisation of aromatic amines in carboxylic acid buffers. *J. Chem. Soc.* **1958,** 70–76.

Huygen, C. (1970). reaction of nitrogen dioxide with Griess type reagents. *Anal. Chem.,* **42,** 407–409.

Huygen, C. and Steerman, P. H. (1971). The determination of nitrogen dioxide in air after absorption in a modified alkaline solution. *Atmosph. Environ.* **5,** 887–889.

Jacobs, M. B. and Hochheiser, S. (1958). Continuous sampling and ultramicrodetermination of nitrogen dioxide in air. *Anal. Chem.* **30,** 426–428.

Johnston, H. S. (1966). *Gas Phase Reaction Rate Theory,* Ronald Press, New York.

Jordan, S. and Bonner, F. T. (1973). Nitrogen and oxygen exchange between nitric oxide and aqueous solution of nitric acid. *Inorg. Chem.* **12,** 1369–1373.

Kaiser, E. W. and Wu, C. H. (1977a). Measurement of the rate constant of the reaction of nitrous acid with nitric acid. *J. Phys. Chem.* **81,** 187–190.

Kaiser, E. W. and Wu, C. H. (1977b). A kinetic study of the gas phase formation and decomposition reactions of nitrous acid. *J. Phys. Chem.* **81,** 1701–1706.

Kalatzis, E. and Ridd, J. H. (1966). Nitrosation, diazotisation, and deamination. Part XII. The kinetics of *N*-nitrosation of *N*-methylaniline. *J. Chem. Soc.* (*B*) **1966,** 529–533.

Kameoka, Y. and Pigford, R. L. (1977). Absorption of nitrogen dioxide into water, sulfuric acid, sodium hydroxide, and alkaline sodium sulfite. *Ind. Eng. Chem. Fundam.* **16,** 163–169.

Kobayashi, H., Takezawa, N., and Niki, T. (1977). Removal of nitrogen oxides with aqueous solutions of inorganic and organic reagents. *Environ. Sci. Technol.* **11,** 190–192.

Kobayashi, H., Takezawa, N., and Niki, T. (1980). Removal of nitrogen oxides with organic amines and metal complexes. *Environ. International.* **4,** 273–275.

Komiyama, H. and Inoue, M. (1978). Reaction and transport of nitrogen oxides in nitrous acid solutions. *J. Chem. Eng. Jap.* **11,** 25–32.

Komiyama, H. and Inoue, H. (1980). Absorption of nitrogen oxides into water. *Chem. Eng. Sci.* **35,** 154–161.

Kramers, H., Blind, M. P. P., and Snoeck, E. (1961). Absorption of nitrogen tetroxide by water jets. *Chem. Eng. Sci.* **14,** 115–125.

Kustin, K., Taub, I. A., and Weinstock, E. (1966). A kinetic study of the formation of the ferrous–nitric oxide complex. *Inorg. Chem.* **5,** 1079–1082.

Lee, Y.-N. and Schwartz, S. E. (1981a). Evaluation of the rate of uptake of nitrogen dioxide by atmospheric and surface liquid water. *J. Geophys. Res.* **86,** 11971–11983.

Lee, Y.-N. and Schwartz, S. E. (1981b). Reaction kinetics of nitrogen dioxide with liquid water at low partial pressure. *J. Phys. Chem.* **85,** 840–848.

Levaggi, D. A., Siu, W., Feldstein, M., and Kothny, E. L. (1972). Quantitative separation of nitric oxide from nitrogen dioxide at atmospheric concentration ranges. *Environ. Sci. Technol.* **6,** 250–252.

Levine, S. Z. and Schwartz, S. E. (1982). In-cloud and below-cloud scavenging of nitric acid vapor. *Atmosph. Environ.* **16,** 1725–1734.

Liljestrand, H. M. and Morgan, J. F. (1981). Spatial variations of acid precipitation in southern California. *Environ. Sci. Technol.* **15,** 333–339.

Liss, P. S. (1976). The exchange of gases across lake surfaces. *J. Great Lakes Res.* **2** (Suppl. 1), 88–100; 126.

Margeson, J. H., Beard, M. E., and Suggs, J. C. (1977). Evaluation of the sodium arsenite method for measurement of NO_2 in ambient air. *J. Air Pollut. Control Assoc.* **27,** 553–556.

Markovits, G. Y., Schwartz, S. E., and Newman, L. (1981). Hydrolysis equilibrium of dinitrogen trioxide in dilute acid solution. *Inorg. Chem.* **20,** 445–450.

Martens, H. H., Dee, L. A., Nakamura, J. T., and Jaye, F. C. (1973). Improved phenoldisulfonic acid method for determination of NO_x from stationary sources. *Environ. Sci. Technol.* **17,** 1152–1154.

Martin, L. R., Damschen, D. E., and Judeikis, H. S. (1981). The reactions of nitrogen oxides with SO_2 in aqueous aerosols. *Atmosph. Environ.* **15,** 191–195.

Matteson, M. J. (1979). Capture of atmospheric gases by water vapor condensation on carbonaceous particles, in *Carbonaceous Particles in the Atmosphere,* Conference Proceedings, Berkeley, CA, March 20, 1978; Lawrence Berkeley Laboratory report LBL-9037; CONF-7803101, pp. 150–154.

Matteson, M. J. (1981). Private communication. [The value of H_{NO_2} in Matteson (1979) and Herrmann and Matteson (1980) was given erroneously as 100 mole fraction (aq) atm^{-1} and should have read 0.01 mole fraction (aq) atm^{-1}, i.e., 0.55 M atm^{-1}.]

Meadows, F. L. and Stalker, W. W. (1966). The evaluation of efficiency and variability of sampling for atmospheric nitrogen dioxide. *Am. Ind. Hyg. Assoc. J.* **27,** 559–566.

Merryman, E. L., Spicer, C. W., and Levy, A. (1973). Evaluation of arsenite-modified Jacobs–Hochheiser procedure. *Environ. Sci. Technol.* **7,** 1056–1059.

Moll, A. J. (1966). The rate of hydrolysis of nitrogen tetroxide. Ph.D. Thesis, University of Washington (available from University Microfilms, Ann Arbor, MI, Order No. 66-12,027; *Dissert. Abstr.* **36,** 125).

Mulik, J., Fuerst, R., Guyer, M., Meeker, J., and Sawicki, E. (1974). Development and optimization of twenty-four hour manual methods for the collection and colorimetric analysis of atmospheric NO_2. *Internatl. J. Environ. Anal. Chem.* **3,** 333–348.

Mustafa, M. G., Elsayed, N., Lim, J. S. T., Postlethwait, E., and Lee, S. D. (1979). Effects of nitrogen dioxide on lung metabolism, in *Nitrogenous Air Pollutants: Chemical and Biological Implications,* D. Grosjean, Ed., Ann Arbor Science Publishers, Ann Arbor, MI, pp. 165–178.

Nash, T. (1970a). Absorption of nitrogen dioxide by aqueous solutions. *J. Chem. Soc.* **A1970,** 3023–3024.

Nash, T. (1970b). An efficient absorbing reagent for nitrogen dioxide. *Atmosph. Environ.* **4,** 661–665.

Nash, T. (1979). The effect of nitrogen dioxide and of some transition metals on the oxidation of dilute bisulphite solutions. *Atmosph. Environ.* **13,** 1149–1164.

National Research Council (1977). Nitrogen oxides. National Academy of Sciences, Washington, pp. 31–55; 197–214.

Nunes, T. L. and Powell, R. E. (1970a). The copper(I)-catalyzed reduction of nitric oxide by tin(II) chloride. *Inorg. Chem.* **9,** 1912–1916.

Nunes, T. L. and Powell, R. E. (1970b). Kinetics of the reaction of nitric oxide with sulfite. *Inorg. Chem.* **9,** 1916–1917.

Oblath, S. B., Markowitz, S. S., Novakov, T., and Chang, S. G. (1981). Kinetics of the formation of hydroxylamine disulfonate by reaction of nitrite with sulfites. *J. Phys. Chem.* 1017–1021.

Ohta, S., Okita, T., and Kato, C. (1981). A numerical model of acidification of cloud water. *J. Meteorol. Soc. Jap.* **59,** 892–902.

Orel, A. E. and Seinfeld, J. H. (1977). Nitrate formation in atmospheric aerosols. *Environ. Sci. Technol.* **11,** 1000–1007.

Ottolenghi, M. and Rabani, J. (1968). Photochemical generation of nitrogen dioxide in aqueous solutions. *J. Phys. Chem.* **72,** 593–598.

Peters, L. K. and Carmichael, G. R. (1982). Modeling of transport and chemical processes that affect regional and global distributions of trace species in the troposphere (Chapter 12, present volume).

Peters, M. S. and Holman, J. L. (1955). Vapor- and liquid-phase reactions between nitrogen dioxide and water. *Ind. Eng. Chem.* **47,** 2536–2539.

Peters, M. S., Ross, C. P., and Klein, J. E. (1955). Controlling mechanism in the aqueous absorption of nitrogen oxides. *Am. Inst. Chem. Eng. J.* **1,** 105–111.

Pick, H. (1920). The electrolytic potential of the reaction nitrite $\rightarrow$ nitrate + nitric oxide. Review of the energetic relationships of the most important compounds of nitrogen with oxygen and hydrogen. *Z. Elektrochem.* **26,** 182–196 (in German).

Pryor, W. A. and Lightsey, J. W. (1981). Mechanisms of nitrogen dioxide reactions: Initiation of lipid peroxidation and the production of nitrous acid. *Science* **214,** 435–437.

Ridd, J. H. (1961). Nitrosation, diazotisation and deamination. *Quart. Rev.* **15,** 418–441.

Rigdon, L. P. and Crawford, R. W. (1971). An experimental system for determining nitrogen oxides in air. University of California Lawrence Radiation Laboratory Report UCRL-51057, Livermore, CA.

Saltzman, B. E. (1954). Colorimetric microdetermination of nitrogen dioxide in the atmosphere. *Anal. Chem.* **26,** 1949–1955.

Sada, E., Kumazawa, H., and Butt, M. A. (1979). Single and simultaneous absorptions of lean SO_2 and NO_2 into aqueous slurries of $Ca(OH)_2$ or $Mg(OH)_2$ particles. *J. Chem. Eng. Jap.* **12,** 111–117.

Sato, T., Matani, S., and Okabe, T. (1979). The oxidation of sodium sulfite with nitrogen dioxide. With special reference to analytical methods for nitrogen-sulfur compounds produced in the reaction system. *Nippon Kagaku Kaishi* **1979** (7), 869–878 (in Japanese); see also American Chemical Society 177th National Meeting, Honolulu, HI, April 1, 1979, paper INDE-210, 1979.

Scaringelli, F. P., Rosenberg, E., and Rehme, K. A. (1970). Comparison of permeation devices and nitrite ion as standards for the colorimetric determination of nitrogen dioxide. *Environ. Sci. Technol.* **4,** 924–929.

Schmid, G. (1961). The autocatalytic nature of the cathodic reduction of nitric acid to nitrous acid. III. Mathematical treatment of an autocatalytic electrode reaction of first order. *Z. Elektrochem.* **65,** 531–534 (in German).

Schmid, G. and Bähr, G. (1964). Formation of HNO_2 from NO and HNO_3 at high acid strength. *Z. Physik. Chem.* **41,** 8–25 (in German).

Schmid, G. and Krichel, G. (1964). The autocatalytic nature of the cathodic reduction of nitric acid to nitrous acid IV. The potentiostatic transient. *Ber. Bunsenges.* **68,** 677–688 (in German).

Schmid, H. (1940). Intermediate reactions, in *Handbuch der Katalyse,* Vol. 2, Katalyse in Lösungen, Springer, Vienna, pp. 1–44 (in German).

Schwartz, S. E. and Freiberg, J. (1981). Mass-transport limitation to the rate of reaction in liquid droplets: Application to oxidation of SO_2 in aqueous solutions. *Atmosph. Environ.* **15,** 1129–1144.

Schwartz, S. E. and White, W. H. (1981). Solubility equilibria of the nitrogen oxides and oxyacids in dilute aqueous solution, in *Advances in Environmental Science and Engineering,* Vol. 4. J. R. Pfafflin and E. N. Ziegler, Eds., Gordon and Breach, New York, pp. 1–45.

Seddon, W. A. and Sutton, H. C. (1963). Radiation chemistry of nitric oxide solutions. *Transact. Faraday Soc.* **59,** 2323–2333.

Seel, F., and Knorre, H. (1961). The reaction of nitrite with bisulfite in bisulfite–sulfite buffer solutions. *Z. Anorg. Allg. Chem.* **313,** 70–89 (in German).

Sherwood, T. K., Pigford, R. L., and Wilke, C. R. (1975). The reactions of NO_x with water and aqueous solutions, in *Mass Transfer,* McGraw-Hill, New York, pp. 346–361.

Söderlund, R. and Svennson, B. H. (1976). The global nitrogen cycle, in *Nitrogen, Phosphorus ud Sulfur-Global Cycles,* SCOPE report No. 7, B. H. Svennson and R. Söderlund, Eds., Ecological Bulletins/NFR, Swedish National Science Research Council, Stockholm, pp. 23–73.

Stedman, G. (1979). Reaction mechanisms of inorganic nitrogen compounds. *Adv. Inorg. Radiochem.* **22,** 113–170.

Szabó, Z. G., Bartha, L. G., and Lakatos, B. (1956). The structure and reactions of dinitrogen tetroxide. *J. Chem. Soc.* **1956,** 1784–1795.

Takeuchi, H., Ando, M., and Kizawa, N. (1977). Absorption of nitrogen oxides in aqueous sodium sulfite and bisulfite solutions. *Ind. Eng. Chem. Process Des. Devel.* **16,** 303–308.

Treinin, A. and Hayon, E. (1970). Absorption spectra and reaction kinetics of NO_2, N_2O_3, and N_2O_4 in aqueous solution. *J. Am. Chem. Soc.* **92,** 5821–5828.

Wendel, M. M. and Pigford, R. L. (1958). Kinetics of nitrogen tetroxide absorption in water. *Am. Inst. Chem. Eng. J.* **4,** 249–256.

Wilke, C. R. and Chang, P. (1955). Correlation of diffusion coefficients in dilute solutions. *Am. Inst. Chem. Eng. J.* **1,** 264–270.

2

REACTIONS OF SULFUR(IV) WITH TRANSITION-METAL IONS IN AQUEOUS SOLUTIONS

Robert E. Huie

Chemical Kinetics Division
National Bureau of Standards
Washington, DC 20234

Norman C. Peterson

Department of Chemistry
Polytechnic Institute of New York
Brooklyn, New York 11201

1. INTRODUCTION

The mechanism of the oxidation of S(IV) in aqueous solutions has been of considerable interest recently because of concern over the effects of SO_2 emissions into the atmosphere. The oxidation of S(IV) is of concern not only as a basis for the understanding of what these effects might be, but also because of the importance of S(IV) oxidation on the performance of the most commonly used SO_2 removal technique, lime–limestone flue gas scrubbers.

It has been known since the work of Titoff (1903) that the rate of oxidation of S(IV) in aqueous solution can be reduced to a negligible level by the use of very pure water and that the rates observed otherwise are due to catalysis by metal ions. Furthermore, it was observed that the addition of complexing agents inhibited the reactions. From these observations and the many others made over the last century on S(IV) autoxidation in solution [for references, see Westley (1982)], it is apparent that a crucial step in this oxidation is the interaction of S(IV) with certain metal ions. Despite this basic understanding and the hundreds of papers written on the subject, a universally accepted mechanism for the oxidation of S(IV) in aqueous solution does not exist. Rather than attempt to review the overall process in S(IV) oxidation, particularly in the environment or in flue gas scrubbers, we focus on one central aspect of that problem: the reactions of S(IV) with metal ions in aqueous solution and the associated reactions of possible complexes that might be formed. Furthermore, the concern that these metal ion–S(IV) complexes might be stable in atmospheric droplets and thereby provide a mechanism for the introduction of S(IV) into the lung has been raised [see Hansen et al. (1974) and also Chapter 5 in this volume by Eatough and Hansen]. This suggestion has been disputed, however, by Dasgupta et al. (1979).

The subsequent oxidation reactions of S(IV) complexed with transition metal ions are discussed by Hoffmann and Boyce in Chapter 3 of this volume and thus are only briefly discussed here.

In this chapter we are concerned primarily with the reactions of ferric, cupric, and maganic ions, since these ions appear to be of greatest importance in flue gas scrubbers and in the environment. Interactions with other metal ions, particularly cobaltic, are also discussed, primarily because the stability of their sulfito complexes permits studies that are apparently not feasible for iron, copper, and manganese complexes. The extensive literature on the formation and structure of SO_2 complexes with the transition metals is not, in general, discussed, particularly since a very recent review has appeared on this subject (Ryan, et al., 1981) and since reactions forming these complexes are normally carried out in nonaqueous solvents. Nor, in general, do we discuss the complexes formed by S(IV) ions with reduced metal ions (e.g., ferrous, cuprous, and manganous). Their spectra (Newman and Powell, 1963) and stability constants (Sillén and Martell, 1964, 1971; Smith and Martell, 1976) are reasonably well established.

The oxidation of S(IV) to S(VI) is a two-electron process. Two-equivalent oxidizing agents (e.g., I_2, H_2O_2), almost always convert S(IV) to S(VI) (Higginson and Marshall, 1957). Reagents that are capable of either one or two equivalent changes (e.g., $KMnO_4$) tend to give mixtures of S(V) (dithionate) and S(VI). The metal ions of interest, Fe(III), Cu(II), and Mn(III), are one-equivalent oxidants. The products of their reactions are sulfate, dithionate, or a mixture of both. The stoichiometry of the reaction (defined here as the ratio of oxidant consumed to S(IV) consumed) can range from 1 for complete dithionate production to 2 for complete sulfate production, depending on not only which metal ion is being used, but also on such diverse factors as the initial concentrations, the presence of complexing agents, or the pH of the solution. The relationship of the stoichiometry to various environmental conditions is a key piece of information available for understanding the mechanism of the overall reaction. The stoichiometry is useful because of the stability of dithionate in the presence of various oxidants. Although dithionate is thermodynamically unstable with regard to disproportionation, it is kinetically stable with regard to both disproportionation and oxidation at room temperature. As the temperature is raised, however, the disproportionation reaction becomes more important, and the S(IV) produced can, in turn, be oxidized by any oxidant present (Yost and Russell, 1944).

In addition, important information about the mechanism of S(IV)–metal ion interactions has been derived from kinetic studies. Typically, the rate of loss of the oxidant, or rate of increase of the reduced oxidant, has been monitored. The effect of a variety of environmental parameters on the derived rate constant is a further clue to the elementary steps in the reaction.

In Section 2 we review the experimental results in the order Fe(III), Cu(II), Mn(III), Co(III), and miscellaneous metal ions, including Hg(II), Cu(III), and Ir(IV). Many of the studies have involved ions complexed by organic chelating agents, for which the abbreviations in Table 1 are used. Unless otherwise stated, experiments were reported at 20–25°C. More work has been done on iron systems than any other, although the literature on cobalt systems is increasing rapidly and may be the most well understood. The few attempts reported from the early 1970s to directly study radical

Table 1. Symbols Used for Ligands Employed in Chelation Studies

Ligand	Symbol
Acetoacetic ester	ac
Acetylacetonate	acac
2,2′-Bipyridine	bipy
Ethylenediamine tetraacetic acid	EDTA
Ethylenediamine	en
Tetraglycine	$H_{-3}G_4$
1,10-Phenanthroline	L
Malonic acid	mal
Salicylaldehyde	salald
2,2′,2″-Triaminotriethylamine	tren

products of metal ion–S(IV) reactions by electron spin resonance (ESR) spectroscopy are discussed briefly. In Section 3 we discuss the mechanisms proposed for the oxidation of S(IV) by transition-metal ions in light of all these results. Finally, we suggest some key experiments we feel should be undertaken to develop a clear understanding of the chemistry of these systems.

2. EXPERIMENTAL OBSERVATIONS ON METAL ION–S(IV) REACTIONS

2.1 Reactions of Fe(III) Complexes

2.1.1. $Fe(H_2O)_6^{3+}$ and $Fe(H_2O)_nCl_m^{(3-m)+}$

There have been a large number of papers dating back well over 100 years on the reaction of the ferric ion with S(IV). Much of the nineteenth century work has been discussed by Bassett and Henry (1935) and Bassett and Parker (1951) and is not reviewed here. Although the earliest work used a suspension of ferric hydroxide, the observation that a red complex is formed that slowly decomposes to ferrous ion and dithionate was later confirmed for ferric salts by Meyer (1920). Pinnow (1923) found that low acidity favored the formation of dithionate and that the addition of quinone or hydroquinone favored the formation of sulfate. Albu and von Schweinitz (1932) studied the effect of pH on the reaction of $FeCl_3$ and Na_2SO_3 by adding KOH, HCl, or H_2SO_4. The yield of dithionate typically was about 50%, with some influence of pH and anion (Cl^- or HSO_4^-) indicated.

The difference between the effect of the change in pH and the change in anion produced by adding HCl and H_2SO_4 was further investigated in the work of Bassett and Parker (1951). They found that the addition of hydrochloric acid to a mixture of ferric chloride and S(IV) greatly reduced the amount of dithionate formed but that the effect of sulfuric acid on the yield of dithionate in the ferric sulfate–sulfur(IV) reaction was quite small.

The first systematic study of the effects of the relative amounts of Fe(III) and S(IV) on dithionate production and of a variety of additives on both the rate and mechanism was conducted by Kuz'minykh and Bomshtein (1953) using a 0.05 M H_2SO_4 solution. They found that the proportion of dithionate formed increased with increasing S(IV):Fe(III) ratio. With a ratio of about unity, 50% of the S(IV) became dithionate. The addition of $ZnSO_4$, $CoSO_4$, $NiSO_4$, K_2SO_4, $(NH_4)_2MoO_4$, $CdSO_4$, $K(SbO)C_4H_4O_6$, Na_3AsO_3, $Cr_2(SO_4)_3$, $MgSO_4$, or $MnSO_4$ had no effect on the reaction. The addition of $CuSO_4$, however, accelerated the reaction and reduced the production of dithionate. The effect of Cu(II) on the rate was greater at the later stages of the reaction.

A detailed kinetic study of this effect of Cu(II) on the products of the reaction of Fe(III) with S(IV) was carried out by Higginson and Marshall (1957) at a hydrogen ion concentration of 0.08 M. They also investigated the effect of added Fe(II). A tenfold increase in Fe(II) decreased the rate of loss of Fe(III) by a factor of somewhat less than 10 but decreased the rate of formation of dithionate by a factor of more than 10. At the same time, the rate of sulfate formation was decreased by only a factor of about 4. A change in the S(IV):Fe(III) ratio from 2.2 to 10 increased the dithionate:sulfate formation rate ratio from 2.4 to 4.9. With the addition of Cu(II), both the rate and the "instantaneous stoichiometry" [the ratio of the rate of formation of Fe(II) to the rate of loss of S(IV)] increased. Increase of the Cu(II) concentration from 2.5×10^{-5} M to 1.5×10^{-2} M in a mixture of 8.0×10^{-3} M Fe(III), 2.0×10^{-3} M Fe(II), and 2.0×10^{-2} M H_2SO_3 increased the instantaneous stoichiometry from 1.25 to 1.99 and the rate of increase of Fe(II) from 2.2×10^{-8} M^{-1} s^{-1} to 1.0×10^{-6} M^{-1} s^{-1}. The rates were measured typically with about 14% of the limiting reagent consumed.

The formation of a red complex when Fe(III) and S(IV) are mixed has been noted since at least the middle of the nineteenth century. Meyer (1920) considered it to be a complex acid, $H_3[Fe(SO_3)_3]$, or its ferric salt. By photometric titration at 458 nm, Danilczuk and Swinarski (1961) confirmed that a complex with three sulfites is the most stable over the pH range 2–3. A change in pH from 2 to 3 causes the absorption to increase by a factor of over 2 for a constant concentration of Fe(III). This reflects the involvement of hydrogen ions in the equilibria associated with the formation of the complexes:

$$Fe^{3+} + HSO_3^- \rightleftharpoons FeSO_3^+ + H^+ \qquad [1]$$

$$FeSO_3^+ + HSO_3^- \rightleftharpoons Fe(SO_3)_2^- + H^+ \qquad [2]$$

$$Fe(SO_3)_2^- + HSO_3^- \rightleftharpoons Fe(SO_3)_3^{3-} + H^+. \qquad [3]$$

To obtain information about the kinetic behavior of the complex, a series of experiments was performed in which the absorption of a series of solutions were measured immediately after mixing and after 5 minutes at several ratios of Fe^{3+} to SO_3^{2-}. The total concentrations were also varied. These data were interpreted to suggest that it is the $Fe(SO_3)_3^{3-}$ complex that undergoes the redox reaction, not the free ion.

Pollar et al. (1961) have carried out further studies of the influence of ferrous ion and cupric ion on both the rate of the reduction of ferric by S(IV) and the product mix resulting from the reaction. The experiments were carried out in 0.5 M sulfuric acid with the use of chromatography to separate ionic species prior to analysis. Second-order plots of the data with no ferrous or cupric added showed considerable curvature. The initial portion of the plots, however, yields an average second-order rate constant of 2.7×10^{-4} $M^{-1}\ s^{-1}$. The runs carried out with the addition of ferrous ion gave results similar to those obtained by Higginson and Marshall (1957). Also, they reported that the addition of cupric sulfate resulted in the production of less dithionate and more sulfate.

Most of the kinetic experiments discussed thus far were done under conditions where dithionate production might be expected to be important: comparable concentrations of Fe(III) and S(IV). Karraker (1963), however, used a large excess of Fe(III) to simplify the reaction and avoid the production of dithionate. The reaction was monitored by measuring the amount of Fe(II) produced. The general behavior of the reaction was similar to that previously discussed—the reaction was slowed by the addition of Fe(II) or more acid. The reaction rate was found to be increased by saturating the solution with air. This is a particularly interesting observation since, under other conditions, Fe(II) and S(IV) are known to undergo an induced oxidation (Richter, 1931) where both are oxidized. An induced oxidation would result in a reduction of the rate of loss of Fe(III). Karraker's results imply that oxygen produces an intermediate capable of reducing Fe(III).

The most thorough series of experiments on the reaction of Fe(III) with S(IV) was carried out by Carlyle and co-workers. The initial work was a kinetic study of the aquation of the complex formed between S(IV) and Fe(III) (Carlyle, 1971). Under the conditions of these experiments, only the $FeSO_3^+$ complex was important. The measured rates depended on the acidity and the bisulfate concentration but were independent of added Fe^{2+}. An acid-independent term was also indicated. These results suggest that the mechanism consists of three steps:

$$FeSO_3^+ \rightleftharpoons Fe^{3+} + SO_3^{2-} \quad [4]$$

$$FeSO_3^+ + H^+ \rightleftharpoons Fe^{3+} + HSO_3^- \quad [5]$$

$$FeSO_3^+ + HSO_3^- \rightleftharpoons Fe^{3+} + HSO_3^- + SO_3^{2-}; \quad [6]$$

any hydrolysis of Fe^{3+} is ignored. They estimated the formation constant $[FeSO_3^+][H^+]/[Fe^{3+}][HSO_3^-]$ to be ≤ 0.4.

The study of the second step in the reaction, the electron transfer between S(IV) and Fe(III), was carried out primarily with S(IV) in great excess

(Carlyle and Zeck, 1973). The $FeSO_3^+$ concentration was monitored directly by absorption spectrophotometry over the range 320–443 nm. In addition to the numerical data, some of their qualitative observations were quite interesting. For one thing, they found that reaction mixtures containing excess Fe(II) or Cu(II) gave unusual rate behavior that could be prevented by pretreating the mixtures with S(IV). Furthermore, the absorbance increased under some conditions, indicating that the amount of Fe(III) was increasing. Although this phenomenon was not explained, apparently one cause was the exposure of the solution to air. This would suggest that induced oxidation of Fe(II) to Fe(III) could occur under their conditions.

The stoichiometry results were in line with those reported by other workers. The kinetic measurements, with S(IV) in great excess, could be fit reasonably well by the expression

$$\frac{-d[\mathrm{Fe(III)}]}{dt} = k'[\mathrm{Fe(III)}]^2, \tag{1}$$

where

$$k' = \frac{a[\mathrm{HSO_3^-}]}{[\mathrm{Fe(II)}][\mathrm{H^+}]} + \frac{b[\mathrm{HSO_3^-}]^2/[\mathrm{H^+}]^2}{c + [\mathrm{SO_2}]} \tag{2}$$

and $a = (1.7 \pm 0.2) \times 10^{-6}\ \mathrm{M\ s^{-1}}$
$b = (5.7 \pm 0.9) \times 10^{-5}\ \mathrm{M\ s^{-1}}$
$c = 0.19 \pm 0.04\ \mathrm{M}$

Although this equation gave a reasonable fit to the rate data taken with S(IV) in excess, rate constants measured with Fe(III) in excess were about a factor of 10 greater than predicted.

With Cu(II) added, the rate of the reaction with S(IV) in excess could be expressed

$$\frac{d[\mathrm{Fe(III)}]}{dt} = k''[\mathrm{Fe(III)}], \tag{3}$$

where

$$k'' = \frac{[\mathrm{Cu(II)}][\mathrm{HSO_3^-}]/[\mathrm{H^+}]}{m[\mathrm{Fe(II)}] + n[\mathrm{Cu(II)}] + p[\mathrm{Cu(II)}]^2/[\mathrm{H^+}]}, \tag{4}$$

$m = 14.1 \pm 0.85$ s, $n = 2.16 \pm 0.095$ s, and $p = 0.83 \pm 0.075$ s. Although this expression did not work as well with Fe(III) in excess, the results were typically within a factor of 2 of the predictions.

Zeck and Carlyle (1974) also carried out a study of the Cu(II) catalyzed reaction of chloroiron(III) with S(IV). An attempt to measure the rate in the absence of copper was unsuccessful, possibly because of trace amounts of molecular oxygen in conjunction with an impurity. The addition of molecular oxygen to these solutions was found to reduce the rate of loss of Fe(III). In a 1 M Cl^- solution and in the presence of Cu(II) the rate of loss of Fe(III) could be represented by the equation

$$\frac{-d[\mathrm{Fe(III)}]}{dt} = k'[\mathrm{Fe(III)}], \tag{5}$$

where

$$k' = \frac{a[\mathrm{S(IV)}]}{[\mathrm{H^+}]^2} \tag{6}$$

and $a = (4.0 \pm 0.14) \times 10^{-3}\ \mathrm{M\ s^{-1}}$. When Fe(II) was present initially, the value of k' was

$$k' = \frac{a[\mathrm{S(IV)}]}{[\mathrm{H^+}]^2(1 + b[\mathrm{Fe^{2+}}])}, \tag{7}$$

with $a = 4.0 \times 10^{-3}\ \mathrm{M\ s^{-1}}$ and $b = 20\ \mathrm{M^{-1}}$.

Dasgupta et al. (1979) observed the ferric–sulfur(IV) reaction in an acidic perchlorate medium. They found in contrast to other work that at a 1:1 ratio of Fe(III):S(IV) (0.005 M) and a pH of 1, conversion of the sulfur to dithionate was quantitative. In systems containing chloride ion, an increase in the ratio of $[\mathrm{Cl^-}]$: $[\mathrm{Fe^{3+}}]$ was found to cause a decrease in the yield of dithionate ion; in 1 M NaCl solutions no dithionate was formed. They also determined that in the absence of complexing anions, the reaction was second order with regard to whichever reagent was limiting. An interesting observation from this work was that the overall reaction rate increased with added chloride ion up to a point and then decreased.

2.1.2. $Fe(CN)_6^{3-}$

The reaction of one other iron compound with S(IV) has been studied by several groups: hexacyanoferrate(III). Its chemistry is of particular interest since it is stable in alkali solutions and since the cyano ligand is inert to substitution. Higginson and Marshall (1957) determined the stoichiometry of the reaction of 0.1 M solutions of $K_3Fe(CN)_6$ and Na_2SO_3. The value ranged from 1.2 to 1.3 over the pH range 7.0–12.3. The reaction was slow and the rate varied by a factor of less than 2 over this pH range. Veprek-Siska and Wagnorova (1965), however, measured a stoichiometry of 2 over the pH range 11.98–5.33, using twentyfold excess of $Fe(CN)_6^{3-}$. Further, at pH 6.37, there was apparently no change in this stoichiometry over a range of concentration ratios $[\mathrm{Fe(CN)_6^{3-}}]/[\mathrm{SO_3^{2-}}] = 0.5$–5.

The only other kinetic study on the reaction of $Fe(CN)_6^{3-}$ with S(IV) was carried out by Swinehart (1967). Rate measurements were made by using a seven- to fortyfold excess of S(IV) and in the presence of ethylenediaminetetracetic acid (EDTA). First-order rate constants were derived from the relationship

$$\frac{-d[\mathrm{Fe(CN)_6^{3-}}]}{dt} = k'[\mathrm{Fe(CN)_6^{3-}}]. \tag{8}$$

The values of k' were found to depend not only on the concentration of S(IV), but also on the nature and concentration of the cation present in solution,

$$k' = k[\mathrm{S(IV)}][\mathrm{M}^+]. \tag{9}$$

The enthalpy of activation for the reaction was also somewhat different for different cations: 11.7 kcal mol^{-1} for Na^+ and 12.6 kcal mol^{-1} for K^+. A particularly interesting result in this study is that there is a decrease in the rate constant when H_2O is replaced by D_2O, $k_{H_2O}/k_{D_2O} = 1.4 \pm 0.1$ for both K^+ and Na^+.

The observation that the addition of different electrolytes to the solution can have an effect beyond a simple ionic strength effect was confirmed by Veprek-Siska and Hasnedl (1968). They assert, however, that the effect is due to impurity copper ions in the electrolyte and can be removed by the addition of sufficient EDTA. Using atomic absorption spectroscopy to measure the concentration of copper, they obtained a good linear relationship between the measured rate and the amount of copper in the various solutions. The solution containing KF was somewhat anomalous, suggesting that fluoride ion can complex the cupric ion and prevent it from being an effective catalyst.

To explain their observations, Veprek-Siska and Wagnerova (1965) used a mechanism involving the displacement of a CN^- ligand by SO_3^{2-} with the formation of a pentacyanosulfitoferrate(III) complex. This mechanism would suggest that the measured rate constant would be affected by the addition of CN^- and would require that the addition of $^{14}CN^-$ result in some labeled CN^- in the product $Fe(CN)_6^{4-}$. Swinehart (1967) tested the effect of added CN^- on the rate, and Wilberg et al., (1965) carried out the reaction in the presence of $^{14}CN^-$. In neither study was there any indication that ligand displacement was important.

In a spectrophotometric study of the reaction of hexacyannoferrate(III) with S(IV), Lancaster and Murray (1971) observed spectral changes characteristic of the formation of a long-lived intermediate. By using ^{35}S-labeled sulfite and removing the product sulfate and unreacted sulfite by reaction with strontium, they were able to show that the intermediate incorporates a sulfur-containing group. It was observed that the introduction of various additives [Cu^{2+}, NaCl, and $Fe(CN)_6^{4-}$] affected the formation of the intermediate but did not affect the intermediate once it was formed.

The rate of formation and hydrolysis of the intermediate were measured over the temperature ranges 25–40°C and 30–39°C. Enthalpies of activation of 12.0 kcal mol^{-1} for the formation reaction and 17.0 kcal mol^{-1} for the hydrolysis reaction were derived, suggesting that Veprek-Siska and Wagnorova (1965) and Swinehart (1967) were both measuring the formation reaction.

These results and kinetic experiments carried out with various amounts of added $Fe(CN)_6^{4-}$ led to the following reaction mechanism (Murray, 1974):

$$Fe(CN)_6^{3-} + SO_3^{2-} \rightleftharpoons Fe(CN)_6^{4-} + SO_3^{-} \quad [7]$$

$$Fe(CN)_6^{3-} + SO_3^{-} \rightarrow Fe(CN)_5(CN \cdot SO_3)^{4-} \quad [8]$$

$$Fe(CN)_5(CN \cdot SO_3)^{4-} + H_2O \rightarrow Fe(CN)_6^{4-} + SO_4^{2-} + 2H^+. \quad [9]$$

2.1.3. *Reactions of Other Ferric Complexes*

The reactions of two other iron complexes with S(IV) have been studied kinetically: pentacyanoaquoferrate(III) and tris(1,10-phenanthroline) ferrate(III). The reaction of $Fe(CN)_5H_2O^{2-}$ with SO_3^{2-} (Veprek-Siska et al., 1966a) resulted not only in the formation of SO_4^{2-}, but also in the stable complex $Fe(CN)_5SO_3^{5-}$. No dithionate was detected, and the overall mechanism of the reaction was determined to be

$$2Fe(CN)_5H_2O^{2-} + 3SO_3^{2-} + 2OH^- \rightarrow 2Fe(CN)_5SO_3^{5-} + SO_4^{2-} + 3H_2O. \quad [10]$$

The ferrous product, $Fe(CN)_5SO_3^{5-}$, was not oxidized by pentacyanoferrate(III). In the early part of the reaction the stoichiometry was 2; as the reaction proceeded, this value changed to $\frac{2}{3}$. Veprek-Siska et al., proposed that the initial step in the reaction was a slow ligand substitution by sulfite

$$Fe(CN)_5H_2O^{2-} + SO_3^{2-} \rightarrow Fe(CN)_5SO_3^{4-} + H_2O, \quad [11]$$

followed by the rapid oxidation of the complex and the hydrolysis of the resulting sulfato complex

$$Fe(CN)_5SO_3^{4-} + Fe(CN)_5H_2O^{2-} \rightarrow Fe(CN)_5SO_3^{3-} + Fe(CN)_5H_2O^{3-} \quad [12]$$

$$Fe(CN)_5SO_3^{3-} + 2H_2O \rightarrow Fe(CN)_5H_2O^{3-} + H_2SO_4. \quad [13]$$

The stable ferrous–sulfite complex is formed in the slow reaction

$$Fe(CN)_5H_2O^{3-} + SO_3^{2-} \rightarrow Fe(CN)_5SO_3^{5-} + H_2O. \quad [14]$$

The kinetics of this last reaction were also measured independently.

The most unusual aspect of the kinetics of the reaction of pentacyanoferrate(III) with S(IV) in comparison to other S(IV) oxidations, is that the rate constant exhibits a maximum at about pH = 5. The rate constant at that point is about 0.4 $M^{-1} s^{-1}$. A similar pH dependence was observed for reaction [14].

The reaction of tris (1,10-phenanthroline) ferrate(III) with S(IV) is somewhat surprising in that the rate equation contains a quadratic dependence on $[HSO_3^-]$ (Carlyle, 1972):

$$\frac{-d[FeL_3^{3+}]}{dt} = \left(k_1 + \frac{k_2}{[H^+]} + k_3[HSO_3^-]\right)[HSO_3^-][FeL_3^{3+}] \quad (10)$$

Although this might suggest the formation of an intermediate that is reactive toward HSO_3^-, no spectral evidence for such an intermediate was observed. The stoichiometry of the reaction was two and no dithionate was found.

The addition of either tris (1,10-phenanthroline) ferrate(II) or dioxygen slowed the reaction significantly. Unlike most other reactions of ferric complexes with S(IV), Cu(II) was not found to catalyze the reaction in perchlorate medium and only slightly in a sulfate medium. The results were interpreted in terms of an outer-sphere oxidation of S(IV) giving a S(V) radical, with the subsequent fast reaction of S(V) with FeL_3^{3+}

$$FeL_3^{3+} + S(IV) \rightarrow FeL_3^{2+} + S(V), \quad [15]$$

$$FeL_3^{3+} + S(V) \rightarrow FeL_3^{2+} + S(VI). \quad [16]$$

The observed term in the rate expression second order in S(IV) was considered to arise from either a reaction mechanism involving $S_2O_5^{2-}$ or a reaction mechanism involving an outer-sphere Fe(III)–S(IV) or Fe(II)–S(V) complex reacting with a second S(IV). Inhibition of the reaction by FeL_3^{2+} or O_2 was thought to arise from reactions involving the S(V) radical.

The products of the reactions of several other ferric complexes with S(IV) have been reported (Veprek-Siska et al., 1966). The complexes $Fe(CN)_5NH_3^{2-}$ and $Fe(bipy)_3^{3+}$ gave only sulfate, whereas $Fe(C_2O_4)_3^{3-}$, $Fe(mal)_3^{3-}$, $Fe(H_2P_2O_7)_3^{3-}$, $Fe(acac)_3$, and $Fe(EDTA)^-$ all gave mixtures of sulfate and dithionate.

2.2. Reactions of Cu(II) Complexes

Although most of the interest in the interaction of cupric salts with S(IV) has centered around the role of Cu(II) as a catalyst in the Fe(III)–S(IV) system or as a catalyst in the oxidation of S(IV) by dioxygen, there have been a few studies of the direct reaction of the cupric ion with S(IV). Much of the early interest in the direct reaction between Cu(II) and S(IV) centered about finding a way to prepare Cu(I) [Keller and Wycoff (1946) and references cited therein]. One early investigator (Baubigny, 1912), however, reported that up to 70% of the sulfite in a solution of sodium sulfite and cupric sulfate was oxidized to dithionate.

In 1925 Reinders and Vles observed that in an ammonia solution of unspecified concentration, sulfite did not reduce cupric ion at ordinary temperature but at 50°C the reaction was noticeable and at 100°C it was rapid. Later, Albu and von Schweinitz (1932) noted that cupric ion forms a colored complex with sulfite in acid solutions but reacts to precipitate CuOH and to form dithionate and sulfate in alkaline solutions. The slowness of the reaction in acid in the absence of added ferric ion was confirmed by Kuz'minykh and Bomshtein (1953). Basset and Parker (1951) found that solutions of cupric chloride changed color on saturation with sulfur dioxide and that a slow subsequent reaction took place. The reaction primarily yielded sulfate. Ammonium sulfite has been reported to quantitatively reduce cupric ions to metallic copper at 170°C (Okabe et al., 1960).

The first kinetic study on the reaction between Cu(II) and S(IV) was reported by Zeck and Carlyle (1974). In 1 M hydrochloric acid and at 25°C the reaction was described by the rate law

$$\frac{-d[\mathrm{Cu(II)}]}{dt} = \frac{[\mathrm{Cu(II)}][\mathrm{S(IV)}]}{1700[\mathrm{H^+}] + 2900[\mathrm{H^+}]^2} \tag{11}$$

The precision of the results, however, was not good, and consequently subtle deviations might have been missed. It was found that addition of Cu(I) strongly retarded the reaction. The influence of Cu(I), however, was not simple, as the measured rate passed through a minimum as the Cu(I) concentration was increased. Complex formation between Cu(I) and S(IV) was also observed, but the equilibrium constant did not appear to be large.

Sulfite complexes of Cu(I) are reported to be very stable, with stepwise formation constants of $K_1 = 10^{7.85}$, $K_2 = 10^{0.85}$, and $K_3 = 10^{0.66}$ (Sillén and Martell, 1964). Copper(I) chloride complexes are also known to be very stable, with

$$\beta_2 = 10^{5.19} = \frac{[\mathrm{CuCl_2^-}]}{[\mathrm{Cu^+}][\mathrm{Cl^-}]^2} \tag{12}$$

(Smith and Martell, 1976). Thus it seems likely that the small equilibrium constant found by Zeck and Carlyle is for the replacement of chloride ion by sulfite ion by

$$\mathrm{CuCl_2^-} + \mathrm{SO_3^{2-}} = \mathrm{CuClSO_3^{2-}} + \mathrm{Cl^-} \qquad [17]$$

This result is consistent with the stability of the mixed ligand complex being less than that of the chloride complex.

Veprek-Siska and Lunak (1974) reported that Cu(II) was immediately and quantitatively reduced by sulfite ion to Cu(I). We infer that the solution is alkaline when this occurs and that chloride ion is absent.

Although the reaction of cupric chloride with S(IV) in acidic solution is not very fast, Dorfman et al. (1977) found that it could be accelerated immensely by the addition of iodide ion. The same is true of cupric bromide (Dorfman, et al., 1978). In the cupric chloride system, even in the presence of iodide ion, the reaction was very slow at chloride ion concentrations of less than 1 M.

The kinetic behavior seemed quite different from that reported (Zeck and Carlyle, 1974) for the cupric chloride reaction with no iodide present. The addition of Cu(I) caused the rate to increase, and this phenomenon rendered the system autocatalytic. Furthermore, Dorfman et al. found that the reaction rate increased with an increase in $[\mathrm{H^+}]$, unlike the inverse dependence noted by Zeck and Carlyle.

The sulfur-containing products of a series of cupric complexes have been reported by Veprek-Siska et al. (1966). The reactants $\mathrm{Cu(H_2O)_4^{2+}}$,

$Cu(H_2P_2O_7)_2^{2-}$, and $Cu(NH_3)_4^{2+}$ all gave mixtures of dithionate and sulfate. The reactants $Cu(C_2O_4)_2^{2-}$, $Cu(mal)_2^{2-}$, $Cu(acac)_2$, $Cu(ac)_2$, and $Cu(bipy)_2^{2+}$ all gave only sulfate; $Cu(en)_3^{2+}$, $Cu(L)_2^{2+}$, Cu(EDTA), and $Cu(CN)_4^{2-}$ were unreactive.

2.3. Reactions of Mn(III) Complexes

There have been only a few reports on the reaction of S(IV) with complexes of Mn(III), even though manganese has been known for some time to catalyze the oxidation of S(IV) (Grodzovskii, 1935; Hoather and Goodeve, 1934). The stoichiometry [Mn(III) : S(IV) consumption ratio] of the reaction of trispyrophosphatomanganate(III) with S(IV) was reported to be 1.24 at pH 0.5, 1.19 at pH 5, and 1.03 at pH 9 (Higginson and Marshall, 1957).

For the same reaction, Brown and Higginson (1972) found the stoichiometry to decrease with increasing pH or with increasing S(IV). Other Mn(III) complexes that have been studied include $Mn(C_2O_4)_3^{3-}$, $Mn(mal)^{3-}$, $Mn(acac)_3$, and $MnEDTA^-$ (Veprek-Siska, et al., 1966). These all gave mixtures of sulfate and dithionate. The complex $Mn(CN)_6^{3-}$ was unreactive toward S(IV).

Recently, a kinetic study of the reaction of Mn(III) in an acid perchlorate medium was carried out (Siskos, et al., 1982). The reaction was found to be very fast and showed little dependence on added Mn(II). The rate constant showed a strong inverse dependence on hydrogen ion concentration with the measured rate increasing by a factor of 5 as the perchloric acid concentration is lowered from 5.02 to 2.31 M. At 5.92 M perchloric acid and over the temperature range 15.4–34.9°C, the rate constant was

$$k = 1.1 \times 10^7 \exp\left(\frac{-1970}{T}\right) \text{M}^{-1}\,\text{s}^{-1}. \tag{13}$$

The stoichiometry of the reaction was 1 within experimental error.

2.4. Reactions of Co(III) Complexes

2.4.1. *Reactions of Co(III) Complexes with S(IV)*

The reactions of Co(III) complexes with S(IV) are of interest primarily because of the formation of stable sulfite complexes that can be isolated and whose decomposition and redox reactions can be studied separately from the starting materials. In addition, there have been two reports on the stoichiometry and products of the overall reactions of some of these complexes.

Higginson and Marshall (1957) studied the reactions of $Co_2(SO_4)_3$, $K_3Co(CO_3)_3$, and $CoN(CH_2CO_2)_3$ with a sodium sulfite solution. At pH 5, $Co(CO_3)_3^{3-}$ gave stoichiometries in the range 1.02–1.12; at pH 0.5,

$Co(H_2O)_6^{3+}$ [from $Co_2(SO_4)_3$] gave stoichiometries in the range 1.04–1.37 and $CoN(CH_2CO_2)_3$ gave a stoichiometry of 1.26 in a slow reaction.

Furthermore, Veprek-Siska et al. (1966b) have reported that $Co(L)_3^{3+}$ and $Co(salald)_3^{3+}$ produce mixtures of dithionate and sulfate but that $Co(C_2O_4)_3^{3-}$, $Co(Mal)_3^{3-}$, and $Co(ac)_3$ oxidize S(IV) only to sulfate.

These reactions may involve formation of a sulfito complex as discussed in Section 2.4.2.

2.4.2. *Decomposition Reactions of Sulfite Complexes of Co(III)*

Cobalt(III) forms complexes inert to substitution, whereas Co(II) complexes are labile, equilibrating rapidly with the solvent. A number of stable Co(III) complexes have been prepared that include sulfite as a ligand. The sulfur-bonded sulfite ligand tends to have a labilizing influence on the ligands in the position trans to the sulfito group in the octahedral complex. Complexes of Co(III) in which the trans labilizing effect of sulfite has been studied include the ligands ammonia, ethylenediamine, cyanide, and dimethylglyoximate. These studies recently have been reviewed by van Eldik and Harris (1980).

The kinetics of the intramolecular redox decomposition of some of these Co(III) sulfito complexes has been studied in detail. This internal electron transfer reaction is thermodynamically favored but kinetically limited. The intramolecular redox decomposition of sulfito pentammine–cobalt(III), however, has not been studied because of the rapid replacement of the *trans*-NH_3 by H_2O, a reaction facilitated by the sulfito ligand as cited above.

Murray and Stranks (1969) were able to observe the internal oxidation–reduction of *trans*-$Co(en)_2SO_3OH_2^+$. The products of the reaction are Co(II) and dithionate ion. This reaction has appreciable rates only at elevated temperatures. At 73°C the rate is conveniently measurable; the first-order rate constants ranged from 0.73×10^{-4} to $3.7 \times 10^{-4}\ s^{-1}$ for $[H^+]$ in the range 0.024–0.95 M. Increase in the concentration of perchloric acid above 2 M caused the rate to diminish, an effect ascribed to the decreasing activity of the solvent, water. The activation parameters for the internal redox reaction are $\Delta H\ddagger = 32.8$ kcal mol^{-1} and $\Delta S\ddagger = 9$ cal K^{-1} mol^{-1}.

The decomposition reaction, in the absence of scavengers, leads exclusively to dithionate. However, the addition of Fe(II) eliminates dithionate production and results in the formation of Fe(III) and S(IV). The formation of Fe(III) was not quite stoichiometric, but the departure could easily be accounted for by secondary reaction of Fe(III) with S(IV). The addition of oxygen to a solution with Fe(II) decreased the amount of Fe(III) produced. In the absence of Fe(II) and in the presence of O_2, sulfate was the only product.

The complex ion *trans*-tetraammine aquosulfito Co(III) undergoes an intramolecular redox reaction (Thacker et al., 1974). The rate was observed at $[H^+]$ from 0.01 to 1.0 M. The products of the reaction included sulfite, sulfate, and dithionate ions. The sulfito ligand is thought to be S bonded in the complex and, in contrast to the decomposition of the O bonded sulfito ligand (below), forms dithionate as a product.

The activation parameters for the reaction are $\Delta H\ddagger = 26.5$ kcal mol^{-1} and $\Delta S\ddagger = 12.5$ cal K^{-1} mol^{-1} in an ionic medium of 1.0 M lithium perchlorate. It was proposed that the rate-determining step is an isomerization in which the sulfito becomes trans to an NH_3 ligand. The (sulfito–O) pentammine Co(III) ion is formed by the rapid addition of SO_2 to the oxygen of the $Co(NH_3)_5OH^{2+}$ ion. At pH values of 3.4–6.5 the intramolecular redox reaction occurs at a measurable rate, with the stoichiometry

$$2Co(NH_3)_5OH^{2+} + SO_2 \rightarrow 2Co^{2+} + 2NH_4^+ + 8NH_3 + SO_4^{2-} \quad [18]$$

The activation parameters for the intramolecular redox reaction are $\Delta H\ddagger = 27$ kcal mol^{-1} and $\Delta S\ddagger = 23$ cal K^{-1} mol^{-1}. Although the proposed reaction mechanism involves generation of the radical ion SO_3^-, no dithionate was observed as a product of the reaction.

The complex ion aquo (sulfito-O) (2,2′,2″-triamino triethylamine) Co(II) was found to undergo intramolecular electron transfer (El-Awady and Harris, 1981). The rate of the intramolecular redox reaction was studied spectrophotometrically after allowing $Co(tren)(OH_2)_2^{3+}$ to react with S(IV) in a buffered aqueous solution. The O-bonded sulfito complex, which is rapidly formed, and slowly reacts according to the stoichiometry

$$2[Co(tren)(OH_2)(OSO_2)]^+ + 11H_2O \rightarrow 2Co(H_2O)_6^{2+} + 2tren + HSO_4^- + HSO_3^-. \quad [19]$$

The redox reaction was observed below pH 5.7, but above pH 7.2 only the replacement of the coordinated water molecule by sulfite was observed. The disulfito complex ion formed may have one sulfito S-bonded and the other one O-bonded. Activation parameters for the intramolecular redox reaction are $\Delta H\ddagger = 24$ kcal mol^{-1} and $\Delta S\ddagger = 8$ cal K^{-1} mol^{-1}.

In contrast to the above results, Dash et al. (1981) have prepared the O-bonded (αβS)–(sulfito) (tetraethylenepentamine) Co(III) ion. This complex ion does not exhibit the intramolecular electron transfer reaction but does slowly isomerize to the S-bonded sulfito complex ion.

2.5. Reactions of Other Metal Complexes

2.5.1. $Ir(Cl)_6^{2-}$

Reactions of a few other metal ions with S(IV) have also been investigated. The reaction of hexachloroiridate(IV) ion has been of interest since it is both substitution inert and appears unlikely to engage in any outer sphere complexation by S(IV). Initial work (Brown and Higginson, 1972) found a stoichiometry of 1.87 at pH 4.40 and 1.81 at pH 6.53. They also found that the product of the reaction at pH 1.6 was 20% pentachloroiridate(III) and 80% hexachloroiridate(III). They suggested that the monoaquo complex is formed by hydrolysis of hexachloroiridate(III). Stapp and Carlyle (1974), however, measured stoichiometries that averaged 1.39 over the pH range 3.7–4.7. The rate of loss of the iridium complex could be expressed as

$$\frac{-d[IrCl_6^{2-}]}{dt} = \frac{0.024[IrCl_6^{2-}][HSO_3^-]}{[H^+]} \tag{14}$$

The addition of Cu(II) had little effect on the rate, but dissolved oxygen slowed the reaction by a factor of about 2.

2.5.2. *Copper(III) Tetraglycine*

The reaction of S(IV) with the Cu(III) tetraglycine complex ion was studied by Anast and Margerum (1981). The complex $Cu(III)(H_{-3}G_4)^-$ can be prepared from the Cu(II) tetraglycine complex by electrolysis on a graphite electrode. The rate of the oxidation of S(IV) by the Cu(III) complex was studied at pH values of 4.5–9.5 in 0.1 M $NaClO_4$. The stoichiometry of this reaction was observed to be

$$2Cu(III)(H_{-3}G_4)^- + SO_3^{2-} + 2OH^- \rightarrow 2Cu(II)(H_{-3}G_4)^{2-} + SO_4^{2-} + H_2O. \qquad [20]$$

Traces of oxygen were reported to influence the observed stoichiometry. This effect was accounted for by the induced oxidation of the Cu(II) complex associated with the sulfite–oxygen reaction summarized below.

The rate law for the reaction in the absence of O_2 was found to be

$$-\frac{d}{dt}[Cu(III)(H_{-3}G_4)^-] = k[Cu(III)(H_{-3}G_4)^-][S(IV)], \tag{15}$$

where

$$k = \frac{k_{HB}K_H[H^+] + k_B}{1 + K_H[H^+]}, \tag{16}$$

$k_{HB} = 2.3 \times 10^3\ M^{-1}\ s^{-1}$, and $k_B = 6.9 \times 10^4\ M^{-1}\ s^{-1}$. The equilibrium constant for the protonation of SO_3^{2-}, K_H, obtained from the pH variation in the rate is $K_H = 6.3 \times 10^6\ M^{-1}$, compared to a recommended value of $1.7 \times 10^7\ M^{-1}$ (Parker et al., 1981). Above pH 5.8 the rate is slowed by $Cu(II)(H_{-3}G_4)^{2-}$, requiring a dependence of

$$\frac{[SO_3^{2-}]}{k'_{obs}} = A[Cu(II)(H_{-3}G_4)^{2-}] + B[Cu(II)(H_{-3}G_4)^{2-}]^2 \tag{17}$$

where $A = 8.01 \times 10^{-6}$ M s, $B = 1.42 \times 10^3$ s, and k'_{obs} is the rate constant for the second-order kinetics observed under these conditions, defined by

$$-\frac{d}{dt}[Cu(III)(H_{-3}G_4)^-] = k'_{obs}[Cu(III)(H_{-3}G_4)^-]^2. \tag{18}$$

The proposed mechanism of reaction [20] is

$$Cu(III)(H_{-3}G_4)^- + SO_3^{2-} \underset{k_{-1}}{\overset{k_1}{\rightleftharpoons}} Cu(II)(H_{-3}G_4)^{2-} + SO_3^-, \qquad [21]$$

$$Cu(III)(H_{-3}G_4)^- + SO_3^- \underset{k_{-2}}{\overset{k_2}{\rightleftharpoons}} Cu(II)(H_{-3}G_4)^{2-} + SO_3, \qquad [22]$$

$$SO_3 + H_2O \overset{k_3}{\rightarrow} H_2SO_4, \qquad [23]$$

$$H_2SO_4 + 2OH^- \xrightarrow{\text{fast}} 2H_2O + SO_4^{2-}. \qquad [24]$$

The steady-state concentration of SO_3^- in this reaction is evidently too small to give an appreciable rate for the formation of dithionate ion by a process such as $2SO_3^- \rightarrow S_4O_6^{2-}$.

The rate law corresponding to the preceding mechanism leads to the steady-state expression

$$-\frac{d[Cu(III)(H_{-3}G_4)^-]}{dt} = \frac{2k_1k_2k_3}{D}[Cu(III)(H_{-3}G_4)^-]^2[SO_3^{2-}] \qquad (9)$$

where

$$D = k_2k_3[Cu(III)(H_{-3}G_4)^-] + k_{-1}k_3[Cu(II)(H_{-3}G_4)^{2-}] + k_{-1}k_{-2}[Cu(II)(H_{-3}G_4)^{2-}]^2,$$

and where, for simplification, the two protonation steps have not been included. The rate constants obtained from this study are

$$k_1 = 3.7 \times 10^4\ M^{-1}\,s^{-1},$$

$$\frac{k_2}{k_{-1}} = 1.66,$$

$$\frac{k_{-2}}{k_3} = 177\ M^{-1}.$$

With added oxygen the authors observed the oxidation of $Cu(II)(H_{-3}G_4)^{2-}$ to $Cu(III)(H_{-3}G_4)^-$ in the presence of sulfite. This induced sulfite oxidation occurs to a maximum extent at pH 8. The induced reaction was found to be autocatalytic, having a first-order dependence on the product concentration:

$$\frac{d}{dt}[Cu(III)(H_{-3}G_4)^-] = k_{obs}[Cu(III)(H_{-3}G_4)^-]. \qquad (20)$$

the reaction involving oxygen undergoes oscillations, notably in the concentration of the intermediates. The explanation offered is that the autocatalytic regeneration of $Cu(III)(H_{-3}G_4)^-$ occurs somewhat independently of the factors governing the rate of reduction of $Cu(III)(H_{-3}G_4)$ in a way that the two processes can become out of phase with each other, producing oscillations. Intermediates involved in the aerobic reaction were suggested to be SO_5^- and SO_5^{2-}.

A mechanism for the autocatalytic reaction occuring in oxygenated solutions was proposed as follows:

$$\mathrm{Cu(III)(H_{-3}G_4)^- + SO_3^{2-} \underset{k_{-1}}{\overset{k_1}{\rightleftharpoons}} Cu(II)(H_{-3}G_4)^{2-} + SO_3^-} \quad [25]$$

$$\mathrm{Cu(III)(H_{-3}G_4)^- + SO_3^- \underset{k_{-2}}{\overset{k_2}{\rightleftharpoons}} Cu(II)(H_{-3}G_4)^{2-} + SO_3} \quad [26]$$

$$\mathrm{SO_3 + H_2O \overset{k_3}{\rightarrow} H_2SO_4 \xrightarrow[\text{fast}]{2OH^-} SO_4^{2-}} \quad [27]$$

$$\mathrm{SO_3^- + O_2 \overset{k_4}{\rightarrow} SO_5^-} \quad [28]$$

$$\mathrm{Cu(II)(H_{-3}G_4)^{2-} + SO_5^- \overset{k_5}{\rightarrow} Cu(III)(H_{-3}G_4)^- + SO_5^{2-}} \quad [29]$$

$$\mathrm{Cu(II)(H_{-3}G_4)^{2-} + SO_5^{2-} + H_2O \overset{k_6}{\rightarrow} Cu(III)(H_{-3}G_4)^- + SO_4^{2-} + OH + OH^-} \quad [30]$$

$$\mathrm{Cu(II)(H_{-3}G_4)^{2-} + OH \xrightarrow{\text{fast}} Cu(III)(H_{-3}G_4)^- + OH^-.} \quad [31]$$

The stoichiometry for this mechanism corresponds to

$$\mathrm{2Cu(II)(H_{-3}G_4)^{2-} + SO_3^{2-} + O_2 \xrightarrow{Cu(III)} 2Cu(III)(H_{-3}G_4)^- + SO_4^{2-} + 2OH^-.} \quad [32]$$

2.5.3. *Hg(III) Complexes*

Unlike other oxidized metal ions, the mercuric ion forms stable complexes with S(IV) at room temperature. This property has been used to advantage in the West-Gaeke method for measuring SO_2 in the atmosphere (West and Gaeke, 1956). The complex of S(IV) with Hg(II) is not only available for subsequent reaction with *p*-rosaniline but is resistant to air oxidation (Dasgupta et al., 1979). Photometric studies indicate that the most stable form of the complex is $Hg(SO_3)_2^{2-}$ and also that SO_3^{2-} and I^- form mixed complex ions with Hg^{2+} (Czakis-Sulikowska, 1966). Mixed complexes with Cl^- have also been reported (Graire, 1924). The sodium salt $Na_2[Hg(SO_3)_2]\cdot H_2O$ has been determined to have mercury–sulfur bonds by x-ray analysis (Nyberg and Cynkier, 1972).

Although the complex formed from the reaction of S(IV) with mercuric chloride is very stable with respect to electron transfer at room temperature and, therefore, appears anomalous when compared to other metal ion–S(IV) systems, the complex does undergo a redox reaction at higher temperatures, leading to mercurous chloride and sulfate (Divers and Shimidzu, 1886).

2.6. ESR Studies

Two studies of the radicals produced from metal ion-sulfur(IV) interactions by means of electron spin resonance (ESR) spectroscopy have been reported as well as an unsuccessful attempt to detect S(V) in the $Fe(CN)_6^{3-}$ + S(IV) reaction (Swinehart, 1967). Norman and Story (1971) found that the reaction of Ti(III) and S(IV) at pH 2 gave a product with a singlet in the ESR spectrum having a *g* value of 2.0058. This spectrum was assigned to the S(III) radical anion SO_2^- arising from the reaction

$$Ti(III) + HSO_3^- \rightarrow Ti(IV) + OH^- + SO_2^- . \qquad [33]$$

A spectrum consisting of a singlet with a *g* value of 2.0030 was found by mixing Ti(III), EDTA, hydrogen peroxide, and sodium sulfite. This spectrum was assigned to the S(V) radical anion SO_3^-. The suggested mechanism for the formation of this radical involved the production of hydroxyl radicals by the reaction of Ti(III) with H_2O_2 and the subsequent one-electron oxidation of SO_3^{2-} by the hydroxyl radical. These authors also found that this singlet at $g = 2.0030$ was obtained by the reaction of Ce(IV) with S(IV) at pH 2. Ozawa et al. (1971) also used the reaction of Ce(IV) with S(IV), but at pH 1, to obtain an ESR spectrum of S(V). A spectrum with a *g* value of 2.0022 and a linewidth of 2.4 G was obtained. The signal was observed to decay slowly, with a half-life of approximately 70 ms.

These results can be compared with ESR measurements of S(V) produced by photolysis of sulfite solutions (Chawla et al., 1973) and by reaction of peroxides with sulfite (Flockhart et al., 1971). In the former case a singlet with a *g* value of 2.00307 and a linewidth of 0.07 G was obtained, and in the latter case a *g* value of 2.0033 was obtained. These ESR results are of interest primarily in that they show that S(V) radicals can be monitored by this technique. It might also be possible to derive information on the products of S(V) reactions or on complexes formed in S(V) reactions with metal ions by use of ESR. Unfortunately, the two experiments in which metal ion production of S(V) was observed obtained different results for the *g* value, and in no case has there been a study of the reaction of S(V) with metal ions by use of ESR.

3. DISCUSSION

3.1. Mechanisms for Oxidation of S(IV) by Transition-Metal Ions

There are two fundamental classes of reaction mechanisms generally considered for the reactions of oxidized transition-metal ions with S(IV) species: (1) those that involve bimolecular reactions of complexes of either S(IV) with the oxidized transition-metal ion or S(V) with the reduced metal ion; and (2) those in which bimolecular reactions of the metal complexes are not important and free S(V) radicals are formed either by inner-sphere or outer-sphere electron transfer reactions.

The first type of mechanism arose from early studies on the reduction of ferric ions by S(IV). Meyer (1920) proposed that the production of dithionate was a consequence of a rearrangement of the ferric salt of the Fe(III)–S(IV) complex:

$$Fe^{3+} + Fe(SO_3)_3^{3-} \rightleftharpoons Fe[Fe(SO_3)_3] \qquad [34]$$

$$Fe[Fe(SO_3)_3] \rightarrow 2Fe^{2+} + SO_3^{2-} + S_2O_6^{2-}. \qquad [35]$$

Similar reactions were considered for manganese, cobalt, and ruthenium.

Bassett and co-workers (Bassett and Henry, 1935; Bassett and Parker, 1951) modified this mechanism to involve the reaction of two sulfite complexes; for example,

$$2Fe(SO_3)^+ \rightarrow 2Fe^{2+} + S_2O_6^{2-}. \qquad [36]$$

In addition, this interaction could lead to S(VI):

$$2Fe(SO_3)^+ \rightarrow 2Fe^{2+} + SO_3^{2-} + SO_3. \qquad [37]$$

Furthermore, it was felt that the increase in S(VI) production on addition of chloride ion could be explained by the increased likelihood of such reactions as

$$FeCl_2(SO_3)^- + FeCl_4^- \rightarrow 2Fe^{2+} + SO_3 + 6Cl^-. \qquad [38]$$

Support for a mechanism involving only metal ion–S(IV) complexes has been provided by Veprek-Siska and co-workers [see particularly Veprek-Siska et al. (1966b)] in their studies of the differences in product distribution between reactions of substitution-inert and substitution-labile complexes. They observed that substitution-inert metal ion complexes formed S(VI) products, whereas substitution-labile complexes formed mixtures of S(VI) and S(V) products. They proposed that sulfite substitutes into the substitution-inert complex to form a S-bonded sulfito complex (reaction [39]) and into substitution-labile complexes to form a bidentate O-bonded complex (reaction [40]):

$$MX_n^{m+} + SO_3^{2-} \rightarrow X_{n-1}M\text{—}SO_3^{(m-2)+} + X \qquad [39]$$

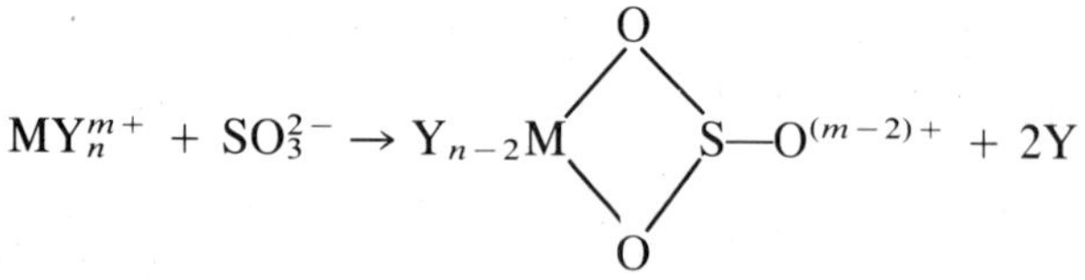

[40]

Here X and Y represent substitution-inert and substitution-labile ligands, respectively. The subsequent fate of the S-bonded complex is oxidation

$$X_{n-1}M\text{—}SO_3^{(m-2)+} + MX_n^{m+} \rightarrow X_{n-1}M\text{—}SO_3^{(m-1)+} + MX_n^{(m-1)+} \qquad [41]$$

and intramolecular electron transfer to form a sulfato complex. This complex is then labile to substitution by the original ligand X:

$$X_{n-1}M{-}SO_3^{(m-1)+} + X \rightarrow MX_n^{(m-1)+} + SO_3. \qquad [42]$$

As discussed in Section 2.1.2, this mechanism does not appear likely, at least for the reaction of $Fe(CN)_6^{3-}$. For substitution-labile complexes, Veprek-Siska et al. suggested that the O-bonded sulfito complexes disproportionate to the reduced metal ions and dithionate

$$2Y_{n-2}M \left\langle \begin{array}{c} O \\ \\ O \end{array} \right\rangle S{-}O^{(m-2)+} + 4Y \rightarrow 2MY_n^{(m-1)+} + S_2O_6^{2-} \qquad [43]$$

Although in many of the metal ion–sulfur (IV) reactions we have been discussing, sulfur(IV)–containing metal complexes have been observed, these complexes are labile with respect to substitution, and it is difficult to separate the metal complex formation equilibria from the electron-transfer steps. Since formation constants of these sulfite complexes are not well known, it is not possible to describe the distribution of metal species in solution with confidence. This then leads to a fundamental uncertainty in the relation of the rate equation derived from kinetic measurements and the actual mechanism. For example, complexes of Fe(III) with sulfite are observed in solution, and the formation equilibria appear to be established very fast with respect to the rate of the redox process (Basolo and Pearson, 1967; Carlyle, 1971; Danilczuk and Swinarski, 1961). In all cases where rapid equilibria are associated with a concurrent rate process, terms in the rate equation can, of course, be interpreted in alternative, indistinguishable ways. Thus the rate processes

$$Fe^{3+} + Fe(SO_3)_2^{2-} \rightarrow \text{products}, \qquad [44]$$

$$2FeSO_3 \rightarrow \text{products}, \qquad [45]$$

and

$$Fe^{3+} + FeSO_3^{+} + SO_3^{2-} \rightarrow \text{products} \qquad [46]$$

all would lead to the same rate dependence on S(IV) and Fe(III). The last, a termolecular process, would be regarded as unlikely but nevertheless is indistinguishable from the others.

By contrast, when the metal is Co(III) the complexes containing the S-bonded sulfito ligand are very stable. Cobalt (III), in general, forms complexes inert to substitution, with the notable exception of the ligand in the position trans to the sulfito ligand. In those intramolecular redox reactions involving oxidation of the sulfito ligand, it is clear that the electron-transfer process involves a metal complex and is an inner-sphere reaction. It seems possible that in other systems where metal-sulfite complexes can form, they may undergo similar intramolecular electron-transfer systems discussed above.

The fact that there are abundant examples of dithionate production in those S(IV) oxidations in which the metal ion undergoes substitition rapidly could also be interpreted to suggest that the S(V) intermediate is stabilized by way of a metal complex providing a pathway to dithionate formation having a low activation energy. The possible role of S(V) radicals coordinated to metal ions has been discussed by Brown and Higginson (1967) and Carlyle and Zeck (1973). However, there is no conclusive evidence on the formation of such complexes.

The second major type of mechanism, involving the production of free radicals, arose from the early work on the photooxidation of S(IV) solutions by O_2 (Bäckström, 1927; Baly and Bailey, 1922; Mason and Mathews, 1926). Bäckström showed that the oxygen reaction was a chain process with a chain length of 50,000 per quantum. Haber and co-workers (Frank and Haber, 1931; Haber, 1931; Haber and Wansbrough-Jones, 1932) interpreted these results as being due to the formation and subsequent reactions of a S(V) free radical

$$\mathrm{S(IV)} + h\nu \rightarrow \mathrm{S(V)} + e^-. \qquad [47]$$

The similarities between the light-induced and metal-catalyzed autoxidations, including the influence of inhibitors, prompted consideration of a similar mechanism for reactions where the S(V) radical is produced by the one-electron oxidation of S(IV) by a trace metal ion. For example,

$$\mathrm{Cu^{2+} + SO_3^{2-} \rightleftharpoons Cu^+ + SO_3^-}. \qquad [48]$$

Dithionate is then formed by the recombination of the S(V) radicals

$$\mathrm{2SO_3^- \rightarrow S_2O_6^{2-}}, \qquad [49]$$

and sulfate is formed in secondary reactions with oxygen,

$$\mathrm{HSO_3 + O_2 \rightarrow HSO_5} \qquad [50]$$

$$\mathrm{HSO_5 + SO_3^{2-} + H_2O \rightarrow 2SO_4^{2-} + OH + 2H^+}. \qquad [51]$$

The chain is continued by the hydroxyl radical

$$\mathrm{OH + SO_3^{2-} \rightarrow OH^- + SO_3^-}. \qquad [52]$$

It should be noted that Veprek-Siska and Lunak (1974, 1978) have argued against reaction [48], as the Cu(I) ion is known to form stable complexes. They find that the kinetics demand a rate constant impossibly high to account for the reaction of O_2 and SO_3^- in the copper-catalyzed autoxidation of S(IV).

One of the most important features of the free-radical mechanism is the ease with which it can explain the influence of added Fe(II) or Cu(II), as in the mechanism described by Higginson and Marshall (1957):

$$Fe(III) + S(IV) \rightarrow Fe(II) + S(V) \quad [53]$$

$$Fe(II) + S(V) \rightarrow Fe(III) + S(IV) \quad [54]$$

$$2S(V) \rightarrow S(V) \cdot S(V)(\text{dithionate}) \quad [55]$$

$$Fe(III) + S(V) \rightarrow Fe(II) + S(VI) \quad [56]$$

$$Cu(II) + S(V) \rightarrow Cu(I) + S(VI) \quad [57]$$

$$Cu(I) + Fe(III) \rightarrow Cu(II) + Fe(II). \quad [58]$$

It is this mechanism, with various modifications, that has been used to explain most of the kinetic results (Carlyle and Zeck,, 1973; Karraker, 1963; Pollard, et al., 1961). This mechanism predicts that the observed rate of loss of Fe(III), in the absence of Cu(II), will depend on the square of the Fe(III) concentration and will have a mixed dependence on S(IV) [see Eqs. (1) and (2)]. On the addition of sufficient Cu(II), the reaction should become first order in Fe(III) [see Eq. (3)]. The discussion of the free-radical mechanisms outlined above suggests that the mechanisms of autoxidation and for metal ion oxidations are essentially similar. It is important to note, however, that Anast and Margerum (1981) have demonstrated for the Cu(III) tetraglycine system essentially different reactions with and without added oxygen. Evidently there are steps in the two mechanisms that differ, are not coupled, and lead to an actual oscillating reaction. The radical mechanism outlined above for autoxidation and oxidation by metal ions must be regarded as probable and consistent with the main features of the rate studies but subject to further experimental tests.

In those reactions involving metal complexes that are inert to substitution, the electron-transfer step can be described as outer sphere, and those one-electron oxidants must lead to formation of intermediate S(V) free radicals. Almost all the reactions studied in this class produce S(VI) as the exclusive sulfur product.

The reaction of $Cu(III)(H_{-3}G_4)^-$ studied by Anast and Margerum (1981) is evidently an outer-sphere oxidation. In that reaction, however, no dithionate is formed. The explanation offered is that the concentration of S(V) radicals is very low because of the rapid reaction of SO_3^- with $Cu(III)(H_{-3}G_4)^-$ associated with the high oxidation potential (0.63 V) of the Cu(III) complex ion. On this basis they estimate a formal potential for the process

$$SO_3^- + e^- = SO_3^{2-} \quad [59]$$

of 0.89 V. By contrast, the Mn(III)–S(IV) system quantitatively produces dithionate ion despite the fact that it has a higher oxidation potential (E^0 = 1.5 V) (Diebler and Sutin, 1964).

Carlyle and Zeck (1973) suggested a mechanism consistent with experiments in the absence of Cu(II) in which an inner sphere Fe(III)–S(IV) complex undergoes electron transfer:

$$Fe^{3+} + H_2O = FeOH^{2+} + H^+ \quad \text{rapid equilibrium} \qquad [60]$$

$$Fe^{3+} + HSO_3^- = FeSO_3^+ + H^+ \quad \text{rapid equilibrium} \qquad [61]$$

$$FeSO_3^+ + H_2O = Fe^{2+} + HSO_4^{2-} + H^+ \quad \text{rapid equilibrium (in absence of copper)} \qquad [62]$$

$$Fe^{3+} + HSO_4^{2-} \rightarrow Fe^{2+} + HSO_4^- \qquad [63]$$

$$FeOH^{2+} + HSO_4^{2-} \rightleftharpoons Fe(OH)_2SO_3 \qquad [64]$$

$$Fe(OH)_2SO_3 + SO_2 \rightarrow Fe^{2+} + S_2O_6^{2-} + H_2O \qquad [65]$$

This mechanism features an Fe(III)–S(V) complex that leads to dithionate. It is also assumed that the S(V) free radical is completely hydrolyzed, forming HSO_4^{2-}. This mechanism is not adequate to account for the rate in solutions containing excess iron. They suggested that a process such as

$$Fe^{3+} + FeSO_3^+ + H_2O \rightarrow 2Fe^{2+} + SO_4^{2-} + 2H^+ \qquad [66]$$

becomes important when Fe(III) is in excess. This suggestion appears inconsistent with the experimental observation by Dasgupta et al. (1979) that the reaction is second order in the *limiting reagent,* i.e., S(IV), under these conditions.

In the presence of Cu(II) the following additional steps are suggested by Carlyle and Zeck (1973):

$$Cu^{2+} + HSO_4^{2-} \rightarrow Cu^+ + HSO_4^- \qquad [67]$$

and

$$Cu^+ + Fe^{3+} \rightarrow Cu^{2+} + Fe^{2+} \quad \text{fast.} \qquad [68]$$

This mechanism includes features of the prototype free-radical mechanism and also includes reasonable features of the complex ion chemistry. Notice that the formation of dithionate ion does not require free-radical chemistry, but rather the reaction of an Fe(III)–S(V) complex with SO_2, forming $S_2O_6^{2-}$. This suggestion is in concordance with the frequent examples of dithionate formation from labile oxidant complexes.

3.2. Suggestions for Further Work

Although there have been a large number of experimental studies on the transition-metal ion–S(IV) reaction and several mechanisms proposed to explain the results, the data available are not sufficient to make a definitive choice from among these mechanisms. There are two basic reasons for this. First, there have been few duplicate studies in different laboratories on the same system, and where there have been, the results are not in complete agreement. This appears to be related to the influence of impurities, including oxygen. Second, there have been almost no measurements on the rates and mechanisms of the elementary reactions that make up the proposed overall reaction mechanisms.

The effect of oxygen on the reaction is very important, both for the information it gives about the mechanism and because oxygen is a likely contaminant in an experiment. Yet, in those cases where the effect of oxygen on the rate has been investigated, the results have not always been consistent (see Section 2.1.1), leading even to an increase or decrease in the apparent rate of the Fe(III) reaction.

One of the most important types of experimental data we have discussed is the reaction stoichiometry, defined here as the ratio of reactants consumed. Related, but not exactly the same, are the relative amounts of sulfate and dithionate produced. It was the relationship of the product mix and the reaction stoichiometry to the substitution stability of the ligands on the metal complex that led Veprek-Siska and co-workers to propose a nonradical mechanism for the reaction and subsequently for the autoxidation and even the photoxidation of sulfite (Veprek-Siska and Lunak, 1978). Yet, in the few cases where stoichiometry measurements have been made in different laboratories, they do not always agree. For example, the hexacyanoferrate(III) reaction has been variously reported to have a stoichiometry of 1.32 (Higginson and Marshall, 1957) or 2 (Veprek-Siska et al., 1966b). The discrepancy in the hexachloroiridate(IV) reaction, 1.84 (Brown and Higginson, 1972) and 1.39 (Stapp and Carlyle, 1974), is somewhat less, but still disturbing. Anast and Margerum (1981) have found that traces of oxygen can lead to an error in the measured stoichiometry of the Cu(III)tetraglycine reaction. It is possible that other systems suffer from similar complications. It is apparent, then, that careful measurements of the stoichiometry of several reactions are needed. These experiments should be carried out under very well defined conditions, with oxygen excluded. The effect of trace impurities, including oxygen, should be investigated. In addition, these measurements should include both the ratio of the reactants consumed and the relative amounts of sulfate and dithionate produced, particularly if there is the possibility of forming a stable sulfite complex.

Although such studies of the reaction stoichiometry and further rate measurements are warranted, they would still not provide the key information needed to allow a reasonably unambiguous understanding of the mechanism. This information must come from studies of the chemistry of the proposed intermediates. The intermediate that warrants immediate attention is the S(V) radical.

There have been some published studies of S(V) reactions; some that were concerned with the detection of S(V) from metal ion–S(IV) systems by ESR were discussed earlier. Others have used flash photolysis or pulse radiolysis to generate S(V) radicals from various precursors and have used absorption spectrophotometry to monitor the S(V) concentration (Eriksen, 1974; Hayon, et al., 1972; Zagorski, et al., 1971). Although these studies have been extremely useful, particularly in relation to the behavior of the S(V) radical in the presence of O_2, some of the observations need confirmation. Eriksen (1974), using the change in electrical conductivity as a probe of product formation, determined that the self-reaction of SO_3^- involved not only recombination, but also disproportionation:

$$2SO_3^- \rightarrow S_2O_6^{2-} \qquad [69]$$

$$\rightarrow SO_3^{2-} + SO_3 \qquad [70]$$

$$SO_3 + H_2O \rightarrow 2H^+ + SO_4^{2-} \qquad [71]$$

The relative importance of the two processes appeared to be pH dependent, with sulfate formed at pH = 10 and a mixture of sulfate and dithionate formed at pH = 4. Obviously, these results should be confirmed, preferably with the direct determination of the reaction products. Such experiments could be carried out by steady-state photolysis or radiolysis. These experiments are particularly crucial since dithionate production seems to be more important in metal ion systems at high pH than at low pH.

Beyond the question of the reactions of S(V) with itself or with O_2 is the question of S(V) reactivity toward transition-metal ions. In several proposed mechanisms, either, or both, oxidation or reduction of a metal ion by S(V) is important. It should be possible to directly measure the rates of these reactions. The first priority should go to the Fe(II) and Cu(II) reactions

$$\text{Fe(II)} + \text{S(V)} \rightarrow \text{Fe(III)} + \text{S(IV)} \qquad [72]$$

$$\text{Cu(II)} + \text{S(V)} \rightarrow \text{Cu(I)} + \text{S(VI)}. \qquad [73]$$

These reactions could be studied by flash photolysis or pulse radiolysis following both S(V) and the metal ions spectrophotometrically. Of course, the effect of O_2 on these systems would be particularly relevant to an understanding of the autoxidation mechanism.

In addition to redox reactions, experiments should be carried out to investigate the possibility that S(V) forms somewhat stable complexes with transition-metal ions. If possible, the subsequent reactions of these complexes should be investigated.

If the rates and mechanisms of S(V) reactions are known, it would be possible to ascertain the viability of the proposed mechanisms for the sulfur(IV)–metal ion reactions. The use of computer modeling, in conjunction with studies of the temporal dependence of the reactants and products, would allow remaining gaps to be filled in. An extension of these studies to systems with O_2 present would then be tractable. We assert that until measurements are made on some of the key elementary reactions suggested as intermediates in S(IV) oxidation, the mechanisms cannot be verified and are not particularly useful for modeling flue gas scrubbers or the transformation of SO_2 in atmospheric droplets.

ACKNOWLEDGMENT

This work was supported in part by the Morgantown Energy Technology Center, U.S. Department of Energy under contract DE-A121-80MC14004.

REFERENCES

Albu, H. W. and von Schweinitz, H. D. (1932). Autooxidation V. Formation of dithionate by the oxidation of aqueous sulfite solutions. *Ber. Deutsch. Chem. Ges.* **B65,** 729–737.

Anast, J. M. and Margerum, D. W. (1981). Trivalent copper catalysis of the autoxidation of sulfite. Kinetics and mechanism of the copper(III/II) tetraglycine reactions with sulfite. *Inorg. Chem.* **20,** 2319–2326.

Bäckström, H. L. J. (1927). The chain theory of negative catalysis. *J. Am. Chem. Soc.* **49,** 1460–1472.

Baly, E. C. C. and Bailey, R. A. (1922). The equilibria in aqueous solutions of the alkali metal bisulphites. *J. Chem. Soc.* **121,** 1813–1821.

Basolo, F. and Pearson, R. G. (1967). Substitution reactions of octahedral complexes, in *Mechanisms of Inorganic Reactions. A Study of Metal Complexes in Solution.* Wiley, New York, Chapter 3, pp. 124–246.

Bassett, H. and Henry, A. J. (1935). The formation of dithionate by the oxidation of sulphurous acid and sulphites. *J. Chem. Soc.* 914–929.

Bassett, H. and Parker, W. G. (1951). The oxidation of sulphurous acid. *J. Chem. Soc.* 1540–1560.

Baubigny, M. H. (1912). Research on the process of formation of dithionic acid from the action of alkali sulfite on the salts of copper. *Comptes Rend.* **154,** 701–703.

Bretsznajder, S., Kotowska, W., and Piskorski, J. (1956). Reaction process between ferric sulfate and sulfur dioxide in solutions containing aluminium sulfate. *Rocz. Chem.* **30,** 411–430; *Chem. Abstr.* **50,** 16507 (1956).

Brown, A. and Higginson, W. C. E. (1972). The oxidation of hydrazine in aqueous solution by complex ions. *J. Chem. Soc. Dalton* 166–170.

Carlyle, D. W. (1971). A kinetic study of the aquation of sulfitoiron(III) ion. *Inorg. Chem.* **10,** 761–764.

Carlyle, D. W. (1972). Electron transfer between sulfur(IV) and tris (1,10-phenanthroline) iron(III) ion in aqueous solution. *J. Am. Chem. Soc.* **94,** 4525–4529.

Carlyle, D. W. and Zeck, O. F. (1973). Electron transfer between sulfur(IV) and hexaaquoiron(III) ion in aqueous perchlorate solution. Kinetics and mechanisms of uncatalyzed and copper(II)-catalyzed reactions. *Inorg. Chem.* **12,** 2978–2983.

Chawla, O. P., Arthur, N. L., and Fessenden, R. W. (1973). An electron spin resonance study of the photolysis of aqueous sulfite solutions. *J. Phys. Chem.* **77,** 772–776.

Czakis-Sulikowska, D. M. (1966). Examination of the mixed complexes of mercury(II). VIII. Spectrophotometric examination of the system: $Hg(ClO_4)_2$–NaI–Na_2SO_3–H_2O. *Rocz. Chem.* **40,** 1393–1400.

Danilczuk, E. and Swinarski, A. (1961). On the complexes $[Fe(III)(SO_3)_n]^{3-2n}$. *Rocz. Chem.* **35,** 1563–1572.

Dasgupta, P. K., Mitchell, P. A., and West, P. W. (1979). Study of transition metal ion–S(IV) systems. *Atmosph. Environ.* **13,** 775–782.

Dash, A. C., El-Awady, A. A., and Harris, G. M. (1981). Kinetics and mechanism of the reactions of sulfito complexes in aqueous solution. 3. Formation, acid-catalyzed decomposition, and intromolecular isomerization of oxygen-bonded (αβS)–(sulfito) (tetraethylenepentamine) cobalt(III) ion and the hydrolysis of its sulfur-bonded analogue. *Inorg. Chem.* **20,** 3160–3166.

Diebler, H. and Sutin, N. (1964). The kinetics of some oxidation–reduction reactions involving manganese(III). *J. Phys. Chem.* **68,** 174–180.

Divers, E. and Shimidzu, T. (1886). Mercury sulphites and the constitution of sulphites. *J. Chem. Soc.* **49,** 533–590.

Dorfman, Y. A., Rogoza, Z. I., and Kashnikova, L. V. (1977). Reduction of copper(II) by sulfur dioxide in presence of chloride and iodide ions. *Zhr. Prikl. Khim.* **50,** 1709–1714; English translation *J. Appl. Chem. USSR* **50,** 1639–1643.

Dorfman, Y. A., Rogoza, Z. I., and Kashnikova, L. V. (1978). Reduction of copper(II) by sulfur dioxide in presence of bromide and iodide ions. *Zh. Prikl. Khim.* **51,** 1457–1461; English translation *J. Appl. Chem. USSR* **51,** 1373–1482 (1978).

El-Awady, A. A. and Harris, G. M. (1981). Kinetics and mechanism of the reactions of sulfito complexes in aqueous solution. 4. Intramolecular electron-transfer and sulfito ligand addition reactions of aquo (sulfito-O) 2,2′,2″-triaminotriethylamine cobalt(III) ion. *Inorg. Chem.* **20,** 4251–4256.

Eriksen, T. E. (1974). pH effects on the pulse radiolysis of deoxygenated aqueous solutions of sulphur dioxide. *J. Chem. Soc. Faraday I,* **70,** 208–215.

Flockhart, B. D., Ivin, K. J., Pink, R. C., and Sharma, B. D. (1971). The nature of the radical intermediate in the reactions between hydroperoxides and sulphur dioxide and their reaction with alkene derivatives: Electron spin resonance study. *Chem. Commun.* 339–340.

Frank, J. and Haber, F. (1931). On the theory of the catalysis by heavy metal ions in aqueous solution and particularly to the autooxidation of sulfite. *Sitzungber. Preuss. Akad. Wiss. Phys. Math. Kl.* 250–256.

Graire, A. (1924). On the reaction of alkali bisulfites with mercuric chloride. *Comptes Rend.* **178,** 1819–1822.

Grodzovskii, M. K. (1935). I. Mechanism of the catalytic oxidation of sulfur dioxide in a solution of manganese salts. *Zh. Fiz. Khim.* **6,** 478–496; through *Chem. Abstr.* **30,** 2090 (1936).

Haber, F. (1931). Autooxidation. *Naturwissenschaften* **19,** 450–455.

Haber, F. and Wansbrough-Jones, O. H. (1932). The effect of light on oxygen free and oxygen containing sulfite solutions. *Z. Phys. Chem.* **B18,** 103–123.

Hansen, L. D., Eatough, D. J., Whiting, L., Barthlomew, C. H., Cluff, C. C., Izatt, R. M., and Christensen, J. J. (1974). Transition metal–SO_3^{2-} complexes. A postulated mechanism for the synergistic effects of aerosols and SO_2 on the respiratory tract, in *Trace Substances in Environmental Health,* Vol. 8, D. D. Hemphill, University of Missouri Press, Columbia, MO, pp. 393–397.

Hayon, E., Treinin, A., and Wilf, J. (1972). Electronic spectra, photochemistry and autoxidation mechanism of the sulfite–bisulfite–pyrosulfite systems. The SO_2^-, SO_3^-, SO_4^-, and SO_5^- radicals. *J. Am. Chem. Soc.* **94,** 47–57.

Higginson, W. C. E. and Marshall, J. W. (1957). Equivalence changes in oxidation–reduction reactions in solution. Some aspects of the oxidation of sulphurous acid. *J. Chem. Soc.* 447–458.

Hoather, R. C. and Goodeve, C. F. (1934). The oxidation of sulphurous acid III. Catalysis by manganous sulfate. *Transact. Faraday Soc.* **30,** 1149–1156.

Karraker, D. G. (1963). The kinetics of the reaction between sulfurous acid and ferric ion. *J. Phys. Chem.* **67,** 871–874.

Keller, R. N. and Wycoff, H. D. (1946). Copper(I) chloride, in *Inorganic Synthesis,* Vol. II, W. C. Fernelius, L. F. Audrieth, J. C. Bailar, H. S. Booth, W. C. Johnson, R. E. Kirk, and W. C. Schumb, Eds., McGraw-Hill, New York, pp. 1–4.

Kuz'minykh, I. N. and Bomshtein, T. B. (1953). Effect of copper on the reaction between sulfur dioxide and ferric sulfate in aqueous solution. *Zh. Priklad. Khim.* **26,** 3–8; through *Chem. Abstr.* **47,** 5832.

Lancaster, J. M. and Murray, R. S. (1971). The ferricyanide–sulphite reaction. *J. Chem. Soc. A,* 2755–2758.

Mason, R. B. and Mathews, J. H. (1926). The effect of ultra-violet light on the oxidation of sodium sulfite by atmospheric oxygen. *J. Phys. Chem.* **30,** 414–420.

Meyer, J. (1920). Oxidation of sulfurous acid by ferric salts. *Ber. Deutsch Chem. Ges.* **B53,** 77–78.

Murray, R. S. (1974). Reinvestigation of the reaction between hexacyanoferrate(III) and sulphite ions. *J. Chem. Soc. Dalton Transact.* 2381–2383.

Murray, R. S. and Stranks, D. R. (1970). Internal oxidation–reduction of trans-$Co(en)_2SO_3OH_2^+$. *Inorg. Chem.* **9,** 1472–1475.

Newman, G. and Powell, D. B. (1963). The infra-red spectra and structures of metal-sulphite compounds. *Spectrochem. Acta* **19,** 213–224.

Norman, R. O. C. and Storey, P. M. (1971). Electron spin resonance studies. Part XXXI. The generation, and some reactions, of the radicals SO_3^-, $S_2O_3^-$, S^-, and SH in aqueous solution. *J. Chem. Soc.* **B,** 1009–1013.

Nyberg, B. and Cynkier, I. (1972). On the crystal structure of $Na_2[Hg(SO_3)_2] \cdot H_2O$. *Acta Chem. Scand.* **26,** 4175–4176.

Okabe, T., Owaku, M., and Hori, S. (1960). The autoxidation of sulfites and related salts in aqueous solution. V. The autoxidation of sulfurous acid. *Nippon Kagaku Zasshi* **81,** 1818–1824; through *Chem. Abstr.* **56,** 9697 (1960).

Ozawa, T., Setaka, M., and Kwan, T. (1971). ESR studies of the sulfite radical ion. *Bull. Chem. Soc. Jap.* **44,** 3473–3474.

Parker, V. B., Staples, B. R., Jobe, T. L., and Neumann, D. B. (1981). Report on some thermodynamic data for desulfurization processes, NBSIR report No. 81-2345.

Pinnow, J. (1923). Reaction between ferric salt and sulfurous acid and its catalysis. *Z. Electrochem.* **29,** 547–552.

Pollard, F. H., Hanson, P., and Nickless, G. (1961). Chromatographic studies on the oxidation of sulphurous acid by ferric ion in aqueous acid solution. *J. Chromatogr.* **5,** 68–73.

Reinders, W. and Vles, S. I. (1925). Reaction velocity of oxygen with solutions of some inorganic salts. III. The catalytic oxidations of sulphites. *Rec. Trav. Chim. Pays-Bas* **44,** 249–268.

Richter, D. (1931). The action of ferrous iron in induced reactions. *Ber. Deustch Chem. Ges* **B64,** 1240–1243.

Ryan, R. R., Kubas, G. J., Moody, D. C., and Eller, P. G. (1981). Structure and bonding of transition metal–sulfur dioxide complexes, in *Structure and Bonding,* Vol. 46, Springer-Verlag, Berlin, pp. 47–100.

Sillén, L. G. and Martell, A. E. (1964). *Stability Constants of Metal-Ion Complexes,* The Chemical Society (London), Special Publication No. 17, pp. 229–232.

Sillén, L. G. and Martell, A. E. (1971). *Stability Constants of Metal-Ion Complexes,* The Chemical Society (London), Suppl. No. 1, Special Publication No. 25, pp. 133, 134.

Siskos, P. A., Peterson, N. C., and Huie, R. E. (1982). The kinetics of the reaction between manganese(III) and sulfur dioxide in aqueous perchlorate solution (manuscript in preparation).

Smith, R. M. and Martell, A. E. (1976). *Critical Stability Constants. Inorganic Complexes,* Vol. 4, Plenum Press, New York, p. 78.

Stapp, E. L. and Carlyle, D. W. (1974). Reduction of hexachloroiridate(IV) ion by sulfur(IV) in acidic aqueous solution. *Inorg. Chem.* **13,** 834–837.

Swinehart, J. H. (1967). The kinetics of the hexacyanoferrate(III)–sulphite reaction. *J. Inorg. Nucl. Chem.* **27,** 2313–2320.

Thacker, M. A., Scott, K. L., Simpson, M. E., Murray, R. S., and Higginson, W. C. E. (1974). Redox decomposition of *trans*-tetraammineaquosulphitocobalt(III) in aqueous solution. *J. Chem. Soc. Dalton* 647–651.

Titoff, A. (1903). Contribution to the knowledge of negative catalysis in a homogeneous system. *Zeit. Phys. Chem. (Leipzig)* **45,** 641–683.

van Eldik, R. and Harris, G. M. (1980). Kinetics and mechanism of the formation, acid-catalyzed decomposition, and intramolecular redox reaction of oxygen-bonded (sulfito) pentaamminecobalt(III) ions in aqueous solution. *Inorg. Chem.* **19,** 880–886.

Veprek-Siska, J. and Hasnedl, A. (1968). Specific effects of electrolytes on the oxidation of sulphite by hexacyanoferrate(III). *Chem. Commun.* 1167–1168.

Veprek-Siska, J. and Lunak, S. (1974). The role of copper ions in copper catalyzed autooxidation of sulfite. *Z. Naturforsch.* **B29,** 689–690.

Veprek-Siska, J. and Lunak, S. (1978). Photocatalytic effects of trace metals. Evidence against a free radical chain mechanism in sulfite autoxidation. *React. Kinet. Cat. Lett.* **8,** 483–487.

Veprek-Siska, J. and Wagnerova, D. M. (1965). One equivalent oxidations. I. Oxidation of sulfite by means of hexacyanoferrate(III). *Collect. Czech. Chem. Commun.* **30,** 1390–1401.

Veprek-Siska, J., Solcova, A., and Wagnerova, D. M. (1966a). One equivalent oxidation. III. Kinetics of the oxidation of sulfite by pentacyanoferrate(III). *Collect. Czech. Chem. Commun.* **31,** 3287–3298.

Veprek-Siska, J., Wagnerova, D. M., and Eckschlanger, K. (1966b). One equivalent oxidation of sulfite by complex ions (II). *Collect. Czech. Chem. Commun.* **31,** 1248–1255.

West, P. W. and Gaeke, G. C. (1956). Fixation of sulfur dioxide as disulfitomercurate(II) and subsequent colorimetric estimation. *Anal. Chem.* **28,** 1816–1819.

Westley, F. (1982). Oxidation of sulfite ion by oxygen in aqueous solution—a bibliography. NBS Special Publication 630, U.S. Government Printing Office, Washington, D.C.

Wiberg, K. B., Maltz, H., and Okano, M. (1965). Mechanism of the ferricyanide oxidation of thiols. *Inorg. Chem.* **7,** 830–831.

Yost, N. M. and Russell, H. (1944). *Systematic Inorganic Chemistry,* Prentice-Hall, New York, pp. 359–360.

Zagorski, Z. P., Sehested, K., and Nielsen, S. O. (1971). Pulse radiolysis of aqueous alkaline sulfite solutions. *J. Phys. Chem.* **75,** 3510–3517.

Zeck, O. F. and Carlyle, D. W. (1974). Electron transfer between sulfur(IV) and chloroiron(III) and chlorocopper(II) ions in aqueous chloride media. Formation of a sulfur dioxide complex with chlorocopper(I). *Inorg. Chem.* **13,** 34–38.

3

CATALYTIC AUTOXIDATION OF AQUEOUS SULFUR DIOXIDE IN RELATIONSHIP TO ATMOSPHERIC SYSTEMS

Michael R. Hoffmann and Scott D. Boyce

Environmental Engineering Science
W. M. Keck Laboratories
California Institute of Technology
Pasadena, California 91125

1. INTRODUCTION

1.1. General Considerations

The fate of SO_2 and NO_x in the atmosphere has been a perplexing problem confronting scientists for more than 100 years. Moreover, with the advent of increased coal utilization on the horizon, concern about the ultimate impact of increased SO_2 and NO_x discharge and their inevitable transformation to acidic sulfates and nitrates on the world ecosystem has grown demonstrably (Brezonik et al., 1980; Liljestrand and Morgan, 1978, 1981; Pack, 1980).

Scientists have been searching for answers to this perplexing problem since the precocious identification of the phenomenon known as "acid rain" by the English climatologist, Robert Angus Smith, in 1872. Dr. Smith, while serving as the General Inspector of the Alkali Works for the Government (England), established an extensive precipitation chemistry network around Great Britain in the 1860s. Results of this study were published in a landmark book, *Air and Rain: The Beginnings of a Chemical Climatology* (Smith, 1872). Some of the general conclusions made by Smith in 1872 were as follows:

> Rain contains sulphates in larger proportion to the chlorides than is found in sea-water.
>
> Sulphates increase inland before large towns are reached.
>
> Sulphates rise very high in large towns because of the amount of sulphur in the coal used as well as of decomposition.
>
> When the sulphuric acid increases more rapidly than the ammonia the rain becomes more acid.
>
> Free acids are found with certainty where combustion or manufacturers are the cause.
>
> When the air has so much acid that two to three grains are found in a

gallon of rain water, or forty parts in a million, there is no hope for vegetation.

As a rule rain is not acid far from towns. If it is, artificial circumstances must be suspected.

In the intervening 108 years our general views on the nature and origin of acid rain have not been altered significantly, although the problem was generally ignored prior to its simultaneous rediscovery by Barrett and Brodin (1955) in Fenno-Scandia, Gorham (1955) in the English Lake District, and Houghton (1955) in fog and cloud water in New England.

Fundamental questions about the multifarious pathways for transformation of SO_2 and NO_x in the atmosphere remain to be answered before a complete mechanistic and dynamic description of the complex SO_2 and NO_x reaction network can be provided. It is clear today that SO_2 and NO_x can be oxidized to sulfate and nitrate aerosols in the gas phase or by reaction in liquid water containing aerosol droplets. The latter reactions may be homogeneous (i.e., involving only dissolved components) or heterogeneous (i.e., involving reaction with or on solids suspended within the liquid aerosol droplets).

1.2. Field-Oriented Studies

Field studies indicate that the relative contribution of gas-phase and aqueous-phase pathways for SO_2 oxidation varies, depending on a variety of climatological conditions such as relative humidity and the intensity of incident solar radiation. In recent years the role of aqueous conversion mechanisms has received increased attention.

Until recently, most discussions of SO_2 oxidation in the atmosphere focused primarily on homogeneous gas-phase oxidation by free radicals OH·, HO_2·, and RO_2· (Calvert et al., 1978). This mechanism was supported by the relatively low SO_2 conversion rates observed at night in studies of power plant plumes. However, more recent studies (Burrows et al., 1979; Graham et al., 1979) have indicated that the rates of oxidation of sulfur dioxide by HO_2· and by organic peroxides are slower than previously thought, leaving OH· reaction the only apparent route for gas-phase oxidation. However, the oxidation rate of sulfur dioxide by the hydroxyl radical is apparently not sufficiently rapid to entirely account for observed ambient sulfate formation rates (Middleton et al., 1980; Möller, 1980). This is especially true in cases where rapid oxidation of sulfur dioxide has been observed when the humidity is high. For these reasons, it is now believed that condensed-phase, homogeneous, and heterogeneous reactions may be predominantly responsible for SO_2 oxidation in clouds (Hegg and Hobbs, 1981) or in conditions of high humidity. Additional field data show the formation of H_2SO_4 and HNO_3 at night, indicating oxidation processes other than reaction with OH· (Cass and Shair, 1980). Newman (1981) arrived at similar conclusions about the relative contributions of gas-phase and condensed-phase pathways for the oxidation of SO_2 in power plant and smelter plumes.

Additional evidence for the role of aqueous-phase oxidation pathways for the conversion of SO_2 to sulfate has been reported by Cass and Shair (1980), Smith and Jeffrey (1975), Cox (1974), Gartrell et al. (1963), Wilson and McMurry (1981), and McMurry et al. (1981). Most notably, Smith and Jeffrey (1975) and Cass (1977) found that the rate of aerosol sulfate formation apparently increased with increasing relative humidities, where submicron aerosol particles are typically in droplet rather than crystalline form. Wilson and McMurry (1981) observed that particle growth rates were proportional to particle diameter for aerosol formation in the Great Smoky Mountains under humid conditions. They interpreted these results to be indicative of dominance of aerosol growth by aqueous-phase oxidation. Cass and Shair (1980) have reported nighttime conversion rates (5.8% h^{-1}) for SO_2 in the Los Angeles sea breeze–land breeze circulation system to be statistically indistinguishable from typical daytime conversion rates (5.7% hr^{-1}) for the month of July in the Los Angeles Basin.

1.3. Laboratory Studies

Laboratory studies have focused on factors affecting the bulk aqueous-phase autoxidation rate of $SO_2 \cdot H_2O$ such as photoassisted catalysis, trace metal catalysis, and pH. Unfortunately, the numerous studies have provided conflicting experimental results on reaction rates, rate laws, and, in particular, pH dependencies. Other experimental work has focused on the possible roles of O_3 and H_2O_2 as aqueous-phase oxidants (Larson et al., 1978; Martin and Damschen, 1981; Penkett et al., 1979).

Hydrogen peroxide has been suggested by Middleton et al. (1980) as a principal oxidant of S(IV) in aqueous aerosols based on theoretical estimates of the rates of alternative reaction pathways. Kok (1978a, 1978b, 1980) and Campbell et al. (1979) have reported that significant concentrations of H_2O_2 and related radical species $HO_2\cdot$ ($2HO_2\cdot \rightarrow H_2O_2 + O_2$) and $OH\cdot$ ($2OH\cdot \rightarrow H_2O_2$) are present in urban atmospheres and plumes. Using the experimental results of Penkett et al. (1979) on the rate of oxidation of S(IV) by H_2O_2, Middleton and co-workers concluded that under daytime conditions aqueous-phase oxidation by H_2O_2 can be a primary contributor to sulfate aerosol production and that under nighttime conditions, without photochemical reactions, catalytic and noncatalytic oxidation on wetted aerosols become important pathways. Similar computations have been made by Möller (1980), who has concluded that aqueous-phase oxidation pathways involving metal catalyzed autoxidations, ozone, and hydrogen peroxide are the most important atmospheric reactions in the production of SO_4^{2-} from SO_2. Using a simplified reaction network model based on literature values for rate constants and rate laws, Möller calculated that SO_2 was removed from the atmosphere by gas-phase oxidation (9%), aqueous-phase oxidation (35%), dry deposition (45%), and wet deposition (11%).

On the basis of laboratory-determined rates, catalytic autoxidation of SO_2 by transition metals dissolved in submicron water aerosols has been suggested (Beilke and Gravenhorst, 1978; Cheng et al., 1971; Dasgupta et al., 1979; Freiberg, 1979; Fuzzi, 1978; Hegg and Hobbs, 1978; Kaplan et al., 1981; Larson et al., 1978) as a potentially important non-photolytic pathway for the rapid formation of sulfuric acid in relatively humid atmospheres.

In view of the likelihood of transition-metal catalyzed oxidation of S(IV) as an important process in aqueous atmospheric aerosols, we undertake a review of this process. Emphasis is given to mechanisms and kinetics of reaction. A review with emphasis on complexation of S(IV) with transition-metal ions is given by Huie and Peterson in Chapter 2 of this volume.

2. CATALYTIC OXIDATION PROCESSES IN THE AQUEOUS PHASE

2.1. Trace-Metal Catalysis

Historically, trace-metal catalysis was an inexplicable nuisance phenomenon for the classical solution-phase kineticist. Impurities in solvents and reagents frequently resulted in nonreproducible rates of chemical reactions and unusual empirical rate laws. Trace-metal concentrations in the range of 0.1–10 nM have been shown to influence the rates of some peroxide reactions (Edwards, 1965). For example, the decomposition of peroxymonosulfuric acid, a suggested intermediate (Larson et al., 1978) in the autoxidation of SO_2 dissolved in aqueous aerosols, is accelerated significantly by nanomolar amounts of some divalent first-row transition metals (Ball and Edwards, 1958). Similarly, the ferrous ion promotes the oxidation of a wide variety of aromatic substrates by hydrogen peroxide (Walling, 1975). Other chemical reactions involving oxidation–reduction, hydrolysis, decarboxylation, transamination, and bromination have been shown to be sensitive to trace-metal catalysis (Hoffmann, 1980).

Historically, the word "catalyst" has been defined in many ways. Ostwald (1902) originally defined a catalyst as a substance that changes the speed of a chemical reaction without undergoing a chemical change itself. Bredig (1909) expanded the definition of a catalyst to include substances that may be chemically altered but are not involved in a whole-number stoichiometric relationship with reactants and products. Bell (1971) subsequently defined a catalyst for a homogeneous reaction to be a substance that appears in the rate expression with a reaction order greater than its stoichiometric coefficient. Another common notion is that a catalyst lowers the activation energy of a given reaction. The catalyst in a homogeneous or heterogeneous system will alter the reaction mechanism to one having a lower activation energy without changing the point of equilibrium for a reaction as predicted by the law of microscopic reversibility (Edwards, 1965; Frost and Pearson, 1961). Since the catalyst is involved in the rate-determining step and consequently in the molecular composition of the activated complex (King, 1955), a discussion of catalytic activity requires consideration of detailed reaction mechanisms. Therefore, the role of transition-metal catalysis in the autoxidation of SO_2 is discussed in terms of postulated reaction mechanisms that are consistent with kinetic observations.

Transition metals that have several stable oxidation states are frequently active as catalysts for electron transfer or redox reactions. Catalysis in many redox reactions results from a special combination of relative rates when an active metal exists in more than one oxidation state (King, 1955). The catalytically active metal can be oxidized and reduced cyclically at a rapid rate with a concomitant catalytic influence on a slower reaction. A general characteristic of trace-metal catalysis is the occurrence of nonreproducible reaction rates and unusual empirical rate laws.

The term "autoxidation" refers to the general reactions of oxidizable materials with molecular oxygen although earlier definitions were more restrictive. In general, reactions of triplet ground-state oxygen with singlet-spin-state reductants are slow in the absence of catalytic influences because they involve a change in spin multiplicity and because a significant amount of bond deformation or alteration of a permanent nature may occur. Autoxidations are often accelerated by trace metals such as Cu(II), Co(II), Fe(II), Mn(II), Ni(II), and many of their coordination complexes.

The autoxidation of H_2S and HS^- in both fresh and seawater systems has been shown to be sensitive to trace-metal catalysis by Mn(II), Cu(II), Fe(II), Ni(II), and Co(II) (Chen and Morris, 1972; Krebs, 1929). The kinetics of the autoxidation of sulfide has been studied by a number of investigators under slightly different experimental conditions. These results have been summarized by Hoffmann (1980). From a comparison of the reported half-lives as measured by different investigators, it is apparent that the reaction is sensitive to a variety of catalytic influences in addition to trace metals. These include general base, microbial, and surface catalysis. The reported half-lives varied from days to minutes under similar concentration conditions, and the observed kinetic orders varied from one investigator to another.

2.2. Interrelationship Between Inorganic and Organic Oxidations in Aerosols

Given the complicated chemical composition of aqueous atmospheric aerosols, interactions between the primary inorganic and organic constituents may be inevitable. For example, the presence of trace metals and the occurrence of H_2O_2, $OH\cdot$, and $HO_2\cdot$ may be interrelated phenomena. This notion can be illustrated by the following mechanistic sequence suggested by Gray (1969) and other investigators (Fieser and Fieser, 1967, 1969; Hewitt, 1970) to explain the copper-catalyzed oxidation of ascorbic acid and phenols where Cu(I) is involved in the activation of dioxygen (Basolo et al., 1975) by direct inner-sphere complexation:

$$Cu^{+} + O_2 \overset{\beta_1}{\rightleftharpoons} CuO_2^{+}, \qquad [1]$$

$$CuO_2^{+} + H^{+} \overset{k_1}{\rightarrow} Cu^{+2} + HO_2\cdot, \qquad [2]$$

$$Cu^{+} + HO_2\cdot \overset{k_2}{\rightarrow} Cu^{+2} + HO_2^{-}, \qquad [3]$$

$$H^{+} + HO_2^{-} \xrightleftharpoons{1/K_{a1}} H_2O_2. \qquad [4]$$

Either $HO_2\cdot$, H_2O_2, or both produced by this sequence of reactions might be responsible for the subsequent oxidation reactions.

In aqueous-phase atmospheric systems, Cu(I) could be generated by photoinduced reduction of Cu(II) (Langford and Carey, 1975; Langford et al., 1973):

$$Cu^{2+} \overset{h\nu}{\rightarrow} {}^{*}Cu^{2+} \overset{H_2O}{\rightarrow} Cu^{+} + OH\cdot + H^{+}. \qquad [5]$$

Similarly, Co(I) and Fe(II) could be generated by photoinduced reduction of Co(II) and Fe(III).

Certain trace metals such as Fe(II)–Fe(III) can play a dual role in both catalyzing the production of H_2O_2 by a sequence of reactions such as that above and catalyzing the subsequent decomposition of H_2O_2 to produce $OH\cdot$ and $HO_2\cdot$. This hypothetical sequence of reactions could proceed according to the classic "Fenton's reagent" reaction (Walling, 1975) for the decomposition of H_2O_2, which evidently proceeds by chain reaction as follows:

Initiation

$$Fe^{2+} + H_2O_2 \rightarrow Fe(OH)^{2+} + OH\cdot \qquad [6]$$

or

$$Fe^{3+} + H_2O_2 \rightarrow Fe^{2+} + H^{+} + HO_2\cdot \qquad [7]$$

Propagation

$$OH\cdot + H_2O_2 \rightarrow H_2O + HO_2\cdot \qquad [8]$$

$$HO_2\cdot + H_2O_2 \rightarrow O_2 + H_2O + OH\cdot \qquad [9]$$

Termination

$$Fe^{3+} + HO_2\cdot \rightarrow Fe^{2+} + H^{+} + O_2 \qquad [10]$$

If the sequence of reactions [1]–[5] and/or [6]–[10] were to take place in atmospheric aerosols containing dissolved S(IV), the oxidizing species might well play a role in the oxidation of S(IV).

An alternative pathway for the formation of aqueous free radicals is the catalytic autoxidation of organic species. A transition metal or transition-metal complex may catalyze the autoxidation of trace-level hydrocarbons or other organics present in aqueous aerosols by a radical chain reaction in which dioxygen is directly activated by complexation with a transition metal. Complexation of molecular oxygen with a transition metal lowers the activation energy for a direct reaction with a substrate (Sheldon and Kochi, 1973). For a generalized hydrocarbon RH, a set of possible reactions would be as follows:

$$M^{2+} + O_2 \rightleftharpoons M^{3+}\text{—}O_2^{-}\cdot \quad [11]$$

$$M^{3+}\text{—}O_2^{-}\cdot + RH \rightarrow M^{2+} + R\cdot + HO_2\cdot \quad [12]$$

$$M^{3+}\text{—}O_2^{-}\cdot + RH \rightarrow M^{3+}\text{—}O_2^{-}H + R\cdot \quad [13]$$

$$M^{3+}\text{—}O_2^{-}\cdot + RH \rightarrow M^{3+} + R^{-} + HO_2\cdot \quad [14]$$

$$M^{3+}\text{—}O_2^{-}\cdot + RH \rightarrow M^{3+} + R\cdot + HO_2^{-} \quad [15]$$

$$HO_2^{-} + H^{+} \rightleftharpoons H_2O_2. \quad [16]$$

In commercial autoxidations M^{2+} is frequently Co(II) or an associated organometallic complex. For example, Uri (1956) proposed a mechanism for Co(II) stearate catalysis of the autoxidation of methyl linoleate that is consistent with the generalized sequence given above:

$$Co(II) + O_2 \overset{\beta_1}{\rightleftharpoons} Co(II)\text{—}O_2 \quad [17]$$

$$Co(II) + RH \overset{\beta_2}{\rightleftharpoons} Co(II)RH \quad [18]$$

$$Co(II)O_2 + Co(II)RH \xrightarrow{k_1} Co(II) + Co(III)R + HO_2\cdot \quad [19]$$

$$Co(II)O_2\cdot RH \xrightarrow{k_2} Co(III)R + HO_2\cdot \quad [20]$$

Similar catalytic mechanisms have been proposed for the autoxidation of benzylthiol (Dance et al., 1974), cumene and acrolein (Hara et al., 1975), 2-mercaptoethanol (Mass et al., 1976), 2,6-di-*t*-butyl phenol (Kothari and Tazuma, 1976), cysteine, thioglycollic acid and butyl mercaptan (Kundo and Kejer, 1968; Kundo et al., 1967; Simonov et al., 1973) catalyzed by Co(II) complexes.

Even though the above reactions are plausible under controlled laboratory conditions, the question as to their applicability in aerosol reactions remains to be answered. In this regard, however, we note that oxidizable organic species that could participate in reactions such as those indicated above have been identified in rain and in atmospheric aerosol samples. Lunde et al. (1977) have identified a wide range of organic micropollutants in rain over Norway. They were able to identify significant numbers of alkanes, polycyclic aromatic hydrocarbons, phthalic acid esters, fatty acids and fatty acid esters, substituted phenols, and benzoic acids. Similar identifications have been reported by Likens et al. (1982) for rainwater in the United States, and by Knights (1980) for aerosol samples obtained in Los Angeles.

Once generated *in situ*, H_2O_2 would be expected to react rapidly with HSO_3^- to produce HSO_4^- (Mader, 1958). Reactions of reduced sulfur compounds such as SO_2, H_2S, and $(R_2N)_2CS$ with H_2O_2 are classified as fast reactions in aqueous systems (Hoffmann, 1977; Hoffmann and Edwards, 1975, 1977).

Hoffmann and Edwards (1975) reported the following two-term rate law for the oxidation of HSO_3^- by H_2O_2 over the pH range of 4–8:

$$-\frac{d[\mathrm{S(IV)}]}{dt} = [\mathrm{S(IV)}][\mathrm{H_2O_2}]\frac{[\mathrm{H^+}]}{[\mathrm{H^+}] + K_{a2}}(k[\mathrm{H^+}] + k'[\mathrm{HA}]) \qquad (1)$$

where $K_{a2} = [SO_3^{2-}][H^+]/[HSO_3^-]$, $[S(IV)] = [HSO_3^-] + [SO_3^{2-}]$, and HA represents a weak acid. The second term in the rate law is an example of general acid catalysis (Bell, 1971). A mechanism consistent with this rate law was proposed:

$$\mathrm{H^+ + SO_3^{2-} \underset{}{\overset{1/K_{a2}}{\rightleftharpoons}} HSO_3^-} \qquad [21]$$

$$\mathrm{HSO_3^- + H_2O_2 \overset{K}{\rightleftharpoons} {}^-O_2SO{-}OH + H_2O} \qquad [22]$$

$$\mathrm{{}^-O_2S{-}O{-}OH + HA \overset{k}{\rightarrow} H_2SO_4 + A^-} \qquad [23]$$

where HA in reaction [23] represents either H^+ or a suitable weak acid. According to this mechanism, the reaction occurs by means of nucleophilic displacement by H_2O_2 on HSO_3^- to form a peroxomonosulfurous acid intermediate, $^-O_2SOOH$, which then undergoes a rate-determining rearrangement to HSO_4^- assisted by H_3O^+ or HA. The rate constants k and k' have been calculated to be 2.7×10^8 and $1.3 \times 10^3\ M^{-2}\ s^{-1}$ at pH 6.4 and 12.0°C. These results were later confirmed by the work of Penkett et al. (1979) down to pH 4.5, although they pointed out that at low pH the first term in the rate law involving specific acid catalysis should dominate the second term involving general acid catalysis by any suitable proton donor. These results were confirmed also by the work of Martin and Damschen (1981), who extended the pH range down to 1.0. They concluded that their results verified the validity of the Hoffmann–Edwards (1975) mechanism over a much broader range of pH.

3. CATALYTIC AUTOXIDATION OF AQUATED SULFUR DIOXIDE

3.1. Introduction

Numerous attempts have been made to characterize the kinetics and mechanisms of the metal-catalyzed autoxidation of sulfur dioxide. However, studies of the reaction of aquated sulfur dioxide as $SO_2 \cdot H_2O$, HSO_3^- or SO_3^{2-} and oxygen have been characterized by inconsistent reaction rates, rate laws, and pH dependences (Beilke and Gravenhorst, 1978; Hegg and Hobbs, 1978; Hoffmann, 1980). In this section postulated mechanisms and their resulting theoretical rate expressions are examined critically. In general, the hypothetical mechanisms have been broken down into three categories; free-radical chain mechanisms involving a sequence of one-electron transfer steps following a thermal initiation, polar mechanisms involving complexation and two-electron transfer steps, and photoassisted mechanisms. These mechanisms and their rate expressions and the empirical rate laws are compared for consistency.

In the past the kinetics and mechanisms of the catalytic autoxidation of sulfite have been the focus of attention of chemical and civil engineers interested in mass-transfer characteristics of aeration devices, of chemical kineticists interested in photoinduced chain reactions and their metal-catalyzed counterparts, and of environmental chemists and engineers interested in atmospheric conversion processes. Consequently, the methods and conditions employed in the study of this fundamental reaction have varied considerably. This lack of uniformity should be kept in mind when making cross-comparisons. Finally, we note that results obtained by these diverse interest groups have seldom been cross-referenced in their respective literatures.

In the studies to be discussed the reduction of oxygen proceeds according to the following stoichiometry:

$$SO_2 \cdot H_2O + \tfrac{1}{2}O_2 \rightarrow H_2SO_4 \qquad [24]$$

Here and generally throughout the discussion we do not distinguish between the several S(IV) and S(VI) species. The equilibria governing these distributions are quite rapid, and hence these distributions are governed simply by solution pH. In any rigorous analysis the contribution of the several S(IV) species, $SO_2 \cdot H_2O$, HSO_3^-, and SO_3^{2-} must be evaluated.

3.2. Free-Radical Reactions

Redox reactions sensitive to homogeneous trace-metal catalysis exhibit reaction rate laws that are often first order in one reactant and zero order in the other. The autoxidations of SO_2 and H_2S in aqueous solution fall into this general category of redox reactions. For example, oxygen concentrations in water are lowered to near zero levels in mass-transfer studies of aeration equipment by the addition of SO_3^{2-} and an appropriately soluble Co(II) salt. In one experimental study of the catalytic effect of hexaquo Co(II) (Chen and Barron, 1972) a nonintegral rate law with a zero-order dependence on oxygen was observed:

$$\frac{-d[SO_2 \cdot H_2O]}{dt} = \nu_M = k_M[\mathrm{Co(II)}]^{1/2}[SO_3^{2-}]^{3/2}. \qquad (2)$$

Numerous other studies of reaction [24] catalyzed by transition-metal cations have been reported. The rate laws derived from these studies are summarized in Table 1. These rate laws are compared below to predictions based on assumed mechanisms. However, some generalization may be made. First, we would note the considerable lack of agreement among these investigators as to the order of reaction in the several species. Most often the various investigators concur that the reaction order in oxygen is zero. However, it should be pointed out that an oxygen dependence in many cases was overlooked by using pseudo-order conditions in oxygen and that in other cases mass-transfer limitations may have resulted in an apparent zero-order oxygen dependence. The majority of investigators report a first-order metal-ion dependence. Finally, a caveat to this comparison is that few investigators used similar analytical and kinetic methodologies and even fewer used identical concentration and pH ranges. As a result, the lack of agreement is somewhat understandable and may suggest that parallel mechanistic pathways are followed that will result in complicated multiterm rate laws. In addition, in the above analysis the role of pH and the speciation of sulfite has been ignored for simplicity.

In addition to the above studies of the metal-ion-catalyzed reaction, we call attention to a study of the noncatalyzed reaction (Matsuura et al., 1969) in which a three-halves reaction order in sulfite ion was reported,

$$\nu = k[SO_3^{2-}]^{3/2}[H_3O^+]^2, \tag{3}$$

as well as an unusual second-order dependence on the hydronium ion concentration.

A frequently proposed mechanism for the thermal initiation of a free-radical chain mechanism based on the experimental work of Bäckström

Table 1. Empirical Rate Laws Reported by Various Investigators for the Metal-Catalyzed Autoxidation of SO_2[a]

M^{n+}	α	β	γ	Reference
Mn^{2+}	2	0	0	Coughanowr and Krause (1965)
Mn^{2+}	≤ 1	≤ 1	0	Matteson et al. (1969)
Fe^{3+}	1	1	0	Brimblecombe and Spedding (1974)
Fe^{3+}	1	1	0	Freiberg (1974)
Fe^{3+}	1	1–2	?	Fuzzi (1978)
Co^{3+}	$\frac{1}{2}$	$\frac{2}{2}$	0	Chen and Barron (1972)
Co^{3+}	$\frac{1}{2}$	?	2	Sawicki and Barron (1973)
Co^{2+}	$\frac{1}{2}$	$\frac{3}{2}$	0	Bengtsson and Bjerle (1975)
Co^{2+}	1	1	1	Yagi and Inoue (1962)
Co^{2+}	2	1	1	Davies et al. (1969)
Co^{2+}	$\frac{1}{2}$	?	1	Linek and Mayrhoferova (1970)
Cu^{2+}	1	1	0	Fuller and Crist (1941)
Cu^{2+}	$\frac{1}{2}$	$\frac{3}{2}$	0	Barron and O'Hern (1966)

[a] Symbols α, β, and γ represent the reaction orders for the reaction rate given by the generalized rate expression $-d[SO_3^{2-}]/dt = k\,[M^{n+}]^{\alpha}\,[SO_3^{2-}]^{\beta}\,[O_2]^{\gamma}$.

(1934) is as follows:

Initiation	$M^{n+} + SO_3^{2-} \xrightarrow{k_1} M^{(n-1)+} + SO_3^{-}\cdot$	[25]
Propagation	$SO_3^{-}\cdot + O_2 \xrightarrow{k_2} SO_5^{-}\cdot$	[26]
	$SO_5^{-}\cdot + SO_3^{2-} \xrightarrow{k_3} SO_5^{2-} + SO_3^{-}\cdot$	[27]
Oxidation	$SO_5^{2-} + SO_3^{2-} \xrightarrow{k_4} 2SO_4^{2-}$	[28]
Termination	$2SO_3^{-}\cdot \xrightarrow{k_5} S_2O_6^{2-}$	[29]
	or	
	$SO_3^{-}\cdot + SO_5^{-}\cdot \xrightarrow{k_6} S_2O_6^{2-} + O_2$	[30]
	or	
	$SO_5^{-}\cdot + SO_5^{-}\cdot \xrightarrow{k_7} S_2O_8^{2-} + O_2.$	[31]

In this mechanism the initiation step involves a one-electron-reducible metal ion as the catalytic initiator [for a truly closed catalytic sequence, $M^{(n-1)+}$ must be oxidized back to M^{n+}]; the propagation steps involve $SO_3^{-}\cdot$, $SO_5^{-}\cdot$, and SO_5^{2-} (peroxymonosulfite ion) as reactive intermediates; the oxidation step is invoked for stoichiometric reasons; and the termination steps result in the production of dithionate, a frequently observed but low-yield product (Higginson and Marshall, 1957).

Theoretical rate expressions for the overall stoichiometric reaction (i.e., the summation of reactions [25]–[28] can be derived readily from the appropriate differential equations describing the total rate of appearance or disappearance of each species involved in the mechanism, provided the steady-state approximation is used. For example, assuming that the chain mechanism involves reactions [25]–[28] and [29] as the termination step, the following differential equations can be written:

$$\frac{-d[SO_3^{2-}]}{dt} = k_1[SO_3^{2-}][M^{n+}] + k_3[SO_3^{2-}][SO_5^{-}\cdot] + k_4[SO_5^{2-}][SO_3^{2-}] \quad (4)$$

$$\frac{d[SO_3^{-}\cdot]}{dt} = k_1[SO_3^{2-}][M^{n+}] - k_2[SO_3^{-}\cdot][O_2] + k_3[SO_3^{2-}][SO_5^{-}\cdot] - 2k_5[SO_3^{-}\cdot]^2 \quad (5)$$

$$\frac{d[SO_5^{-}\cdot]}{dt} = k_2[SO_3^{-}\cdot][O_2] - k_3[SO_5^{-}\cdot][SO_3^{2-}] \quad (6)$$

$$\frac{d[SO_5^{2-}]}{dt} = k_3[SO_5^{-}\cdot][SO_3^{2-}] - k_4[SO_5^{2-}][SO_3^{2-}]. \quad (7)$$

By the steady-state approximation for the radical ions, we set

$$\frac{d[SO_3^{-}\cdot]}{dt} = \frac{d[SO_5^{-}\cdot]}{dt} = \frac{d[SO_5^{2-}]}{dt} = 0. \quad (8)$$

From these equations the following relationships are evident:

$$k_3[SO_5^- \cdot][SO_3^{2-}] = k_4[SO_5^{2-}][SO_3^{2-}] \tag{9}$$

$$k_2[SO_3^- \cdot][O_2] = k_3[SO_5^- \cdot][SO_3^{2-}] \tag{10}$$

$$2k_5[SO_3^- \cdot]^2 = k_1[SO_3^{2-}][M^{n+}] \tag{11}$$

$$[SO_3^- \cdot] = \left(\frac{k_1}{2k_5}\right)^{1/2} [M^{n+}]^{1/2}[SO_3^{2-}]^{1/2}. \tag{12}$$

Substitution of Eqs. (9)–(12) into Eq. (4) gives the overall rate of disappearance of sulfite and by mass-balance considerations, the rate of appearance of sulfate:

$$\frac{-d[SO_3^{2-}]}{dt} = \frac{d[SO_4^{2-}]}{dt}$$

$$= k_1[SO_3^{2-}][M^{n+}] + k_2\left(\frac{2k_1}{k_5}\right)^{1/2}[M^{n+}]^{1/2}[SO_3^{2-}]^{1/2}[O_2]. \tag{13}$$

Since, according to Bäckström (1934), the sulfite chain reaction has a chain length of approximately 5×10^4, the long-chain-length approximation can be invoked (i.e., $v_p/v_i \gg 1$ or the ratio of the rate of propagation to the rate of initiation is very large), and Eq. (13) can be rewritten as

$$\frac{-d[SO_3^{2-}]}{dt} \simeq k[M^{n+}]^{1/2}[SO_3^{2-}]^{1/2}[O_2] \tag{14}$$

where $k = k_2(2k_1/k_5)^{1/2}$. Different theoretical rate expressions can be obtained if different assumptions are made about the termination step. In addition, the steady-state and the long-chain-length approximations for a free-radical, closed-sequence catalytic process (Boudart, 1968; Frost and Pearson, 1961) can be used to simplify the algebraic manipulations. Theoretical rate expressions obtained with these procedures are listed in Table 2.

Comparison of the theoretically derived rate expressions in Table 2 with the empirical rate laws summarized in Table 1 shows only a few instances of agreement. The empirical rate laws observed by Barron and O'Hern (1966), Chen and Barron (1972), and Bengtsson and Bjerle (1975) at constant

Table 2. Theoretical Rate Expressions Obtained from Bäckström (1934) Mechanism for Various Termination Steps[a]

Termination Step	Rate Coefficient, k	α	β	γ
A. k_5	$k_2(2k_1/k_5)^{1/2}$	$\frac{1}{2}$	$\frac{1}{2}$	1
B. k_6	$(2k_1k_2k_3/k_6)^{1/2}$	$\frac{1}{2}$	1	$\frac{1}{2}$
C. k_7	$k_3(2k_1/k_7)^{1/2}$	$\frac{1}{2}$	$\frac{3}{2}$	0

[a] Symbols α, β, and γ represent the reaction orders for the reaction rate given by the generalized rate expression $-d[SO_3{}^{2-}]/dt = k\,[M^{n+}]^{\alpha}\,[SO_3{}^{2-}]^{\beta}\,[O_2]^{\gamma}$.

pH are in general agreement with rate expression B in Table 2. However, later results reported by Sawicki and Barron (1973) using a thin-film reactor system contradicted earlier results reported by Barron and co-workers (Barron and O'Hern, 1966; Chen and Barron, 1972). We have already pointed out that the zero order in oxygen reported in many of the studies may be experimental artifact. All the rate expressions in Table 2 show a one-half-order metal-ion dependence, and expressions A and B show a nonzero order in oxygen. In contrast, the majority of investigators listed in Table 1 report a first-order metal ion dependence and a zero-order oxygen dependence. Additionally, it should be noted that for initiation of a free-radical sequence by means of reaction [25], metal catalysts must have an available lower oxidation state and a reduction potential compatible with the S(V)/S(IV) couple [+0.89 V; Anast and Margerum (1981)]. Few metal couples that have been employed meet this requirement. They are Co(III)–Co(II) and Cu(III)–Cu(II), with Fe(III)–Fe(II) as a borderline case. Copper(II), Ni(II), and Mn(II) do not meet this requirement and should be expected to proceed by a different mechanism. Despite all these considerations, many of the investigators listed in Table 1 cite the Bäckström (1934) mechanism as a logical sequence of elementary reactions that is adequate to explain their empirical observations. Clearly, in many cases it is inadequate.

Alternative free-radical mechanisms have been postulated by Hayon et al. (1972) and Schmittkunz (1963). Hayon et al. (1972) observed the formation of the sulfate radical ion $SO_4^- \cdot$ during the flash photolysis and pulse radiolysis of oxygenated sulfite solutions and found no firm evidence for $SO_5^- \cdot$. If the extreme experimental conditions employed by Hayon et al. (1972) are representative also of the more moderate energetic conditions anticipated in thermal reactions, the following mechanism can be postulated:

Initiation

$$M^{n+} + SO_3^{2-} \xrightarrow{k_1} M^{(n-1)+} + SO_3^- \cdot \quad [32]$$

Propagation

$$SO_3^- \cdot + O_2 \xrightarrow{k_2} SO_5^- \cdot \quad [33]$$

$$SO_5^- \cdot + SO_3^{2-} \xrightarrow{k_3'} SO_4^- \cdot + SO_4^{2-} \quad [34]$$

$$SO_4^- \cdot + SO_3^{2-} \xrightarrow{k_4'} SO_4^{2-} + SO_3^- \cdot \quad [35]$$

Termination

$$SO_5^- \cdot + SO_5^- \cdot \xrightarrow{k_7} S_2O_6^{2-} + 2O_2 \quad [36]$$

or

$$SO_4^- \cdot + SO_4^- \cdot \xrightarrow{k_8} S_2O_6^{2-} + O_2 \quad [37]$$

or

$$SO_4^- \cdot + SO_4^- \cdot \xrightarrow{k_8'} S_2O_8^{2-} \quad [38]$$

The mechanism is similar in many respects to the Bäckström mechanism with the exception of the inclusion of the sulfate radical ion as a reactive propagation intermediate.

Once again, theoretical rate expressions can be derived from the mechanism [32]–[38] by use of the steady-state approximation, the long-chain-length assumption, and the assumption that the rate of initiation is equal to rate of termination at steady state (i.e., $v_i = v_t$ when $d[SO_3^- \cdot]/dt = d[SO_4^- \cdot]/dt = d[SO_5^- \cdot]/dt = 0$). The resulting rate expressions are listed in Table 3. Regardless of the termination step, the mechanism proposed by Hayon et al. (1972) yields a single rate expression in which there is a half-order dependence on the metal ion concentration and a three-halves-order dependence on sulfite. It is seen that this mechanism gives rise to a dependence on species concentrations identical to that of the Bäckström mechanism with termination step k_7 (Table 2).

The mechanism proposed by Schmittkunz (1963) postulates two separate chain propagation sequences resulting from two different initiation steps. The first sequence is essentially the same as the Hayon et al. mechanism and is not discussed further here. The second mechanism, which involves $O_2^- \cdot$, $HO_2 \cdot$, $OH \cdot$, and $H \cdot$ as chain carriers, is as follows:

$$\textit{Initiation} \quad M^{x+}L_3^{n-} + O_2 \xrightarrow{k_1'} M^{(x-1)+}L_3^{n-} + O_2^- \cdot \qquad [39]$$

$$\textit{Acid–base equilibrium} \quad O_2^- \cdot + H^+ \overset{K_a^{-1}}{\rightleftharpoons} HO_2 \cdot \qquad [40]$$

$$\textit{Propagation} \begin{cases} HO_2 \cdot + SO_3^{2-} \xrightarrow{k_9} SO_4^{2-} + OH \cdot & [41] \\ OH \cdot + SO_3^{2-} \xrightarrow{k_{10}} SO_4^{2-} + H \cdot & [42] \\ H \cdot + O_2 \xrightarrow{k_{11}} HO_2 \cdot & [43] \end{cases}$$

$$\textit{Termination} \quad 2HO_2 \cdot \xrightarrow{k_{12}} H_2O_2 + O_2 \qquad [44]$$

where $M^{n+}L_3^{n-}$ is a metal (M)–ligand (L) complex of appropriate charge $x - n$ and K_a is the acid dissociation constant for the hydroperoxyl radical/superoxide acid–base equilibrium. The rate expression obtained for this mechanism is

$$-\frac{d[SO_3^{2-}]}{dt} = k_9 \left(\frac{2k_1'}{k_{12}}\right)^{1/2} [M^{x+}L_3^{n-}]^{1/2}[SO_3^{2-}][O_2]^{1/2}. \qquad (15)$$

This mechanism is seen to predict reaction orders (Table 3) identical to those of the Bäckström mechanism with termination step k_6 (Table 2). It is thus clear that these mechanisms cannot be distinguished solely by determination of reaction order.

In all the cases treated pH dependences in terms of the acid–base chemistry of $SO_2(aq)$ have been ignored. A complete analysis valid over broad pH ranges requires consideration of the roles of $SO_2 \cdot H_2O$, HSO_3^-, $SO_2 \cdot H_2O^+$, and $HSO_3 \cdot$ as reactive species. Larson et al. (1978) utilized a hybrid mechanism involving two distinct chain propagation sequences with superimposed acid–base equilibria to obtain a rate expression that agreed moderately well with their empirical rate law.

Table 3. Theoretical Rate Expressions Obtained from the Hayon et al. (1972) and Schmittkunz (1963) Mechanisms[a]

Mechanism	Termination Step	Rate Coefficient, k	α	β	γ
Hayon et al. (1972)	k_7	$k_3'\ (2k_1/k_7)^{1/2}$	$\frac{1}{2}$	$\frac{3}{2}$	0
Hayon et al. (1972)	k_8 or k_8'	$k_4'\ (2k_1/k_8)^{1/2}$	$\frac{1}{2}$	$\frac{3}{2}$	0
Schmittkunz (1963)	k_{12}	$k_9\ (2k_1'/k_{12})^{1/2}$	$\frac{1}{2}$	1	$\frac{1}{2}$

[a] Symbols α, β, and γ represent the reaction orders for the reaction rate given by the generalized rate expression $-d[SO_3^{2-}]/dt = k\,[M^{n+}]^{\alpha}\,[SO_3^{2-}]^{\beta}\,[O_2]^{\gamma}$.

3.3. Polar Mechanisms

Nonradical, polar mechanisms for the metal-catalyzed autoxidation of $SO_2(aq)$ have been proposed by Bassett and Parker (1951), Matteson et al. (1969), and Freiberg (1974). A common feature of the mechanisms proposed by these investigators is inner-sphere complexation of the catalytic metal by sulfite as a prelude to electron transfer. Few metal–sulfite stability constants have been determined because of the thermodynamic instability of sulfite toward oxidation in oxic systems; however, those that have been reported (Sillén and Martell, 1964; Smith and Martell, 1976) are significantly larger than the corresponding constants for metal sulfate complexes, as shown in Table 4. Other common features of these mechanisms include

Table 4. Comparison of Stability Constants for Metal Sulfito and Sulfato Complexes at 25.0°C

Metal	Ionic Strength, (M)	Sulfito Complex	log β	Ionic Strength, (M)	Sulfato Complex	log β
Ag^+	0	$AgSO_3^-$	5.6	0	$AgSO_4^-$	1.3
Cd^{2+}	1.0	$Cd(SO_3)_2^{2-}$	4.2	1.0	$Cd(SO_4)_2^{2-}$	1.6
Hg^{2+}	1.0	$Hg(SO_3)_2^{2-}$	24.1	0.5	$Hg(SO_4)_2^{2-}$	2.4
Ce^{3+}	0	$CeSO_3^+$	8.0	0	$CeSO_4^+$	3.6
Fe^{3+}	0.1	$FeSO_3^+$	18.1[b]	0	$FeSO_4^+$	4.0
Fe^{2+}				0	$FeSO_4^0$	2.2

[a] From Smith and Martell (1976) except as indicated.
[b] Evaluated by the sequence

Reaction	log K
$Fe^{3+} + SO_2{\cdot}H_2O \rightleftharpoons FeSO_3^+ + 2H^+$	9.7[c]
$H^+ + HSO_3^- \rightleftharpoons SO_2{\cdot}H_2O$	1.6
$H^+ + SO_3^{2-} \rightleftharpoons HSO_3^-$	6.8
$Fe^{3+} + SO_3^{2-} \rightleftharpoons FeSO_3^+$	18.1

[c] Hansen et al. (1976).

binding of dioxygen by the metal-sulfito complex and electron transfer by means of successive two-electron transfers. The latter process is to be contrasted with the series of one-electron transfer, chain-propagation steps treated in Section 3.1.

In a study of Mn(II)-catalyzed oxidation of S(IV) Bassett and Parker (1951) found that Mn^{2+} salts had a pronounced catalytic effect on the rate of sulfite autoxidation with almost complete suppression of dithionate formation and that the nature of the Mn^{2+} counterion (i.e., SO_4^{2-} or Cl^-) strongly influenced the reaction rate as a result of competitive complexation; a similar catalytic behavior was observed for Co(II) and Ni(II) salts. On the basis of these observations they proposed the following polar mechanism:

$$Mn^{2+} + SO_3^{2-} \underset{k_{-1}}{\overset{k_1}{\rightleftharpoons}} MnSO_3^0, \qquad [45]$$

$$MnSO_3^0 + SO_3^{2-} \underset{k_{-2}}{\overset{k_2}{\rightleftharpoons}} Mn(SO_3)_2^{2-}, \qquad [46]$$

$$Mn(SO_3)_2^{2-} + O_2 \underset{k_{-3}}{\overset{k_3}{\rightleftharpoons}} Mn(SO_3)_2O_2^{2-}, \qquad [47]$$

$$Mn(SO_3)_2O_2^{2-} \xrightarrow{k_4} Mn^{2+} + 2SO_4^{2-} \quad \text{(slow)}. \qquad [48]$$

A theoretical rate expression can be derived from this sequence of elementary reactions by use of the vector method of King and Altman (1956). In this procedure the mechanism is written in a cyclic form that represents the closed nature of the catalytic cycle for the overall reaction as shown in Figure 1. The number of intermediate complexes is readily identified, and the concentration of each form of the catalyst can be shown to be proportional to the sums of the terms that are obtained from the elementary reaction steps that individually or in sequence lead to the specific form of the catalyst. The number of rate constants (or products of individual rate constants and concentrations of species in excess) must be one less than the number of species in a particular cycle. The steady-state approximation is applied to the resultant equations for each reactive intermediate, and the final rate

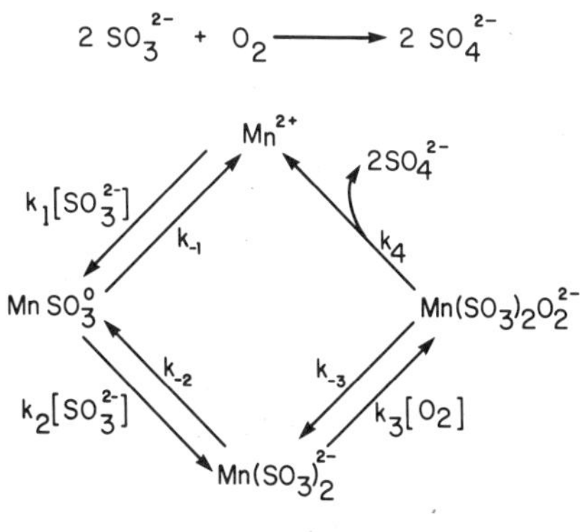

Figure 1. Reaction mechanism for the manganese-catalyzed autoxidation of sulfite (Bassett and Parker, 1951) expressed in cyclic form as a closed catalytic sequence. Inner-sphere complexation precedes electron transfer.

expression is obtained by substitution into a designated rate-limiting step. By the use of this procedure, the following rate expression is obtained:

$$\nu = -\frac{d[SO_3^{2-}]}{dt} = \frac{k_4[Mn^{2+}]_0[SO_3^{2-}]^2[O_2]}{D} \tag{16}$$

where

$$D = K_A + K_B[O_2] + K_C[SO_3^{2-}] + K_D[SO_3^{2-}][O_2] + K_E[SO_3^{2-}]^2 + [SO_3^{2-}]^2[O_2].$$

Here

$$K_A = K_1^{-1}K_2^{-1}\left(K_3^{-1} + \frac{k_4}{k_3}\right)$$

$$K_B = K_1^{-1}\frac{k_4}{k_2}$$

$$K_C = K_2^{-1}\left(K_3^{-1} + \frac{k_4}{k_3}\right)$$

$$K_D = \frac{k_4}{k_2} + \frac{k_4}{k_1}$$

$$K_E = K_3^{-1} + \frac{k_4}{k_3}$$

and $[Mn^{2+}]_0$ is the initial manganese concentration. K_1, K_2, and K_3 are the equilibrium constants for reactions [45], [46], and [47], respectively. This cumbersome expression can be simplified for certain limiting cases if assumptions are made about the magnitude of individual terms, as shown in Table 5. It should be emphasized, however, that these limiting cases are idealizations and that generally intermediate cases would be expected.

From laboratory studies of S(IV) oxidation in aqueous aerosols Matteson et al. (1969) reported that the following rate law was adequate to account for their experimental observations under certain conditions:

$$\nu_0 = -\left(\frac{d[SO_2 \cdot H_2O]}{dt}\right)_0 = \frac{k[Mn^{2+}]_0[SO_2 \cdot H_2O]_0}{k_1[SO_2 \cdot H_2O]_0 + k'[Mn^{2+}]_0 + K}. \tag{17}$$

Here the concentration of S(IV) is expressed as $[SO_2 \cdot H_2O]$ to reflect the condition of high acidity employed in the study. The rate expression (17) can be reduced to

$$\nu_0 = k''[Mn^{2+}]_0[SO_2]_0 \tag{18}$$

for the initial rate when $K \gg k_1[SO_2 \cdot H_2O]_0 + k'[Mn^{2+}]_0$. Indication of such behavior based on aerosol conversion experiments was observed by Cheng et al. (1971) and Kaplan et al. (1981). The rate law of Eq. (18) exhibits the same form as the limiting case D shown in Table 5.

Table 5. Rate Expressions for Limiting Cases of the Bassett–Parker (1951) Mechanism[a,b]

Dominant Term in Denominator	α	β	γ
A	1	2	1
B	1	2	0
C	1	1	1
D	1	1	0
E	1	0	1
$[SO_3^{2-}]^2\,[O_2]$	1	0	0

[a] $$\nu = \frac{d[SO_3^{2-}]}{dt} = \frac{k_4[Mn^{2+}]_0\,[SO_3^{2-}]^2\,[O_2]}{K_A + K_B[O_2] + K_C[SO_3^{2-}] + K_D[SO_3^{2-}][O_2] + K_E[SO_3^{2-}]^2 + [SO_3^{2-}]^2[O_2]}$$

[b] Symbols α, β, and γ represent the reaction orders for the reaction rate given by the generalized rate expression $-d[SO_3^{2-}]/dt = k\,[M^{n+}]^{\alpha}\,[SO_3^{2-}]^{\beta}\,[O_2]^{\gamma}$.

Freiberg (1975) has proposed a similar catalytic cycle involving the inner-sphere complexation of sulfite as a prelude to electron transfer. His hypothetical mechanism, which was postulated in an effort to obtain an empirical rate law consistent with experimental data for Fe^{3+}-catalyzed oxidation of S(IV) (Karraker, 1963; Neytzell de Wilde and Taverner, 1958) is as follows:

$$SO_2 \cdot H_2O \overset{K_{a1}}{\rightleftharpoons} H^+ + HSO_3^-, \quad [49]$$

$$HSO_3^- \overset{K_{a2}}{\rightleftharpoons} H^+ + SO_3^{2-}, \quad [50]$$

$$Fe(OH)^{2+} + H^+ \overset{K_h}{\rightleftharpoons} Fe^{3+} + H_2O, \quad [51]$$

$$Fe^{3+} + HSO_3^- \overset{\beta_1}{\rightleftharpoons} Fe(HSO_3)^{2+}, \quad [52]$$

$$Fe(HSO_3)^{2+} + SO_3^{2-} \underset{\text{slow}}{\overset{\beta_2}{\rightleftharpoons}} Fe(SO_3)_2H^0, \quad [53]$$

$$Fe(SO_3)_2H^0 + O_2 \overset{k_3}{\rightleftharpoons} Fe(SO_3)_2HO_2^0, \quad [54]$$

$$H_2O + Fe(SO_3)_2HO_2^0 \overset{k_4}{\rightarrow} Fe(OH)^{2+} + 2HSO_4^-. \quad [55]$$

This mechanism is similar to that proposed by Bassett and Parker (1951) (reactions [45]–[48]; Figure 1) for the manganese-catalyzed oxidation.

A rate law with no apparent dependence on oxygen concentration was postulated as a limiting case of this mechanism:

$$\nu = k'[Fe^{3+}][SO_2 \cdot H_2O]^2[H^+]^{3-}; \quad (19)$$

the inverse third-order dependence on $[H^+]$ results from expressing S(IV) concentrations in terms of $SO_2 \cdot H_2O$. This rate law is identical to that proposed by Bassett and Parker in the limiting case of high $[O_2]$ (see Table 5, case B).

In addition to a sequence of two-electron transfers involving oxygen as the oxidant, oxidation by Fe^{3+} by one-electron transfer may also occur;

$$Fe(SO_3)_2H^0 + Fe^{3+} + H_2O \xrightarrow{k_6} 2Fe^{2+} + HSO_4^- + HSO_3^- + H^+. \quad [56]$$

However, in view of the extremely slow rate of oxidation of Fe(II) to Fe(III) at low pH (Sung and Morgan, 1981), a loss of apparent catalytic activity would be predicted if reaction [56] were a predominant pathway in the Fe(III)–O_2–SO_2 system, unless Fe(II) also exhibits catalytic activity in a fashion similar to Mn(II), Co(II), Ni(II), and Cu(II).

Brimblecombe and Spedding (1974) and Fuzzi (1978) also studied the Fe(III)-catalyzed autoxidation of dissolved sulfur dioxide. At pH 5 Brimblecombe and Spedding reported that the empirical rate law was second order overall and first order each in [Fe(III)] and [S(IV)], whereas Fuzzi (1978) reported a change in the S(IV) reaction order from first order to second order when pH $\geq$ 5. This later result agrees in part with Freiberg's observations of a second-order S(IV) dependence, although the pH ranges in these two studies were not comparable. In addition, results over the pH range 4–8 may not be representative of homogeneous catalysis by soluble monomeric or polymeric iron species such as $Fe(OH)_n^{(3-n)+}$ or $Fe_m(OH)_n^{(3m-n)+}$; rather, because of the formation of amorphous hydroxide and oxyhydroxide solids such as $Fe(OH)_3$ and FeOOH, these results may be more representative of heterogeneous catalysis on colloid surfaces.

3.4. Photoassisted Catalysis in Aqueous Systems

3.4.1. Homogeneous Processes

The catalytic effect of light on the autoxidation of dissolved SO_2 has been acknowledged for years as an alternative mode of initiation of the chain reaction sequence given in reactions [28]–[34] (Alyea and Bäckström, 1929; Bäckström, 1934; Dogliotti and Hayon, 1967; Haber and Wansbrough-Jones, 1931; Hayon et al., 1972; Lunak and Veprek-Siska, 1976; Mason and Mathews, 1926; Mathews and Dewey, 1912; Mathews and Weeks, 1917). Ultraviolet irradiation of sulfite–bisulfite systems is thought to produce an aquated electron (Dogliotti and Hayon, 1967; Hayon et al., 1972) and a sulfite radical ion,

$$SO_3^{2-} \xrightarrow{h\nu} SO_3^{-}\cdot + e_{aq}^-. \quad [57]$$

A transient electronic absorption maximum at λ = 700 nm has been assigned to the solvated electron and a similar transient maximum at λ = 260 nm has been assigned to the $SO_3^-\cdot$ radical ion. The aquated electron decays by reaction with O_2 or HSO_3^- to produce a superoxide ion or sulfite ion and a hydrogen atom, respectively. However, only Lunak and Veprek-Siska (1976) have systematically explored the possibility of a metal-catalyzed photooxidation of sulfite even though the phenomenon of homogeneous metal-assisted photooxidation–reduction processes has been reported previously for Fe(III) and Cu(II)–ligand systems (Baker et al., 1980; Balzani and Carassiti, 1970; Langford et al., 1973; Langford and Carey, 1975; Lockhart and Blakeley, 1975; Miles and Brezonik, 1981).

The preliminary work of Lunak and Veprek-Siska has shown a distinct photoassisted catalysis by Fe(III) at wavelengths higher than the cutoff for absorption by sulfite and bisulfite ion in aqueous solution. Quantum yields were shown to be dependent on the wavelength of irradiation and the concentration of Fe(III) in the system. They have proposed that a quantum of light is absorbed by a ferric–sulfito complex, thereby initiating the catalytic cycle through photoreduction of Fe(III) to Fe(II). In the absence of added Fe(III), the photochemical autoxidation of sulfite above $\lambda = 300$ nm did not proceed to a noticeable extent. A mechanism consistent with their findings would be as follows:

$$\mathrm{Fe(OH)}_n^{(3-n)+} + \mathrm{HSO_3^-} \rightleftharpoons \mathrm{Fe(OH)}_n\mathrm{SO}_3^{(1-n)+} + \mathrm{H}^+ \quad [58]$$

$$\mathrm{Fe(OH)}_n\mathrm{SO}_3^{(1-n)+} + \mathrm{HSO_3^-} \rightleftharpoons \mathrm{Fe(OH)}_n(\mathrm{SO_3})_2^{(n+1)-} + \mathrm{H}^+ \quad [59]$$

$$\mathrm{Fe(III)(OH)}_n(\mathrm{SO_3})_2^{(n+1)-} \xrightarrow{h\nu} {}^*\mathrm{Fe(II)(OH)}_n(\mathrm{SO_3})(\mathrm{SO_3}\cdot)^{(n+1)-} \quad [60]$$

$$^*\mathrm{Fe(II)(OH)}_n(\mathrm{SO_3})(\mathrm{SO_3}\cdot)^{(n+1)-} \rightleftharpoons \mathrm{Fe(II)(OH)}_n\mathrm{SO}_3^{n-} + \mathrm{SO_3^-}\cdot \quad [61]$$

$$\mathrm{Fe(III)(OH)}_n(\mathrm{SO_3})_2^{(n+1)-} + \mathrm{O_2} \rightleftharpoons \mathrm{Fe(III)(OH)}_n(\mathrm{SO_3})_2\mathrm{O}_2^{(n+1)-} \quad [62]$$

$$\mathrm{Fe(III)(OH)}_n(\mathrm{SO_3})_2\mathrm{O}_2^{(n+1)-} \xrightarrow{h\nu} {}^*\mathrm{Fe(III)(OH)}_n(\mathrm{SO_4})_2^{(n+1)-} \quad [63]$$

$$2\mathrm{H}^+ + {}^*\mathrm{Fe(III)(OH)}_n(\mathrm{SO_4})_2^{(n+1)-} \rightleftharpoons \mathrm{Fe(III)(OH)}_n^{(3-n)+} + 2\mathrm{HSO_4^-}. \quad [64]$$

After photoinitiation, autoxidation would proceed by a free-radical chain after binding of dioxygen and followed by a photoassisted inner-sphere electron transfer as shown in reactions [62]–[64].

3.4.2. *Heterogeneous Processes*

A primary source of iron, manganese, zinc, titanium, and other metals found in atmosphere aerosols is the combustion of coal. These metals will be released to the atmosphere in their highest oxidation states in the form of metal oxides such as Fe_2O_3, MnO_2, ZnO, and TiO_2. Taylor and Flagan (1981) and Ouimette and Flagan (1981) have found direct evidence for significant mass fractions of Fe_2O_3 in submicrometer particle size fractions of both fly ash from a coal-fired combustor and ambient aerosol. The presence of such material in atmospheric aerosol suggests the possibility of heterogeneous reactions (i.e., involving solids suspended in aqueous solution) occurring in the ambient atmosphere. These processes are examined in this section.

Frank and Bard (1977) reported a heterogeneous photoassisted autoxidation of HSO_3^- on Fe_2O_3 (hematite) surfaces. Irradiation at 400 nm of an oxygenated bisulfite solution with suspended Fe_2O_3 particles resulted in a rapid conversion of S(IV) to S(VI), whereas no conversion of sulfite to sulfate was observed in the absence of light. A sequence of stoichiometric reactions that offers one possible explanation for photoassisted catalysis on a metal oxide (semiconductor) surface is as follows:

$$Fe_2O_3 + 2h\nu \xrightarrow{\text{surface}} (Fe_2O_3) + 2p^+ + 2e^- \qquad [65]$$

$$H_2O + HSO_3^- + 2p^+ \xrightarrow{\text{surface}} HSO_4^- + 2H^+ \qquad [66]$$

$$O_2 + 2e^- + 2H^+ \xrightarrow{\text{surface}} H_2O_2 \qquad [67]$$

$$H_2O_2 + HSO_3^- \xrightarrow{\text{solution}} HSO_4^- + H_2O. \qquad [68]$$

Here p^+ is a charge vacancy created by the photoassisted excitation of an electron from the valence band of the metal oxide to the conduction band. Excited electrons in the conduction band and positive holes in the valence band migrate to the surface (presumably without recombination), where they react with oxygen and bisulfite, respectively, to produce hydrogen peroxide as the reduced product and bisulfate as the oxidized product.

Similar but less effective photocatalytic effects have been reported by Frank and Bard (1977) for other semiconductor materials such as TiO_2, CdS, and ZnO. However, results for CdS and ZnO are obscured by the fact that at low pH (~pH 3) these solids dissolve to a certain extent, and thus any apparent photocatalytic effects may be due to dissolved metal species rather than to inherent properties of the semiconductor material.

Faust and Hoffmann (1982) have studied the photocatalytic autoxidation of HSO_3^- on synthetic Fe_2O_3 suspensions at pH 2.5 and find that the presence of Fe_2O_3 particles results in a slight rate enhancement as compared to the corresponding homogeneous photocatalytic system.

Childs and Ollis (1980) have addressed some of the fundamental considerations that are required in order to ascertain whether a heterogeneous photoreaction is truly photocatalytic according to reactions [65]–[68] or is stoichiometrically photochemical. In a stoichiometric photochemical reaction, the observed conversion of sulfite to sulfate in irradiated systems (Frank and Bard, 1977) may result from any of the following processes: direct photoactivation of an adsorbed sulfite on the surface; surface-prompted reaction of sulfite photoactivated in solution; noncatalytic reaction of the Fe_2O_3 surface with the adsorbed sulfite; or homogeneous catalysis by Fe(III)–Fe(II) leached from the Fe_2O_3 surface. The possibility of such processes remains to be resolved in the case of iron-catalyzed systems.

4. HOMOGENEOUS AND HETEROGENEOUS CATALYSIS BY METAL–LIGAND COMPLEXES

4.1. Kinetics of Homogeneous Catalysis by Co(II), Fe(II), and Mn(II) Tetrasulfophthalocyanine

Boyce et al. (1983) have studied kinetics and mechanisms of the autoxidation of dissolved surfur dioxide over the pH range of 4.5–11.0 in the presence of Co(II), Fe(II), Mn(II), Cu(II), Ni(II), and V(IV)–4,4′,4″,4‴-tetrasulfophthalocyanine (TSP) complexes. Phthalocyanine ligands are macrocyclic tetrapyrrole compounds that readily form square planar complexes with metal atoms located in the plane of the ring as shown in Figure 2. Certain metal phthalocyanine complexes such as cobalt tetrasulfophthalocyanine are known to actively bind dioxygen and to serve as reversible oxygen carriers (Abel et al., 1971; Jones et al., 1979). Because of their relationship to naturally occurring organic macromolecules, as well as their high stability, catalytic specificity and oxidaselike activity, metal phthalocyanine complexes are suitable models for study of the catalytic effects of trace metals in aqueous systems (Hoffmann and Lim, 1979).

Kinetic data were obtained by monitoring the rate of disappearance of sulfite [total S(IV)] spectrophotometrically and the rate of disappearance of dissolved oxygen potentiometrically; pseudo-first-order spectrophotometric results (Figure 3) obtained at 213 nm were linear for more than three half-lives. Similarly, pseudo-zero-order potentiometric results were linear also

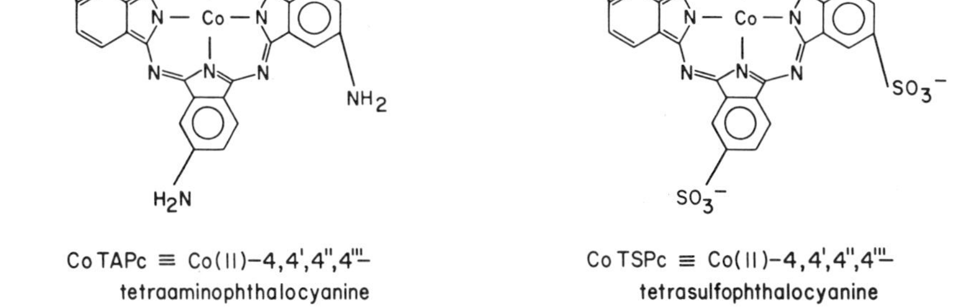

Figure 2. Cobalt phthalocyanine complexes employed as catalysts for reaction of S(IV) with oxygen. Catalysts may be employed in solution or in a heterogeneous suspension on silica gel particles.

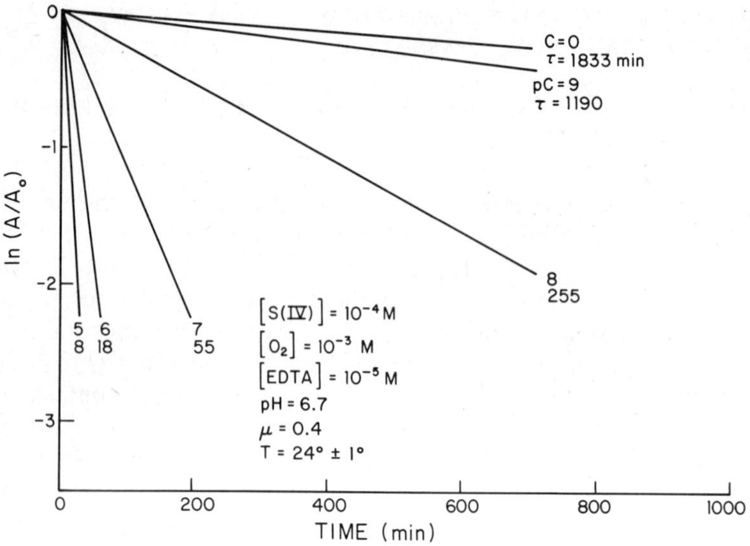

Figure 3. Pseudo-first-order plots of ln(absorbance at time = *t*/absorbance at time *t* = 0) for reaction of S(IV) with oxygen: C = [Co(II)TSP]; $pC = -\log_{10}[C]$; τ = reaction half-life; pH = 6.7; ionic strength = 0.4 M; EDTA was present. From Boyce et al. (1983).

with catalyst reaction orders of 0.5 at pH 6.7 and 1.0 when the pH exceeded 9.4. Other factors affecting the observed reaction rates reported by Boyce et al. (1983) were ionic strength, light, catalyst age, the nature of the central metal, and competitive complexation.

The observed rate law for Co(II)TSP at pH 6.7 was as follows:

$$\frac{-d[\mathrm{S(IV)}]}{dt} = k'[\mathrm{Co(II)TSP}]^{1/2}[\mathrm{S(IV)}] \tag{20}$$

and at pH 9.4

$$\frac{-d[\mathrm{S(IV)}]}{dt} = k''[\mathrm{Co(II)TSP}][\mathrm{S(IV)}] \tag{21}$$

where $k' = 1.62\ \mathrm{M}^{-1/2}\ \mathrm{s}^{-1}$ and $k'' = 1.83 \times 10^3\ \mathrm{M}^{-1}\ \mathrm{s}^{-1}$. In the presence of competitive complexing agents such as EDTA (10 μM) and cyanide (10 μM) k'' was reduced to $3.3 \times 10^2\ \mathrm{M}^{-1}\ \mathrm{s}^{-1}$ and $2.9 \times 10^2\ \mathrm{M}^{-1}\ \mathrm{s}^{-1}$, respectively, whereas addition of the free-radical trap mannitol (10 μM) reduced the observed rate constant to a lesser extent.

Independent spectral evidence over the wavelength range 600–700 nm indicated the existence of a monomeric and dimeric form of the Co(II) catalyst. Addition of S(IV) to this previously established equilibrium resulted in a shift in the visible spectrum that was indicative of a corresponding shift in the monomer–dimer equilibrium in favor of the monomeric form. The maximum absorbance of the spectral peak due to the monomer was shown to be a direct function of $[S(IV)]_0$. This result was interpreted in terms of the formation of a discrete sulfite Co(II)TSP complex. However, Boyce et al. (1983) also observed that the reaction exhibited sensitivity to light as shown in Figure 4. In the absence of light the reaction proceeded at a noticeably slower rate. This result suggests that metal catalysis may proceed by way of both light-dependent and light-independent pathways.

On the basis of experimental observations and kinetic data reported by Boyce et al. (1983), two distinct mechanisms can be postulated to describe the detailed molecular processes involved in the homogeneous catalytic autoxidation of sulfite: a two-electron transfer, bisubstrate complexation pathway and a one-electron transfer, chain reaction sequence. Each of these mechanistic possibilities is presented and examined for consistency with experimental observations.

4.2. Two-Electron, Bisubstrate Complexation Mechanism

The catalytic activity of metal phthalocyanines in aqueous solution was documented initially by Cook (1938) for the decomposition of H_2O_2 and the

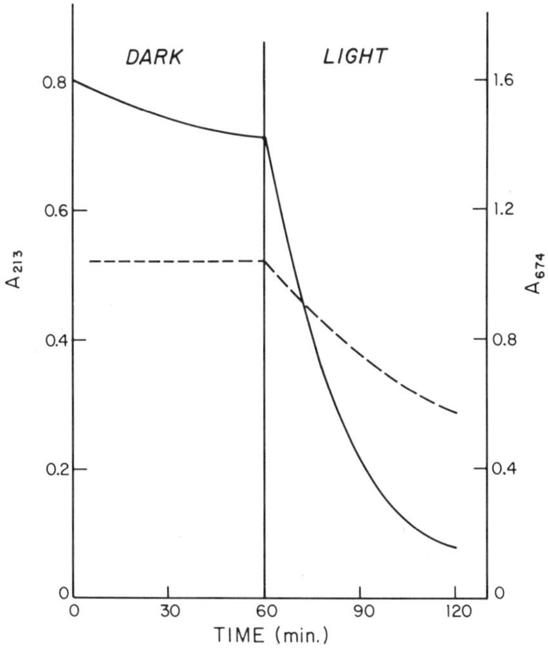

Figure 4. Effect of fluorescent room light on the catalytic autoxidation of sulfite. Solid line (left scale) refers to the absorbance of SO_3^{2-}; dashed line (right scale) refers to absorbance of all monomeric Co(II) complexes. $[S(IV)] = 10^{-4}$ M; $[O_2] = 10^{-3}$ M; $[Co(II)TSP] = 10^{-6}$ M; pH = 6.7; ionic strength = 0.4 M; $[EDTA] = 10^{-5}$ M; $T = 24°C$. From Boyce et al. (1983).

oxidation of HI and later by Wagnerova and co-workers for the autoxidation of hydroxylamine (Wagnerova et al., 1974), hydrazine (Wagnerova et al., 1973), ascorbic acid (Wagnerova et al., 1978), and cysteine (Dolansky et al., 1976) and by Hoffmann and Lim (1979). These investigators suggested that the catalytic behavior of cobalt, iron, and manganese phthalocyanine complexes was similar to that of an oxidase enzyme. A well-known characteristic of enzymatic reactions is the variability of reaction orders for catalyst and substrate. Under certain conditions the order in substrate concentration can vary between zero and one, but, most likely, a nonintegral order will be observed. To interpret the observed kinetic behavior of sulfite within this framework, Boyce et al. (1983) developed a bisubstrate model for catalytic activity. The mechanism postulated to account for the observed kinetic behavior is the ordered ternary-complex mechanism (Laidler and Bunting, 1973), which is depicted symbolically in Figure 5 and consists of the following reactions:

$$(\mathrm{Co(II)TSP})_2^{4-} \overset{K_d}{\rightleftharpoons} 2\mathrm{Co(II)TSP}^{2-}, \quad [69]$$

$$\mathrm{Co(II)TSP}^{2-} + \mathrm{SO}_3^{2-} \underset{k_{-0}}{\overset{k_0}{\rightleftharpoons}} \mathrm{SO_3Co(II)TSP}^{4-} \quad [70]$$

$$\mathrm{SO_3Co(II)TSP}^{4-} + \mathrm{O}_2 \underset{k_{-1}}{\overset{k_1}{\rightleftharpoons}} \mathrm{SO_3Co(III)TSPO}_2^{4-}\cdot \quad [71]$$

$$\mathrm{SO}_3^{2-} + \mathrm{SO_3Co(III)TSPO}_2^{4-}\cdot \underset{k_{-2}}{\overset{k_2}{\rightleftharpoons}} \mathrm{SO_3Co(III)TSPO_2{\cdot}SO}_3^{6-} \quad [72]$$

$$\mathrm{H_2O} + \mathrm{SO_3Co(III)TSPO_2{\cdot}SO}_3^{6-} \underset{h\nu}{\overset{k_3}{\rightarrow}} \mathrm{SO_3Co(II)TSPO}_2^{6-} + \mathrm{H_2SO_4} \quad [73]$$

$$\mathrm{SO_3Co(II)TSPO}_2^{6-} + 2\mathrm{H}^+ \overset{k_4}{\rightarrow} \mathrm{SO_3Co(II)TSP}^{4-} + \mathrm{H_2O_2} \quad [74]$$

$$\mathrm{H_2O_2} + \mathrm{SO}_3^{2-} \overset{k_5}{\rightarrow} \mathrm{SO}_4^{2-} + \mathrm{H_2O}. \quad [75]$$

Reaction [70] represents the formation of the reactive monomeric catalytic center, $SO_3Co(II)TSP^{4-}$ (C in Figure 5) from the dominant dimeric form of the catalyst in solution. This sequence of events is consistent with observed spectral changes that indicate a shift in the dimer–monomer equilibrium on complexation of the added substrate in an axial coordination position on Co(II) either above or below the plane of the phthalocyanine complex. Complexation of SO_3^{2-} by Co(II)TSP in an axial coordination site enhances the subsequent complexation of dioxygen as written in reaction [71]. Carter et al. (1974) have studied the reversible binding of dioxygen by Co(II) complexes of the general form Co(II)(L)B, where L is a quadridentate planar ligand and B is an axial ligand. They conclude that as the π-electron donating ability of B increased, the electron density on the cobalt atom would be enhanced and that this, in turn, results in a greater π-bonding electron flow from cobalt to oxygen. Rollman and Chan (1971) report that the imidazole

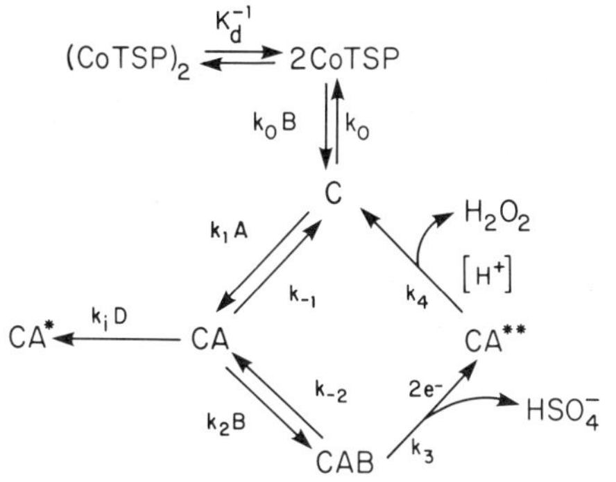

$A \equiv O_2$
$B \equiv S(IV)\ \{HSO_3^-,\ SO_3^{2-}\}$
$C \equiv {}^{=}SO_3\text{-}Co(II)\,TSP$
$D \equiv$ inhibitor {mannitol}
$CA \equiv {}^{=}SO_3\text{-}Co(III)\,TSPO_2^-$
$CA^{**} \equiv {}^{=}SO_3\text{-}Co(II)\,O_2^{2-}$

Figure 5. Mechanism proposed for the Co(II)TSP-catalyzed autoxidation of sulfite (two-electron transfer polar reaction) expressed symbolically in cyclic form.

complex of Co(II)TSP rapidly complexes O_2, whereas the pyridine complex does not. Similar enhancements of the binding of dioxygen by Co(II)TSP when complexed by an appropriate axial ligand have been reported by Cookson et al. (1977), Przywarska-Boniecka and Fried (1976), and Hoffmann and Lim (1979). In addition, studies of the catalytic properties of heme have shown that an axial ligand trans to dioxygen affects both the formation constant and the reversibility of oxygen transfer (Basolo et al., 1975; Jones et al., 1979).

Electron spin resonance studies on Co(II)TSP–O_2 complexes have been shown to be consistent with the formulation of the complex as Co(III)TSP$O_2^-\cdot$ (Cookson et al., 1977; Rollman and Chan, 1971). The intermediate complex given in reaction [71] is considered to be a mixed-ligand Co(III) complex with a superoxide ion $O_2^-\cdot$ and a sulfite ion bound trans to one another in axial coordination positions. In the proposed mechanism the O_2 adduct reacts with an additional SO_3^{2-} ion to form a ternary complex as indicated in reaction [72]. This complex, in turn, undergoes a rate-limiting two-electron transfer from the second bound sulfite to the coordinated dioxygen–Co(II) system to form a SO_3-bound Co(II)–peroxide complex. The attached SO_3 hydrolyzes rapidly to give a coordinated H_2SO_4, which readily dissociates from the complex. After protonation the coordinated O_2^{2-} is released as H_2O_2. This intermediate H_2O_2 reacts with another molecule of SO_3^{2-} to form SO_4^{2-} by means of an additional two-electron transfer in the final step.

In support of this mechanism we note that Schutter and Beelen (1981) have directly observed the formation and accumulation of H_2O_2 as an intermediate reduction product of O_2 in the Co(II)TSP catalyzed autoxidation of 2-mercaptoethanol in water. Also, Davies et al. (1969) have reported on the basis of ^{18}O tracer experiments that approximately half of the oxygen transferred to sulfite comes from a superoxo complex, $(NH_3)_5Co(III)–O_2^-–Co(II)NH_3^{5+}$; the other half presumably comes from water. A compatible conclusion as to the origin of oxygen in oxidized sulfite was reported recently by Holt et al. (1981), who have shown conclusively that the ^{18}O content of the product sulfate for metal-catalyzed autoxidations of SO_3^{2-} is linearly related to the ^{18}O content of water and that at least three of the four oxygen atoms in the sulfate products are isotopically controlled by the solvent water. (The remaining oxygen atom originated from O_2.) Finally, Yatsimirskii et al. (1977) have found strong evidence for complexation of sulfite by the Co(II)–dioxygen complex, Co_2(L-histidine)O_2, as a prelude to electron transfer from S(IV) to O_2. In total, these results are consistent with the mechanism postulated in reactions [69]–[75].

The observed photocatalytic effect (Figure 4) may be due to absorption of light at $\lambda > 620$ nm by the reactive intermediate $SO_3Co(III)TSPO_2{\cdot}SO_3^{6-}$ to form a bound singlet oxygen complex $SO_3Co(II)TSP^1O_2SO_3^{6-}$, which has a more favorable spin symmetry for rapid electron transfer. Cox and Whitten (1979) have shown that various metalloporphyrins give rise to singlet oxygen when irradiated at appropriate wavelengths.

In the above mechanism the active catalytic center is the complex $SO_3Co(II)TSP^{4-}$, which is designated as C in Figure 5. The catalytic cycle begins and ends with this complex. Steps leading to C are assumed to be in rapid equilibrium and are ignored in the initial development of a rate expression. In the catalytic cycle there are three intermediates, $SO_3Co(III)TSPO_2^{4-}{\cdot}$, $SO_3Co(III)TSPO_2{\cdot}SO_3^{6-}$, and $SO_3Co(II)TSPO_2^{6-}$, and correspondingly three steady-state equations. With these assumptions a rate law was derived by use of the method of King and Altman (1956), which is based on a standard determinant procedure used for solving a system of nonhomogeneous linear equations obtained from steady-state considerations,

$$\nu = \frac{d[SO_4^{2-}]}{dt} = \frac{k'[SO_3Co(II)TSP^{4-}]_T[SO_3^{2-}][O_2]}{K_A + K_B[O_2] + K_C[SO_3^{2-}] + [O_2][SO_3^{2-}]}, \qquad (22)$$

where $k' = \dfrac{k_3k_4}{k_3 + k_4}$

$$K_A = \frac{k_4(k_{-1}k_{-2} + k_{-1}k_{-3})}{k_1k_2(k_3 + k_4)}$$

$$K_B = \frac{k_4(k_{-2} + k_3)}{k_2(k_3 + k_4)}$$

$$K_C = \frac{k_3k_4}{k_1(k_3 + k_4)}$$

and $[SO_3Co(II)TSP^{4-}]_T$ represents the total catalyst concentration,

$$[SO_3Co(II)TSP^{4-}]_T = [SO_3Co(II)TSP^{4-}] + [SO_3Co(III)TSPO_2^{4-}\cdot] + [SO_3Co(III)TSPO_2\cdot SO_3^{6-}] + [SO_3Co(II)TSPO_2^{6-}]. \quad (23)$$

Equation (22) must be modified to account for rapid equilibria that preceded the catalytic cycle involving $SO_3Co(II)TSP^{4-}$ as the active center. The concentration of this active center can be expressed in terms of the dimer dissociation constant and the formation constant for the initial sulfito complex,

$$K_d = \frac{[Co(II)TSP^{2-}]^2}{[(Co(II)TSP)_2^{4-}]}, \quad (24)$$

$$\beta = \frac{[SO_3Co(II)TSP^{4-}]}{[Co(II)TSP^{2-}][SO_3^{2-}]}. \quad (25)$$

Equations (24) and (25) can be combined to give

$$[SO_3Co(II)TSP^{4-}] = \beta K_d^{1/2}[(Co(II)TSP)_2^{4-}]^{1/2}[SO_3^{2-}]. \quad (26)$$

The expression in Eq. (26) can be further reduced since $[SO_3^{2-}]_0 \gg [(Co(II)TSP)_2^{4-}]_0$. This condition allows the approximation that $[SO_3^{2-}]$ is relatively constant with respect to the amount tied up in the complex $SO_3Co(II)TSP^{4-}$, such that

$$[SO_3Co(II)TAP^{4-}]_T \simeq K'[(Co(II)TSP)_2^{4-}]^{1/2}, \quad (27)$$

where $K' = \beta K_d^{1/2}[SO_3^{2-}]$ (i.e., K' is a pseudoequilibrium constant). Equation (27) can then be substituted into Eq. (22) to give an approximate rate expression:

$$v = \frac{k'K'[(Co(II)TSP)_2^{4-}]^{1/2}[SO_3^{2-}][O_2]}{K_A + K_B[O_2] + K_C[SO_3^{2-}] + [O_2][SO_3^{2-}]}. \quad (28)$$

Equation (28) can be simplified for the experimental conditions $[O_2] \gg [SO_3^{2-}]$ such that $k_B[O_2] \gg K_A$, $K_C[SO_3^{2-}]$, yielding

$$v \simeq \frac{k'K'[(Co(II)TSP)_2^{4-}]^{1/2}[SO_3^{2-}]}{K_B + [SO_3^{2-}]}. \quad (29)$$

Two extremes can be considered for Eq. (29). If $K_B \gg [SO_3^{2-}]$,

$$v \simeq k'\beta K_d^{1/2}[(Co(II)TSP)_2^{4-}]^{1/2}[SO_3^{2-}]^2 \quad (30)$$

whereas if $[SO_3^{2-}] \gg K_B$,

$$v \simeq k'\beta K_d^{1/2}[(Co(II)TSP)_2^{4-}]^{1/2}[SO_3^{2-}]. \quad (31)$$

The theoretical rate expressions for the limiting cases given in Eqs. (30) and (31) can be compared with the experimentally observed rate laws for pH 6.7 and 9.4 [Eqs. (20) and (21)]. Since the dimer is the dominant catalyst species in solution at pH 6.7, a one-half-order dependence on total added catalyst is consistent with the assumption that the monomer is actually the active form. Nonintegral reaction orders arise frequently in polar reactions when the principal reactive species is derived from the dissociation of a dimer (Frost and Pearson, 1961). At pH 9.4 a first-order dependence on the catalyst concentration is observed. This result is consistent with this kinetic formulation if the dimeric form of the catalyst is no longer the dominant species. Cookson et al. (1977) see evidence for a shift in the monomer–dimer equilibrium toward the monomer with an increase in pH. With an increase in pH, the remaining solvent molecules in the coordination sphere of $Co(II)TSP^{2-}$ are replaced by hydroxyl groups, resulting in an increase in stability of the monomeric species $HOCo(II)TSP^{3-}$, $(HO)_2Co(II)TSP^{4-}$, and $HOCo(III)\ TSPO_2^{3-}\cdot$. If this is, in fact, the situation, the prior equilibrium can be neglected. Consequently, the theoretical rate expression would show a first-order dependence in catalyst concentration as indicated in Eq. (22).

The mechanism proposed above appears to explain the kinetic and spectral data over a broad range of conditions. The extremes of Eq. (29)–(31) explain both the apparent shifts in the reaction order of sulfite to values greater than 1 as reported by some investigators (Table 1) and the apparent zero-order dependence on oxygen.

A variant of the above mechanism occurs when SO_3^{2-} reacts with $SO_3Co(III)TSPO_2^{4-}\cdot$ without the formation of a ternary complex of sufficiently long life to be kinetically significant. This type of mechanism was originally suggested by Theorell and Chance (1951); the resulting form of the theoretical rate expression is identical to Eq. (22) except that the constants K_A, K_B, and K_C are given by different combinations of rate constants. This second alternative would be consistent with the observed ionic strength dependence (Boyce et al., 1983), which indicates that two negatively charged species are interacting in a rate-determining step such as reaction [72].

For verification of the applicability of Eq. (28) to experimental data and to obtain actual rate constants, a double-reciprocal analysis of initial rate data is required. Equation (28) can be rearranged to give

$$\frac{1}{v_0} = \frac{1 + K_A/[O_2][SO_3^{2-}] + K_B/[SO_3^{2-}] + K_C/[O_2]}{V_0}, \tag{32}$$

where $V_0 = (k_3k_4/k_3 + k_4)\beta K_d^{1/2}[SO_3^{2-}][(Co(II)TSP)_2^{4-}]^{1/2}$ at pH 6.7. A plot of v_0^{-1} versus $[O_2]_0^{-1}$ at constant $[SO_3^{2-}]$ should be linear, and conversely, a plot of v_0^{-1} versus $[SO_3^{2-}]_0^{-1}$ at constant $[O_2]$ should be linear. These linear relationships will yield values for the four kinetic parameters V_0, K_A, K_B, and K_C.

4.3. A One-Electron Transfer Chain Reaction

A second, although less likely, mechanism for the catalytic action of Co(II)TSP is shown in Figure 6. In this mechanistic sequence, the ternary

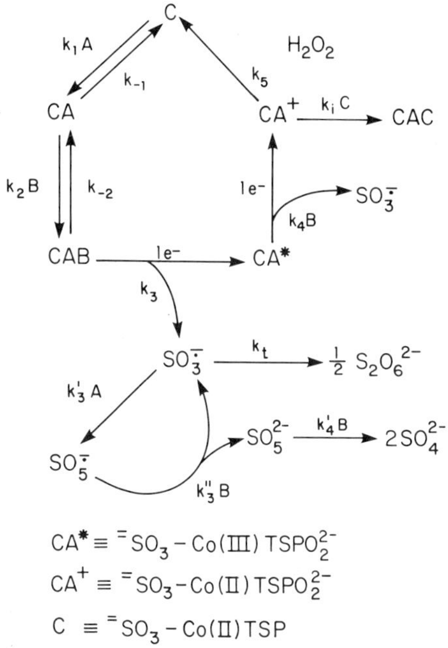

Figure 6. Alternative reaction mechanism for the Co(II)TSP-catalyzed autoxidation of sulfite involving a one-electron transfer radical pathway in which the complex, $SO_3Co(II)TSPO_2^{4-}\cdot$, acts as the primary chain-reaction initiator.

dioxygen–sulfite complex (C) initiates a chain reaction identical in nature to the mechanism given previously in reactions [25]–[31] or [33]–[38]. An additional one-electron transfer to the reduced form of the ternary complex would produce H_2O_2 as an intermediate reduction product. If the complexation steps are assumed to be rapidly attained prior equilibria, the theoretical rate expression would be expected to be of the form

$$\nu = k'[M^{n+}]^{1/2}[SO_3^{2-}]^{3/2}, \tag{33}$$

where in this case $[M^{n+}] = [SO_3Co(III)TSPO_2^{4-}\cdot]$ and $[SO_3Co(III)\text{-}TSPO_2^{4-}\cdot] \simeq \beta K_d^{1/2}[(Co(II)TSP)_2^{4-}]^{1/2}[SO_3][O_2]$. Substitution of these relationships into Eq. (33) gives a rate expression that is second order in sulfite and first order in oxygen. These kinetic dependences are not observed; therefore, the one-electron transfer mechanism is an unlikely candidate.

Alternative one-electron transfer or two electron pathways involving the μ-superoxo complex of Co(II)TSP can be considered as suggested by Davies et al. (1969) for $(NH_3)_5Co(III)O_2Co(II)(NH_3)_5^{5+}$ and by Yatsimirskii et al. (1977) for (L-histidine)$_2$Co(III)O$_2$Co(II)(L-histidine)$_2$. In this mechanism the μ-superoxo complex $(CoTSP)_2O_2^{4-}$ is the active catalytic center and reacts as follows:

$$(CoTSP)_2O_2^{4-} + SO_3^{2-} \xrightarrow{\text{slow}} SO_4^{2-} + Co(II)TSP + Co(IV)TSPO \quad [76]$$

$$Co(IV)TSPO + SO_3^{2-} \longrightarrow SO_4^{2-} + Co(II)TSP \quad [77]$$

or

$$(CoTSP)_2O_2^{4-} + SO_3^{2-} \xrightarrow{\text{slow}} Co(III)TSPO_2Co(II)TSP^{5-} + SO_3^{-}\cdot \quad [78]$$

$$Co(III)TSPO_2Co(II)TSP^{5-} + SO_3^{-}\cdot \longrightarrow Co(II)TSPO_2^{4-} + SO_3 + Co(II)TSP^{2-} \quad [79]$$

$$SO_3 + H_2O \longrightarrow H_2SO_4 \quad [80]$$

$$Co(II)TSPO_2^{4-} + 2H^+ \longrightarrow Co(II)TSP^{2-} + H_2O_2. \quad [81]$$

These mechanistic pathways, although chemically viable, seem to be less consistent with kinetic observations than does the bisubstrate complexation mechanism. For example, these mechanisms predict a second-order dependence on the Co(II)TSP monomer or a first-order dependence on the corresponding dimer and first-order dependence on oxygen. These conditions were not observed by Boyce et al. (1983).

4.4. Alternative Mechanisms

Anast and Margerum (1981) have reported that Cu(III) tetraglycine reacts rapidly with sulfite in two reversible one-electron steps to give Cu(II) tetraglycine and aquated SO_3 as follows:

$$Cu(III)(H_{-3}G_4)^- + SO_3^{2-} \underset{k_{-1}}{\overset{k_1}{\rightleftharpoons}} Cu(II)(H_{-3}G_4)^{2-} + SO_3^{-}\cdot \quad [82]$$

$$Cu(III)(H_{-3}G_4)^- + SO_3^{-}\cdot \underset{k_{-2}}{\overset{k_2}{\rightleftharpoons}} Cu(II)(H_{-3}G_4)^{2-} + SO_3 \quad [83]$$

$$SO_3 + H_2O \rightarrow 2H^+ + SO_4^{2-}. \quad [84]$$

Anast and Margerum (1981) propose that in the presence of oxygen the sulfite radical anion reacts with oxygen to give the peroxymonosulfite radical (reaction [26]), which then oxidizes the Cu(II) back to Cu(III):

$$SO_3^{-}\cdot + O_2 \overset{k_4}{\rightarrow} SO_5^{-}\cdot \quad [85]$$

$$Cu(II)(H_{-3}G_4)^{2-} + SO_5^{-}\cdot \rightarrow Cu(III)(H_{-3}G_4)^- + SO_5^{2-} \quad [86]$$

$$2Cu(II)(H_{-3}G_4)^{2-} + SO_5^{2-} \rightarrow 2Cu(III)(H_{-3}G_4)^- + SO_4^{2-} + OH^-. \quad [87]$$

This mechanism is an example of an induced catalytic reaction in which the Cu(III)–Cu(II) tetraglycine couple catalyzes the autoxidation of sulfite to sulfate; the latter reaction can be described by the simple overall stoichiometry

$$2SO_3^{2-} + O_2 \xrightarrow{Cu(III)(H_{-3}G_4)^-} 2SO_4^{2-}. \quad [88]$$

Anast and Margerum (1981) obtained the following rate law from reactions [82]–[87]:

$$\frac{d[SO_3^{2-}]}{dt} = \frac{k_1k_2k_3[Cu(III)(H_{-3}G_4)^-]^2[SO_3^{2-}]}{C} \quad (34)$$

where

$$C = k_2k_3[\mathrm{Cu(III)(H_{-3}G_4)^-}] + k_{-1}k_3[\mathrm{Cu(II)(H_{-3}G_4)^{2-}}] + k_1k_{-2}[\mathrm{Cu(II)(H_{-3}G_4)^{2-}}]^2.$$

Values for k_1, k_2/k_{-1}, and k_{-2}/k_3 were reported to be 3.7×10^4 M^{-1} s^{-1}, 1.66 M^{-1}, and 177 M^{-1}, respectively. In this reaction system the oxidation of SO_3^{2-} by O_2 promotes an induced oxidation of Cu(II) to Cu(III) by an autocatalytic process in which there is a net gain in $Cu(III)(H_{-3}G_4)^-$ when $Cu(II)(H_{-3}G_4)^{2-}$, O_2, and SO_3^{2-} are present initially. Such a process could take place by the sequence [82]–[85]–[86]–[87], which would have stoichometry

$$SO_3^{2-} + 2Cu(II)(H_{-3}G_4) + O_2 \rightarrow SO_4^{2-} + Cu(III)(H_{-3}G_4). \qquad [89]$$

This reaction would take place in parallel with the overall reaction [88].

An induced oxidation mechanism of this type represents a viable conceptual alternative to a free-radical chain sequence (reactions [25]–[31]) even though $SO_3^- \cdot$ and $SO_5^- \cdot$ are postulated as reactive intermediates.

4.5. Heterogeneous Catalysis by Hybrid Complexes

Solid-supported analogs of Co(II) tetrasulfophthalocyanine and tetraminophthalocyanine (Figure 2) have been synthesized and examined for catalytic activity toward the autoxidation of sulfite (Boyce et al. 1983). Such solid-supported catalysts can be used as models for examining changes in catalytic behavior when a homogeneous catalyst is chemically anchored or physically adsorbed to a solid surface.

Kinetic data in these heterogeneous reaction systems were obtained by monitoring changes in dissolved oxygen concentration as a function of time as illustrated in Figure 7. The most effective hybrid catalyst appears to be the Co(II)TAP complex covalently linked through a peripheral amino group of the phthalocyanine ring. Attachment achieved by direct complexation of the surface-bound imidazole to the Co(II) of the ring system results in lower catalytic activity, and apparent inactivation occurs after one half-life or less. However, when the activity of imidazole-bonded Co(II)TSP is normalized with respect to the surface concentration of the covalently bonded Co(II)TAP, there appear to be no ostensible differences in net activity, although some of the activity of the imidazole–Co(II)TSP system may be due to the dissociation of the bound complex to give active Co(II)TSP in solution.

McCord and Fridovich (1969) reported that dimethylsulfoxide (DMSO) was an effective catalyst for the autoxidation of sulfite. Since DMSO was used in the preparation of Co(II)TAP, Boyce et al. (1983) examined its catalytic effect as shown in Figure 7; however, the effectiveness of DMSO as a catalyst is significantly less than the hybrid catalysts when normalized with respect to concentration.

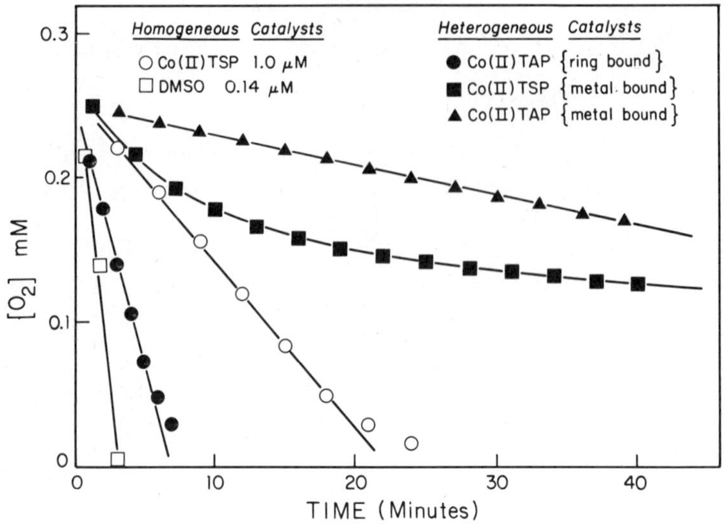

Figure 7. Comparison of catalytic activity of various solid supported cobalt phthalocyanine complexes toward autoxidation of sulfite at pH 6.7; $[S(IV)]_0 = 1.0$ mM and $[O_2]_0 = 0.25$ mM. From Boyce et al. (1983).

In general, the rates of reactions catalyzed by hybrid catalysts tend to be slower than those catalyzed by the corresponding homogeneous form of the catalyst. However, there are reports of improved catalytic activity and of longer catalyst lifetimes when homogeneous catalysts are anchored to solid supports (Hartley and Vezey, 1977; Pittman et al., 1975; Robinson, 1976). In a hybrid catalyst system the reaction is constrained to take place on the surface of the catalyst; but in the case of catalyst molecules attached to pore wall surfaces of the solid support, substrate accessibility may limit reactivity. In contrast, in a homogeneous system all the catalyst molecules are available as catalytic centers. Therefore, the homogeneous catalysts are potentially more efficient in terms of the absolute amount of a catalyst needed to promote a reaction to a given extent (Hartley and Vezey, 1977). For a hybrid catalytic system, the catalytic activity will depend on the solid-phase properties such as surface area, porosity, and the degree of cross-linkage in polymeric supports.

5. SUMMARY AND CONCLUSIONS

Three distinctly different reaction pathways for the catalytic autoxidation of dissolved S(IV) have been discussed. These are free-radical chain mechanisms involving initiation by reducible metal ions or light, polar pathways involving inner-sphere complexation of metal and substrate, and photo-assisted pathways involving homogeneous or heterogeneous metal catalysts.

Free-radical mechanisms proposed by Bäckström (1934), Schmittkunz (1963), Hayon et al. (1972) and others are often inadequate to explain experimental observations. Reaction orders in metal, S(IV), and oxygen predicted by these mechanisms are either zero or nonintegral, whereas most empirical reaction orders are integral. For initiation of a free-radical reaction according to the proposed mechanisms, a suitable one-electron oxidant is required. Few metals meet this criterion.

As suggested previously by Bassett and Parker (1951), Matteson et al. (1969), Freiberg (1975), and Boyce et al. (1983), certain metal-catalyzed autoxidations of sulfite proceed through the formation of discrete inner-sphere complexes between the reductant, SO_3^{2-}, and the catalyst as a prelude to electron transfer. As noted in Section 3.2, few metal sulfite stability constants have been determined. Determination of formation constants for metal sulfite complexes in aqueous solution should be an area of active research.

Additional experimental evidence examined herein suggests that the binding and subsequent activation of dioxygen plays a significant role in the catalytic cycle. From this perspective, the most effective catalysts should involve complexes of Fe(II)–Fe(III), Mn(II)–Mn(III), Co(II)–Co(III), and V(III)–V(IV), in which the central metal is reversibly oxidized and reduced on complexation by oxygen and/or sulfite. Inhibition of catalytic activity by strong chelating agents and complexing agents supports this conclusion.

Certain metal-catalyzed reactions of sulfite may be enhanced by a photoassisted pathway as shown by Boyce et al. (1983) for Co(II) and by Lunak and Veprek-Siska (1976) for Fe(III). This apparent coupling of photolytic and metal-catalyzed processes may help to explain relative differences between nighttime and daytime SO_2 conversion rates. More work is needed in this area in order to extend and apply this concept to aerosol systems.

In atmospheric aqueous aerosol systems, important factors to consider are the nature and roles of dissolved organic molecules that can act as competitive complexing agents for metals. For example, the liquid-phase autoxidation of benzaldehyde produces benzoic acid which can act as a suitable complexing agent [e.g., $pK_{a1} = 3.97$, $\log \beta_{11} = 1.51$ for $Cu(C_7H_6O_2)^+$] and a similar oxidation of 2-hydroxybenzaldehyde to 2-hydroxybenzoic acid produces even a stronger potential ligand [e.g., $pK_{a1} = 2.78$, $\log \beta_{11} = 10.13$ for $Cu(C_7H_6O_3)^+$. The presence of complexing agents of this type will accelerate the dissolution of Fe_2O_3 and MnO_2, which are the likely sources of soluble iron and manganese in aerosol systems. As shown recently by Cohen et al. (1981), the catalytic activity of soot-derived aerosols correlates well with the total iron released to the liquid phase.

Results discussed in this chapter indicate that H_2O_2 is a major intermediate reduction product of the catalyzed oxidation of various sulfur compounds and oxidizable organic molecules. As such, metal-catalyzed autoxidation reactions in liquid aerosols may be a significant source of atmospheric peroxide as reported by Kok (1981). However, accumulation of H_2O_2 would not occur to an appreciable extent in liquid aerosols that contain or are exposed to high concentrations of SO_2 (Hoffmann and Edwards, 1975).

Preliminary research on hybrid organometallic catalysts has shown that attachment of homogeneous catalysts to solid surfaces results in a negligible loss of catalytic activity for the autoxidation of dissolved S(IV).

Improved catalytic ability, elimination of recovery problems, and longer catalyst lifetimes may be achieved with supported organometallic catalysts (Pittman et al., 1975; Robinson, 1976). Mass et al. (1976) and Schutten and Beelen (1981) have reported an enhanced catalytic autoxidation of mercapto-ethanol by Co(II) tetraminophthalocyanine attached to cross-linked poly-acrylamide and polyvinylamine, respectively. Autoxidation of Klaus plant effluents or SO_2 in flue gases by supported organometallic complexes such as the hybrid phthalocyanines (Boyce et al. 1983) may provide an alternative method for SO_2 emission control. Figure 8 illustrates a hypothetical countercurrent reactor with fixed beds of solid-supported catalyst that could be used for SO_2 scrubbing with the production of H_2SO_4 as an alternative to

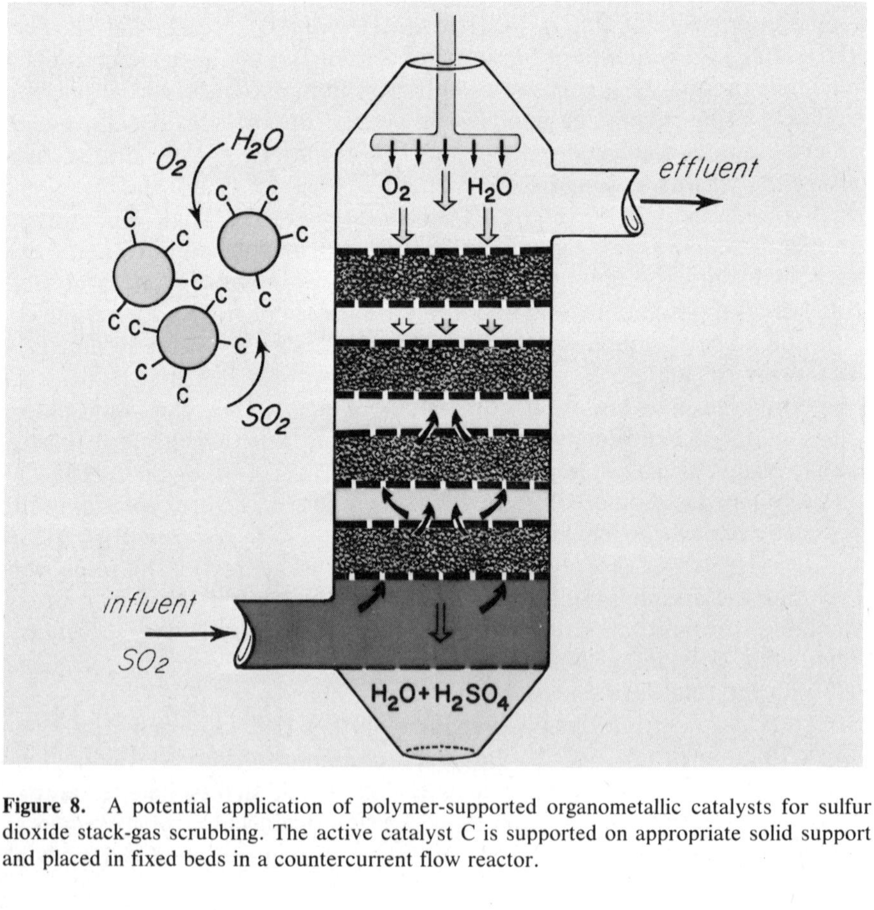

Figure 8. A potential application of polymer-supported organometallic catalysts for sulfur dioxide stack-gas scrubbing. The active catalyst C is supported on appropriate solid support and placed in fixed beds in a countercurrent flow reactor.

limestone slurry scrubbers, which produce an unusable solid reaction product.

ACKNOWLEDGMENTS

The authors acknowledge gratefully the financial support of the U.S. Environmental Protection Agency (EPA) (Grant no. R808086-01) and President's Fund/Sloan Foundation Grant Administered by the California Institute of Technology. They are also indebted to their colleagues in the air pollution field, Drs. P. McMurry, G. Cass, R. Flagan, G. McRae, and J. Seinfeld, for their heuristic discussions on the subject of aerosol chemistry.

REFERENCES

Abel, E. W., Pratt, J. M., and Whelan, R. (1971). Formation of a 1:1 oxygen adduct with the cobalt(II)–tetrasulphophthalocyanine complex. *Chem. Commun.* 449–450.

Alyea, H. N. and Bäckström, L. J. (1929). The inhibition action of alcohols on the oxidation of sodium sulfite. *J. Am. Chem. Soc.* **51,** 90–109.

Anast, J. M. and Margerum, D. W. (1981). Trivalent copper catalysis of the autoxidation of sulfite. Kinetics and mechanism of the copper(III/II) tetraglycine reactions with sulfite. *Inorg. Chem.* **20,** 2319–2326.

Bäckström, H. (1934). Der Kettenmechanisms bei der Autoxydation von Natrium-sulfirlösungen. *Z. Phys. Chem.* **25B,** 122–138.

Baker, A. D., Casadavell, A., Gafney, H. D., and Gellender, M. (1980). Photochemical reactions of tris(oxalato) iron(III). *J. Chem. Educ.* 314–315.

Balzani, V. and Carassiti, V. (1970). *Photochemistry of Coordination Compounds,* Academic Press, New York.

Ball, D. L., and Edwards, J. O. (1958). The catalysis of the decomposition of Caro's acid. *J. Phys. Chem.* **62,** 343–345.

Barron, C. H. and O'Hern, H. A. (1966). Reaction kinetics of sodium sulfite by the rapid-mixing method. *Chem. Eng. Sci.* **21,** 397–404.

Barrett, E. and Brodin, G. (1955). The acidity of Scandinavian precipitation. *Tellus* **7,** 251–257.

Basolo, F., Hoffman, B. M., and Ibers, J. A. (1975). Synthetic oxygen carriers of biological interest. *Acc. Chem. Res.* **8,** 384–392.

Bassett, H. and Parker, W. G. (1951). The oxidation of sulphurous acid. *J. Chem. Soc.,* 1540–1560.

Beilke, S., and Gravenhorst, G. (1978). Heterogeneous SO_2-oxidation in the droplet phase. *Atmos. Environ.* **12:**231–239.

Bell, R. P. (1971). *Acid and Base Catalysis,* 2nd ed., Cornell University Press, Ithaca, NY.

Bengtsson, S. and Bjerle, I. (1975). Catalytic oxidation of sulphite in dilute aqueous solutions. *Chem. Eng. Sci.* **30,** 1429–1435.

Boudart, M. (1968). *The Kinetics of Chemical Processes,* Prentice-Hall, Englewood-Cliffs, NJ, pp. 246.

Boyce, S. D., Hoffmann, M. R., Hong, P. A. and Moberly, L. M. (1983). Catalysis of the autoxidation of aquated sulfur dioxide by homogeneous and heterogeneous transition metal complexes, in *Acid Precipitation: SO_2, NO, and NO_2 Oxidation Mechanisms—*

Atmospheric Considerations. J. G. Calvert, Ed. Ann Arbor Science Publishers, Ann Arbor MI (in press).

Bredig, G. (1909). *Ulmann's Enzyke Technol. Chem.* **6,** 670.

Brezonik, P. L., Edgerton, E. S., and Hendry, C. C. (1980). Acid precipitation and sulfate deposition in Florida. *Science* **208,** 1027–1029.

Brimblecombe, P., and Spedding, D. J. (1974). The catalytic oxidation of micromolar aqueous sulphur dioxide. *Atmosph. Environ.* **8,** 937–945.

Burrows, J. P., Cliff, D. I., Harris, G. W., Thrush, B. A., and Wilkinson, J. P. T. (1979). Atmospheric reactions of the HO_2 radical studied by laser magnetic resonance spectroscopy. *Proc. Roy. Soc.* (*Lond.*) **A368,** 463–481.

Calvert, J. G., Su, F., Bottenheim, J. W., and Strausz, O. P. (1978). Mechanisms of homogeneous oxidation of sulfur dioxide in the troposphere. *Atmosph. Environ.* **12,** 197–226.

Campbell, M. J., Sheppard, J., and Crown, D. (1979). Measurement of hydroxyl radical as part of STATE program. *Geophys. Res. Lett.* **6,** 175.

Carter, J. J., Rillema, D. P., and Basolo, F. (1974). Oxygen carrier and redox properties of some neutral cobalt chelates. *J. Am. Chem. Soc.* **96,** 392–400.

Cass, G. R. (1977). Methods for sulfate air quality management with applications to Los Angeles. Ph.D. Thesis, Environmental Engineering Sciences, California Institute of Technology, Pasadena, CA.

Cass, G. R. and Shair, F. H. (1980). Transport of sulfur oxides within the Los Angeles sea breeze/land breeze circulation system, *Proceedings of the 2nd Joint Conference on Applications of Air Pollution Meteorology,* American Meteorological Society, pp. 320–327.

Chen, T. I., and Barron, C. H. (1972). Some aspects of the homogeneous kinetics of sulfite oxidation. *Ind. Eng. Chem. Fundam.* **11:** 466–470.

Cheng, R. T., Corn, M., and Frohliger, J. O. (1971). Contributions to the reaction kinetics of water-soluble aerosols and SO_2 in air at ppm concentrations. *Atmosph. Environ.* **5,** 987–1008.

Chen, K. Y. and Morris, J. C. (1972). Oxidation of sulfide by O_2: Catalysis and inhibition. *J. Sanit. Eng. Div. ASCE* **98,** 215–227.

Childs, L. P. and Ollis, D. F. (1980). Is photocatalysis catalytic? *J. Catal.* **66,** 383–390.

Cohen, S., Chang, S., Markowitz, S., and Novakov, T. (1981). Role of fly ash in catalytic oxidation of S(IV) slurries. *Environ. Sci. Technol.* **15,** 1498–1502.

Cook, A. H. (1938). Catalytic properties of the phthalocyanine. Part II. Oxidase properties. *J. Chem. Soc.* 1761–1780.

Cookson, D. J., Smith, T. D., Boas, J. F., Hicks, P. R., and Pilbrow, J. R. (1977). Electron spin resonance study of the autoxidation of hydrazine, hydroxylamine and cysteine catalyzed by the cobalt(II) chelate complex of 3,10,17,24-tetrasulphophthalocyanine. *J. Chem. Soc.* 109–14.

Coughanowr, D. R. and Krause, F. E. (1965). The reaction of SO_2 and O_2 in aqueous solutions of $MnSO_4$. *Ind. Eng. Chem. Fundam.* **4,** 61–66.

Cox, G. S. and Whitten, D. G. (1979). Interaction of porphyrin and metalloporphyrin excited states with molecular oxygen. Energy transfer versus electron-transfer quenching mechanisms in photooxidations. *Chem. Phys. Lett.* **67,** 511–515.

Cox, R. A. (1974). Particle formation from homogeneous reactions of sulfur dioxide and nitrogen dioxide. *Tellus* **26,** 235–240.

Dance, I. G., Conrad, R. C., and Cline, J. E. (1974). Mechanism of cobalt dithiolene complex catalysis of thiol autoxidation of acidic acetonitrile solution. *Chem. Commun.* 13–14.

Davies, R. A., Hagopian, K. E., and Sykes, A. G. (1969). Kinetic and oxygen-18 tracer studies on the reaction of sulphite with the superoxo- complex $(NH_3)_5Co{\cdot}O_2{\cdot}Co(NH_3)_5^{5+}$ in aqueous media. *J. Chem. Soc.* **(A)**, 623–629.

Dogliotti, L. and Hayon, E. (1967). Flash photolysis study of sulfite, thiocyanate and thiosulfate ions in solution. *J. Phys. Chem.* **72,** 1800–1807.

Dolansky, J., Wagnerova, D. M., and Veprek-Siska, J. (1976). Autooxidation of cysteine catalyzed by cobalt(II) tetrasulphophthalocyanine models of oxidases V. *Collect. Czech. Chem. Commun.* **41,** 2326–2332.

Edwards, J. O. (1965). *Inorganic Reaction Mechanisms,* Benjamin, New York.

Faust, B. C. and Hoffmann, M. R. (1982). Unpublished results.

Fieser, M. and Fieser, L. F. (1967, 1969). *Reagents for Organic Synthesis,* Vols. 1 and 2, Wiley, New York.

Frank, S. N. and Bard, A. J. (1977). Heterogeneous photocatalytic oxidation of cyanide and sulfite in aqueous solution at semiconductor powders. *J. Phys. Chem.* **81,** 1484–1488.

Freiberg, J. (1974). Effects of relative humidity and temperature on iron-catalyzed oxidation of SO_2 in atmospheric aerosols. *Environ. Sci. Technol.* **8,** 731–734.

Freiberg, J. (1975). The mechanism of iron catalyzed oxidation of SO_2 in oxygenated solutions. *Atmosph. Environ.* **9,** 661–672.

Frost, A. A. and Pearson, R. G. (1961). *Kinetics and Mechanism,* Wiley, New York.

Fuller, E. C. and Crist, R. H. (1941). The rate of oxidation of sulfite ions by oxygen. *J. Am. Chem. Soc.* **63,** 1644–1650.

Fuzzi, S. (1978). Study of iron(III) catalyzed sulphur dioxide oxidation in aqueous solution over a wide range of pH. *Atmosph. Environ.* **12,** 1439–1442.

Gartrell, J. E., Thomas, J. W., and Carpenter S. B. (1963). Atmospheric oxidation of SO_2 in coal-burning power plant plumes. *Am. Ind. Hyg. Assoc. Quart.* **24,** 113–120.

Gorham, E. (1955). On the acidity and salinity of rain. *Geochim. Cosmochim. Acta* **7,** 231–239.

Graham, R. A., Winer, A. M., Atkinson, R., and Pitts, J. N., Jr. (1979). Rate constants for the reaction of HO_2 with HO_2, SO_2, CO, N_2O, *trans*-2-butene, and 2,3-dimethyl-2-butene at 300 K. *J. Phys. Chem.* **83,** 1563–1566.

Gray, R. D. (1969). The kinetics of oxidation of copper(I) by molecular oxygen in perchloric acid solutions. *J. Am. Chem. Soc.* **91,** 56.

Haber, F. and Wansbrough-Jones, O. H. (1931). Über die Einwirkung des Lichtes auf sauerstoffreie and sauerstoffhaltige Sulfitlösung. *Z. Phys. Chem.* **B18,** 103–123.

Hansen, L. D., Whiting, L., Eatough, D. J., Jensen, T. E., and Izatt, R. M. (1976). Determination of Sulfur (IV) and Sulfate in Aerosols by Thermometric Methods. *Anal. Chem.* **48,** 634–638.

Hara, T., Ohkatusu, Y., and Osa, T. (1975). Catalytic activity of metal phthalocyanines in autoxidation reactions. *Bull. Chem. Soc. Jap.* **48,** 85–89.

Hartley, F. R. and Vezey, P. N. (1977). Supported transition metal complexes as catalysts. *Adv. Organomet. Chem.* **15,** 189–234.

Hayon, E., Treinin, A., and Wilf, J. (1972). Electronic spectra, photochemistry, and autoxidation mechanism of the sulfite–bisulfite–pyrosulfite systems. The SO_2^-, SO_3^-, and SO_5^- radicals. *J. Am. Chem. Soc.* **94,** 47–57.

Hegg, D. A. and Hobbs, P. V. (1978). Oxidation of sulfur dioxide in aqueous systems with particular reference to the atmosphere. *Atmosph. Environ.* **12,** 241–253.

Hewitt, D. G. (1970). The oxidation of 2,4-di-*t*-butylphenol by Cu–amine complexes. *J. Chem. Soc.* **D,** 227–228.

Higginson, W. C. E. and Marshall, J. W. (1957). Equivalence changes in oxidation–reduction reactions in solution: Some aspects of the oxidation of sulphurous acid. *J. Chem. Soc.* 447–458.

Hoffmann, M. R. (1980). Trace metal catalysis in aquatic environments. *Environ. Sci. Technol.* **14,** 1061–1066.

Hoffmann, M. R. (1977). Kinetics and mechanism of the oxidation of hydrogen sulfide by hydrogen peroxide in acidic solution. *Environ. Sci. Technol.* **11,** 61–66.

Hoffmann, M. R. and Edwards, J. O. (1975). Kinetics and mechanism of the oxidation of sulfur dioxide by hydrogen peroxide in acidic solution. *J. Phys. Chem.* **79,** 2096–2098.

Hoffmann, M. R. and Edwards, J. O. (1977). Kinetics and mechanisms of the oxidation of thiourea and *N,N*-dialkylthioureas by hydrogen peroxide. *Inorg. Chem.* **16,** 3333–3338.

Hoffmann, M. R. and Lim, B. C. H. (1979). Kinetics and mechanism of the oxidation of sulfide by oxygen: Catalysis by homogeneous metal–phythalocyanine complexes. *Environ. Sci. Technol.* **13,** 1406–1413.

Holt, B. D., Kumar, R., and Cunningham, P. T. (1981). Oxygen-18 study of the aqueous phase oxidation of sulfur dioxide. *Atmosph. Environ.* **15,** 557–566.

Houghton, H. G. (1955). On the chemical composition of fog and cloud water. *J. Meteorol.* **12,** 355–357.

Jones, R. D., Summerville, D. A., and Basolo, F. (1979). Synthetic oxygen carriers related to biological systems. *Chem. Rev.* **79,** 139–179.

Kaplan, D. J., Himmelblau, D. M., and Kanaoka, C. (1981). Oxidation of sulfur dioxide in aqueous ammonium sulfate aerosols containing manganese as a catalyst. *Atmosph. Environ.* **15,** 763–773.

Karraker, D. G. (1963). The kinetics of the reaction between sulphurous acid and ferric ion. *J. Phys. Chem.* **67,** 871–874.

King, E. L. (1955). Catalysis in homogeneous reactions in a liquid phase, in *Catalysis: Fundamental Principles,* P. H. Emmett, Ed., New York, Reinhold, Vol. II, Part 2.

King, E. L. and Altman C. (1956). A schematic method of deriving the rate laws for enzyme catalyzed reactions. *J. Phys. Chem.* **60,** 1375–1378.

Knights, R. L. (1980). Analysis of particulate organic air pollutants by high resolution mass spectrometry, in *The Character and Origins of Smog Aerosols,* G. M. Hidy, P. K. Mueller, D. Grosjean, B. R. Appel, and J. J. Wesolowski, Eds., *Adv. Environ. Sci. Technol.* **9,** 237–251.

Kok, G. L. (1980). Measurements of hydrogen peroxide in rainwater. *Atmosph. Environ.* **14,** 653–656.

Kok, G. L. (1981). Measurements of hydrogen peroxide in rain water. *EOS Transact. Am. Geophys. Union.* **45,** 884.

Kok, G. L., Holler, T. P., Lopez, M. B., Nachtrieb, H. A., and Yuan, M. (1978a). Chemiluminescent method for determination of hydrogen peroxide in the ambient atmosphere. *Environ. Sci. Technol.* **12,** 1072–1076.

Kok, G. L., Darnall, K. R., Winer, A. M., Pitts, J. N., Jr., and Gay,B. W. (1978b). Ambient air measurements of hydrogen peroxide in the California south coast basin. *Environ. Sci. Technol.* **12,** 1077–1080.

Kothari, V. M. and Tazuma, J. J. (1976). Selective autoxidation of some phenols using salcomines and metal phthalocyanines. *J. Catal.* **41,** 180–189.

Krebs, H. A. (1929). Über der Wirkung der Schweimetalle auf die Autoxydation der Alkalisulfide und des Schwefelwasserstoffs. *Biochem. Z.* **204,** 343–346.

Kundo, N. N. and Kejer, N. P. (1968). Mechanism of the catalytic action of cobalt tetrasulfophthalocyanine. *Russ. J. Phys. Chem.* **42,** 707–711.

Kundo, N. N., Kejer, N. P., Glazneva, G. V., and Mamseva, E. K. (1967). Catalytic properties of phthalocyanines in cysteine oxidation. *Kinet. Catal.* **8,** 1119–1124.

Laidler, K. J. and Bunting, P. S. (1973). *The Chemical Kinetics of Enzyme Action,* Clarendon Press, Oxford.

Langford, C. H. and Carey, J. H. (1975). The charge transfer photochemistry of the hexaaquoiron(III) ion, the chloropentaaquoiron(III) ion, and the μ-dihydroxo dimer explored with *tert*-butyl alcohol scavenging. *Can. J. Chem.* **53,** 2430–2435.

Langford, C. H., Wingham, M., and Sastri, V. S. (1973). Ligand photooxidation in copper(II) complexes of nitrilotriacetic acid. *Environ. Sci. Technol.* **7,** 820–822.

Larson, T. V., Horike, N. R., and Halstead, H. (1978). Oxidation of sulfur dioxide by oxygen and ozone in aqueous solution: A kinetic study with significance to atmospheric rate processes. *Atmosph. Environ.* **12,** 1597–1611.

Likens, G. E., Edgerton, S. S., and Galloway, J. N. (1982). The composition and deposition of organic carbon in precipitation. *Tellus* (in press).

Liljestrand, H. M. and Morgan, J. J. (1978). Chemical composition of acid precipitation in Pasadena, California. *Environ. Sci. Technol.* **12,** 1271 (1978).

Liljestrand, H. M. and Morgan, J. J. (1981). Spatial variations of acid precipitation in Southern California. *Environ. Sci. Technol.* **15,** 333–338.

Linek, V. and Mayrhoferova, J. (1970). The kinetics of oxidation of aqueous sodium sulfite solution. *Chem. Eng. Sci.* **25,** 787–800.

Lockhart, H. B., Jr. and Blakeley, R. V. (1975). Aerobic photodegradation of Fe(III(–(ethylenedinitrilo) tetraacetate (ferric EDTA). *Environ. Sci. Technol.* **12,** 1035–1038.

Lunak, S. and Veprek-Siska, J. (1976). Photochemical autooxidation of sulfite catalyzed by iron(III) ions. *Coll. Czech. Chem. Commun.* **41,** 3495–3503.

Lunde, G., Gether, J., Gjos, N., and Lande, M. S. (1977). Organic micropollutants in precipitation in Norway. *Atmosph. Environ.* **11,** 1007–1014.

McCord, J. M. and Fridovich, I. (1969). The utility of superoxide dismutase in studying free radical reactions. *J. Biol. Chem.* **244,** 6056–6063.

McMurry, P. H., Rader, D. J., and Smith, J. L. (1981). Studies of aerosol formation in power plant plumes. I. Parametrization of conversion rate for dry, moderately polluted ambient conditions. *Atmosph. Environ.* **15,** 2315–2329.

Mader, P. M. (1958). Kinetics of the hydrogen peroxide–sulfite reaction in solution. *J. Am. Chem. Soc.* **80,** 2634–2639.

Martin, R. L. and Damschen, D. E. (1981). Aqueous oxidation of sulfur dioxide by hydrogen peroxide at low pH. *Atmosph. Environ.* **15,** 1615–1622.

Mason, R. B. and Mathews, J. H. (1926). The effect of ultra-violet light on the oxidation of sodium sulfite by atmospheric oxygen. *J. Phys. Chem.* **30,** 414–420.

Mass, T., Kuijer, M., and Zwart, J. (1976). Activation of cobalt–phthalocyanine catalyst by polymer attachment. *Chem. Commun.* 87–88.

Mathews, H. J. and Dewey, L. H. (1912). A quantitative study of some photochemical effects produced by ultra-violet light. *J. Phys. Chem.* **17,** 211–218.

Mathews, H. J. and Weeks, M. E. (1917). The effect of various substances on the photochemical oxidation of solutions of sodium sulfite. *J. Am. Chem. Soc.* **39,** 635–647.

Matsuura, A., Harada, J., Akehata, T., and Shirai, T. (1969). Rate of ammonium sulfite oxidation in aqueous solution. *J. Chem. Eng. Jap.* **2,** 199–203.

Matteson, J. J., Stöber, W., and Luther, H. (1969). Kinetics of the oxidation of sulfur dioxide by aerosols of manganese sulfate. *Ind. Eng. Chem. Fundam.* **8,** 677–684.

Middleton, P., Kiang, C. S., and Mohnen, V. A. (1980). Theoretical estimates of the relative importance of various urban sulfate aerosol production mechanisms. *Atmosph. Environ.* **14,** 463–472.

Miles, C. J. and Brezonik, P. L. (1981). Oxygen consumption in humic-colored waters by a photochemical ferrous–ferric catalytic cycle. *Environ. Sci. Technol.* **15,** 1089–1095.

Möller, D. (1980). Kinetic model of atmospheric SO_2 oxidation based on published data. *Atmosph. Environ.* **14,** 1067–1076.

Newman, L. (1981). Atmospheric oxidation of sulfur dioxide: A review as viewed from power plant and smelter plume studies. *Atmosph. Environ.* **15,** 2231–2239.

Neytzell de Wilde, F. G. and Taverner, L. (1958). Experiments relating to the possible production of an oxidizing acid leach liquor by auto-oxidation for the extraction of uranium. *2nd U.N. International Conference on the Peaceful Uses of Atomic Energy Proceedings,* Vol. 3, pp. 303–317.

Ostwald, W. (1902). Über Katalysé. *Physik. Z.* **3,** 313–322.

Ouimette, J. R. and Flagan, R. C. (1981). Chemical species contributions to light scattering by aerosols and a remote site, *ACS Symposium No. 167. Atmospheric Aerosol: Source/Air Quality Relationships,* E. S. Macias and P. K. Hopke, Ed., American Chemical Society, Washington, DC, pp. 125–156.

Pack, D. H. (1980). Precipitation chemistry patterns: A two-network data set. *Science* **208,** 1143–1145.

Penkett, S. A., Jones, B. M. R., Brice, K. A., and Eggleton, A. E. J. (1979). The importance of atmospheric ozone and hydrogen peroxide in oxidizing sulfur dioxide in cloud and rainwater. *Atmosph. Environ.* **13,** 123–137.

Pittman, C. U., Smith, L. R., Hanes, R. M. (1975). Catalytic reactions using polymer-bound vs. homogeneous complexes of nickel, rhodium and ruthenium. *J. Am. Chem. Soc.* **97,** 1742–1760.

Przywarska-Boniecka, H. and Fried, K. (1976). The influence of additional ligands on autooxidation of cobalt and iron 4,4′,4″,4‴-tetrasulfonated phthalocyanines. *Roczniki Chemii* **50,** 43–52.

Robinson, A. L. (1976). Homogeneous catalysis(II): Anchored metal complexes. *Science* **194,** 1261–1263.

Rollman, L. D. and Chan, S. I. (1971). Electron spin resonance studies of low-spin cobalt(II) complexes. Base adducts of cobalt phthalocyanine. *Inorg. Chem.* **10,** 1978–1982.

Sawicki, J. E. and Barron, C. H. (1973). On the kinetics of sulfite oxidation in heterogeneous systems. *Chem. Eng. J.* **5,** 153–159.

Schmittkunz, H. (1963). Chemilumineszenz der Sulfitooxidation. Dissertation of the Naturwissenschaftliche Fakultät der Universität Frankfurt.

Schutten, J. H. and Beelen, T. P. M. (1981). The role of hydrogen peroxide during the autoxidation of thiols promoted by bifunctional polymer-bonded cobalt phthalocyanine catalysts. *J. Molec. Catal.* **10,** 85–97.

Sheldon, R. A. and Kochi, J. K. (1973). Mechanisms of metal-catalyzed oxidation of organic compounds in the liquid phase. *Oxidation Combust. Rev.* **5,** 135–242.

Sillén, L. G. and Martell, A. E. (1964). *Stability Constants of Metal–Ion Complexes,* Chemical Society, London, Special Publications 17 and 25.

Simonov, A. D., Kejer, N. P., Kundo, N. N., Mamseva, E. I., and Glazneva, G. V. (1973). The catalytic properties of sulfonated cobalt phthalocyanines in the oxidation of cysteine and hydrogen sulfide. *Kinet. Catal.* **14,** 864–868.

Smith, R. A. (1872). *Air and Rain: The Beginning of a Chemical Climatology,* Longmans Green, London, pp. 671.

Smith, F. B. and Jeffrey, G. H. (1975). Airborn transport of sulphur dioxide from the U.K. *Atmosph. Environ.* **9,** 643–659.

Smith, R. M. and Martell, A. M. (1976). *Critical Stability Constants,* Vol. 4, *Inorganic Complexes,* Plenum Press, New York.

Sung, W. and Morgan, J. J. (1981). Kinetics and products of ferrous iron oxygenation in aqueous systems. *Environ. Sci. Technol.* **14,** 561–568.

Taylor, D. D. and Flagan, R. C. (1981). Aerosols from a laboratory pulverized coal combustor. *ACS Symposium Series No. 167. Atmospheric Aerosol: Source/Air Quality Relationships*, E. S. Macias and P. K. Hopke, Eds., American Chemical Society, Washington, DC, pp. 157–172.

Theorell, H. and Chance, B. (1951). Studies on liver alcohol dehydrogenase II. The kinetics of the compound of horse liver alcohol dehydrogenase and reduced diphosphopyridine nucleotide. *Acta Chemica Scand.* **5,** 1127–1144.

Uri, N. (1956). Metal ion catalysis and polarity of environment in the aerobic oxidation of unsaturated fatty acids. *Nature* **177,** 1177–1178.

Wagnerova, D. M., Schwertnerova, E., and Veprek-Siska, J. (1973). Autoxidation of hydrazine catalyzed by tetrasulfophthalocyanines. *Collect. Czech. Commun.* **38,** 756–764.

Wagnerova, D. M., Schwertnerova, E., and Veprek-Siska, J. (1974). Autoxidation of hydroxylamine catalyzed by Co(II) tetrasulfophthalocyanine. Models of oxidases. *Collect. Czech. Chem. Commun.* **39,** 3036–3047.

Wagnerova, D. M., Blanek, J., Smettan, G., and Veprek-Siska, J. (1978) Catalyzed autoxidation of ascorbic acid: Evidence of a ternary complex. *Collect. Czech. Chem. Commun.* **43,** 2105–2110.

Walling, C. (1975). Fenton's reagent revisited. *Acc. Chem. Res.* **12,** 125–131.

Wilson, J. C. and McMurry, P. H. (1981). Studies of aerosol formation in power plant plumes—I. Secondary aerosol formation in the Navajo generating station plume. *Atmosph. Environ.* **15,** 2329–2339.

Yagi, S. and Inoue, H. (1962). The absorption of oxygen into sodium sulfite solution. *Chem. Eng. Sci.* **17,** 411–421.

Yatsimirskii, K. B., Bratushko, I. Yu., and Zatsny, I. L. (1977). Kinetics and mechanism of the reduction of molecular oxygen coordinated in the complex Co_2(L-histidine)$_4O_2$ by sodium sulphite in aqueous solution. *Zh. Neorgan. Khimii* **22,** 1611–1616.

4

ROLE OF CARBON PARTICLES IN ATMOSPHERIC CHEMISTRY

Shih-Ger Chang and Tihomir Novakov

Lawrence Berkeley Laboratory
University of California
Berkeley, California 94720

1. INTRODUCTION

Carbonaceous particles in the atmosphere consist of two major components—graphitic or black carbon (sometimes referred to as *elemental* or *free* carbon) and organic material. The latter can be either directly emitted from sources (primary organics) or produced by atmospheric reactions from gaseous precursors (secondary organics). For the sake of clarity, we define soot as the total primary carbonaceous particulate material, i.e., the sum of black carbon and primary organics. Black carbon is a chemically and catalytically active material that can be an effective carrier for other toxic air pollutants through their adsorptive capability. The chemical, adsorptive, and catalytic behaviors of black carbon particles depend very much on their crystalline structure, surface composition, and electronic properties. This chapter discusses these properties and examines their relevance to atmospheric chemistry.

The assessment of the chemical role of black carbon in the atmosphere in general, and in photochemical environments such as Los Angeles in particular, began with an empirical assessment of the black carbon concentrations. The presence of soot in the atmosphere of industrial cities was once obvious but has become less so more recently. Improvements in combustion technology and the use of better-grade fuels have led to the virtual elimination of visible smoke emissions. The emphasis of air pollution control thus has shifted away from controlling primary particulate emissions toward controlling gaseous emissions, especially in view of the concept of Los Angeles–type photochemical smog, which was believed to contain neither smoke nor fog. According to such a view, the haze over the Los Angeles Air Basin on polluted days is due almost entirely to the photochemical conversion of certain invisible gases to light-scattering particles consisting of sulfates, nitrates, and secondary organics but almost no soot. However, our studies, as recapitulated below, have clearly demonstrated that soot is ubiquitous not only in urban atmospheres but also in remote regions such as the Arctic (Rosen et al., 1981). Los Angeles, with its abundant coastal fog, thus contains both components of London-type fog—smoke (or soot) and fog.

2. ABUNDANCE

The methodology that we adopted involved systematic measurements of the ratio of black carbon to total carbon for a large number of samples collected directly from sources, source-dominated environments, and well-aged ambient air (24-h samples) (Hansen et al., 1980). The ambient samples were collected in areas with widely different atmospheric chemical characteristics (e.g., degree of photochemical activity, source composition, geographic location). Measurements of this ratio from a number of source samples give insights into the relative black : total carbon ratio of primary emissions and the source variabilities. Secondary material, does not contain the black component but increases the total mass of carbon, thus reduces the black : total carbon fraction. That is, under high photochemical conditions one would expect this ratio to be significantly smaller than under conditions obviously heavily influenced by sources.

Because of the large number of samples that had to be analyzed in these studies, a fast-throughput optical attenuation method (Rosen et al., 1980) was developed and used for determining black carbon. The validity of the optical attenuation method was checked by performing Raman spectroscopic (Rosen et al., 1978) and optoacoustic (Yasa et al., 1978) measurements on some of the ambient and source samples. Total particulate carbon was determined by a combustion method.

The optical attenuation method compares the transmission of a 633-nm He–Ne laser beam through a loaded filter relative to that of a blank filter. The relationship between the optical attenuation and the loading of black carbon on the filter can be expressed as

$$[\mathrm{C_{black}}] = \frac{1}{K} \times ATN, \tag{1}$$

where $[\mathrm{C_{black}}]$ is the carbon black loading (μg/cm^2) and $ATN = -100 \ln(I/I_0)$ (I and I_0 are the transmitted light intensities for the loaded filter and the filter blank, respectively). The proportionality constant K has been shown to have an average value of 20 ± 1 cm^2/μg (Hansen et al., 1980).

Besides black carbon, particulate material contains organic carbon, which is not optically absorbing. The total amount of particulate carbon is the sum of these two components:

$$[\mathrm{C_{tot}}] = [\mathrm{C_{black}}] + [\mathrm{C_{org}}].$$

We define specific attenuation σ as the attenuation per unit loading of total carbon:

$$\sigma \equiv \frac{ATN}{[\mathrm{C_{tot}}]}. \tag{2}$$

Since the optical attenuation is due only to black carbon, substitution of Eq. (1) gives

$$\sigma = K \times \frac{[\mathrm{C_{black}}]}{[\mathrm{C_{tot}}]}. \tag{3}$$

Therefore, for K known, the determination of the specific attenuation of a given sample gives an estimate of black carbon as a fraction of the total carbon in a sample. It is seen from (3) that K represents the specific attenuation of black carbon alone.

Examination of the specific attenuation and black carbon fraction from a variety of combustion sources shows that these quantities do not vary greatly source to source. Examples of these quantities are shown for a number of sources in Table 1.

In view of the relative constancy of σ from various sources, it is possible in principle to ascertain the percentage of soot (i.e., primary carbonaceous material) in ambient particles from the ratio of the specific attenuation of

Table 1. Specific Attenuation σ and black carbon (BC) (Percentage of Total Carbon) of Source Samples

Source	Number of Samples	Average σ		Highest σ		Lowest σ	
		($cm^2\ \mu g^{-1}$)	BC (%)[a]	($cm^2\ \mu g^{-1}$)	BC (%)[a]	($cm^2\ \mu g^{-1}$)	BC (%)[a]
Parking garage	12	5.4	27	7.7	39	2.25	11
Diesel	6	5.6	28	5.7	29	3.5	18
Scooter	9	5.1	26	6.1	31	4.2	21
Tunnel	63	6.3	32	12.5	63	3.7	19
Natural gas	6	2.6	13	3.3	17	1.9	10

Source: Novakov (1981).

[a] Evaluated by Eq. (3).

Table 2. Mean Specific Attenuation of Ambient Samples

Site	Number of Samples	$\bar{\sigma}$ ($cm^2\ \mu g^{-1}$)	Standard Deviation (%)	Soot[a] (%)
New York, NY	211	5.69	24	97
Gaithersburg, MD	155	4.72	32	81
Argonne, IL	221	4.35	38	74
Berkeley, CA	513	4.28	34	73
Anaheim, CA	444	3.99	43	68
Fremont, CA	461	3.74	33	64
Denver, CO	42	3.47	43	59

Source: Novakov (1981).

[a] Evaluated by Eq. (4) for $\sigma_{source} = 5.8\ cm^2\ \mu g^{-1}$.

ambient samples to the average value characteristic of major primary sources (Novakov, 1981):

$$\frac{[\text{Soot}]}{[\text{C}]} = \frac{\sigma_{ambient}}{\sigma_{source}}. \tag{4}$$

Table 2 lists the mean specific attenuation of ambient samples (weekends excluded) in order of decreasing σ and soot fractions obtained by use of Eq. (4) and $\sigma_{source} = 5.8 \pm 1.5\ cm^2\ \mu g^{-1}$ (Novakov, 1981). On the basis of this estimate, the New York City carbonaceous aerosol is essentially entirely primary soot. A different value of σ_{source} would certainly change the estimated soot percentage. However, New York City's average soot content would nevertheless remain the highest, irrespective of the actual numerical value of σ_{source}. It is logical that samples from this location have the highest soot content because the site represents a heavily traveled street canyon that is strongly influenced by primary carbonaceous material. The Fremont and Anaheim samples have the smallest soot content on the average, as might be expected, because both locations represent receptor sites that would be influenced to a greater extent by aged, secondary material.

These results demonstrate that black carbon is certainly a major fraction of ambient particulate carbon at all locations studied. These findings thus suggest that there is a catalytically active material that is present in the atmosphere in concentrations sufficiently high to warrant the assessment of its role in heterogeneous atmospheric chemistry.

3. COMPOSITION

The diameter of black carbon particles varies from 50 Å or even smaller to several thousand angstroms. X-ray diffraction studies (Hofmann and Wilm, 1936) have shown that each particle is made up of a large number of crystallites 20–30 Å in diameter. Each crystallite consists of several carbon layers with a graphitic hexagonal structure and has defects, dislocations, and discontinuities in the layer planes and thus contains high concentrations of unpaired electrons that constitute the active sites. Carbon atoms located at these sites show strong tendencies to react with other molecules because of residual valencies. During particle formation, interactions of air, water, flue gas, etc., with carbon particles occur, resulting in the incorporation of oxygen, hydrogen, and traces of nitrogen into the structure (Mattson and Mark, 1971).

Nearly every type of oxygen-containing functional group known in organic chemistry has been postulated to exist on the carbon surface; examples are shown in Figure 1. The functional groups most often suggested are

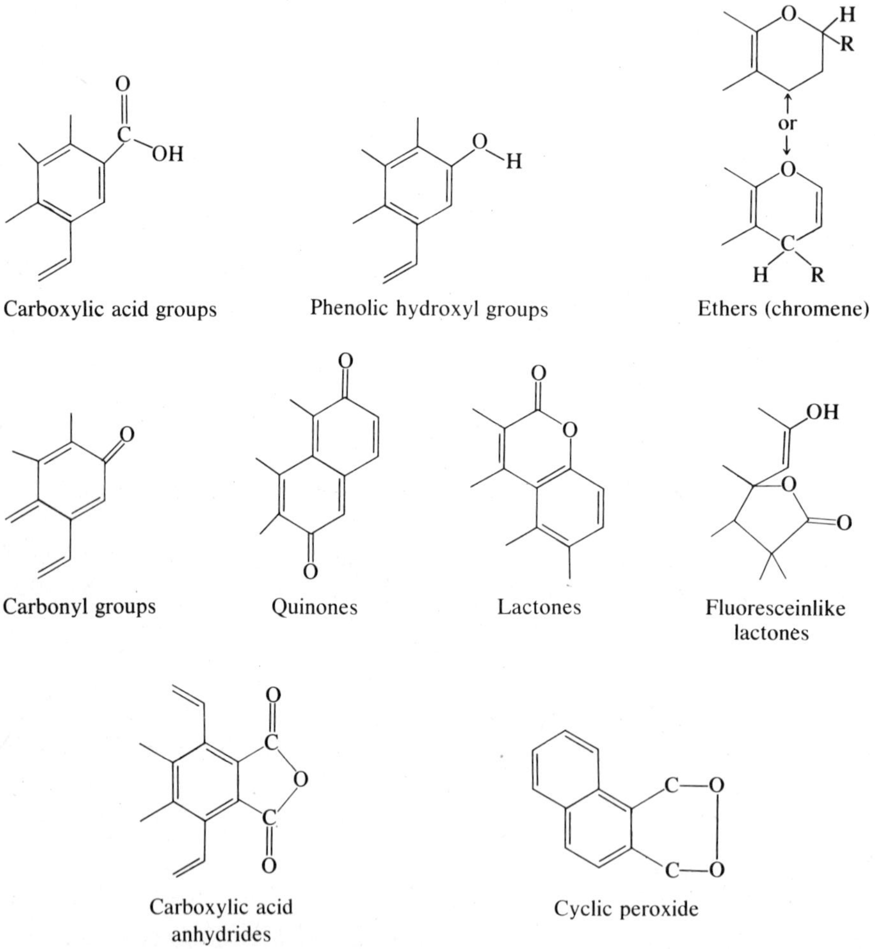

Figure 1. Oxygen-containing functional groups on elemental carbon particle surfaces. (From Chang et al., 1982; reprinted with permission of Plenum Press.)

carboxyl groups, phenolic hydroxyl groups, and quinone carbonyl groups (Boehm, 1966; Coughlin and Ezra, 1968; Garten and Weiss, 1957; Puri, 1966, 1970; Smith, 1959; Zarif'yanz et al., 1967). Less often suggested are ether, peroxide, and ester groups in the forms of normal and fluoresceinlike lactones (Garten et al., 1957), carboxylic acid anhydrides (Boehm et al., 1964), and cyclic peroxide (Puri, 1962). The relative amounts of these complexes and their structure depends on the thermal history of carbon particles (Hart et al., 1967; Laine et al., 1963; Palmer and Cullis, 1965; Weller and Young, 1948). Little is known about the structure of surface nitrogen species, although the capability of fixation of nitrogen (Emmett, 1948) in carbon particles and the promoting effect of the catalytic activity of nitrogenous carbon (Larsen and Walton, 1940) have been observed.

On the basis of both structural and functional considerations, black carbon particles should be regarded as a complex three-dimensional organic polymer with the capability of transferring electrons, rather than merely as an amorphous form of elemental carbon.

4. ACIDITY

Depending on the thermal history, black carbon particles may possess either acidic or basic character. Because of this property, soot may influence the pH of atmospheric water droplets in which it is suspended (Chang and Novakov, 1975a). It has been shown that activation of elemental carbon by exposure to O_2 at temperatures of 200–400°C produces an acidic type. By contrast, activation of carbon at high temperatures either in pure CO_2 or under vacuum, followed by exposure to oxygen at room temperature, results in a basic type.

The acidic character can be explained by the dissociation of acidic oxygen functional groups such as carboxyl and hydroxyl in aqueous solution. The nature of the basic character has been a topic of considerable discussion and controversy (Frumkin et al., 1931; Mattson and Mark, 1971; Rivin, 1963; Schilow et al., 1930; Steenberg, 1944). The presence of basic sites in the form of surface oxides has been proposed by Schilow et al. (1930) and Garten and Weiss (1957) among others, to account for the chemisorption of acids. The latter authors suggested that the oxides were in the form of a chromenelike structure, which would result in the formation of carbonium ion after neutralization with acid. The presence of carbonium cationic sites on the surface of acid-treated carbon was confirmed by Rivin (1963), but it could not be established whether the basic sites are due to the presence of the chromenelike surface oxides or to the inherent polynuclear aromatic structure of the carbon particles. Frumkin et al. (1931) proposed an electrochemical theory in which the adsorption of the acids by carbon is determined by the electrical potential at the carbon solution interface and by the capacity of the double layer. According to Steenberg (1944), adsorption of acids involved primary adsorption of protons by physical force and secondary adsorption of anions in the diffuse double layer. In contrast, Mattson and Mark (1971) attributed the neutralization of acids at high acid concentrations to the primary adsorption of the anions and secondary adsorption of the protons.

5. CHEMICAL ACTIVITY

We have investigated the reaction between black carbon particles and NH_3 in both oxidizing (Chang and Novakov, 1975b) and reducing (Novakov and Chang, 1977) atmospheres. The first set of experiments involved the exposure of combustion-produced soot to NH_3 in air. The nature of nitrogen species thus formed was studied with the aid of electron spectroscopy for chemical analysis (ESCA). Soot particles for these experiments were generated by a premixed propane–oxygen flame. The exposure of soot particles to NH_3 was done under two different experimental conditions: in a static regime, with propane soot collected on a silver membrane filter subsequently exposed to the reactant gas at ambient temperature; and in a flow system, by introducing the reactant gas downstream from the propane-oxygen flame, i.e., while the soot particles were still at high temperature.

The ESCA spectra of the nitrogen (1*s*) region of soot samples prepared in these ways are shown in Figures 2 and 3. Interaction of NH_3 with "cold" soot particles can result in ammoniumlike species (Figure 2). However, as seen from Figure 3, NH_3 interacting with "hot" soot particles produces additional species with ESCA peaks of somewhat lower binding energies, designated N_x. Ammonium in these samples is probably produced on soot particles after they have been collected on the filter and cooled.

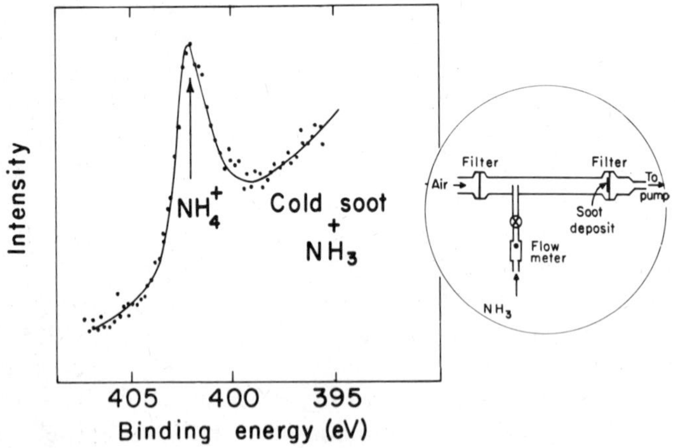

Figure 2. Nitrogen (1*s*) ESCA spectrum of cold soot particles exposed to NH_3. The setup used for exposure is also shown. (From Chang and Novakov, 1975b; reprinted with permission of Pergamon Press.)

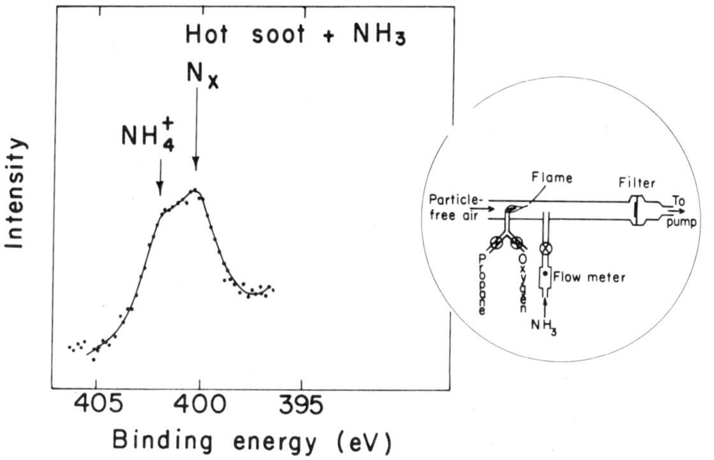

Figure 3. Nitrogen (1*s*) ESCA spectrum of hot soot particles exposed to NH_3. The experimental arrangement used for sample preparation is also shown. (From Chang and Novakov, 1975b; reprinted with permission of Pergamon Press).

Using ESCA to analyze ambient particulates, Novakov and co-workers (1972, 1973) observed, in addition to commonly occurring nitrate and ammonium, two reduced nitrogen species with N(1*s*) binding energy corresponding to the N_x surface species produced under laboratory conditions. The chemical equivalency of the ambient and synthetic N_x species is demonstrated by their thermal behavior. The experimental procedure was to measure ESCA spectra at gradually increasing sample temperatures. The results of such measurements for an ambient particulate sample collected in Pomona, California during a moderate smog episode (October 24, 1972) and for a sample prepared by NH_3–hot soot interaction are shown in Figures 4 and 5.

The spectrum of the ambient sample (Figure 4) at 25°C shows the presence of NO_3^-, NH_4^+, and N_x. At 80°C the entire nitrate peak is lost, accompanied with a similar loss in intensity of the ammonium peak. The shaded portion of the ammonium peak in the 25°C spectrum represents the ammonium fraction volatilized between 25 and 80°C. It appears, therefore, that the nitrate in this sample is mainly in the form of ammonium nitrate. At 150°C the only nitrogen species remaining is N_x. The ammonium fraction still present at 80°C but absent at 150°C is associated with some ammonium compound more stable than NH_4NO_3, possibly ammonium sulfate. At 250°C the appearance of another peak, labeled N_x', is seen. This peak continues to increase at 350°C. The total peak areas of spectra recorded at 150, 250, and 350°C remain constant, indicating that the N_x component is transformed into N_x' by heating in vacuum.

The N_x species produced by surface reactions of hot soot with NH_3 have the same kind of temperature dependence as those in ambient samples (see

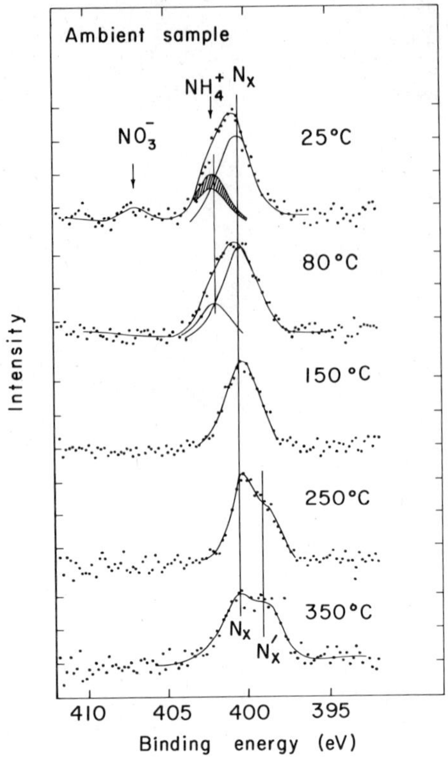

Figure 4. Nitrogen (1*s*) ESCA spectrum of an ambient sample as measured at 25, 80, 150, 250, and 350°C. (From Chang and Novakov, 1975b; reprinted with permission of Pergamon Press.)

Figure 5). The spectrum taken at room temperature shows that most nitrogen species in this sample are of the N_x type. Heating the sample in vacuum to 150°C does not influence the line shape or intensity. At 250°C, however, the formation of N_x' is evident. Further transformation of N_x to N_x' occurred at 350°C.

Both ambient and synthetic N_x' species will remain unaltered even if the temperature is lowered back to room temperature if the sample remains in vacuum. However, if the sample is removed from vacuum and exposed to moisture, N_x' will be transformed back to the original N_x compound. It can be concluded that N_x' species are produced by dehydration of N_x.

These results indicate that the thermal behavior of nitrogen species of the N_x and N_x' type observed in ambient pollution particulates is identical to that of the reduced nitrogen species produced by reactions at elevated temperature of ammonia with finely divided carbon or soot. On the basis of these experimental results, N_x was assigned to a mixture of amines and amides and N_x', to nitrile.

The foregoing results, together with structural and functional considerations, suggest a possible formation mechanism for these compounds. Since the soot particle surface was in contact with air and flue gas prior to the interaction with NH_x, it should have been covered with surface oxygen complexes. By using the most often mentioned surface oxygen–carbon functional groups (i.e., carboxyl groups and phenolic hydroxyl groups), we can describe, in analogy with organic chemistry, some possible reactions of NH_3 and soot leading to the formation of amides, amines, nitriles, and ammonium–salt-like compounds associated with the black carbon component of soot particles.

At low temperatures black carbon covered with surface carboxyl or phenolic groups may act as a Brønsted acid when interacting with NH_3. Carboxyl ammonium or phenolic ammonium salts will be formed as the result of proton exchange. Ammonia may also be physically adsorbed by hydrogen bonding to surface OH or COOH groups. At elevated temperatures the carboxyl group is electrophilic and has the tendency to accept an electron pair from the basic species in the process of coordination. The nucleophilic substitution reaction of NH_3 with carboxylic acid yields an amide, which may dehydrate and become a nitrile on further heating. Carboxyl and phenolic hydroxyl ammonium salts may dehydrate at elevated temperature to produce amides and/or nitriles and amines, respectively. These processes are illustrated in Figure 6.

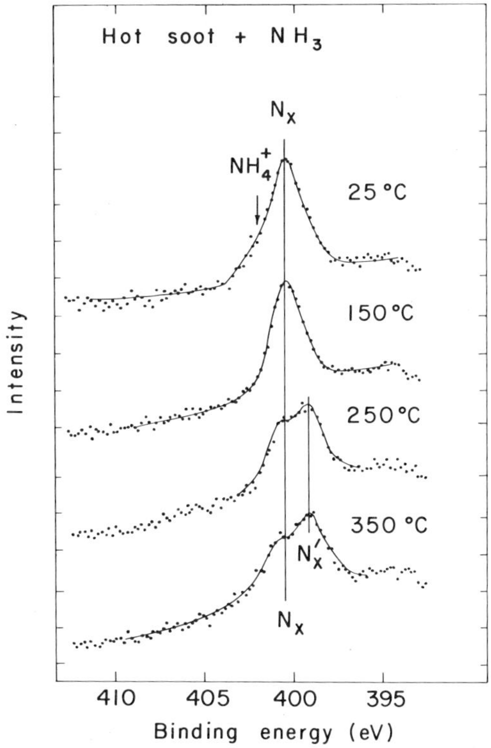

Figure 5. Nitrogen (1*s*) ESCA spectrum of (hot) soot sample exposed to NH_3, as measured at 25, 150, 250, and 350°C. (From Chang and Novakov, 1975b; reprinted with permission of Pergamon Press.)

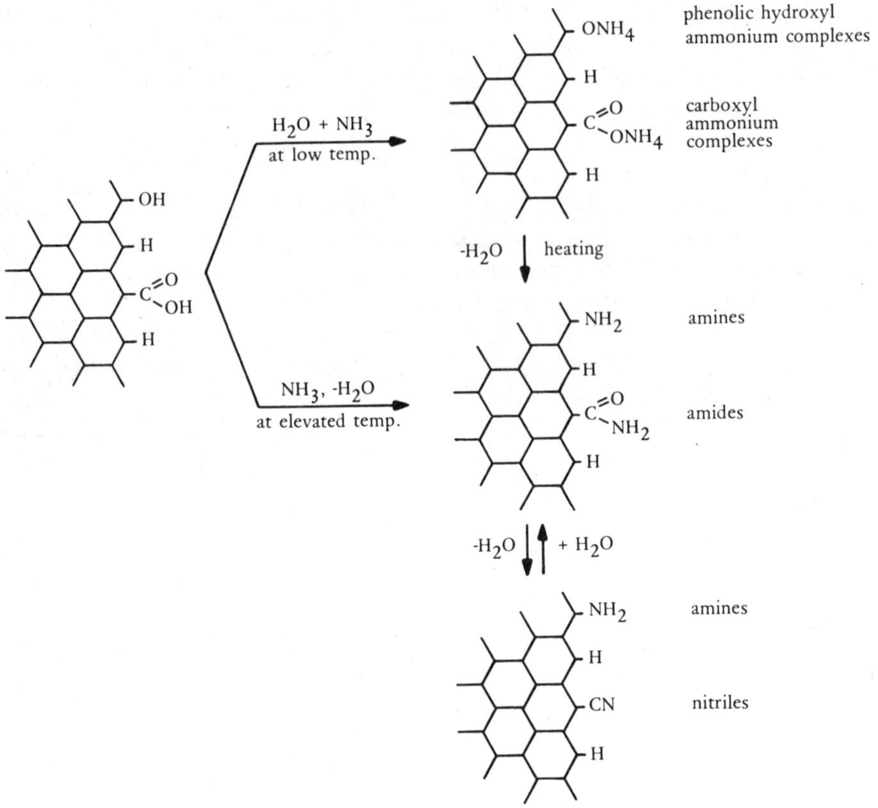

Figure 6. Possible reaction paths of NH_3 with surface-oxidized black carbon. (From Chang and Novakov, 1975b; reprinted with permission of Pergamon Press.)

The photoelectron spectroscopic results indicate that the amides and amines correspond to the N_x species. These appear as broad peaks indicating the presence of more than one chemical species. Nitriles formed from amides by dehydration on heating correspond to the N_x' species. We have established the reversibility of the

$$N_x \underset{+H_2O}{\overset{-H_2O}{\rightleftharpoons}} N_x'$$

process. The carboxyl ammonium and phenolic hydroxyl ammonium salts produced by NH_3 chemisorption correspond to the volatile ambient ammonium species.

The stability of ambient particulate nitrogen in water has been studied by combining ESCA measurements with determination of total nitrogen by proton activation (Gundel et al., 1979). Results with samples from several locations (Berkeley, Los Angeles, and St. Louis) indicate that (1) a large fraction of N_x (85%) originally present in ambient particulate matter can be removed by water extraction; and (2) more NH_4^+ appears in the extract than was present on the untreated sample and less N_x appears in the extract than was present on the untreated sample. The N_x deficiency in the extract matches the surplus in NH_4^+. The former behavior may be attributed to the presence of water-soluble stoichiometric compounds such as amines and of surface species such as amides and nitrile that can undergo hydrolysis. The latter may be attributed to the hydrolysis of amide and nitrile groups. These processes may be responsible for the disagreement (Brosset, 1980) of chemical composition in so-called black episodes from those predicted from the equilibrium phase diagram.

A further set of experiments (Novakov and Chang, 1977) involved the grinding of a purified-grade POCO graphite in NH_3 in the absence of air at room temperature. The concentration of nitrogen relative to carbon was determined by ESCA. The information on the structure of surface nitrogen species was obtained with the aid of Fourier transform infrared spectroscopy. To help in the assignment of vibrational frequencies, infrared spectra of graphite particles after reaction with deuterated ammonia were also obtained.

The grinding reduces particle sizes and creates fresh surfaces. Surface carbon atoms of graphite particles show a strong chemical reactivity because of unsaturation in valency. Figure 7 shows infrared spectra of the graphite particles after extensive grinding in an atmosphere of NH_3 and ND_3 (traces *a* and *c*, respectively); expansions of these spectra are shown in traces *b* and *d*. These infrared spectra suggest the occurrence of dissociative chemisorption of NH_3 on the carbon particle surface. Vibrational frequencies associated with the surface groups C—NH_2, C=N—H, C≡N, and C—H are observed in traces *a* and *b*. The isotope shifts shown in traces *c* and *d*

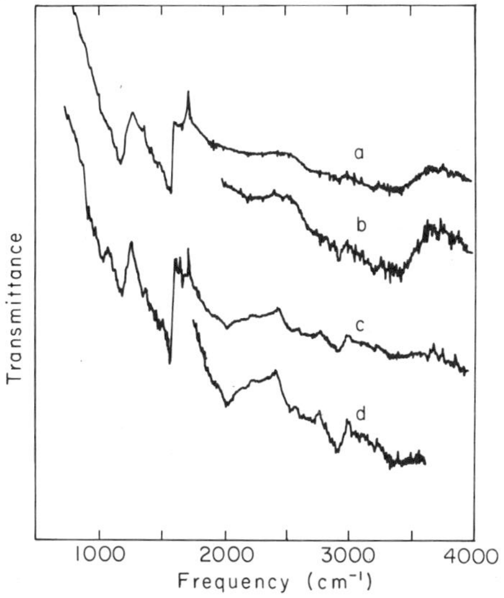

Figure 7. Infrared spectra of the graphite particles after extensive grinding in an atmosphere of NH_3 (*a*) and ND_3 (*c*); *b* and *d* are 2× expansions along the ordinate of *a* and *c*, respectively. (From Chang et al., 1982; reprinted with permission of Plenum Press.)

support these assignments. Surface CNH_2 groups give rise to two bands near 3400 cm^{-1} that are attributed to symmetric and antisymmetric N—H stretching modes. These two bands should shift to 2500 cm^{-1} for CND_2. This shift is shown in traces *c* and *d*. A NH_2 bending mode near 1580 cm^{-1} should shift to about 1200 cm^{-1} for the ND_2 groups. However, a strong band due to the $k = 0$ E_{2g} phonon mode of the graphite lattice (Tuinstra and Koenig, 1970) and/or to a vibrational mode of the aromatic structure of graphite (Friedel and Hofer, 1970) also occurs at about 1580 cm^{-1}. Likewise, the C—N stretching mode vibrates at approximately 1200 cm^{-1} and appears in both the C—NH_2 and the C—ND_2 surface groups.

In addition to the bands for the NH_2 group mentioned above, we detected surface nitrogen groups indicating the dissociation of more than one bond in a molecule of ammonia. A band between 1600 and 1700 cm^{-1} could be assigned to immines (C═NH and C═N—C), a weak band at 2300 cm^{-1} to nitrile (C≡N), and a band at 2180 cm^{-1} to isocyanide (—N^+≡C^-).

The evidence of the dissociative chemisorption of ammonia on carbon particle surfaces is supported also by the appearance of the C—D stretching band at 2050 cm^{-1}. However, the assignment of the C—H stretching is ambiguous because the C—H stretching is near 2900 cm^{-1}, where a vibrational band appears on both NH_3 and ND_3 samples. This band could be the overtone and/or combination bands resulting from the strong absorption band at 1300–1600 cm^{-1}. There is also a band, possibly of the same nature, at 2700 cm^{-1} in both samples.

6. CATALYTIC ACTIVITY

Black carbon particles are effective catalysts (Coughlin, 1969) for many different types of reaction, including oxidation–reduction, halogenation, hydrogenation–dehydrogenation, dehydration, polymerization, and isomerization. Table 3 lists a few reactions catalyzed by carbon that could have direct bearing on atmospheric chemistry. It is difficult to assess the importance of all these reactions in the atmosphere at this time because useful rate equations have not been determined. One reaction system that has been extensively studied and may be of particular atmospheric importance is the catalytic oxidation of SO_2 on carbon in aqueous suspensions. Studies of this reaction and its atmospheric importance are reviewed in this section.

The oxidation of SO_2 on carbon particles was studied by photoelectron spectroscopy (ESCA) by Novakov et al. (1974). It was found that under some conditions a significant amount of sulfate can be produced by the catalytic action of carbon particles. Although these early experiments were qualitative, it was possible to conclude the following:

1. The reaction product is in the +6 oxidation state (i.e., sulfate).
2. Soot-catalyzed oxidation of SO_2 is more efficient at higher humidity.

Table 3. Some Reactions Catalyzed by Carbon

Reactions	References
1. $SO_2 + \frac{1}{2}O_2 \rightarrow SO_3$	Novakov et al. (1974), Chang et al. (1979), Brodzinsky et al. (1980), Chang et al. (1981)
2. $SO_2 + NO_2 \rightarrow SO_3 + NO$	Cofer et al. (1980), Britton and Clarke (1980)
3. $SO_2 + O_3 \rightarrow SO_3 + O_2$	Cofer et al. (1981)
4. $NO + \frac{1}{2}O_2 \rightarrow NO_2$	Rao and Hougen (1952)
5. $2H_2O_2 \rightarrow 2H_2O + O_2$	Bente and Walton (1943)
6. $CO + Cl_2 \rightarrow COCl_2$	Dulou (1945)
7. $SO_2 + Cl_2 \rightarrow SO_2Cl_2$	Dulou (1945)
8. $HCOOH \nearrow H_2O + CO$; $HCOOH \searrow H_2 + CO_2$	Stumpp (1965)
9. 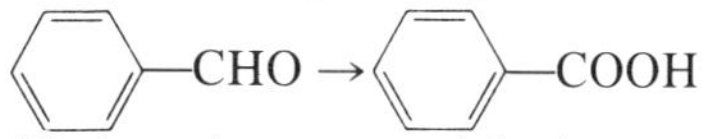	Gundel (1979)
10. Hydroquinone → quinhydrone → quinone	Bente and Walton (1943)

3. The oxygen in air plays an important role in SO_2 oxidation.
4. Black carbon–catalyzed oxidation exhibits a saturation effect.
5. Sulfur dioxide can be oxidized on other types of graphitic carbonaceous particle, such as ground graphite particles and activated carbon.

Results from the experiments with combustion-produced soot particles are essentially similar to those obtained for activated carbon by Davtyan and Tkach (1961) and Siedlewski (1965).

Soot-catalyzed SO_2 oxidation can proceed by two mechanisms: a "dry" mechanism, in the presence of water, and a "wet" mechanism, when the soot particles are covered by a liquid water layer. The wet mechanism is much more efficient than the dry and is applicable to situations in plumes, clouds, fogs, and the ambient atmosphere when the aerosol particles are covered with a liquid water layer. The dry mechanism is expected to operate in stacks or under conditions of low relative humidity.

6.1. Dry Mechanism of SO_2 Oxidation

A description of the dry mechanism was given by Yamamoto et al. (1972), who studied the reaction kinetics on dry activated carbon in the presence of O_2 and H_2O vapor. The rate of reaction was found to be first order with respect to SO_2, provided the partial pressure of SO_2 was less than 10^{-4} atm, and depended on the square root of the concentration of O_2 and H_2O vapor. The activation energy in the temperature range 70–150°C was found to vary from −4 to −7 kcal mol^{-1} (i.e., reaction rate decreasing with increasing temperature), depending on the origin of the activated carbon. Initially the reaction occurs on the surface of both micropores and macropores, and the rate is constant for a given activated carbon until the amount of accumulated H_2SO_4 reaches about 10% by weight of the carbon. Beyond that amount the rate gradually decreases with the reaction time until the micropore volume is saturated by H_2SO_4. The reaction continues only on the macropores at a constant, but much slower, rate.

According to Yamamoto et al. (1972), a rate expression (valid until the amount of H_2SO_4 formed reaches 10% by weight of the carbon) for the activated carbon used can be written as follows:

$$\frac{d[H_2SO_4]}{dt} = [C]P_{SO_2}P_{O_2}^{1/2}P_{H_2O}^{1/2}(k_{micro} + k_{macro})e^{-E_a/RT},$$

where E_a is the activation energy, R is the universal gas constant; T is the absolute temperature; t is the time in seconds; [C] and [H_2SO_4] are the respective concentrations in μg/m^3; P_{SO_2}, P_{O_2}, and P_{H_2O} are the respective partial pressures in atmospheres; and k_{micro} and k_{macro} are the rate constants on the surface micropores and macropores, respectively, in atm^{-2} s^{-1}. For the specific carbon employed, with physical properties as follows—micropore volume, 0.4 ml/g; micropore radius, 7.4 Å; macropore volume, 0.4 ml/g; macropore radius, 2.2 μm; BET surface area, 894 m^2/g; particle diameter, 0.25 mm—the values reported were $k_{micro} = 4.2 \times 10^2$ atm^{-2} s^{-1} and $k_{macro} = 22.9$ atm^{-2} s^{-1}.

The dry mechanism is relatively inefficient because the reaction product remains on the carbon surface and acts as the catalyst poison.

6.2. Wet Mechanism of SO_2 Oxidation

The situation is entirely different when soot (or black carbon) is covered with a layer of liquid water and the catalytic oxidation occurs at the solid–liquid interface: there is constant regeneration of active sites because the reaction product is soluble in water and thus leaves the soot surface. Such reactions were studied in detail by Chang et al. (1979, 1981) and Brodzinsky et al. (1980), who used both combustion soots and activated carbons.

The kinetics of reaction was studied by batch (flask) experiments, from which a rate law was derived. This rate law has been confirmed by fog chamber studies (Benner, 1980). The flask experiments were performed by using suspensions of commercially available activated carbons as well as suspensions of combustion-produced soots.

Figure 8 shows typical reaction curves of the oxidation of S(IV) in aqueous suspensions of soot particles collected from acetylene and natural gas flames.

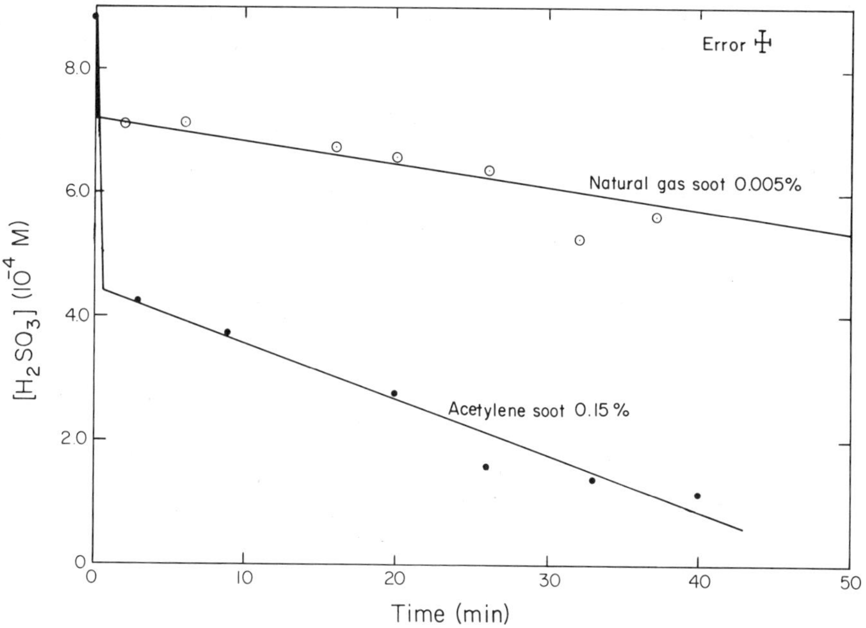

Figure 8. S(IV) concentration as a function of time for acetylene and natural gas soot suspensions. (Reprinted with permission from Brodzinsky et al., 1980. Copyright 1980 American Chemical Society.)

The reaction occurs in two steps. The initial disappearance of S(IV) is so fast that its rate could not be followed by the analytical techniques used. The second step is characterized by a much slower reduction of S(IV). The results obtained with these combustion-produced soots were reproduced (Figure 9) by suspensions of similar concentrations of a commercially available activated carbon (Nuchar C-190, a trademark of West Virginia Pulp and Paper Co.). Figure 9 also shows a mass balance between the S(IV) consumed and the sulfate produced. At a constant temperature, the amount of S(IV) oxidized by the rapid first step process was found to be proportional to the carbon particle concentration.

The reaction of the second step was found to have the following characteristics:

1. The reaction rate is independent of pH (pH < 7.6); therefore, $SO_2 \cdot H_2O$, HSO_3^-, and SO_3^{2-} are indistinguishable in terms of oxidation on the carbon surfaces.
2. The reaction is first order with respect to the concentration of carbon particles.
3. The activation energy of the reaction is ~8.5 kcal mol^{-1}, being slightly different for different carbons.
4. The reaction rate has a complex dependence on the concentration of

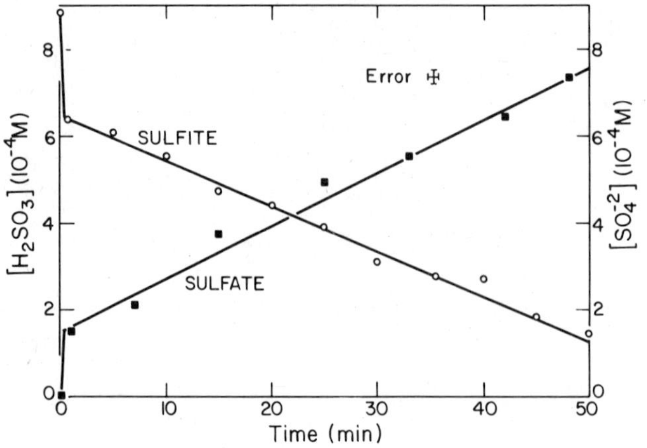

Figure 9. S(IV) and SO_4^{2-} concentrations as a function of time for a 0.16%-by-weight activated carbon suspension. (Reprinted with permission from Brodzinsky et al., 1980. Copyright 1980 American Chemical Society.)

S(IV), ranging between a second- and zero-order reaction as the S(IV) concentration increases.

5. The reaction rate has a complex dependence on the concentration of dissolved O_2, with the order of reaction between zeroth and first.

Figure 10 shows the effective rate of reaction (normalized carbon concentration, room temperature–20°C, and air) as a function of the sulfurous acid concentration for the activated carbons studied (see also Table 4). From the Nuchar C-190 curve, the rate of reaction is second order with respect to S(IV) below 10^{-7} M, moves through a first-order reaction around 5×10^{-6}

Table 4. Summary of Kinetic Data for Catalytic Oxidation of SO_2 by Various Elemental Carbon Particles in Aqueous Suspension[a]

Kinetic Data \ Elemental Carbon	Nuchar C-190 (WESVACO)	Nuchar SN (WESVACO)	Aktivkohle (MERCK)
A (mol g^{-1} s^{-1})	0.88	0.16	2.5
E_a (kcal mol^{-1})	8.8	8.1	8.8
K_1 (M^{-1})	2.1×10^3	7.4×10^3	4.4×10^3
α (M^{-2})	2.4×10^{12}	4.9×10^8	9.5×10^{11}
β (M^{-1})	1.2×10^7	3.0×10^5	3.7×10^7

[a] Reaction rate equation:

$$\frac{d[SO_4^{-2}]}{dt} = Ae^{-E_a/RT}[C_x]\left\{\frac{K_1[O_2]}{1 + K_1[O_2]}\right\}\left\{\frac{\alpha[S(IV)]^2}{1 \times \beta[S(IV)] + \alpha[S(IV)]^2}\right\}$$

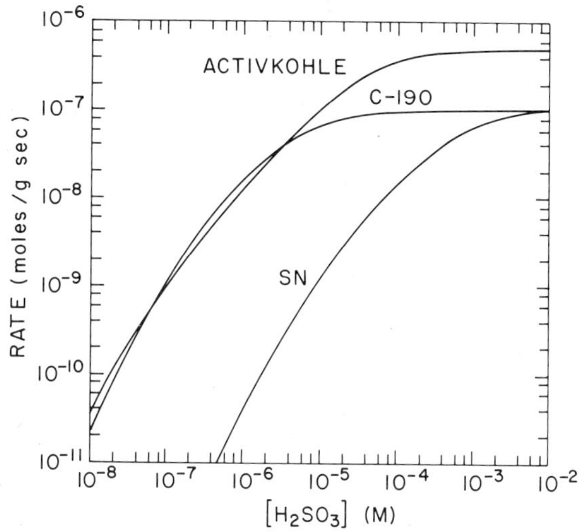

Figure 10. Effective rate of oxidation of S(IV) catalyzed on various activated carbon particles versus S(IV) concentration. (From Chang et al., 1982; reprinted with permission of Plenum Press.)

M, and becomes independent of S(IV) concentrations above 10^{-4} M. The other curves are seen to be similar in their behavior.

On the basis of the experimental results, we propose the following reaction mechanism:

$$C_x + O_2(aq) \underset{k_{-1}}{\overset{k_1}{\rightleftharpoons}} C_x{\cdot}O_2 \qquad [1]$$

$$C_x{\cdot}O_2 + S(IV) \underset{k_{-2}}{\overset{k_2}{\rightleftharpoons}} C_x{\cdot}O_2{\cdot}S(IV), \qquad [2]$$

$$C_x{\cdot}O_2{\cdot}S(IV) + S(IV) \underset{k_{-3}}{\overset{k_3}{\rightleftharpoons}} C_x{\cdot}O_2{\cdot}2S(IV), \qquad [3]$$

$$C_x{\cdot}O_2{\cdot}2S(IV) \overset{k_4}{\rightarrow} C_x + 2S(VI), \qquad [4]$$

where C_x = soot, $O_2(aq)$ = dissolved oxygen molecule, S(IV) = sulfite species, and S(VI) = sulfate species.

Reaction [1] indicates that dissolved oxygen is adsorbed on the soot particle surface to form an activated complex. This adsorbed oxygen complex then oxidizes the S(IV) to form sulfate according to Reactions [2]–[4]. If it is assumed that the reaction follows the condition of Langmuir adsorption equilibrium (Clark, 1970), the rate of acid formation is

$$\frac{d[S(VI)]}{dt} = 2k_4[C_x]\left(\frac{K_1[O_2]}{1 + K_1[O_2]}\right)\left(\frac{K_2[S(IV)]}{1 + K_2[S(IV)]}\right)\left(\frac{K_3[S(IV)]}{1 + K_3[S(IV)]}\right) \qquad (5)$$

where $K_1 = k_1/k_{-1}$, $K_2 = k_2/k_{-2}$, and $K_3 = k_3/k_{-3}$.

The experimental results yielded the following rate law for this reaction:

$$\frac{d[\mathrm{S(VI)}]}{dt} = k[\mathrm{C}_x]\left(\frac{K_1[\mathrm{O}_2]}{1 + K_1[\mathrm{O}_2]}\right) f[\mathrm{S(IV)}] \qquad (6)$$

where $f[\mathrm{S(IV)}] = (\alpha[\mathrm{S(IV)}]^2/1 + \beta[\mathrm{S(IV)}] + \alpha[\mathrm{S(IV)}]^2)$. Here $[\mathrm{C}_x]$ = grams of carbon particles per liter, $[\mathrm{O}_2]$ = moles of dissolved oxygen per liter, and $[\mathrm{S(IV)}]$ = total moles of S(IV) per liter. Values of the several constants are presented in Table 4.

The dependence of the rate of formation of sulfate on the partial pressure of $SO_2((P_{SO_2})$ in the atmosphere can be obtained from Eq. (6). The relationship of $f[\mathrm{S(IV)}]$ with P_{SO_2} and pH is shown in Figures 11 and 12. The function $f[\mathrm{S(IV)}]$ and in turn the production rate of sulfate (because the rate is linearly proportional to $f[\mathrm{S(IV)}]$ decrease as the pH decreases at a given P_{SO_2}. The magnitude of change in $f[\mathrm{S(IV)}]$ per unit pH change is much larger at a lower P_{SO_2}. Also, $f[\mathrm{S(IV)}]$ and the rate depend only slightly on P_{SO_2} under most atmospheric conditions when P_{SO_2} is 1–10 ppb and the pH is 5–6; $f[\mathrm{S(IV)}]$ increases only 10% and twofold at pH 6 and 5, respectively, when P_{SO_2} increases from 1 to 10 ppb. However, $f[\mathrm{S(IV)}]$ depends strongly on P_{SO_2} when the pH is low.

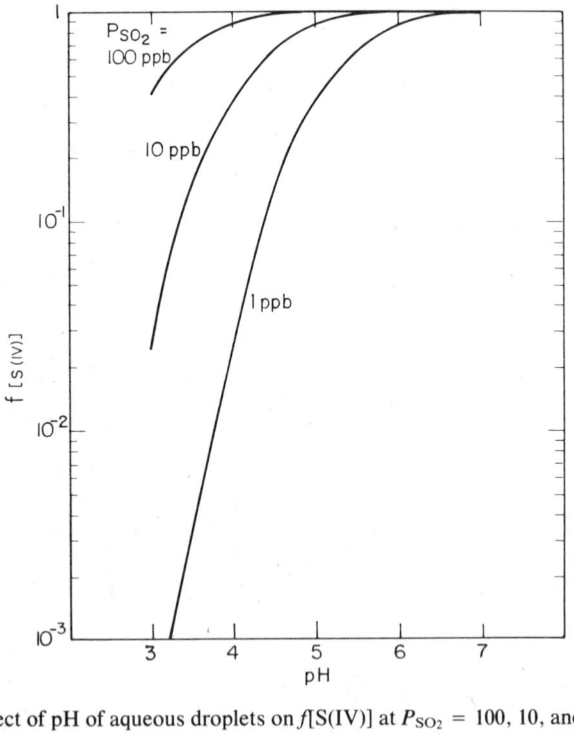

Figure 11. Effect of pH of aqueous droplets on $f[\mathrm{S(IV)}]$ at P_{SO_2} = 100, 10, and 1 ppb; $f[\mathrm{S(IV)}]$ exhibits the dependence of the soot catalyzed aqueous oxidation rate of S(IV) on S(IV) concentration. (From Chang et al., 1981; reprinted with permission of Pergamon Press.)

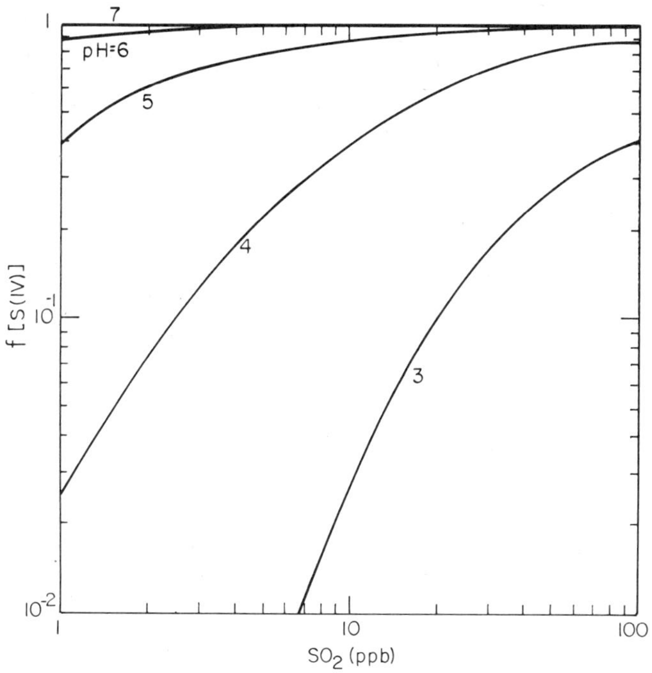

Figure 12. Effect of partial pressure of SO_2 on f[S(IV)] at various values of pH. (From Chang et al., 1981; reprinted with permission of Pergamon Press.)

The catalytic oxidation of sulfurous acid on carbon particles of different origins shows the same kinetic behavior. However, the rate constants of several different types of carbon particle were found to differ from type to type, as indicated in Table 4. In principle, the reaction rate should be proportional to the concentration of active sites on the carbon particles, rather than to the concentration of carbon particles. The number of active sites per unit mass of carbon particles varies with each type and is not necessarily proportional to the surface area. Siedlewski (1965) has shown by means of the electron paramagnetic resonance (EPR) method, that free electrons on carbon particles can serve as active centers for the adsorption of oxygen molecules and for the oxidation of SO_2. The concentration of free electrons is related to the origin and thermal history of the carbon particles.

It is thus impractical to formulate a generally applicable rate constant for atmospheric soot particles because these particles may arise from the combustion of different types of fossil fuel under different combustion conditions and thus possess a different catalytic activity.

6.3. Relative Importance of Some Wet SO_2 Oxidation Mechanisms

We have carried out a box-type calculation (Chang et al., 1979, 1981) to compare the relative importance of sulfate production mechanisms by black carbon particles with other mechanisms involving liquid water. The species considered included SO_2, CO_2, $H_2O(l)$, and air; oxidizing agents O_2, O_3, and HNO_2; and catalysts Fe^{3+}, Mn^{2+} (as discussed by Hoffmann and Boyce in Chapter 3 of this volume), and black carbon. The role of NH_3 in these reactions was also investigated. The kinetic expressions employed in this evaluation were obtained from the following studies: oxygen, Beilke et al. (1975); ozone, Erickson et al. (1977); iron, Freiberg (1975); manganese, Matteson et al. (1969); nitrous acid, Chang et al. (1981), Oblath et al. (1981); black carbon, Brodzinsky et al. (1980).

All the oxidation mechanisms considered except Mn^2 are pH dependent. Most of these mechanisms have lower oxidation rates at a lower pH, but some are more sensitive to the change in pH than others. However, the HNO_2 mechanism shows a larger oxidation rate when the solution is more

Table 5. Equilibria and Equilibrium Constants Involved in Aqueous-Phase Oxidation of S(IV) by Various Mechanisms

Equilibrium	Equilibrium Constant	Value[a]	Reference
$H_2O = H^+ + OH$	K_w	1.0(−14)[b]	McKay (1971)
$CO_2(aq) + H_2O = HCO_3^- + H^+$	K_{1c}	4.45(−7)	McKay (1971)
$HCO_3^- = CO_3^{2-} + H^+$	K_{2c}	4.68(−11)	McKay (1971)
$NH_3(aq) + H_2O = NH_4^+ + OH^-$	K_a	1.77(−5)	McKay (1971)
$SO_2(aq) + H_2O = HSO_3^- + H^+$	K_{1s}	7.7(−2)	McKay (1971)
$HSO_3^- = SO_3^{2-} + H^+$	K_{2s}	6.24(−8)	McKay (1971)
$HSO_4^- = SO_4^{2-} + H_2O$	K_{3s}	1.2(−2)	Hodgman (1958)
$HNO_2(aq) = H^+ + NO_2^-$	K_n	5.1(−4)	Schwartz and White (1981)
$Fe^{3+} + HSO_3^- + Fe(HSO_3)^{2+}$	K_{1f}		
$Fe(HSO_3)^{2+} + SO_3^{2-} = Fe(HSO_3)(SO_3)$	K_{2f}		
$CO_2(g) = CO_2(aq)$	H_c	3.4(−2)	McKay (1971)
$NH_3(g) = NH_3(aq)$	H_a	5.7 (1)	McKay (1971)
$SO_2(g) = SO_2(aq)$	H_s	1.24	McKay (1971)
$HNO_2(g) = HNO_2(aq)$	H_n	4.9(1)	Schwartz and White (1981)
$O_2(g) = O_2(aq)$	H_{O_2}	1.08(−3)	Hodgman (1958)
$O_3(g) = O_3(aq)$	H_{O_3}	1.23(−2)	Mellor (1946)

[a] Concentrations in mol/liter, M; gas pressures in atm; water activity taken as unity; $T = 25°C$.
[b] The notation 1.0(−14) represents 1.0×10^{-14}.

Table 6. Reactions and Rate Constants Involved in Aqueous-Phase Oxidation of S(IV) by Various Mechanisms

Reaction	Rate Constant, 25°C	Units	Reference
$SO_2 \cdot H_2O \underset{-1}{\overset{1}{\rightleftharpoons}} HSO_3^- + H^+$	$k_1 = 3.4(6)$	s^{-1}	Eigen et al. (1961)
	$k_{-1} = 2(8)$	$M^{-1}\ s^{-1}$	
$SO_3^{2-} + \frac{1}{2}O_2 \rightarrow SO_4^{2-}$	$k_2 = 1.7(-3)^a$	s^{-1}	Beilke et al. (1975)
$SO_2 \cdot H_2O + OH^- \underset{-3}{\overset{3}{\rightleftharpoons}} HSO_3^- + H_2O$	$k_3 = 2.9(5)$	$M^{-1}\ s^{-1}$	Beilke et al. (1975)
	$k_{-3} = 2.3(-7)$	s^{-1}	
$HSO_3^- + O_3 \xrightarrow{4} HSO_4^- + O_2$	$k_4 = 1.1(5)$	$M^{-1}\ s^{-1}$	Erickson et al. (1977)
$SO_3^{2-} + O_3 \xrightarrow{5} SO_4^{2-} + O_2$	$k_5 = 7.4(8)$	$M^{-1}\ s^{-1}$	Erickson et al. (1977)
$Fe(HSO_3)(SO_3) + O_2 + H_2O \xrightarrow{6} Fe(OH)^{2+} + 2HSO_4^-$	$K_{1f}K_{2f}k_6 = 5.6(12)$	$M^{-3}\ s^{-1}$	Freiberg (1975)
$Fe(HSO_3)(SO_3) + Fe^{3+} + H_2O \xrightarrow{7} 2Fe^{2+} + HSO_4^- + HSO_3^- + H^+$	$K_{1f}K_{2f}k_7 = 3.6(11)$	$M^{-3}\ s^{-1}$	Freiberg (1975)
$Mn^{2+} + SO_2(aq) \underset{-8}{\overset{8}{\rightleftharpoons}} X$	$k_8 = 4.0(3)$	$M^{-1}\ s^{-1}$	Matteson et al. (1969)
	$k_{-8} = 1.7\ (-1)$	s^{-1}	
$X \underset{-9}{\overset{9}{\rightleftharpoons}} Mn^{2+} + H^+ + HSO_4^-$	$k_9 = 3.7(-3)$	s^{-1}	Matteson et al. (1969)
	$k_{-9} = 1.18(-1)$	$M^{-2}\ s^{-1}$	
$2H^+ + NO_2^- \xrightarrow{10} NO^+ + H_2O$	$k_{10} = 8(5)$	$M^{-2}\ s^{-1}$	Oblath et al. (1981)
$H^+ + NO_2^- + HSO_3^- \xrightarrow{11} NOSO_3^- + H_2O$	$k_{11} = 3.8(3)$	$M^{-2}\ s^{-1}$	Oblath et al. (1981)
$NO_2^- + 2HSO_3^- \xrightarrow{12} ON(SO_3)_2^{3-} + H_2O$	$k_{12} = 9(-4)$	$M^{-2}\ s^{-1}$	Oblath et al. (1981)
$C_x + O_2 + 2S(IV) \xrightarrow{13} C_x + 2S(VI)$	k_{13}		Brodzinsky et al. (1980)

[a] For $P_{O_2} = 0.2$ atm.

acidic. The following initial conditions were used in the calculation: liquid water, 0.05 g m^{-3}; SO_2, 0.01 ppm; O_3, 0.05 ppm; and CO_2, 311 ppm. Concentrations of particulate iron and manganese of 250 ng m^{-3} and 20 ng m^{-3}, respectively, were assumed. However, since only 0.13% of the total iron and 0.25% of the manganese are water soluble, according to Gordon et al. (1975), the concentrations employed in the kinetic evaluations were correspondingly decreased. The concentrations of black carbon and HNO_2 were taken as 10 μg m^{-3} and 8 ppb, respectively. The latter corresponds to 25 ppb of NO and 50 ppb of NO_2 at equilibrium conditions. For NH_3, a concentration of 5 ppb was used; this is higher than the highest equilibrium partial pressure of NH_3 over the United States as calculated by Lau and

Charlson (1977). Tables 5–7 list the equations describing the equilibria and oxidation rates used in this comparative study.

The following assumptions were made in the calculations:

1. The size of liquid water drops suspended inside the box is so small that the absorption rate of gaseous species (SO_2, NH_3, and HNO_2) is governed by chemical reaction and not by mass transport.
2. There is no mass transfer of any species into the box during the reaction; therefore, the SO_2 (and NH_3 or HNO_2) is depleted with time. The mass balance of the SO_2, CO_2, NH_3, and HNO_2 is always maintained (i.e., $-\Delta[SO_2(g)] = \Delta[SO_2 \cdot H_2O] + \Delta[HSO_3^-] + \Delta[SO_3^{-2}] + \Delta[HSO_4^-] + \Delta[SO_4^{-2}]$; $-\Delta[CO_2(g)] = \Delta[CO_2 \cdot H_2O] + \Delta[HCO_3^-] + \Delta[CO_3^{-2}]$; $-\Delta[NH_3(g)] = \Delta[NH_3 \cdot H_2O] + \Delta[NH_4^+]$; and $-\Delta[HNO_2(g)] = \Delta[HNO_2] + \Delta[NO_2^-] + 2\Delta[N_2O(g)]$; all units are in moles).
3. The growth of liquid water droplets due to the vapor pressure lowering effect of the sulfuric acid formed in the droplets is neglected.

The rate of sulfate production was determined by a calculation scheme involving a combination of equilibrium and kinetic steps. Equilibrium be-

Table 7. Rate Expressions Describing Aqueous-Phase Oxidation of S(IV)

Mechanism	Reaction Rate Law, M s^{-1}	Reference
O_2	$\dfrac{k_2 K_{2s}(k_1 + k_3 K_w/[H^+]) H_s P_{SO_2}}{k_{-1}[H^+]^2 + k_{-3}[H^+] + K_{2s}k_2}$	Beilke et al. (1975)
O_3	$(k_4 K_{1s}/[H^+] + k_5 K_{1s} K_{2s}/[H^+]^2) H_{O_3} H_s P_{O_3} P_{SO_2}$	Erickson et al. (1977)
Fe^{3+}	$K_{1f} K_{2f} K_{1s}^2 K_{2s}(2k_6 H_{O_2} P_{O_2} + k_7[Fe^{3+}]) \times H_s^2 P_{SO_2}^2/[H^+]^3$	Freiberg (1975)
Mn^{2+} [a]	$k_9[X] - k_{-9}[S(VI)]^2([Mn^{2+}] - [X])$ where $[X] = \dfrac{H_s k_8 P_{SO_2}[Mn^{2+}]}{H_s(k_8 P_{SO_2} + v_L RT[Mn^{2+}]) + k_{-8} + k_9}$	Matteson et al. (1969)
HNO_2	$(k_{10}[H^+]^2 + k_{11} K_{1s} H_s P_{SO_2} + k_{12} K_{1s}^2 H_s^2 P_{SO_2}^2/[H^+]^2)[NO_2^-]$	Oblath et al. (1981)
Black carbon[b]	$k_{13}[C_x][O_2]^{0.7} \dfrac{\alpha[S(IV)]^2}{1 + \beta[S(IV)] + \alpha[S(IV)]^2}$	Brodzinsky et al. (1980)

[a] v_L = liquid water volume fraction; R = universal gas constant (atm M^{-1} K^{-1}); T = temperature (K).

[b] $k_{13} = 1.2 \times 10^{-4}$ $M^{0.3}$ (g/liter)$^{-1}$ s^{-1}; $\alpha = 1.5 \times 10^{12}$ M^{-2}; $\beta = 3.06 \times 10^6$ M^{-1}.

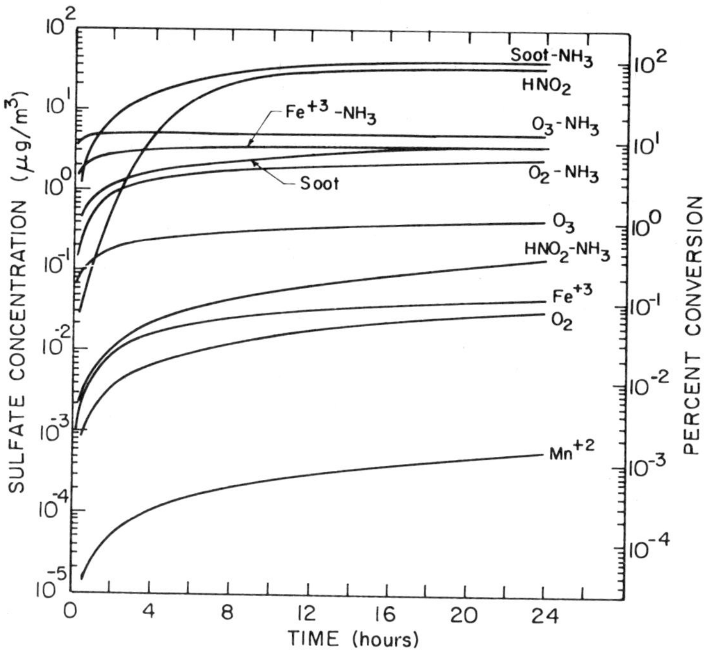

Figure 13. Comparison of relative importance of various sulfate production mechanisms involving liquid water based on a box-type calculation. The following initial conditions were used in the calculation: P_{SO_2} = 0.01 ppm; P_{CO_2} = 311 ppm; P_{NH_3} = 5 ppb; P_{O_3} = 0.05 ppm; P_{HNO_2} = 8 ppb; $[Fe^{3+}] = 1.2 \times 10^{-7}$ M; $[Mn^{2+}] = 1.8 \times 10^{-8}$ M; soot = 10 $\mu g\ m^{-3}$; and liquid water = 0.05 $g\ m^{-3}$. (From Chang et al., 1981; reprinted with permission of Pergamon Press.)

tween SO_2 in the gas phase and S(IV) in the droplet is established several orders of magnitude faster than oxidation of S(IV) to sulfate (Beilke and Gravenhorst, 1978). Similar assumptions were made regarding NH_3 and CO_2. At the outset of the calculation, therefore, gases are taken to be in equilibrium with the dissolved solutes. Then the formation of sulfate proceeds by the given time-dependent production rate. The increase in the sulfate level in the small time step Δt causes the reduction in pH of the solution, which in turn disturbs the equilibrium between the droplet and its surrounding gaseous environment. More gases are dissolved in the droplet to maintain the equilibrium. At the same time, these gases are depleted in the surrounding atmosphere. After each calculation, the time step is adjusted and the process is repeated until a 24-h period is completed. The results are shown in Figure 13.

Figure 13 indicates that O_3, black carbon, and HNO_2 can be important mechanisms for sulfate aerosol formation. In general, the O_3 mechanism is more important under high pH and/or photoactivity conditions when the concentration of O_3 is high, whereas both carbon and HNO_2 processes are more important when the lifetime of fog or clouds is long and the pH of the droplets is low. Both carbon and HNO_2 processes can be dominant processes close to sources and in heavily polluted urban areas, where the concentrations of soot and $NO-NO_2$ are high and the pH of aqueous droplets is low.

The rate constant for oxidation of S(IV) by black carbon particles varies, depending on the nature and history of particle production as discussed previously. In a fog-chamber study, Benner (1980) recently found that the reaction rate of soot particles from a natural gas diffusion flame can be considerably faster than the reaction rate employed in these calculations. This variability suggests that further kinetic studies of S(IV) oxidation by soot from different types of fuel should be conducted.

ACKNOWLEDGMENT

This work was supported by the Assistant Secretary for the Environment, Office of Health and Environmental Research, Pollutant Characterization and Safety Division of the U.S. Department of Energy under Contract No. W-7405-ENG-48 and by the National Science Foundation under Contract No. ATM 80-13707.

REFERENCES

Beilke, S. and Gravenhorst, G. (1978). Heterogeneous SO_2 oxidation in the droplet phases. *Atmosph. Environ.* **12,** 231–239.

Beilke, S., Lamb, D., and Müller, J. (1975). On the uncatalyzed oxidation of atmospheric SO_2 by oxygen in an aqueous system. *Atmosph. Environ.* **9,** 1083–1090.

Benner, W. H. (1980). Private communication, Lawrence Berkeley Laboratory.

Bente, P. F. and Walton, J. H. (1943). The catalytic activity of activated nitrogenous carbons. *J. Phys. Chem.* **47,** 133–148.

Boehm, H. P. (1966). Chemical identification of surface groups. *Adv. Catal. Relat. Subj.* **16,** 179–274.

Boehm, H. P. Diehl, E., Heck, W., and Sappok, R. (1964). Surface oxides of carbon. *Angew. Chem. Internatl. Ed. Engl.* **3,** 669–677.

Britton, L. G. and Clarke, A. G. (1980). Heterogeneous reactions of sulphur dioxide and SO_2-NO_2 mixtures with a carbon soot aerosol. *Atmosph. Environ.* **14,** 829–839.

Brodzinsky, R., Chang, S. G., Markowitz, S. S., and Novakov, T. (1980). Kinetics and mechanism for the catalytic oxidation of sulfur dioxide on carbon in aqueous suspensions. *J. Phys. Chem.* **84,** 3354–3358.

Brosset, C. (1980). Equilibrium composition of aerosols generated from sulphuric and nitric acids, water and ammonia. Paper presented at the Chemical Institute of Canada Annual Meeting, Ottawa, Ontario, June 8–11, 1980.

Chang, S. G. and Novakov, T. (1975a). *Infrared and Photoelectron Spectroscopic Study of SO_2 Oxidation on Soot Particles,* Lawrence Berkeley Laboratory Report LBL-4446.

Chang, S. G. and Novakov, T. (1975b). Formation of pollution particulate nitrogen compounds by NO–soot and NH_3–soot gas particle surface reactions. *Atmosph. Environ.* **9,** 495–504.

Chang, S. G., Brodzinsky, R., Toossi, R., Markowitz, S. S., and Novakov, T. (1979). Catalytic oxidation of SO_2 on carbon in aqueous suspensions. *Proceedings of the Conference on Carbonaceous Particles in the Atmosphere,* Lawrence Berkeley Laboratory Report LBL-9037, pp. 122–130.

Chang, S. G., Toossi, R., and Novakov, T. (1981). The importance of soot particles and nitrous acid in oxidizing SO_2 in atmospheric aqueous droplets. *Atmosph. Environ.* **15,** 1287–1292.

Chang, S. G., Brodzinsky, R., Gundel, L. A. and Novakov, T. (1982). Chemical and Catalytic Properties of Elemental Carbon, in *Particulate Carbon: Atmospheric Life Cycle*, Wolff, G. T. and Klimisch, R. L. Eds., Plenum Press, New York pp. 159–181.

Clark, A. (1970). *The Theory of Adsorption and Catalysis,* Academic Press, New York.

Cofer, III, W. R., Schryer, D. R., and Rogowski, R. S. (1980). The enhanced oxidation of SO_2 by NO_2 on carbon particulates. *Atmosph. Environ.* **14,** 571–575.

Cofer, III, W. R., Schryer, D. R., and Rogowski, R. S. (1981). The oxidation of SO_2 on carbon particles in the presence of O_3, NO_2, and N_2O. *Atmosph. Environ.* **15,** 1281–1286.

Coughlin, R. W. (1969). Carbon as adsorbent and catalyst. *Ind. Eng. Chem. Res. Devel.* **8,** 12–23.

Coughlin, R. W. and Ezra, F. S. (1968). Role of surface acidity in the adsorption of organic pollutants on the surface of carbon. *Environ. Sci. Technol.* **2,** 291–297.

Davtyan, O. K. and Tkach, Yu. A. (1961). The mechanism of oxidation, hydrogenation, and electrochemical oxidation on solid catalysts. II. The catalytic activity of surface "oxides" on carbon. *Russ. J. Phys. Chem.* **35,** 992–998.

Dulou, R. (1945). Catalyse par les adsorbants non métalliques. *Chim. Ind. (Paris)* **54,** 396–403.

Eigen, M., Kustin, K., and Maass, G. (1961). Die Geschwindigkeit der Hydration von SO_2 in wässriger Lösung. *Z. phys. Chem. neue Folge* **30,** 130–136.

Emmett, P. H. (1948). Adsorption and pore-size measurements on charcoals and whetlerites. *Chem. Rev.* **43,** 69–148.

Erickson, R. E., Yates, L. M., Clark, R. L., and McEwen, D. (1977). The reaction of sulfur dioxide with ozone in water and its possible atmospheric significance. *Atmosph. Environ.* **11,** 813–817.

Freiberg, J. (1975). The mechanism of iron catalyzed oxidation of SO_2 in oxygenated solutions. *Atmosph. Environ.* **9,** 661–673.

Friedel, R. A. and Hofer, L. J. E. (1970). Spectral characterization of activated carbon. *J. Phys. Chem.* **74,** 2921–2922.

Frumkin, A., Burstein, R., and Lewin, P. (1931). Über aktivierte Kohle. *Z. Phys. Chem.* **A157,** 442–446.

Garten, V. A. and Weiss, D. E. (1957). The ion- and electron-exchange properties of activated carbon in relation to its behaviour as a catalyst and adsorbent. *Rev. Pure Appl. Chem.* **7,** 69–122.

Garten, V. A., Weiss, D. F., and Willis, J. B. (1957). A new interpretation of the acidic and basic structures in carbon. I. Lactone groups of the ordinary and fluorescein types in carbon. *Aust. J. Chem.* **10,** 295–308.

Gordon, G. E., Davis, D. D., Israel, G. W., Landsberg, H. E., and O'Haver, T. C. (1975). *Atmospheric Impact of Major Sources and Consumers of Energy.* National Science Foundation Report NSF/RA/E-75/189. National Technical Information Service PB262574.

Gundel, L. A. (1979). Private communication, Lawrence Berkeley Laboratory.

Gundel, L. A., Chang, S. G., Clemenson, M. S., Markowitz, S. S., and Novakov, T. (1979). Characterization of particulate amines, in *Nitrogenous Air Pollutants,* D. Grosjean, Ed., Ann Arbor Science, Ann Arbor, pp. 211–220.

Hansen, A. D. A., et al. (1980). The use of an optical attenuation technique to estimate the carbonaceous component of urban aerosols, in *Atmospheric Aerosol Research Annual Report FY-1979,* Lawrence Berkeley Laboratory Report LBL-10735, pp. 8–16–8–21.

Hart, P. J., Vastola, F. J., and Walker, Jr., P. L. (1967). Oxygen chemisorption on well cleaned carbon surfaces. *Carbon* **5,** 363–371.

Hodgman, C. D., Ed. (1958). *Handbook of Chemistry and Physics,* 39th ed., Chemical Rubber Publishing Company, Cleveland, OH.

Hoffman, U. and Wilm, D. (1936). Über die Kristallstruktur von Kohlenstoff. *Z. Elektrochem. Angew. physik. Chem.* **42,** 504–522.

Laine, N. R., Vastola, F. J., and Walker, Jr., P. L. (1963). The importance of active surface area in the carbon oxygen reaction. *J. Phys. Chem.* **67,** 2030–2034.

Larsen, E. C. and Walton, J. H. (1940). Activated carbon as a catalyst in certain oxidation-reduction reactions. *J. Phys. Chem.* **44,** 70–85.

Lau, N. C. and Charlson, R. J. (1977). On the discrepancy between background atmospheric ammonia gas measurements and the existence of acid sulfates as a dominant atmospheric aerosol. *Atmosph. Environ.* **11,** 475–478.

Matteson, M. J., Stöber, W., and Luther, H. (1969). Kinetics of the oxidation of sulfur dioxide by aerosols of manganese sulfate. *Ind. Eng. Chem. Fundam.* **8,** 677–687.

Mattson, J. S. and Mark, Jr., H. B. (1971). *Activated Carbon,* Marcel Dekker, New York, Chapters 3 and 6.

McKay, H. A. C. (1971). The atmospheric oxidation of sulfur dioxide in water droplets in the presence of ammonia. *Atmosph. Environ.* **5,** 7–14.

Mellor, H. W. (1946). *Comprehensive Treatise on Inorganic and Theoretical Chemistry,* Vol. 1, Longmans, London.

Novakov, T. (1973). Chemical characterization of atmospheric pollution particulates by photoelectron spectroscopy. *Proceedings of 2nd Joint Conference on Sensing of Environmental Pollutants,* Instrument Society of America, Pittsburgh, PA, pp. 197–204.

Novakov, T. (1981). Microchemical characterization of aerosols, in *Nature, Aim and Methods of Microchemistry,* H. Malissa, M. Grasserbauer, and R. Belcher, Eds., Springer-Verlag, Vienna, pp. 141–165.

Novakov, T. and Chang, S. G. (1977). *ESCA in Environmental Chemistry.* Lawrence Berkeley Laboratory Report LBL-6323; to be published as a chapter in *Advances in Analytical Chemistry,* D. Natusch, Ed., American Chemical Society.

Novakov, T., Mueller, P. K., Alcocer, A. E., and Otvos, J. W. (1972). Chemical composition of Pasadena aerosol by particle size and time of day. III. Chemical states of nitrogen and sulfur by photoelectron spectroscopy. *J. Colloid Interface Sci.* **39,** 225–234.

Novakov, T., Chang, S. G., And Harker, A. B. (1974). Sulfates in pollution particulates: Catalytic oxidation of SO_2 on carbon particles. *Science* **186,** 259–261.

Oblath, S. B., Markowitz, S. S., Novakov, T., and Chang, S. G. (1981). Kinetics of the formation of hydroxylamine disulfonate by reaction of nitrite with sulfites. *J. Phys. Chem.* **85,** 1017–1021.

Palmer, H. B. and Cullis, C. F. (1965). The formation of carbon from gases, in *Chemistry and Physics of Carbon,* Vol. I, Marcel, Dekker, New York, pp. 265–325.

Puri, B. R. (1962). Surface oxidation of charcoal at ordinary temperatures. *Proceedings of Conference on Carbon, 5th,* Vol. I, Pergamon Press, Oxford, pp. 165–170.

Puri, B. R. (1966). Chemisorbed oxygen evolved as carbon dioxide and its influence on surface reactivity of carbons. *Carbon* **4,** 391–400.

Puri, B. R. (1970). Surface complexes on carbon, in *Chemistry and Physics of Carbon,* Vol. VI, Marcel Dekker, New York, pp. 191–282.

Rao, M. N. and Hougen, O. H. (1952). Catalytic oxidation of nitric oxide on activated carbon. *Chem. Eng. Progr. Symp. Ser.* **48,** 110–124.

Rivin, D. (1963). Hydride-transfer reactions of carbon black. *Proceedings of Conference on Carbon, 5th,* Vol. II, Pergamon Press, Oxford, pp. 199–209.

Rosen, H., Hansen, A. D. A., Gundel, L., and Novakov, T. (1978). Identification of the optically absorbing component of urban aerosols. *Appl. Opt.* **17,** 3859–3861.

Rosen, H., Hansen, A. D. A., Dod, R. L., and Novakov, T. (1980). Soot in urban atmospheres: Determination by an optical absorption technique. *Science* **208,** 741–744.

Rosen, H., Novakov, T., and Bodhaine, B. A. (1981). Soot in the Arctic. *Atmosph. Environ.* **15,** 1371–1374.

Schilow, N., Schatunowskja, H., and Tschmutow, K. (1930). Adsorptionsercheinungen in Lösungen. XX. Über den chemischen Zustand der Oberfläche von aktiver Kohle. *Z. Phys. Chem. (Leipzig)* **A149,** 211–222.

Schwartz, S. E. and White, W. H. (1981). Solubility equilibria of the nitrogen oxides and oxyacids in dilute aqueous solution, in *Advances in Environmental Science and Engineering,* Vol. 4, J. R. Pfafflin and E. N. Ziegler, Eds., Gordon and Breach, New York, pp. 1–45.

Siedlewski, J. (1965). The mechanism of catalytic oxidation on activated carbon: The influence of free carbon radicals on the adsorption of SO_2. *Internatl. Chem. Eng.* **5,** 297–301.

Smith, R. N. (1959). The chemistry of carbon–oxygen surface compounds. *Quart. Rev.* **13,** 287–305.

Steenberg, B. (1944). *Adsorption and Exchange of Ions on Activated Charcoal,* Almquist and Wiksells, Uppsala.

Stumpp, E. (1965). Untersuchungen über den Ameisensäurezerfall an Graphit und an Metallchlorid-Graphitverbindungen. *Z. Anorg. Allgem. Chem.* **337,** 292–300.

Tuinstra, F. and Koenig, J. L. (1970). Raman spectrum of graphite. *J. Chem. Phys.* **53,** 1126–1130.

Weller, S. W. and Young, T. F. (1948). Oxygen complexes on charcoal. *J. Am. Chem. Soc.* **70,** 4155–4162.

Yamamoto, K., Seki, M., and Kawazoe, K. (1972). Absorption of sulfur dioxide on activated carbon in the flue gas desulfurization process. III. Rate of oxidation of sulfur dioxide on activated carbon surfaces. *Nippon Kagaku Kaishi* **6,** 1046–1052.

Yasa, Z., Amer, N., Rosen, H., Hansen, A. D. A., and Novakov, T. (1978). Photoacoustic investigation of urban aerosol particles. *Appl. Opt.* **18,** 2528–2530.

Zarif'yanz, Y. A., Kiselev, V. F., Lezhnev, N. N., and Nikitina, D. V. (1967). Interaction of graphite fresh surface with different gases and vapors. *Carbon* **5,** 127–135.

5

ORGANIC AND INORGANIC S(IV) COMPOUNDS IN AIRBORNE PARTICULATE MATTER

Delbert J. Eatough and Lee D. Hansen

Thermochemical Institute and
Department of Chemistry
Brigham Young University
Provo, Utah 84602

1. Introduction

Epidemiologic and animal toxicologic studies have indicated that reactions between gaseous SO_2 and metal-containing aerosols result in the formation of respiratory irritants (Amdur et al., 1975; Colucci, 1976; Lave and Seskin, 1977; National Academy of Sciences, 1979). Initially it was suggested that sulfate per se was responsible for the observed health effects (Shy and Finklea, 1973; U.S. Environmental Protection Agency, 1975). That now appears unlikely. It is not even clear that the effects can be associated with aerosol acidity (e.g., H_2SO_4) (Colucci, 1976; National Academy of Sciences, 1979; Sachner et al., 1978). These studies point out the importance of understanding in detail the chemical species formed by the reactions of $SO_2(g)$ to form new chemical species in atmospheric aerosols. The known reactions of $SO_2(g)$ to form aerosol sulfur species with sulfur in the +4 oxidation state are the focus of this chapter. A flow diagram of known reactions of $SO_2(g)$ in the atmosphere is given in Figure 1. The transformations indicated

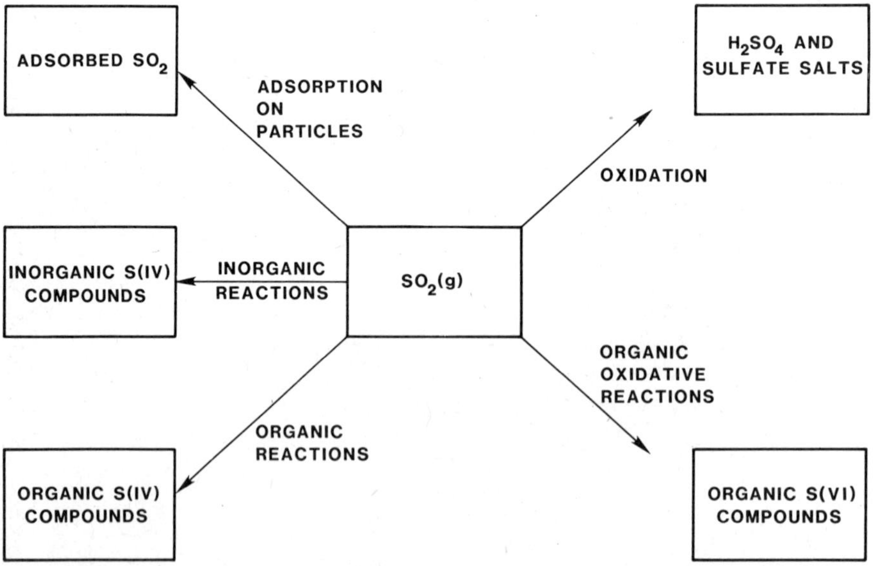

Figure 1. Compounds formed from reactions of $SO_2(g)$ in the atmosphere.

in Figure 1 show the chemical entities that must be studied to understand the reactions of SO_2(g) with aerosols. Knowledge of the rate of conversion of SO_2(g) to aerosol species, the total sulfur content of the aerosol, or even the oxidation state of sulfur is not sufficient since the lifetimes, concentrations, toxicities, etc. of the various species will be highly variable.

The oxidation of SO_2(g) to form inorganic sulfate compounds has been the subject of many studies both in the laboratory (Husar et al., 1978) and in the field (Newman, 1981). Aspects of this chemistry are treated in Chapters 2–4, 6, 7, 11, and 12 of this volume. In contrast, the formation of organic S(VI) compounds in the atmosphere has received little attention. Previous studies have reported the presence of methane sulfonic acid (Panter and Penzhorn, 1980; Penzhorn and Filby, 1976) and monomethyl and dimethyl sulfate (Eatough et al., 1981a; Lee et al., 1980) in atmospheric aerosols. The mechanisms of formation of these compounds are not now known.

No compounds with sulfur in the +5 oxidation state have been shown to be present in the atmosphere. It is possible that dithionate is formed under some conditions; however, this species has not yet been identified. Dithionate chemistry is discussed in relation to inorganic S(IV) chemistry in Section 2.2 as well as in Chapter 2 of this volume (Huie and Peterson).

Known S(IV) compounds resulting from the interaction of SO_2(g) with particulate matter include physically adsorbed SO_2 and inorganic and organic compounds. Only the first two products appear to be important in emissions from smelters. The chemistry occurring in plumes from oil- or coal-fired power plants or in urban atmospheres leads to all three classes of compounds. Procedures that have been used to study the formation of aerosol inorganic S(IV) and organic S(IV) species are described, together with the results obtained from studies in the flue lines, workrooms, and plumes of smelters, in the flue lines and plumes of fossil-fuel fired power plants and in rural and urban environments are described in the present chapter. Work on these compounds that may have relevance to toxicology is outlined.

2. INORGANIC S(IV) COMPOUNDS

2.1. Characterization of Inorganic S(IV) in Flue Dusts and Airborne Particulate Matter

The presence of S(IV) compounds in airborne particulate matter was first suggested by photoelectron spectroscopy studies of aerosol samples (Allegrini and Mattogno, 1978; Novakov et al., 1972). It was not possible from these results to establish whether the S(IV) observed by ESCA analysis represented adsorbed SO_2(g) or a stable sulfite compound. The presence of stable compounds with sulfur in the +4 oxidation state was hypothesized by Hansen et al. (1974, 1975) to explain analytical results obtained from a variety of experiments on size-fractionated samples collected from the workroom environment of a copper smelter. The workroom samples were analyzed by use of the following techniques:

Calorimetry. The calorimetric techniques used for the determination of S(IV) and other reducing agents and of sulfate in the collected aerosol samples have been described by Hansen et al. (1976). The portion of an aerosol sample extractable with 0.1 M HCl and 2.5 (or 5.0) mM $FeCl_3$ was analyzed for S(IV) by a calorimetric redox titration with dichromate. Other dichromate-oxidizable species are each independently detected by this method. Sulfate was determined calorimetrically by precipitation with $BaCl_2$. The samples were extracted at room temperature in an ultrasonic bath for 20 min. All solutions used in the analysis were prepared, stored, and used under an argon atmosphere. The dichromate titration indicated the presence of reducing agents with ΔH_{ox} values of -30 and -24 kcal eq^{-1}. End points observed in the thermograms indicated that the redox reactions being monitored calorimetrically were rapid and stoichiometric. The species titrated with a ΔH_{ox} value of -30 kcal eq^{-1} was assumed to be Fe^{2+}, as that is the value of ΔH_{ox} of this cation with $Cr_2O_7^{2-}$ in 0.1 M HCl. Both the $\Delta G°$ and $\Delta H°$ values for reaction of $Cr_2O_7^{2-}$ with HSO_3^- are more negative than the corresponding values for oxidation of Fe^{2+}. The calorimetric results indicated that HSO_3^- was not present in the solutions analyzed. Another possible reducing species is As(III), whose ΔH_{ox} value is comparable to the -24 kcal eq^{-1} observed; however, the oxidation of As(III) by $Cr_2O_7^{2-}$ in 0.1 M HCl was found to be slow, resulting in an effective ΔH_{ox} value in the titration experiment of below -15 kcal eq^{-1}. In addition, elemental analysis indicated insufficient amounts of arsenic or any other nonsulfur species that could explain the observed calorimetric results. Calorimetric titration of solutions prepared from Fe^{3+} and HSO_3^- solutions indicated the presence of a species that titrated after Fe^{2+} with a ΔH_{ox} value of -24 kcal eq^{-1}. It was thus postulated that the species titrated was a stable Fe(III)–S(IV) complex (Hansen et al., 1974, 1975, 1976).

PIXE. Elements present in the acid extractant solutions with atomic number 17 (chlorine) or greater were determined by proton-induced x-ray emission (PIXE). The methods used for data reduction have been described by Nielsen et al. (1976). The inorganic S(IV) determined by calorimetric titration was present in the same size distribution pattern as most transition metals in the sample and did not follow the size distribution observed for sulfate (Figure 2). This suggests that the inorganic S(IV) compounds and sulfate were not formed by the same mechanism. The concentration of S(IV) in these samples is correlated with the iron concentration for all samples studied (Figure 3). Calorimetric redox titration of these samples and a Moessbauer spectrum (one sample only) showed that no Fe(II) was present in the samples (Hansen et al., 1975). These data also suggest the presence of an Fe(III)–S(IV) compound in the col-

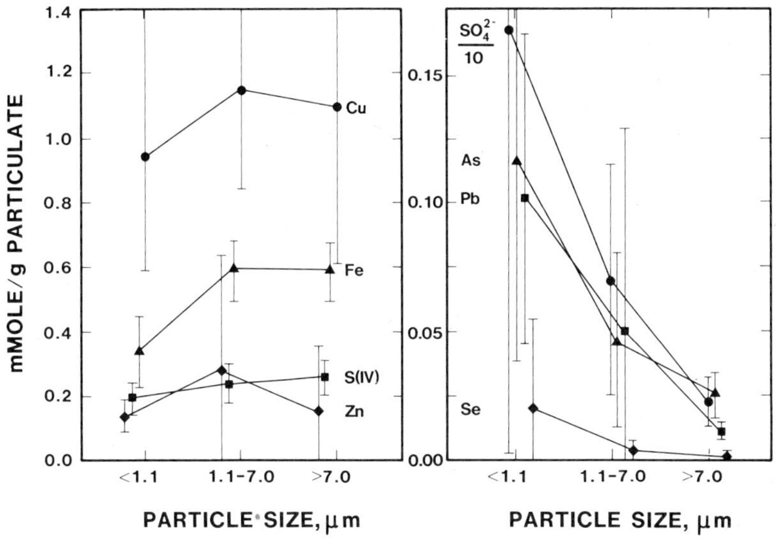

Figure 2. Dependence on particle size of the concentration of various species in suspended particulate matter from the workroom environment of a copper smelter. The error bars are the standard deviations among several collected samples. (From Eatough and Hansen, 1981; reprinted with permission of Alan R. Liss, Inc.)

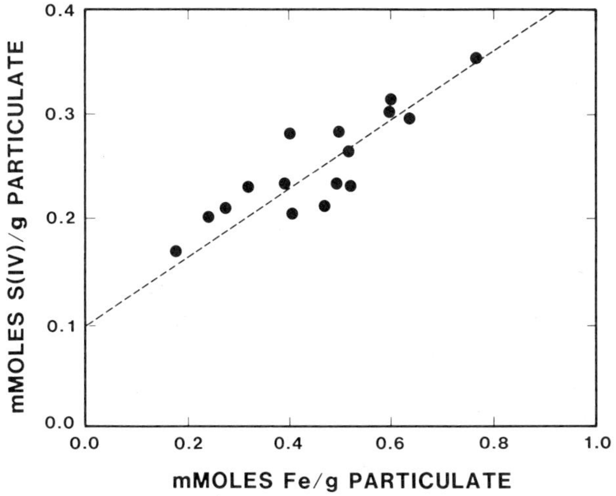

Figure 3. Concentrations of S(IV) and Fe(III) found in suspended particulate matter from the workroom environment of a copper smelter. Taken from Hansen et al. (1975).

lected samples. It was found that PIXE was not reliable for total sulfur determinations because of decomposition and volatilization of sulfur from the PIXE targets (Hansen et al., 1980).

ESCA. Samples for analysis were prepared by transferring collected particulate matter onto a sample platen. The electron spectroscopy for chemical analysis (ESCA) spectra were obtained with a Hewlett-Packard x-ray photoelectron spectrometer. High-resolution data on the energy of the sulfur 2*p* photoelectron lines were obtained to determine S(−II), S°, S(IV), and S(VI) species present on the surface of the sample. The measured peak heights were adjusted for specific line sensitivities to obtain quantitative data as previously described (Eatough et al., 1979).

The ESCA data obtained in analysis of >7 μm particulate matter collected near the reverberatory furnace of the smelter are shown in Figure 4. (Here and throughout this chapter reference to particle size indicates the equivalent aerodynamic diameter.) Analysis of this sample 1 month after collection gave S(VI):S(IV):S(−II) ratios of 1.00:0.98:0.80. After storage of the sample in air for 6 months, reanalysis gave the ratios of 1.00:1.66:0.00. These results suggest that oxidation of sulfide species in smelter aerosols may result in the formation of S(IV), rather than sulfate compounds. This result is not unexpected since the S(IV) fraction of aerosols from smelters apparently exists as transition-metal–S(IV) complexes that are stable toward oxidation by air. Since the S(−II) component of the samples exists as transition-metal sulfides, the formation of transition-metal–S(IV) complexes may be favored over other possible oxidation products (Eatough et al., 1979).

All the above methods have also been applied to analysis of a series of flue line samples collected from both copper and lead smelters (Eatough et al., 1979, 1982). In addition to the calorimetric, PIXE, and ESCA analyses, the following analyses were also performed on these flue line samples:

Ion Chromatography. Ions extractable into water were determined by using ion chromatography (IC). Details of the experimental procedures used have been published (Hansen et al., 1979). Problems associated with use of this analytical technique to analyze for some S(IV) species have also been discussed (Dasgupta et al., 1982; Eatough et al., 1979, 1982; Eatough and Hansen, 1979, 1981; Hansen et al., 1979). The technique can be used to determine physically adsorbed SO_2 or simple sulfite salts but not stable transition-metal–S(IV) complexes.

Colorimetry. Sulfur dioxide thermally released from samples by heating and S(IV) in extract solutions were determined colorimetrically by use of the procedure described by West and Gaeke (1956). In these experiments the evolved gases trapped in Na_2HgCl_4 solutions or material extractable in Na_2HgCl_4 solutions were analyzed. In some experiments the flue dust samples were extracted into 1.0 M HF solution and this solution analyzed colorimetrically by addition of the West–Gaeke reagents.

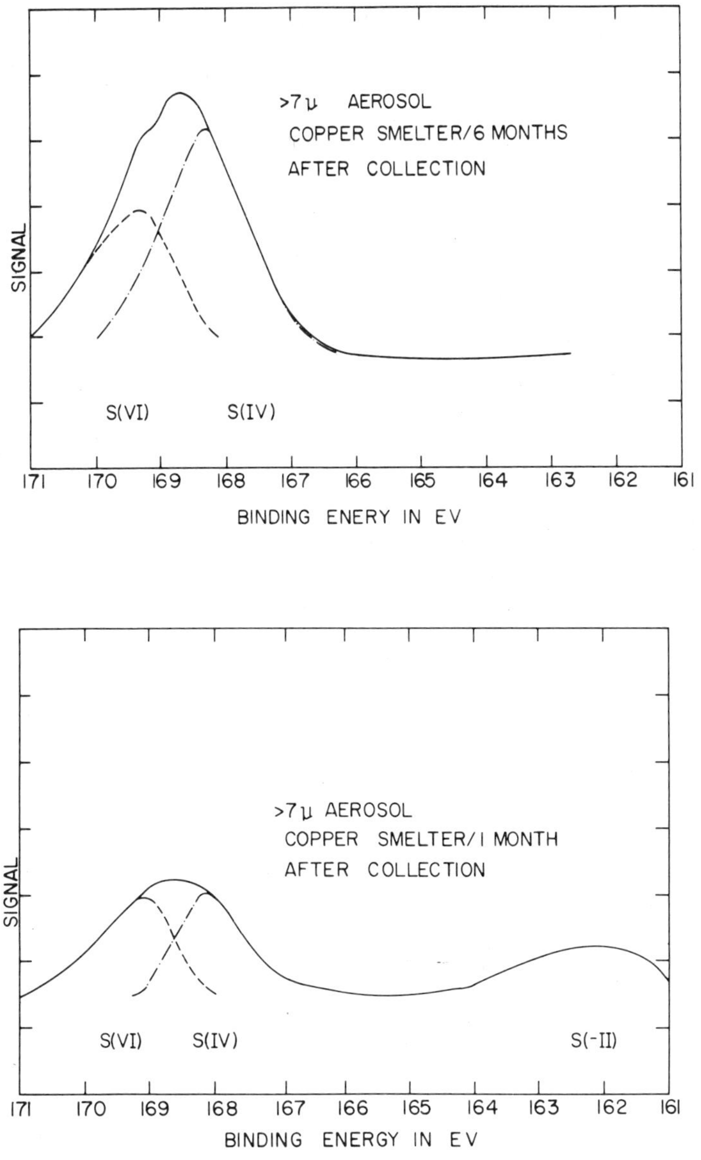

Figure 4. ESCA spectra of >7-μm particulate matter collected in the workroom environment of a copper smelter. (From Eatough and Hansen, 1981; reprinted with permission of Alan R. Liss, Inc.)

Comparative results obtained for the smelter flue line samples (Eatough et al., 1979, 1981b, 1982; Eatough and Hansen, 1981) with the various analytical techniques are summarized in Table 1. Comparable results were found in the determination of total inorganic S(IV) by calorimetric titration with $K_2Cr_2O_7$ in 0.1 M HCl solution with or without added Fe^{3+}, measurement of the increase in sulfate associated with dichromate oxidation, and meas-

Table 1. Comparative Determination of Inorganic S(IV) in Smelter Flue Dust Samples[a]

Analysis Procedure	Range[b]	Mean[b]	Number of Samples
$K_2Cr_2O_7$, titration in 0.1 M HCl, no $FeCl_3$	74–138	114	8
Increase in calorimetrically determined SO_4^{2-} due to $K_2Cr_2O_7$ addition in 0.1 M HCl	37–116	84	6
SO_2(g) evolution at 280°C	104–117	112 ± 6	3
Ion chromatography	0–3	1 ± 1	4
West–Gaeke analysis	0–8	3 ± 3	4
West–Gaeke analysis in 1.0 M HF	13–24	19 ± 8	2

[a] Data are from Eatough et al. (1979, 1981b, 1982) and Eatough and Hansen (1981).

[b] The results are given as a percentage of the value obtained by determination in 0.1 M HCl, 2.5 mM $FeCl_3$ solution by calorimetric redox titration with 5 mM $K_2Cr_2O_7$ in 0.1 M HCl.

urement of SO_2(g) evolved from the samples. Problems associated with the ion chromatographic and the West–Gaeke procedures have been discussed (Dasgupta et al., 1982; Eatough et al., 1979; Eatough and Hansen, 1979; Hansen et al., 1979). We previously suggested that $FeCl_3$ be added to the HCl extractant solution to assure complete formation of the Fe(III)–S(IV) complex which is titrated with a ΔH_{ox} value of -24 kcal eq^{-1} (Hansen et al., 1976); however, results obtained on flue dust samples indicate that if only inorganic S(IV) is to be measured, the $FeCl_3$ may be omitted if the calorimetric redox titration is carried out immediately after extraction. [The addition of $FeCl_3$ to the extract solution is essential if hydrolysis of organic S(IV) is to be followed by measuring the increase in the Fe(III)–S(IV) complex (Eatough et al., 1978a).] The poor results obtained in the West–Gaeke determination are presumably due to slow formation of the unstable chromophore (Eatough et al., 1979; Eatough and Hansen, 1981). As indicated in Table 1, better, but still unsuitable, results are obtained by the West–Gaeke procedure when a competing ligand for Fe(III), such as F^-, is added. The negative results obtained by ion chromatography are due to oxidation of the sulfur in the Fe(III)–S(IV) complex in basic solution (Eatough et al., 1979; Eatough and Hansen, 1979; Hansen et al., 1979; Hilton et al., 1979). However, it would appear that weakly adsorbed SO_2 can be measured by ion chromatographic determination of sulfite (Dasgupta et al., 1982; Eatough et al., 1979; Hansen et al., 1979). Addition of formaldehyde to the aqueous extract solution facilitates stabilization of adsorbed SO_2 for analysis by IC (Dasgupta et al., 1982; Hansen et al., 1979). Calorimetric determination of SO_4^{2-} in 0.1 M HCl extracts of samples before and after addition of $Cr_2O_7^{2-}$ shows that $Cr_2O_7^{2-}$ addition results in an increase in sulfate that is in reasonable agreement with the direct determination of inorganic S(IV) by redox titration. Potential problems with the general application of this method for determination of S(IV) have been discussed (Eatough and Hansen, 1979). Good agreement was also seen between calorimetrically determined S(IV) and SO_2(g) evolution at 280°C (Eatough and Hansen, 1981), as indicated in Table 1. The time course of the SO_2 evolution is shown in Figure 5. The very rapid evolution of 0.07 μmol SO_2(g) in this sample corresponds with the amount of adsorbed SO_2 expected on the basis of ion chromatographic analysis of the sample. This very rapid evolution was seen only in the one sample illustrated. In addition, this was the only sample studied by SO_2(g) evolution in which sulfite was found by ion chromatography. The evolution of 0.58 μmol of SO_2(g) over 3–4 days correlates with the calorimetric determination of inorganic S(IV) in this sample. The final very slow SO_2(g) evolution presumably represents SO_2(g) produced by reactions of sulfides or sulfates in the sample. Although good results are obtained in the determination of S(IV) in smelter flue dust samples by measuring SO_2(g) evolution in the temperature range 180–280°C or by measuring the increase in sulfate on addition of $K_2Cr_2O_7$ in 0.1 M HCl, the techniques are not recommended for routine analysis. The SO_2(g) evolution measurement is tedious and may give erroneous answers in analysis of coal ash or ambient samples because of reduction of sulfate by elemental carbon in the heated sample (Hansen, 1981). The measurement of SO_4^{2-} increase following $Cr_2O_7^{2-}$ oxidation of inorganic S(IV) requires taking differences in the measured sulfate concentrations, leading to significant error if the ratio of inorganic S(IV) to SO_4^{-2} is small. This ratio is usually less than 0.2 and often less than 0.1 in samples collected from the ambient atmosphere and in plumes from coal- and oil-fired boilers (see Section 2.4).

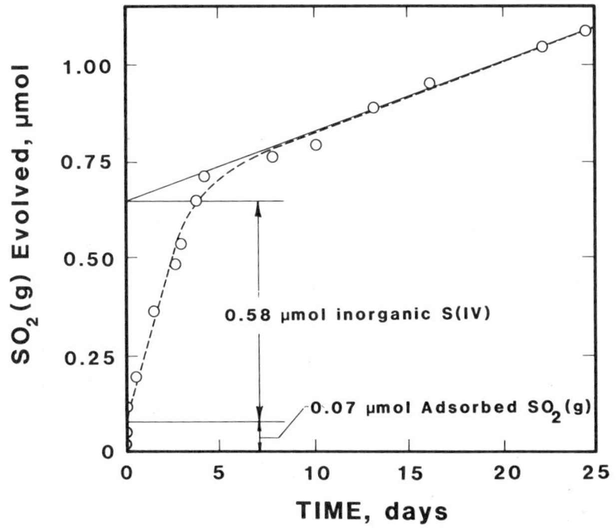

Figure 5. Evolution of SO_2(g) at 280°C from a copper smelter dust sample under a flowing N_2(g) stream as a function of time. (From Eatough and Hansen, 1981; reprinted with permission of Alan R. Liss, Inc.)

The formation of inorganic S(IV) compounds has also been studied in size-fractionated particulate samples collected from the plume of the Utah smelter of Kennecott Copper Corporation (Eatough and Hansen, 1981; Eatough et al., 1982). The samples were collected with Sierra cascade impactors at distances of 4–60 km from the stacks at four sampling stations located on the eastern edge of the Oquirrh mountains at an elevation about 500 m above the smelter. All stations were in the plume when northerly wind flows off the Great Salt Lake placed the plume along the mountains.

Analysis of the collected samples by the calorimetric procedures showed that the chemistry occurring in the plume is more complex than that occurring in the flue line of the same smelter. Whereas the analyses of samples collected inside the smelter are consistent with formation of Fe(III)–S(IV) complexes only, other inorganic complexes appear to be formed in the plume. Calorimetric dichromate titration of acidic extracts of the plume samples showed that extractable Fe^{2+} was not present in the plume aerosol. The Fe(III)–S(IV) complex with $\Delta H_{ox} = -24$ kcal eq^{-1} was routinely seen in these samples, and, in addition, a reducing species with $\Delta H_{ox} = -16$ kcal eq^{-1} was seen in some samples. This latter reducing agent appeared to be converted over a period of several weeks to a species that oxidized with $\Delta H_{ox} = -24$ kcal eq^{-1}. In addition, the concentration of inorganic S(IV) in the collected plume samples was not well correlated with acid-extractable Fe ($r = 0.39$) but was better correlated with the sum of acid extractable Fe + Cu ($r = 0.73$) (see Figure 6). The Fe(III):Cu(?) ratio in the samples varied from 3 to 1. These data suggest the presence of Cu(?)–S(IV) species in the collected samples. It was previously shown

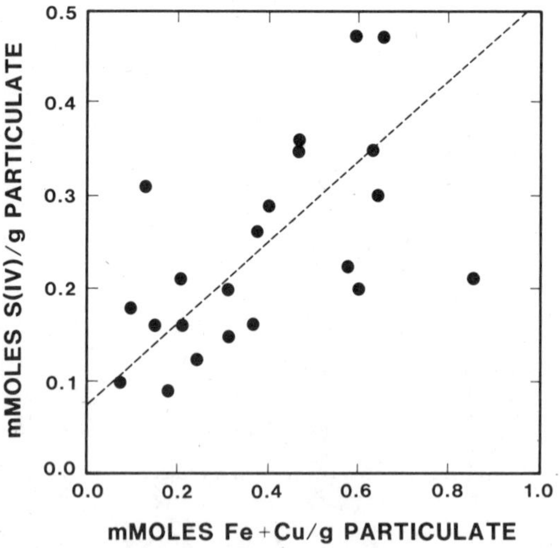

Figure 6. Concentrations of S(IV) and Fe + Cu found in particulate matter in the plume of a copper smelter. (From Eatough and Hansen, 1981; reprinted with permission of Alan R. Liss, Inc.)

Table 2. Comparison of Total to Surface Concentration of Inorganic S(IV) Species in Smelter Flue Line Dust Samples and Particulate Matter from a Copper Smelter Plume[a]

Source Number	Sample Source	Smelter Type	Total S(IV) (μmol g^{-1})	Surface S(IV) (μmol g^{-1})	Surface/ Total
1	Flue line	Cu	138	<80	<0.5
2	Flue line	Cu	163	331	2.0
3	Flue line	Cu	100	164	1.6
4	Flue line	Pb	125	340	2.7
5	Flue line	Cu	100	164	1.6
	Plume, <3 km	Cu	90	70	0.8
	Plume, >30 km	Cu	320	<40	<0.1

[a] Data are from Eatough et al. (1979, 1982).

(Hansen et al., 1974; Hilton et al., 1979) that stable Cu(I) complexes of SO_3^{2-} can be formed. It is also interesting to note that the plot of S(IV) versus Fe(III) in the smelter workroom samples (Figure 3) does not have a zero intercept. This suggests some S(IV) species other than Fe(III)–S(IV) is present. The slope of the plot of S(IV) versus Fe in Figure 3 is 0.3, whereas the slope for the data in Figure 6 is about 0.5. The formulas of the inorganic S(IV) complexes formed in the flue line and plume of the smelter are not known at present.

Since a major motive behind work such as is described in this chapter is to assess the potential health hazards of the collected emissions, it is of interest to compare total concentration with surface concentration of inorganic S(IV) in the samples. Surface concentrations of S(IV) have been determined in smelter flue dust and plume samples by photoelectron spectroscopy (Eatough et al., 1979, 1982). The total analyses represent the total amount of material, noxious and otherwise, that is added to the environment. As the data summarized in Figures 3 and 6 indicate, the total concentration of S(IV) in plume aerosols is comparable to the concentrations found in aerosols in the flue line or in the workroom environment of the smelter. However, when these aerosols are inhaled, it is the particle surface that contacts the tissues of the body, and we might expect toxic effects to correlate with particle surface concentrations, particularly if chemical speciation plays a role. A comparison of total and surface S(IV) concentrations (Eatough et al., 1979, 1982) in the smelter flue dust and plume samples is given in Table 2. In general, the concentration of S(IV) on the surface of the flue line samples is higher than the total concentration. The aged plume samples (Eatough and Hansen, 1981; Eatough et al., 1982) have an average total S(IV) concentration of 320 μmol g^{-1}, about three times that seen in primary aerosol emissions from the same source in spite of the formation of additional particulate sulfate and the mixing in of ambient aerosols. (The total concentration of the conservative tracer, As, was 530 μmol g^{-1} in primary emissions and 24 μmol g^{-1} in the aged plume.) However, the surface concentration of the inorganic S(IV) species decreased from 164 μmol g^{-1} in the primary aerosol to less than measurable (<1% of the sulfate concentration) in the aged plume sample. This difference appears to result mainly from differences in sulfate and S(IV) formation mechanisms in the plume (Eatough et al., 1982). The inorganic S(IV) species are formed by dissolution of $SO_2(g)$ into a liquid droplet containing Fe(III) and Cu(?) (see Section 2.3). This mechanism would presumably lead to comparable total and surface concentrations. The secondary sulfate, however, is believed to be formed principally by homogeneous gas-phase processes that result in the condensation of H_2SO_4 on the particle surface, and thus the surface composition is significantly different from the total composition in the aged plume samples. It is possible that this physical difference may also result in a difference in the toxicities of the aged plume versus primary aerosols.

2.2. Characterization of Laboratory-Generated Inorganic-S(IV)-Containing Aerosols

Further support for the formation of stable Fe(III) and Cu(?) complexes with S(IV) has been obtained from laboratory studies. It has been shown by spectrophotometry (Danilczuk and Swinarski, 1962), kinetics (Carlyle, 1971), and calorimetry (Hansen et al., 1974, 1975, 1976; Eatough and Hansen, 1979; Hilton et al., 1979; Schlesinger et al., 1980) that a complex between Fe^{3+} and HSO_3^- will form in aqueous solution. If we assume that the reaction occurring can be represented by

$$Fe^{3+} + SO_2{\cdot}H_2O = FeHSO_3^{2+} + H^+, \qquad [1]$$

then measurement of the equilibria attained at pH 1 in $FeCl_3$, $NaHSO_3$, and Na_2SO_4 solutions results in a log K value for reaction [1] of 24 ± 2 [calculated from data summarized in Eatough and Hansen (1979)]. Calorimetric results for the titration of HSO_3^- with Fe^{3+} (Hansen et al., 1974) and potentiometric studies of the Fe^{3+}–SO_3^{2-} system (Danilczuk, 1964) also support a large log K value (Hansen et al. 1974). However, other studies have concluded that the equilibrium constant for formation of an Fe^{3+}–HSO_3^- complex is much smaller than that implied from the calorimetric work (Carlyle, 1971; Dasgupta et al., 1979; Hegg and Hobbs, 1978). It should be pointed out that the equilibrium is complex and that the final species probably cannot be represented as a simple $FeHSO_3^{2+}$ complex (Eatough and Hansen, 1979; Hilton et al., 1979; Schlesinger et al., 1980). All the data currently in the literature related to the formation of Fe(III)–S(IV) complexes in aqueous solution can be reconciled by the following reaction scheme (Eatough and Hansen, 1979):

$$Fe^{3+} + HSO_3^- = (I) \qquad [2]$$

$$(I) = (II) \qquad [3]$$

The Fe(III)–S(IV) complex designated as (I) is rapidly formed in solution when Fe^{3+} and HSO_3^- solutions are mixed. The formation of complex (I) results in a red-colored solution. Complex (I) has a ΔH_{ox} value for titration with dichromate in acidic solution of -38 kcal eq^{-1}. Complex (I) is not stable in basic solution, dissociating to give SO_3^{2-}. Over a period of several minutes to a few hours the metastable species (I) is converted to the thermodynamically more favorable state represented by (II). Although the exact nature of the end complex (II) is unknown, it is reasonable to assume that reaction [3] involves hydrolysis, polymerization, hydration, ligand rearrangement, or other similar changes in the complex. The ΔH_{ox} value for (II) is -24 kcal eq^{-1}. Solutions containing (II) are not red colored. The complex is not stable above pH 3–5, undergoing an auto redox process to form Fe^{2+} and SO_4^{2-} at higher pH. Complexes of the form of (II) are probably the predominant inorganic S(IV) species found in particulate matter in the atmosphere.

The earliest experiments reported that have attempted to characterize aerosols containing transition-metal–S(IV) compounds were experiments conducted cooperatively by personnel at Harvard and Brigham Young University (BYU) (Eatough et al., 1978b; Eatough and Hansen, 1981) that reproduced the generation of aerosols containing iron and copper known to produce a "synergistic" response in guinea pigs (Amdur et al., 1975). The results obtained, summarized in Table 3, indicate that both H_2SO_4 and S(IV) species exist in these aerosols. In all cases, except for the $CuSO_4$ aerosols

Table 3. Chemical Composition of Metal Salt Aerosols Exposed to $SO_2(g)$[a]

		Mole Ratio of Species to Total Metal in Aerosol			
Metal Salt	SO_2 (ppm)	S(IV)[b]	Fe(III)[c]	H_2SO_4[d]	H_2O
$CuSO_4$	0	0.00	—	0.00	2.5
$CuSO_4$	4	0.043	—	0.12	2.1
$CuSO_4$	4	0.076	—	0.21	2.8
$CuSO_4$	20	0.021	—	0.00	0.8
$CuSO_4$	40	0.087[e]	—	0.00	−0.2
$CuSO_4$	190	0.052[e]	—	0.00	−0.4
$FeSO_4$	4	0.091	0.248	0.15	1.7
$FeSO_4$	4	0.089	0.349	0.11	2.1

[a] Data are from Eatough and Hansen (1981).

[b] Determined by calorimetric redox titration of acidic extracts of the aerosol as the species that titrates with $Cr_2O_7^{2-}$ with $\Delta H_{ox} = -24$ kcal eq^{-1}.

[c] Determined as the difference between total Fe and Fe^{2+} (determined by calorimetric redox titration) in acidic extracts.

[d] Determined by a combined calorimetric–pH titration of aqueous extracts of the aerosol (Eatough et al., 1977).

[e] The ΔH_{ox} value for the redox titration of the S(IV) species with $K_2Cr_2O_7$ was -12 to -18 kcal eq^{-1}. These species hydrolyzed in 6 weeks to a species that titrated with $\Delta H_{ox} = -24$ kcal eq^{-1}.

exposed to very high SO_2 concentrations, the S(IV) species gave the ΔH_{ox} value expected for titration of S(IV) with $K_2Cr_2O_7$ in the analytical technique used (Hansen et al., 1976), indicating that labile S(IV) complexes are formed. Kinetically inert species appear to be formed, however, if $CuSO_4$ aerosols are exposed to very high concentrations of SO_2.

Hilton et al. (1979) have reported on the preparation of Fe(III)–S(IV) aerosols by the aerosolization of equilibrated solutions prepared by mixing solutions of $Fe_2(SO_4)_3$, H_2SO_4, and Na_2SO_3 or $NaHSO_3$. The composition of the prepared solutions was observed to change over a period of hours to days. These changes could be observed both calorimetrically and colorimetrically. The solutions were aerosolized only after equilibrium had been reached. The aerosols produced from these different solutions generally had S(IV) : Fe(III) ratios of 0.5–1, even when the S(IV) : Fe(III) ratio in the prepared solution was as high as 3. Aerosolization of the solutions also resulted in evolution of labile S(IV) as SO_2(g). The heat of oxidation of the S(IV)-containing species in these aerosols when initially titrated with $K_2Cr_2O_7$ was -24 kcal eq^{-1}, -15 to -18 kcal eq^{-1}, or both. The S(IV) concentration in the aerosol was dependent on several variables, including the S(IV) concentration in the solution that was aerosolized, the S(IV) : Fe(III) ratio in the solution aerosolized, and whether the S(IV) in solution originated from the SO_3^{2-} ion or the HSO_3^- ion. The calorimetric–pH titration data indicated that all the Fe(III) in the aerosol was involved in the formation of the S(IV) complex.

Schlesinger et al. (1980) have reported on the production and characterization of an Fe(III)–S(IV)-containing aerosol prepared by mixing SO_2(g) with a submicrometer-size aerosol of Fe_2O_3. Residual SO_2(g) was removed by using a NaOH-coated denuder tube. Their results corroborate those reported earlier by Hilton et al. (1979). In addition, the aerosol produced by Schlesinger et al. (1980) is better suited for animal toxicology studies. They also performed several innovative analyses that provide additional insights into the chemical nature of the Fe(III)–S(IV) aerosol produced. As was the case in studies reported by Hilton et al. (1979), Schlesinger et al. (1980) found that the solutions prepared from the generated aerosols contained several Fe(III)–S(IV) complexes with varying ΔH_{ox} values. Furthermore, they demonstrated that the equilibria in 0.1 M HCl solutions of the aerosols could be shifted by adding excess $FeCl_3$ or by addition of H_3PO_4 to 5 mM $FeCl_3$, 0.1 M HCl extracts. The addition of H_3PO_4 resulted in oxidation of all sulfur in the aerosol to sulfate and the concurrent production of Fe^{2+}. In experiments in which both the increase in sulfate following $Cr_2O_7^{2-}$ oxidation and the production of Fe^{2+} following addition of H_3PO_4 were determined, Schlesinger, et al. (1980) were able to show that all the nonsulfate sulfur in the generated aerosols was present in the +4 oxidation state. In addition, they performed specific analyses to test for the possible presence of dithionate and found that it was not present. It had been predicted earlier (on the basis of chemistry observed in solutions) by Dasgupta et al. (1979) that, in the absence of chloride, dithionate would be the principal sulfur-containing species in aerosols resulting from the interation of $SO_2(g)$ with iron salts. Schlesinger et al. (1980) were able to demonstrate that formation of dithionate did not occur in aerosols formed from SO_2 and Fe_2O_3. It has been suggested from the results of laboratory studies (Sato et al., 1979) that nitrogen oxide–sulfur dioxide chemistry may lead to the formation of dithionate in aerosols. Specific analysis for dithionate in environmental samples has not been reported. It is clear, however, from the results reported by Schlesinger et al. (1980) that dithionate does not interfere with the calorimetric determination of inorganic S(IV).

The calorimetric determinations performed by Schlesinger et al. (1980) on extracts of their prepared Fe(III)–S(IV) aerosols indicate the complex nature of Fe(III)–S(IV) aerosol chemistry. In solutions prepared either from pure salts or from generated aerosols, species were present that oxidized with ΔH_{ox} values of -31 and -25 kcal eq^{-1}. They correctly identified the species with a ΔH_{ox} value of -31 kcal eq^{-1} as Fe^{2+}. However, Fe^{2+} was shown not to be initially present in the generated aerosols by colorimetric tests. Assignment by Schlesinger et al. (1980) of the species with a ΔH_{ox} value of -25 kcal/eq as uncomplexed sulfite is erroneous for two reasons: (1) the species with a ΔH_{ox} value of -25 kcal eq^{-1} titrated after Fe^{2+}, but any H_2SO_3 or HSO_3^- in the solution would be titrated before Fe^{2+} (Hansen et al., 1976); and (2) the ΔH_{ox} value for the reaction of sulfite in acidic aqueous solution with $Cr_2O_7^{2-}$ is calculated from potentiometric data to be -38.5 kcal eq^{-1} if no inert Cr(III)–SO_4^{2-} complexes are formed and has been measured to be -29.8 kcal eq^{-1} when inert Cr(III)–SO_4^{2-} complexes are formed (Hansen et al., 1976). The species that titrates with a ΔH_{ox} value of -25 kcal eq^{-1} is identical in redox characteristics to the compound referred to as an Fe^{3+}–SO_3^{2-} complex ($\Delta H_{ox} = -24$ kcal eq^{-1}) in the publication by Hansen et al. (1976). With this in mind, the data reported in Schlesinger et al. (1980) are in complete agreement with the results reported in Hilton et al. (1979). The data of Schlesinger et al. (1980) do indicate that the shift in equilibria of Fe(III)–S(IV) complexes with ΔH_{ox} values of less than -24 kcal eq^{-1} to the species that titrates with $\Delta H_{ox} = -24$ kcal eq^{-1} by addition of excess Fe^{3+} also results in the production of some sulfate. Thus the concentrations of the Fe(III)–S(IV) species with ΔH_{ox} values of less than -24 kcal eq^{-1} which may have been present in environmental samples collected in the programs at BYU (Eatough et al., 1982; Eatough and Hansen, 1980; Richter, 1981) have been underestimated.

Additional information on equilibria involving Fe(III)–S(IV) complexes can be deduced from data given in Schlesinger et al. (1980) on the rates of heat generation and sulfate production when an aliquot of an equilibrated solution of 1 M $Fe_2(SO_4)_3$, 1 M Na_2SO_3, and 1 M H_2SO_4 is added to a 0.1 M HCl, 5 mM $FeCl_3$ solution. Possible reactions that might occur can be expressed as

$$H_2SO_3 + 2Fe^{3+} + H_2O = 2Fe^{2+} + HSO_4^- + 3H^+ \quad [4]$$

$$\text{Fe(III)–S(IV)} = Fe^{2+} + HSO_4^- \quad [5]$$

Reaction [5] is not balanced, as the structure of the Fe(III)–S(IV) complex is not known. However, available data suggest that the complex is formally equivalent to a bisulfite complex (Hilton et al., 1979). Data available in the literature (Hansen et al., 1976) allow the calculation of ΔH values for reactions [4] and [5]. The calculated ΔH value for reaction [4] is -17.6 kcal mol^{-1} sulfate. The calculated ΔH value for reaction [5] is -13 kcal mol^{-1} sulfate for the Fe(III)–S(IV) complex with $\Delta H_{ox} = -24$ kcal eq^{-1} and is between -2 and $+1$ kcal mol^{-1} sulfate for the Fe(III)–S(IV) complex with ΔH_{ox} in the range -15 to -18 kcal eq^{-1}. The data due to Schlesinger et al. (1980) indicate that addition of an equilibrated 1 M $Fe_2(SO_4)_3$, 1 M Na_2SO_3, and 1 M H_2SO_4 solution to 0.1 M HCl, 5 mM $FeCl_3$ results in production of sulfate with a measured ΔH value of about -8 kcal mol^{-1} sulfate for the first 2–3 min (~20% of the total sulfate formed), followed by a measured ΔH value of -1 kcal mol^{-1} for further production of sulfate over about a 60-min period. Approximately 40% of the initial S(IV) remained as a stable Fe(III)–S(IV) species after 2.5 h. These data further support the suggestion that Fe(III)–S(IV) complexes exist that are slowly converted to both sulfate (reaction [5]) and a stable Fe(III)–S(IV) complex (species (II) in reaction [3]). The data also prove that the observed increase in sulfate is not due to oxidation of H_2SO_3 by Fe^{3+} as shown in reaction [4].

In summary, laboratory experiments by two independent groups using three different aerosol generation procedures have shown that aerosols can be prepared that contain stable Fe(III)–S(IV) compounds. The data indicate that the chemistry of formation is complex and that metastable species are formed that slowly change with time to produce an Fe(III)–S(IV) compound that is stable for a time period of at least days to weeks. Observations on samples collected in the flue lines, plumes, and workrooms of smelters show the same Fe(III)–S(IV) complexes to exist in these environments (see Section 2.1).

2.3. Mechanism of Formation of Aerosol Inorganic S(IV)

The mechanism of formation of inorganic S(IV) species in ambient particulate matter has been studied in the plumes of a copper smelter (Eatough et al., 1982) and a lead smelter (Richter, 1981) and in the rural northeastern United States (Richter, 1981).

2.3.1. Smelter Plume Studies

Sulfur transformation chemistry was studied in the plume of the Utah smelter of Kennecott Copper Corporation from April to October 1977 (Eatough et al., 1982; Eatough and Hansen, 1981). Data collected at each station included $SO_2(g)$ concentration, low-volume collected total particulate mass, high-volume collected size-fractionated particulate mass, wind velocity and direction, temperature, and relative humidity. Particulate samples were analyzed for S(IV), sulfate, strong acid, anions, cations, and elemental concentrations by use of calorimetric, ion chromatographic, PIXE, ESCA, ion microprobe, and scanning electron microscope (SEM)–ion microprobe techniques. The concentration of arsenic in the particulate matter was used as a conservative plume tracer.

Two distinct metal–S(IV) species similar to those observed in laboratory aerosol experiments (see Sections 2.1 and 2.2) were found in the plume. Aerosol S(IV) species are formed by reactions of SO_2(g) with both ambient and plume-derived aerosol. The extent of reaction is equilibrium controlled and not dependent on kinetics. If adsorption kinetics controlled the rate of incorporation of S(IV) into the particulate matter, the mechanism would be at least second order overall, since it must be first order or higher in both the plume particulate matter and SO_2(g) concentrations, both of which decrease as the plume expands (Freiberg, 1978; Schwartz and Newman, 1978). Thus the rate of formation of aerosol S(IV) would decrease as the plume dilution increased if the extent of reaction were kinetically controlled. The opposite was observed. This is illustrated in Figure 7, where data for the formation of SO_4^{2-} from SO_2(g) [a first-order process (Eatough et al., 1982)] are compared to corresponding data for the formation of inorganic S(IV) in particulate matter associated with the plume.

If the amount of inorganic S(IV) species formed in the aerosol is controlled by a reversible reaction and if the rate of the reaction is fast compared to plume travel time between sampling sites, the concentration of inorganic S(IV) in the particulates will be related by the mass action law to the con-

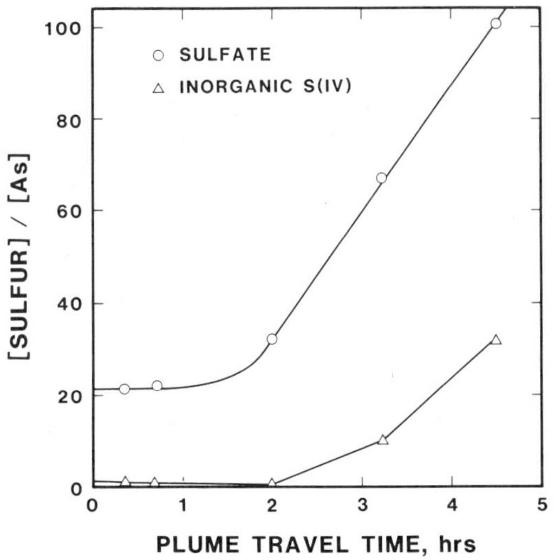

Figure 7. Formation of inorganic S(IV) and sulfate (corrected for background) in plume of a copper smelter relative to conservative tracer arsenic as a function of plume transport time. (Data are from Eatough et al., 1982).

centrations of the chemical species in the reaction. The concentrations of inorganic S(IV) determined to be present in plume-associated aerosols in 13 different sampling sets were shown to be consistent with the mass action expression given in Eq. (1):

$$K_E = \frac{[H^+][S(IV)]}{[Fe + Cu - S(IV)](P_{SO_2})} \tag{1}$$

where the brackets represent the concentration of the indicated species in micromoles per milligram of particulate, and P_{SO_2} refers to the $SO_2(g)$ concentration in parts per million. The value of K_E calculated from the data is 5.2 ± 2.4, indicating that the formation of inorganic S(IV) species in this plume is well characterized by Eq. (1). Elimination of any of the terms in Eq. (1) resulted in a larger relative standard deviation. It should be noted that the mass action expression given in Eq. (1) is formally equivalent to the equilibrium given in reaction [1] (Section 2.2).

The fit of the copper smelter plume S(IV) data to Eq. (1) shows that the stability of aerosol S(IV) to loss of $SO_2(g)$ is dependent on the aerosol acidity. It has been shown (Eatough et al., 1982) that removal of plume aerosols with high acidity from the high $SO_2(g)$ concentrations present in the near-smelter plume resulted in loss of inorganic S(IV) in the collected aerosols. The data obtained on these near-smelter plume samples allowed the equilibrium constant for the reaction occurring in the aqueous droplet aerosol as given by

$$M^{n+} + SO_2{\cdot}H_2O = MHSO_3^{(n-1)+} + H^+ \qquad [6]$$

to be evaluated as log K = 22.9 ± 1.1. The equilibrium constant for reaction [6] where $M^{n+} = Fe^{3+}$ has also been estimated by the authors to be 24 ± 2 from data obtained in the measurement of the equilibria attained at pH 1 in $FeCl_3$ and $NaHSO_3$ solutions (Eatough and Hansen, 1979). It was also shown by Eatough et al. (1982) that the Fe(III)–S(IV) complexes within aerosol droplets reduced As(V) to As(III) during plume transport. The arsenic and sulfur speciation data can be combined to estimate a value of log K for reaction [6]. The resulting value, 23.4 ± 1.3, is in excellent agreement with the other values obtained as outlined above. It should be emphasized again that the structure of the Fe(III)–S(IV) complex is not known. The $MHSO_3^{(n-1)+}$ structure in reaction [6] is intended only to qualitatively describe the species and not to imply a specific structure.

The equilibrium concentration of S(IV) in particulate matter in the copper smelter plume is thus controlled by acidity, the concentration of $SO_2(g)$ and the concentration of soluble Fe(III) and Cu(?) compounds (see Figures 3 and 6). This equilibrium involves the interaction of $SO_2(g)$ with both plume and ambient aerosols. The production of H_2SO_4 and the consequent condensation of water on the particles and solubilization of Fe(III) and Cu(?) compounds may be important contributing factors to the formation of inorganic S(IV) species in this plume even though the increased H^+ would decrease the extent of the formation reaction. The results show that SO_2(g) may be rapidly lost from highly acidic particles if the SO_2 concentration in the atmosphere is markedly changed but that the S(IV) species are quite stable toward SO_2(g) loss in particles of low acidity. The increase in inorganic S(IV) at plume residence times in excess of 3 hours (Figure 7) is due mainly to neutralization of aerosol acidity as the plume travels. At a distance of 40–50 km from the copper smelter stacks the mole ratio of S(IV) to secondary sulfate in the plume varied from 0.1 to 1.5 and averaged 0.7 in eleven sampling sets. The variability was due mainly to changes in the rate of SO_2(g) to sulfate conversion with temperature. Thus formation of aerosol S(IV) species can be as important a chemical process for formation of aerosol sulfur as the formation of sulfate. Qualitatively similar results have been obtained in a more limited study of a lead smelter plume (Richter, 1981).

2.3.2. *Rural Northeast Studies*

A similar dependence of the concentration of aerosol S(IV) species on aerosol acidity has been observed in a series of samples collected during the summertime in the rural northeastern United States (Richter, 1981). This study was conducted during the summers of 1978 and 1979. Sampling stations were operated near Rockport, Indiana; Duncan Falls, Ohio; State College, Pennsylvania; Johnstown, Pennsylvania; and Upton, New York (see Figure 8). Two sampling stations were operated at Rockport: one at the Sulfate Regional Experiment (SURE) station just west of town and the other at a meteorological tower site located north of town. These sites were chosen to study the S(IV) chemistry in background aerosols in the Rockport area and the possible formation of particulate S(IV) compounds in the plume from a metallurgical facility located between the two sites. Samples were collected at these sites during June 1978. During July 1978, sampling equipment was located at the SURE station at Duncan Falls, Ohio and the Multistate Atmospheric Power Production Pollution Study (MAP3S) station located outside State College, Pennsylvania. These sites were chosen to study the effect of high acidity on the S(IV) species as well as their stability during long-range transport. This was a period of intensive sampling for the SURE and MAP3S programs, and this enabled us to make use of the meteorological data obtained in those studies. Three sampling stations were set up in the vicinity of the Homer City and Conemaugh power plants near Johnstown, Pennsylvania during August 1978. Both power plants are at the same elevation, and both have 244-m stacks. Sampling stations were located so as to facilitate collection of both ambient background and coal-fired power plant plume samples. Finally, samples were obtained during May 1979 at Brookhaven National Laboratory, Upton, New York.

The dependence of the aerosol concentration of inorganic S(IV) on aerosol acidity is shown in Figure 9 for data obtained in these northeastern U.S. studies on days when the sampling sites were not influenced by proximate sources and for the data obtained at the copper smelter. The results observed

Figure 8. Location of sampling sites for northeastern U.S. aerosol study.

in the northeastern U.S. study are consistent with the equilibrium formation mechanism given in Eq. (1). This equilibrium expression predicts that an increase in aerosol acidity will result in a decrease in the concentration of inorganic S(IV). The data in Figure 9 show that, at high $[H^+]$, the inorganic S(IV) concentration is always low. At low aerosol acidity the concentration of inorganic S(IV) can be appreciable. The acidity thus creates an upper bound to the possible concentration of inorganic S(IV), as suggested by the dotted line in Figure 9. Inorganic S(IV) will thus not be present in highly acidic aerosols formed when the conversion of SO_2(g) to sulfuric acid is rapid. On the other hand, as $[H^+]$ decreases, the aerosol inorganic S(IV) species will become more stable, and the formation of these species from SO_2(g) may take place in the aerosol. Low $[H^+]$ does not indicate that inorganic S(IV) formation will necessarily take place, as the formation is also dependent on the presence of soluble Fe^{3+} and Cu(?). However, high $[H^+]$ will limit inorganic S(IV) production. The $[H^+]$ thus sets an upper limit on the concentration of inorganic S(IV) that can exist in an aerosol.

It should be noted that there were samples collected in the copper smelter plume that were very acidic and had concentrations of inorganic S(IV) higher than would be expected from the results of the ambient samples collected in the northeastern United States. These smelter plume samples had high concentrations of copper and iron that also favor the formation of inorganic S(IV) as predicted by reaction [6]. However, these high copper and iron concentrations are not typical of ambient aerosols, and the formation of inorganic S(IV) would not be expected in acidic ambient aerosols that have low copper and iron concentrations.

2.4. Occurrence of Aerosol Inorganic S(IV)

The data given in Section 2.3 indicate that the atmospheric concentration of inorganic S(IV) is dependent on the following variables: aerosol acidity; aerosol water concentration; aerosol iron and copper concentrations; and atmospheric $SO_2(g)$ concentration. Concentrations of inorganic S(IV) determined in <3 μm airborne particulate matter are summarized in Table 4. The highest atmospheric concentrations of inorganic S(IV) in aerosols are

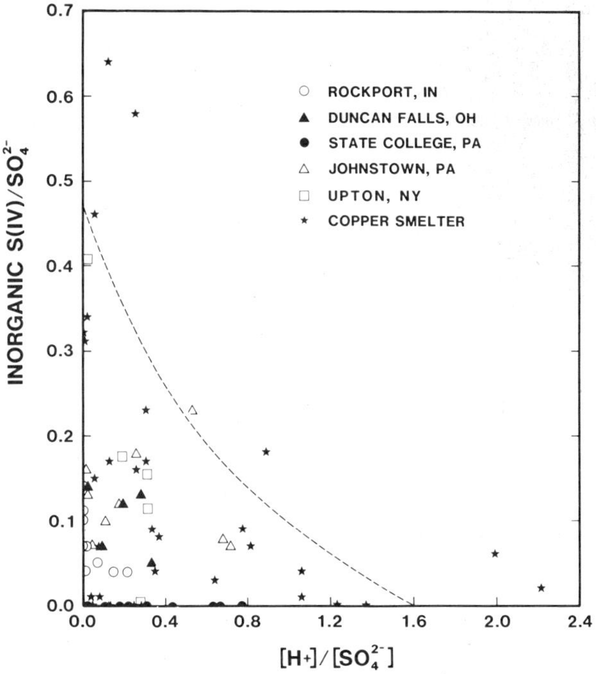

Figure 9. Ratio of inorganic S(IV) to sulfate in <3 μm particles as a function of aerosol acidity. Data are for samples from a northeastern U.S. aerosol study (Hansen, 1981; Richter, 1981) and a copper smelter plume chemistry study (Eatough et al., 1982).

Table 4. Occurrence of Inorganic S(IV) in Airborne Aerosols

Location	Aerosol Source	Number of Samples	Inorganic S(IV) Concentration in <3 μm Aerosols, μg/m^3 Expressed as $FeHSO_3SO_4$			Average Mole Ratio Inorganic S(IV)/SO_4^{2-}	Reference
			Range	Median	Average		
Utah	Copper smelter plume, within 8 km of stack	12	0–19.8	3.9	5.1	0.02 ± 0.03	Eatough et al. (1982)
	Copper smelter plume, 20 km from stack	5	0–9.8	1.3	1.5	0.05 ± 0.06	Eatough et al. (1982)
	Copper smelter plume, 40–60 km from stack	16	0–7.8	3.3	3.4	0.21 ± 0.14	Eatough et al. (1982)
Idaho	Lead smelter plume, 15 and 23 km from stack	2	1.6–2.4	—	2.0	0.12 ± 0.01	Richter (1981)
Utah	Coal-fired power plant plume, within 6 km of stack	7	0–1.4	1.0	0.7	0.39 ± 0.36	Eatough et al. (1981c)
	Coal-fired power plant plume, 20–42 km from stack	9	0–3.2	0.9	1.0	1.01 ± 1.8	Eatough et al. (1981c)

Pennsylvania	Coal-fired power plant plume within 12 km of stack[a]	5	1.4–3.1 (0.2–1.9)[a]	2.4 (1.1)[a]	2.2 (1.1)[a]	0.10 ± 0.05 (0.26 ± 0.27)[a]	Richter (1981)
California	Oil-fired power plant plumes	5	0–0.7	0	0.1	0.04 ± 0.09	Eatough et al. (1981b,c)
Rockport, IN	Rural ambient	9 .	0.9–1.8	1.3	1.4	0.05 ± 0.02	Richter (1981)
Duncan Falls, OH	Rural ambient	5	1.0–2.3	2.0	1.7	0.07 ± 0.03	Richter (1981)
Johnstown, PA	Rural ambient	4	1.4–2.0	1.8	1.7	0.08 ± 0.03	Richter (1981)
State College, PA	Rural ambient	9	0	0	0	0	Richter (1981)
Upton, NY	Rural ambient[b]	10 (4)[b]	0–3.7 (1.2–3.7)[b]	0 (2.5)[b]	1..0 (2..6)[b]	0.09 ± 0.13 (0.22 ± 0.13)[b]	Hansen (1981)
Los Angles basin, CA	Urban air, coastal—high moisture content	2	0.6–1.3	—	0.9	0.01 ± 0.00	Ellis (1982)
	Urban air, dry air mass	4	0	0	0	0	Ellis (1982)
New York, NY	Urban air	3	0–1.4	0.2	0.5	0.08 ± 0.09	Eatough el al. (1978a)

[a] The values in parentheses have been corrected for background aerosol contribution.

[b] The values in parentheses are for 4 of the 10 days when inorganic S(IV) was present in the sampled aerosol.

associated with the plumes of the copper and lead smelters. The increase in inorganic S(IV) relative to sulfate as the plume is transported and the aerosol acidity reduced is evident in the data from the copper smelter. Even though the average atmospheric concentration of $SO_2(g)$ (Eatough et al., 1982) is reduced from 500 ppb in the plume within 8 km of the stack to 20

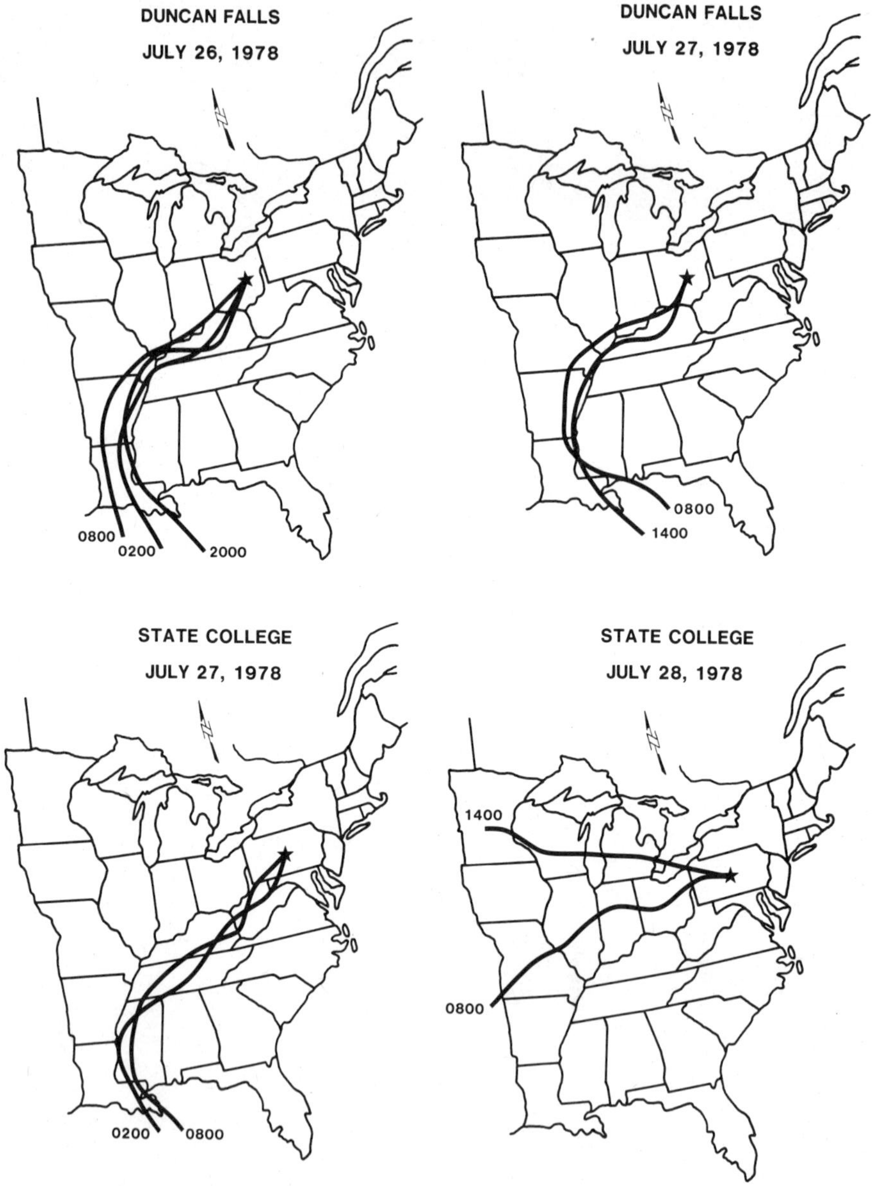

Figure 10. Trajectory maps for State College, Pennsylvania and Duncan Falls, Ohio during July 26–28, 1978.

Table 5. Results of Analysis of Samples Obtained at Duncan Falls and State College During the Times Covered by the Back Trajectories Given in Figure 10

Sampling Site	Date	Time of Sampling	SO_2(g) (ppm)	Relative Humidity (%)	T (°C)	Concentration in <3 μm Particulate Matter, (nmol m^{-3})			
						Inorganic S(IV)	Organic S(IV)	SO_4^{2-}	H^+
Duncan Falls, OH	July 26	09:45–15:45	27	73	30	10	5	113	3
	July 27	00:00–09:00	15	90	29	5	3	81	6
State College, PA	July 27	01:00–07:00	0	95	16	0	0	90	46
	July 28	07:30–13:30	12	97	21	0	8	70	18

ppb 40–60 km from the stack, the concentration of inorganic S(IV) in plume particulate matter is little changed. Concentrations in the plume of the lead smelter show similar behavior.

Inorganic S(IV) concentrations in plume aerosols from coal-fired power plants are similar for the two power plants studied. The data suggest a continuing increase in inorganic S(IV) relative to sulfate in the plume aerosol associated with the power plant in Utah during plume transport. Relatively low concentrations of inorganic S(IV) were seen in aerosols collected in the urban centers or in the plumes of oil fired power plants.

The composition of the particulate matter obtained in the rural areas sampled in Indiana, Ohio, and Pennsylvania was all heavily influenced by coal-fired power plant emissions from the Ohio river valley when the wind was from the south and west. The samples collected at Upton, New York and some samples collected at State College, Pennsylvania when the wind was from the north or east were probably not strongly influenced by coal combustion. Appreciable concentrations of aerosol inorganic S(IV) were seen at all sites except State College. The aerosols collected at State College had the highest acidity. This accounts for the absence of inorganic S(IV) in aerosols at this site. In the study at State College and Duncan Falls, during one time period Providence was kind, and samples were collected at Duncan Falls on one day and then on the following day from the same air mass at State College. Figure 10 gives the results of 3-day back-trajectory calculations at Duncan Falls and State College during this time period. (The calculations were kindly provided by Ronald Meyers of Brookhaven National Laboratory.) The results of analyses of the samples collected at these two sites during this time period are given in Table 5. As the air mass moved from Duncan Falls to State College on July 27–28, the acidity of the aerosol increased and the inorganic S(IV) was lost, probably because of decomposition by acid.

3. ORGANIC S(IV) COMPOUNDS

3.1. Characterization of Organic S(IV) in Airborne Particulate Matter

3.1.1. Background

During the study of inorganic S(IV) chemistry in aerosol samples from the plume of a coal-fired power plant and from New York City we found evidence for a S(IV) compound distinctly different from the inorganic S(IV) compounds identified in smelter plumes (Eatough et al., 1978a). We postulated that an organic S(IV) compound(s) was present in these samples. The evidence that led us to this postulate is presented here. Data that conclusively show the presence of an organic S(IV) compound(s) and, in part, identify the structure of the compound are given in Sections 3.1.2 and 3.1.3.

In 1976 Dr. Joan Daisey of New York University sent us a concentrated acetone extract of airborne particulate matter collected in New York City. After evaporation of the acetone, a fraction of the sample was dissolved in 0.1 M HCl, 2.5 mM $FeCl_3$ solution on a Friday afternoon and subjected to calorimetric determinations for inorganic S(IV) and sulfate (Hansen et al. 1976). No $Cr_2O_7^{2-}$ oxidizable species with ΔH_{ox} values more exothermic than about -8 kcal eq^{-1} were seen in the sample. [As summarized in Sections 2.1 and 2.2, inorganic S(IV) compounds observed in smelter plume samples have ΔH_{ox} values from -15 to -24 kcal eq^{-1}).] The next Monday it was decided to repeat the sulfate determination partly because of some strange features seen in the thermograms from the $BaCl_2$ injection. To our surprise we found it necessary to dilute the solution to run the dichromate titration because of the presence of large amounts of a material that reduced dichromate with a ΔH_{ox} value of -24 kcal eq^{-1}, identical to that observed for the Fe(III)–S(IV) complex which is the basis of our calorimetric determination of inorganic S(IV). Also, the measured concentration of sulfate was much greater on Monday than on Friday. Similar results were later obtained on particulate matter collected in New York City and from the plume of a coal-fired heating plant (Eatough et al., 1978a) when the samples were directly extracted into 0.1 M HCl and 2.5 mM $FeCl_3$ and analyzed immediately after extraction and again after the solutions had sat at room temperature for 1 week or longer. In all these analyses the extracted material had been removed from the insoluble particulate matter by filtration. Furthermore, we found that this increase in Fe(III)–S(IV) with time was not correlated with the extractable iron and copper found in the samples (Figure 11). The observed *change* in inorganic S(IV) frequently exceeded the upper limit of inorganic S(IV) predicted from the study of this chemistry in samples derived from the laboratory (Section 2.2) and from the environment (Section 2.1), as given by the dashed line in Figure 11 (e.g., cf. Figures 3 and 6 to Figure 11). We postulated (Eatough et al., 1978a) that this increase in inorganic S(IV) resulted from the hydrolysis of an organic S(IV) compound(s). On the basis of the limited data then available, our original publication speculated that the organic S(IV) compound(s) (Eatough et al., 1978a) might be addition compounds of H_2SO_3 to aldehydes or to polyketo- or hydroxy-substituted aromatic hydrocarbons.

Several sulfur-containing organic compounds that probably result from SO_x chemistry (Figure 12) have been shown or are expected to be present in the atmosphere. Aerosol-phase methane sulfonic acid (Panter and Penzhorn, 1980) and gas-phase ethylene sulfite (Jones, 1974) have been identified at low concentrations in areas impacted by coal combustion or urban emissions. The related compound, ethylene sulfate, would be expected to be present as an oxidation product formed from ethylene sulfite, but ethylene sulfate has not yet been found. Hydroxymethane sulfonic acid is expected to be formed from H_2SO_3 and formaldehyde. On the basis of the favorable thermodynamics for formation of the compound and the high concentration

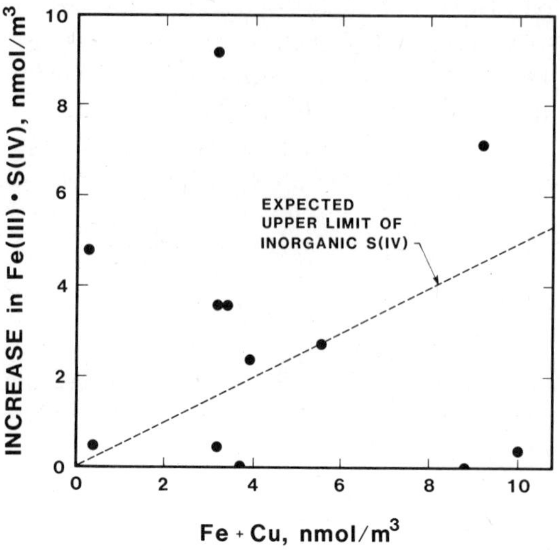

Figure 11. Plot of the *increase* over a 1-week period in the Fe(III)–S(IV) (ΔH_{ox} = −24 kcal eq^{-1}) species titrated in 0.1 M HCl, 2.5 mM $FeCl_3$ extracts of particulate matter collected in New York City *versus* the sum of iron and copper in the sample. The dashed line is the upper limit of inorganic S(IV) expected in the sample (see Figures 3 and 6). (Data are from Eatough et al., 1978a).

of gas-phase formaldehyde usually seen in urban atmospheres, it has been argued that hydroxymethane sulfonic acid or its salts should be formed in the atmosphere (Charlson et al., 1978; Dasgupta et al., 1982). These compounds have also not yet been positively identified, however. Dimethyl and monomethyl sulfate have been shown to be present in particulate matter in the flue line and plumes of coal-fired power plants (Eatough et al., 1981a; Lee et al., 1980).

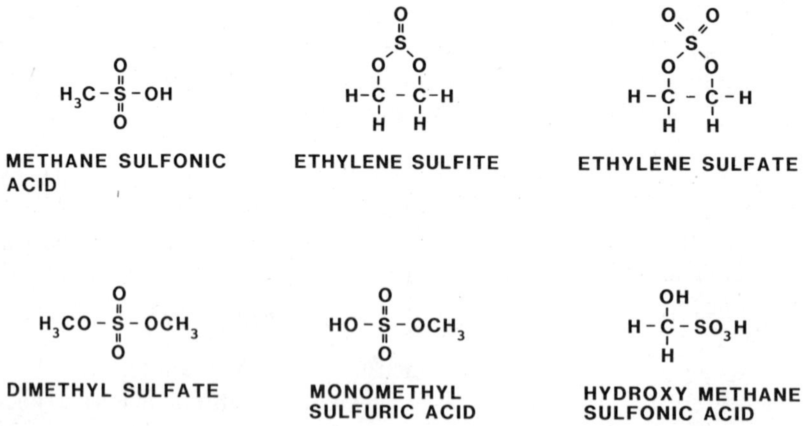

Figure 12. Organic sulfur-containing compounds known or expected to be formed from SO_2(g) in the atmosphere.

None of these compounds has the same chemical properties as the organic S(IV) compound(s), and thus in 1979 we instituted a program designed to determine the structure of the organic S(IV) compound(s) found to be present in atmospheric particulate matter. The data obtained support the presence of an organic S(IV) compound in the collected samples and indicate that the compound will dissociate to give one S(IV)-containing fragment and one two-carbon fragment per molecule. The data obtained to date, however, do not positively identify the structure. The results obtained are summarized in Sections 3.1.2 and 3.1.3.

3.1.2. Evidence for Organic S(IV) Compound(s)

Several different methods of analysis of particulate samples collected from the plumes of coal-fired power plants or from areas heavily impacted by coal-fired boilers indicate that a S(IV) compound distinctly different from inorganic S(IV) is present in the samples (Eatough et al., 1978a, 1981b,c). The results of these analyses are summarized in Tables 6 and 7. The presence of an organic compound with sulfur in the +4 oxidation state is supported by the observation that the compound will slowly evolve S(IV) in either 0.1 M HCl, 2.5 mM $FeCl_3$, or 3 M H_3PO_4 extract solution to give a positive test for S(IV) by calorimetric redox titration or West–Gaeke colorimetric determination, respectively (Table 6). The organic S(IV) compound can be extracted into benzaldehyde, counterextracted into 0.1 M HCl, 2.5 mM $FeCl_3$ and then hydrolyzed to inorganic S(IV), which can be determined by calorimetric redox titration (Table 6). Samples containing the compound will

Table 6. Comparative Determination of Organic S(IV) in Plume Particulate Matter from Coal Combustion[a]

Analysis Procedure	S Determination	Range[b]	Mean[b]	Number of Samples
Increase in SO_4^{2-} in basic solution due to oxidation by Ag_2O, $KMnO_4$, or H_2O_2	SO_4^{2-} by ion chromatography	79–128	96 ± 17	6
Calorimetric determination of S(IV) in benzaldehyde extracts	Hydrolyzable S(IV) by calorimetry	53–139	90 ± 28	6
West–Gaeke method	Hydrolyzable SO_3^{2-}	46–64	55 ± 12	2
SO_2(g) evolution in N_2 atmosphere at 50–150°C	SO_2(g) by West-Gaeke colorimetry	47–100	70 ± 24	5

[a] Data are from Eatough et al. (1981b).

[b] The results are given as a percent of the value determined calorimetrically by the procedure given in Eatough et al. (1978a).

Table 7. Results of Analyses of Three Different Samples of Particulate Matter from the Los Angeles Basin[a]

Sample Preparation	Organic S(IV)[b]	Diethylene Glycol × 2[b]	Ethylene Glycol[b]
Original sample	1.00 ± 0.10	—	—
Sample after extraction with CH_2Cl_2	0.96 ± 0.20	—	—
CH_2Cl_2 extractable	<0.05	<0.05	<0.05
CH_2Cl_2 extracted sample after extraction with CH_3OH	<0.10	—	—
CH_3OH extractable	0.98 ± 0.14	0.70 ± 0.02	0.16 ± 0.13

[a] Data are from Eatough et al. (1981b).

[b] The values are given relative to organic S(IV) determined calorimetrically in the original sample, molar basis.

evolve SO_2(g) at about 100°C (Figure 13). Furthermore, the amount of SO_2(g) evolved agrees quantitatively with the results of titration calorimetric analysis for organic S(IV) in the sample (Table 6). In addition, the compound can be oxidized to sulfate in basic aqueous solution by Ag_2O, $KMnO_4$, or H_2O_2 (Table 6). The results obtained by these various techniques are in reasonable agreement. In contrast, inorganic S(IV) cannot be detected by the normal West–Gaeke method (Eatough et al., 1979; Eatough and Hansen, 1979, 1981), by oxidation in basic solution (Eatough and Hansen, 1979; Hansen et al., 1979; Hilton et al., 1979), or by thermal evolution of SO_2(g) below 180°C (Eatough and Hansen, 1981). The inorganic S(IV) is also not extracted by benzaldehyde (Eatough, 1978c). It is thus established that a

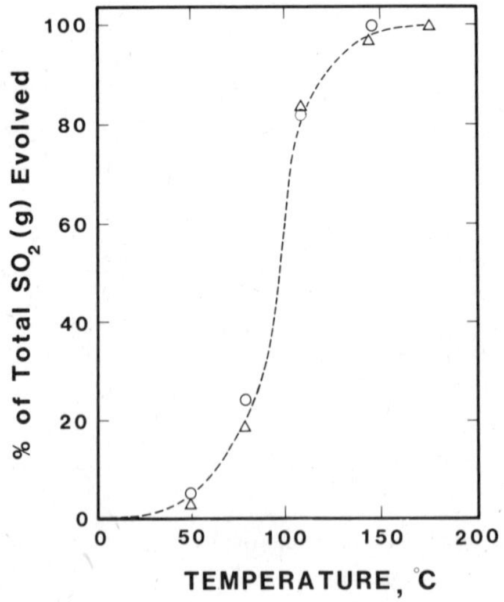

Figure 13. Evolution of SO_2(g) in a flowing N_2(g) stream from two ambient particulate matter samples collected in the Los Angeles Basin as a function of temperature. Data are from Eatough et al. (1981b).

S(IV) compound is present that is quite different from the inorganic S(IV) compounds discussed above.

3.1.3. *Characterization of the Organic S(IV) Compound(s)*

The structure of the organic S(IV) compound(s) is not yet known. Data related to identifying the structure have been obtained from a combination of calorimetric, colorimetric, ion chromatographic, gas chromatographic–mass spectroscopic (GCMS), and nuclear magnetic resonance (NMR) spectroscopic analyses (Eatough et al., 1981b). These data, taken together, conclusively show that the S(IV) compound identified by the techniques given in Section 3.1.2 is organic.

The concentration of the organic S(IV) compound(s) in samples can be determined by extraction of a portion of the sample with 0.1 M HCl, 2.5 mM $FeCl_3$ and determination of hydrolyzable S(IV) by redox titration calorimetry (Eatough et al., 1978a) or by extraction of a portion of the sample with H_2O or methanol and determination by ion chromatography of the increase in SO_4^{2-} after Ag_2O oxidation (Eatough et al., 1981b), see Table 6.

Although most of the organic material in the sample can be removed by extraction for 24 h in a Soxhlet apparatus with methylene chloride, no organic S(IV) compound(s) has ever been found in the CH_2Cl_2 extract. Titration calorimetric analysis of a portion of the sample after the CH_2Cl_2 extraction further shows that the organic S(IV) compound(s) is not affected by the methylene chloride extraction. The results obtained on a series of samples from the Los Angeles basin by this analysis scheme are given in Table 7. Results obtained on other samples also have shown that less than 5–10% of the organic S(IV) compound(s) is removed from the filter by 24-h Soxhlet extraction with diethylether. The samples extracted with CH_2Cl_2 and diethylether were next extracted with methanol. Titration calorimetric analysis of both the resulting methanol solution and the treated filter indicated that the organic S(IV) compound(s) was quantitatively removed by the methanol (Table 7). The resulting methanol solutions have been analyzed by a variety of techniques.

Samples from the Los Angeles Basin and from the plume of a coal-fired boiler were prepared by the above extraction sequence and the resulting solutions concentrated and analyzed by gas chromatography–mass spectroscopy using a hot port injection onto a Tenax packed column or a glass capillary column. The only sulfur-containing compound identified with either column was dimethyl sulfate (Lee et al., 1980) found in low concentrations in the samples from the plume of the coal-fired boiler. Neither SO_2(g) nor SO_3(g) was detected by the gas chromatographic–mass spectroscopic analysis. The only organic compounds found in the methanol extracts in quantities comparable to the organic S(IV) compound(s) were ethylene glycols identified in the Tenax column experiments. Ethylene glycols have also been identified as decomposition products associated with SO_2(g) in direct-probe mass spectroscopic analysis of the methanol-extractable material. The glycol compounds were not seen in capillary-column gas chromatographic analysis of the methanol extracts even though the compounds theoretically would have eluted in this column. Identification of the various peaks was made both by spiking the sample with the indicated compounds and by mass spectral analysis of the eluted material. The ethylene glycols were also not detected in the CH_2Cl_2 extracts of the sample (Table 7) even though they should be extracted in this solvent. These results suggest that the glycols are formed by reactions in the gas chromatograph. Finally, the moles of C_2 units seen as ethylene glycol and diethylene glycol in methanol extracts were equal to the moles of organic S(IV) found by titration calorimetry (Table 7). Thus the data suggest that, under some conditions, the organic S(IV) compound(s) decompose in the gas chromatograph port or on the Tenax column to give ethylene glycols. Any SO_3(g) or SO_2(g) produced would be adsorbed by the Tenax. It has also been shown that oxidation of basic aqueous solutions containing the organic S(IV) compound(s) will produce under some conditions, equimolar amounts of sulfate and an anion identified as oxalate by ion chromatography (Eatough et al., 1981b).

The methanol extract containing the organic S(IV) compound(s) can be further purified by column chromatography on silica gel. Usually in these experiments about 400–800 mg of collected particulate matter is extracted sequentially with CH_2Cl_2 and $(CH_3CH_2)_2O$ in a Soxhlet apparatus for 24–48 h each, followed by batch extraction with CH_3OH. The resulting 100 ml of CH_3OH solution is concentrated to 0.5 ml and the condensed fraction put on a column packed with 15 g of silica gel in methanol. Methanol is used as the mobile phase. The organic S(IV) compound(s) is separated as an orange–red band on the column. The resulting eluted solution does not contain any appreciable amount of any metal ions. The resulting material has been analyzed by direct probe–mass spectroscopy, by capillary-column GCMS, and by ^1H and ^{13}C NMR spectroscopy. Attempts have been made to determine the molecular weight of the compound by using electron-impact, chemical ionization with CH_4 and NH_3 as reagent gases, and field-desorption mass spectral techniques. All these attempts have been unsuccessful. Decomposition products are usually seen at around 90°C (cf. results given in Figure 13). Sublimation experiments in vacuum over the temperature range 60–95°C show that the compound decomposes but does not sublime to any measurable degree over this temperature range (Hansen, 1981). Identified decomposition products in the direct probe experiments include ethylene glycol, diethylene glycol, SO_2, SO_3 (or H_2SO_4), CO_2, and several fragments of apparent composition CH_xSO (x = 1, 2, 3).

Nuclear magnetic resonance data have been obtained in CD_3OD, CD_3SOCD_3, and D_2O. Data on ^{13}C nuclei show that only one type of carbon is present in the compound. The nuclear magnetic resonance peak for this carbon occurs at 74.7 ± 0.8 ppm relative to TMS in CD_3OD and at 75.9 ± 0.2 ppm in CD_3SOCD_3. In the fully coupled spectrum, this peak is split into a triplet (Figure 14), indicating the presence of a methylene carbon. The carbon peak at 75 ppm was the only ^{13}C resonance identified other than those due to solvent carbons in either CD_3OD or CD_3SOCD_3 solvents. The observed shifts for the methylene carbon and the methylene hydrogens are compared to those for several sulfur-containing organic compounds in Table 8. The NMR data suggest that the compound is structurally similar to the cyclic sulfur compounds or to the formaldehyde adduct. However, the cyclic compounds both behave as stable molecular species in gas chromatographic–mass spectroscopic determinations. In addition, hydroxymethane sulfonic acid is not stable as the acid; salts of this compound hydrolyze rapidly (seconds) in basic solution and are identified as free sulfite ion in ion chromatographic analysis of solutions (Dasgupta et al. 1982; Hansen et al. 1979). Solutions containing the organic S(IV) compound(s) found in particulate matter do not give either sulfite or sulfate ions with $NaHCO_3$, $NaHCO_3$–Na_2CO_3, or Na_2CO_3–NaOH eluents. Sulfate is produced only after addition of strong oxidizing agents. No other anion that can be attributed to the organic S(IV) compound(s) is seen by ion chromatography, so the compound is probably not an anion.

Final identification of the structure will require additional spectroscopic data to determine the functionality of the carbon, oxygen, and sulfur atoms in the compound as well as data on highly purified samples to provide molecular weight and composition. Such experiments are currently under way. Final proof of the structure will probably also require synthesis of the compound. Experiments are also underway to examine formation of organic S(IV) compounds in chambers containing SO_2, O_3, and either ethylene or propylene. These experiments are being conducted in cooperation with D. Fox, University of North Carolina and J. Meagher, Tennessee Valley Authority.

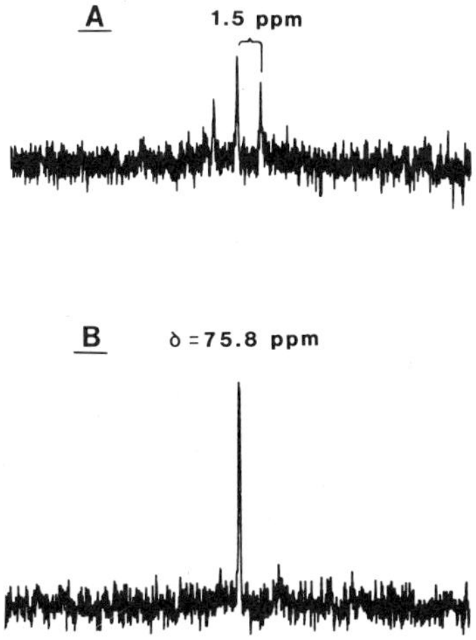

Figure 14. Gated decoupled (*A*) and fully decoupled (*B*) ^{13}C nuclear magnetic resonance spectra of the methylene carbon in the organic S(IV) compound. Data obtained at the University of Utah Nuclear Resonance Spectroscopy Facility by use of a Varian SC 300 spectrometer.

Table 8. Measured and Literature (Sadtler, 1980) Chemical Shifts for ^{13}C and ^{1}H for CH_n Groups Bonded to —S or —O—S Linkages in Solutions of Organic S(IV) From Particulate Matter and in Known Compounds

Compound	Solvent	Chemical Shifts, ppm[a] δ, ^{13}C	δ, ^{1}H
Organic S(IV)	$(CD_3)_2SO$	75.9 ± 0.2	3.92 ± 0.05
	CD_3OD	74.7 ± 0.8	4.31 ± 0.02
	D_2O	No data	4.64
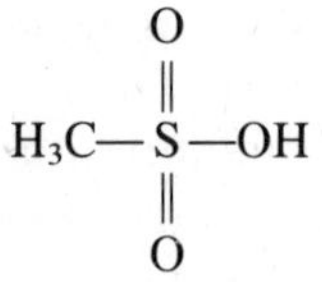	D_2O	~42[b]	2.82
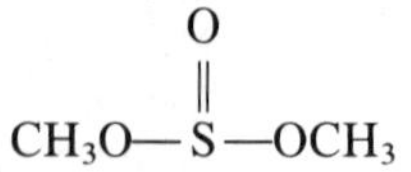	CCl_4	48.4	3.59
$DOCH_2CH_2—SO_3Na$ (β α)	D_2O	53.7 α 57.9 β	3.15 α 3.95 β
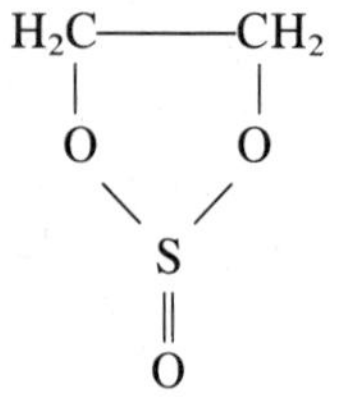	$CDCl_3$	67.7	4.30 , 4.61
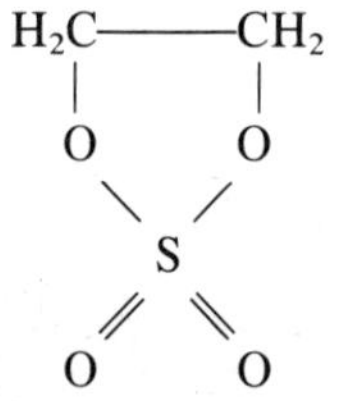	CCl_4	No data	4.68
$DOCH_2—SO_3Na$	D_2O	75.0	4.46
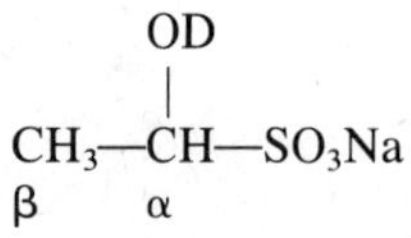	D_2O	81.1 α 17.9 β	4.57 α 1.50 β
$DOCH_2OSONa$	D_2O	84.3	3.82
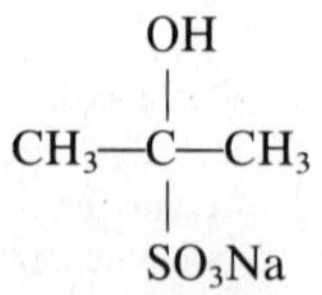	D_2O	86.7	—

[a] Chemical shift values are with respect to δ for TMS = 0.

[b] Estimated from $δ^{13}C$ values from related compounds.

3.1.4. Summary

The data summarized in Sections 3.1.2 and 3.1.3 conclusively show that one or more organic S(IV) compound are present in the atmospheric particulate matter studied. The data do not give a positive identification of the structure of the compound. The solution and gas chromatography–mass spectral data obtained show that the compound can be thermally or oxidatively cleaved to ethylene glycol (or related carboxylic acids) and SO_2 or sulfate. The compound appears to be nonionic. The NMR data indicate that the compound is structurally similar to cyclic organic sulfites or sulfates or to α-hydroxy sulfonic acids. However, it is not identical to any of the previously characterized compounds shown in Figure 12 or Table 8.

3.2. Mechanism of Formation of Aerosol Organic S(IV)

3.2.1. Sources Studied

The formation of organic S(IV) in particulate matter has been studied in emissions from several coal- and oil-fired power plants. Details of these studies have been published (Eatough et al., 1981b,c; Richter, 1981). Samples were obtained from the flue lines and plumes of six coal-fired and three oil-fired power plants, with the method of collection varying at each site. Flue line sampling included obtaining particulate matter with isokinetic sampling and cryogenic trapping of possible gaseous species (Eatough et al., 1981b). Plume samples were obtained from aircraft or ground collection programs. In all cases in which particulate matter was collected on filters, the filters used were HCl-washed quartz fiber filters prepared as previously described (Eatough et al., 1982). Temperature, relative humidity, and concentration of $SO_2(g)$ were measured at all locations but one. Wind speed and direction were also measured to assist in determining plume flow relative to ground- or air-based sampling locations.

Material extractable in 0.1 M HCl, 2.5 mM $FeCl_3$, or solutions of cryogenically trapped materials adjusted to 0.1 M HCl, 2.5 mM $FeCl_3$ were analyzed for inorganic S(IV) (Hansen et al., 1976), organic S(IV) (Eatough et al., 1978a), sulfate (Hansen et al., 1976), and nitrite (Hansen et al., 1977) by calorimetric procedures. Materials extractable in H_2O or the cryogenically trapped solutions were analyzed by ion chromatography for monovalent cations and for anions. The aqueous extracts and trapped solutions were also analyzed for strong and weak acids by measuring the pH and by a combined calorimetric–pH titration technique (Eatough et al., 1977). Selected samples were analyzed for elemental content by PIXE. Sulfur determinations were not done by PIXE for reasons previously outlined (Hansen et al., 1980).

3.2.2. Flue Line Samples

The results of analyses of the collected particles from the flue lines of five coal-fired and one oil-fired power plant (Table 9) indicate that organic S(IV)

Table 9. Composition of 0.1 M HCl-Soluble Material in Primary Emissions (<3 μm) from Coal- and Oil-Fired Power Plants

Location	Source	SO_4^{2-} (μmol g^{-1})[a]	Inorganic S(IV)[b]	Organic S(IV)[b]	H^+ [b]	SO_3(g)[b]	SO_2(g)[b]
Michigan	Coal-A EP[c]	1330	<0.8	<0.8	27.1	530[d]	104,000[d]
Indiana	Coal-B FL[c]	53.4	<0.4	5.6	<2.0	320[d]	31,900[d]
Nevada	Coal-D IL[c]	171	<0.5	<0.5	<0.6	ND[e]	ND
Nevada	Coal-D EP[c]	270	<0.3	<0.3	<0.3	46	ND
Utah	Coal-E WS[f]	ND	9.5	<0.3	<0.1	<1[g]	5,780
Utah	Coal-E EP[f]	ND	<0.3	<0.3	<0.1	128	13,200
Utah	Coal-F FL[c]	191	<1.0	<1.0	<0.5	ND	ND
Mississippi	Oil-G FL[c]	3610	<0.1	<0.1	31.1	1100[d]	33,500[d]

[a] μmol g^{-1} of ash in weighed sample.

[b] Expressed as mole % relative to SO_4^{2-} in collected sample.

[c] Data are from Eatough et al. (1981b). The notations EP, FL, and IL indicate that the sample was obtained from the flue line after an electrostatic precipitator, from the flue line on a plant with no emission control systems, or from the flue line at the inlet to an electrostatic precipitator, respectively.

[d] Data are from Easter et al. (1980).

[e] ND = not determined.

[f] Data are from Eatough et al. (1981c). The notations WS and EP indicate that the sample was collected from the flue line after a wet scrubber or an electrostatic precipitator, respectively.

[g] No SO_3(g) was present, as the stack temperature was below the acid dew point.

was found in samples of particulate matter from the flue line of only one coal-fired plant. Furthermore, no condensable S(IV) species other than SO_2(g) could be shown to be present. On the basis of these results, the concentration of the organic S(IV) compound(s) in primary particulates from coal- or oil-fired power plants, with or without electrostatic precipitator control is less than 0.01 wt.% as S. Therefore, the organic S(IV) found in particulate matter from these sources must be the result of reactions in the plume. It is probable that the organic S(IV) compound(s) is formed by a photochemical mechanism involving SO_2, O_3 and ethylene.

3.2.3. Conversion of SO_2(g) to Organic S(IV) in Plumes

The procedures used by us to describe the reaction kinetics of SO_2(g) in transported plumes has been published (Eatough et al., 1981c). The conversion of SO_2(g) to organic S(IV) in the plumes of oil- and coal-fired power plants is well described by a process that is first order in SO_2(g) (Figure 15). Conservative tracer data were available only for Coal-E and Oil-H samples (Eatough et al., 1981c). Details on data analysis for the other steam plants studied have been given (Eatough et al., 1981b). The resulting k_1 values for the conversion of SO_2(g) to organic S(IV) are given in Table 10.

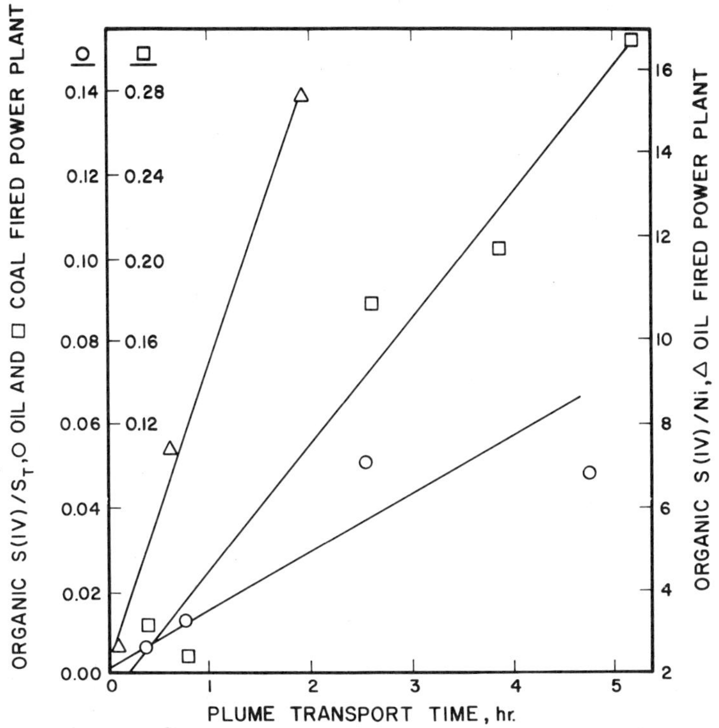

Figure 15. Test for first-order mechanism for conversion of SO_2(g) to the organic S(IV) compound(s). (Data are from Eatough et al., 1981c).

A temperature dependence of the k_1 values for the formation of the organic S(IV) species is not apparent from the results of studies completed to date. Least-squares analysis of ln (k_1) versus $1/T$ (10 data points) for formation of the organic S(IV) gives a poor fit with $r^2 = 0.3$ and a calculated increase in the rate with decreasing temperature. Such a temperature dependence is consistent not with the activation energy of a reaction pathway controlling the reaction, but rather with changes in atmospheric chemistry as the controlling factor. Such effects have been suggested, for example, in the Fe^{3+}–NH_3(g)–SO_2(g) reaction system (Freiberg, 1974), where the chemistry is controlled by the solubility of the gases in a liquid droplet, resulting in an increase in the overall reaction rate with decreasing temperature. Although the details are not known, similar effects could control the formation rate of the organic S(IV) compound(s). The rate of formation is inversely related to the atmospheric water concentration (Table 10 and Figure 16). This may be due to interference with the formation mechanism or to decomposition of the S(IV) species by water. The observed short-term stability of aqueous extracts of the organic S(IV) suggests that water probably affects the formation reaction rather than a decomposition reaction. Such chemistry would be expected if the organic S(IV) compound(s) is structurally related to ethylene sulfite or hydroxymethane sulfonic acid (see Section 3.1) and is formed from the photochemistry of SO_2, O_3, and olefins. The linear relationship depicted in Figure 16 suggests that the inhibition is first order in atmospheric water concentration. Data are currently being analyzed by us on detailed studies at an oil-fired power plant in California and a coal-

Table 10. Calculated k_1 Values for the Conversion of SO_2(g) to Organic S(IV)[a]

Plant[b]	k_1 [% SO_2(g) h^{-1}]	T (°C)	P_{H_2O} (mbar)
Coal-C	0.4	24	23.5
Coal-D	>1.0	8	3.2
Coal-E	0.78	20	11.5
	1.38	25	6.0
	3.01	14	10.0
	2.54	21	3.7
Oil-G	0.6	22	11.9
Oil-H	0.6	17	10.8
	1.0	8	5.6
Oil-I	1.4	15	6.4

[a] Data are from Eatough et al. (1981b,c).

[b] Coal-fired power plant C is located in Pennsylvania. Oil-fired power plants H and I are located in California (Eatough et al., 1981b). No flue line samples were obtained at these sources. Location of other plants is given in Table 9.

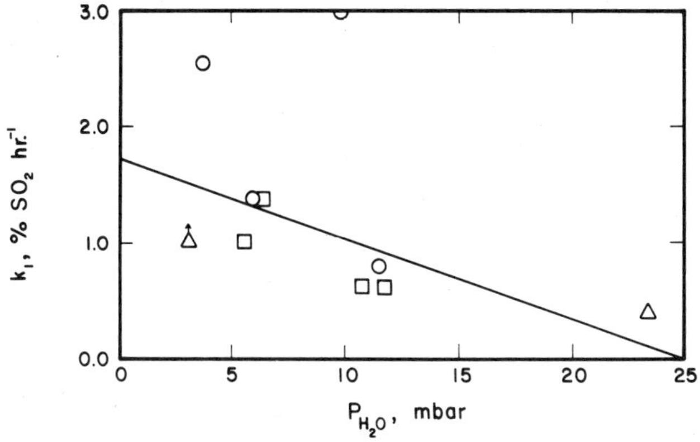

Figure 16. Plot of k_1 for conversion of SO_2(g) to organic S(IV) compound(s) *versus* atmospheric water partial pressure. Data points are from studies at coal-fired power plants, Δ and O, and oil-fired power plants, □. (Data are from Eatough et al., 1981c).

fired power plant in Alabama that will add about 16 more data points to Figure 16. Preliminary results support the indication that the k_1 value is dependent on atmospheric water concentration. The details of the reaction mechanism must await elucidation of the structure of the organic S(IV) compound(s). The data in Figure 16 suggest a maximum value of about 2% per hour for conversion of SO_2(g) to the organic S(IV) compound(s).

3.3. Occurrence of Aerosol Organic S(IV)

The data given in Sections 3.1 and 3.2 indicate that the aerosol concentration of organic S(IV) is dependent on the following variables: atmospheric residence time; atmospheric water content; atmospheric ethylene and propylene concentrations; atmospheric SO_2(g) concentration; and photochemistry. Atmospheric concentrations of organic S(IV) in <3 μm particulate matter that have been measured are summarized in Table 11. In almost all cases where analyses have been performed on size-fractionated material, over 80% of the organic S(IV) is found in <0.5-μm-size particulate matter (Ellis, 1982; Eatough et al., 1981c; Richter, 1981). It has also been shown (Richter, 1981) that organic S(IV) is not stable over the time period of several weeks in samples stored at room temperature.

The data summarized in Table 11 for the coal-fired power plant in Utah and the oil-fired power plant in California show that the formation of organic S(IV) in the plume can compete favorably with the formation of sulfate for plume SO_2(g). The observed ratio of organic S(IV) to sulfate doubles for distant samples as compared to samples obtained close to the stack. The rate of formation of organic S(IV) in these two plumes averages one-half the rate of formation of sulfate. This results in the observed organic

Table 11. Occurrence of Organic S(IV) Compound(s) in Airborne Aerosols

Location	Aerosol Source	Number of Samples	Organic S(IV) Concentration in <3-μm Aerosols in μg/m³.[a]			Average Mole Ratio, Organic S(IV) : SO_4^{2-}	Reference
			Range	Median	Average		
Utah	Coal-fired power plant plume, within 6 km of stack	7	0–0.8	0.6	0.4	0.33 ± 0.19	Eatough et al. (1981c)
	Coal-fired power plant plume, 20–42 km from stack	9	0.1–1.0	0.4	0.3	0.65 ± 0.38	
Utah	Coal-fired boiler plume, next to stack during inversion	3	0.3–0.9	0.6	0.6	0.61 ± 0.31	Eatough et al. (1981a)
Nevada	Coal-fired power plant plume, 4–34 km from stack, ground-based samples	6	0–0.4	0.3	0.3	0.58 ± 0.42	Eatough et al. (1981b)
Pennsylvania	Coal-fired power plant plume, within 12 km of stack	5	0.2–0.8 (0–0.4)[b]	0.4 (0)[b]	0.4 (0.1)[b]	0.05 ± 0.03 (0.04 ± 0.08)[b]	Richter (1981)
California	Oil-fired power plant plume, within 3 km of stack	4	0–0.7	0.2	0.3	0.15 ± 0.15	Eatough et al. (1981c)
	Oil-fired power plant plume, 10–28 km from stack	3	0.3–0.9	0.6	0.6	0.27 ± 0.34	Eatough et al. (1981c)

Mississippi	Oil-fired power plant plume, 15–30 km from stack	1	0.4 (0.4)[b]	—	—	0.08 (0.22)[b]	Eatough et al. (1981b)
Rockport, IN	Rural ambient	9	0.3–1.2	0.7	0.6	0.05 ± 0.03	Richter (1981)
Duncan Falls, OH	Rural ambient	5	0.1–0.6	0.3	0.3	0.03 ± 0.02	Richter (1981)
Johnstown, PA	Rural ambient	4	0.3–1.1	0.6	0.7	0.06 ± 0.01	Richter (1981)
State College, PA	Rural ambient	9 (4)[c]	0–1.6 (0.9–1.6)[c]	0. (1.0)[c]	0.5 (1.1)[c]	0.03 ± 0.05 (0.08 ± 0.04)[c]	Richter (1981)
Upton, NY	Rural ambient	8 (5)[c]	0–1.8 (0.2–1.8)[c]	0.9 (1.6)[c]	0.9 (1.4)[c]	0.14 ± 0.15 (0.22 ± 0.13)[c]	Hansen (1981)
Los Angeles, CA	Urban air, coastal mixed layer, high moisture content	8	0–1.0	0	0.1	0.003 ± 0.006	Farber et al. (1981)
	Urban air, dry air mass	14	0–8.2	1.9	3.5	1.05 ± 1.50	
	Urban air, inversion	13	0–9.0	1.8	3.0	0.60 ± 0.78	
New York, NY	Urban air, winter	3	0.2–1.7	1.1	1.0	1.93 ± 1.2	Eatough et al. (1978a)

[a] Molecular weight assumed to be 112.
[b] Data in parentheses have been corrected for background aerosol contribution.
[c] Data in parentheses include only those days when organic S(IV) was present.

S(IV) : SO_4^{2-} ratio of about 0.5 in aged plume samples from these two sources. In contrast, the rate of formation of sulfate observed in the plume of the coal-fired power plant in Nevada was small because of the low ambient temperature (Eatough et al., 1981c), and in aged plume samples from this source the organic S(IV) : SO_4^{2-} ratio was sometimes greater than 1. The observed organic S(IV) : SO_4^{2-} ratios in the rural and urban samples follow the expected effect of atmospheric H_2O on formation of the organic S(IV) compound(s). The highest organic S(IV) : SO_4^{2-} ratios are seen in dry air masses in the inland Los Angeles Basin or in New York in the winter. The lowest ratios are seen in the east in the summer and in moist air masses in the Los Angeles Basin. The highest observed atmospheric concentrations of the organic S(IV) compound(s) occur in the Los Angeles Basin. Presumably this reflects a combination of high olefin concentration, low water concentration, and high photochemical activity in this area.

As noted in Section 1.4, air mass back trajectories are available (Figure 10) for the sampling periods in which the Duncan Falls and State College samples (Table 5) were obtained. These trajectory calculations can be used to determine the origin of the particulate samples collected as well as any sources that may have influenced their composition. The samples collected on July 26 and 27 at Duncan Falls were from air masses that originated in the Ohio River Valley, where large amounts of coal are burned. The $SO_2(g)$ and hydrocarbons necessary to produce organic S(IV) were thus present in these air masses, and some organic S(IV) was found in both samples. The sample collected on July 27 at State College originated south of the main sources in the Ohio River Valley, and no organic S(IV) was seen in this sample. The other samples collected at State College that contained no organic S(IV) were collected in air masses that originated from the northeastern or northwestern regions, which are free of large numbers of sources. The samples with organic S(IV) all originated from the Ohio River Valley (e.g., see July 28 data in Figure 10 and Table 5). The sample collected at State College on July 28 was from an air mass that was over the Duncan Falls site on July 27. Additional formation of organic S(IV) in the suspended aerosols during transport from Duncan Falls to State College is suggested by the data.

Even though the ratio of organic S(IV) to SO_4^{2-} in aerosols collected over the eastern United States is small, the data suggest that long-range transport and formation of both sulfate and organic S(IV) take place in air masses in this region. Ratios of both sulfate and organic S(IV) to total atmospheric sulfur [($SO_2(g)$ + aerosol sulfate, organic S(IV), and inorganic S(IV)] for all samples where organic S(IV) was seen are given in Figure 17. Least-squares fit of the data presented in Figure 17 gives

$$\frac{[SO_4^{2-}]}{[S_T]} = 0.185 + 7.15\,\frac{[S(IV)_{org}]}{[S_T]}, \qquad (2)$$

with $r^2 = .63$. If the average rate of conversion of $SO_2(g)$ to sulfate in the

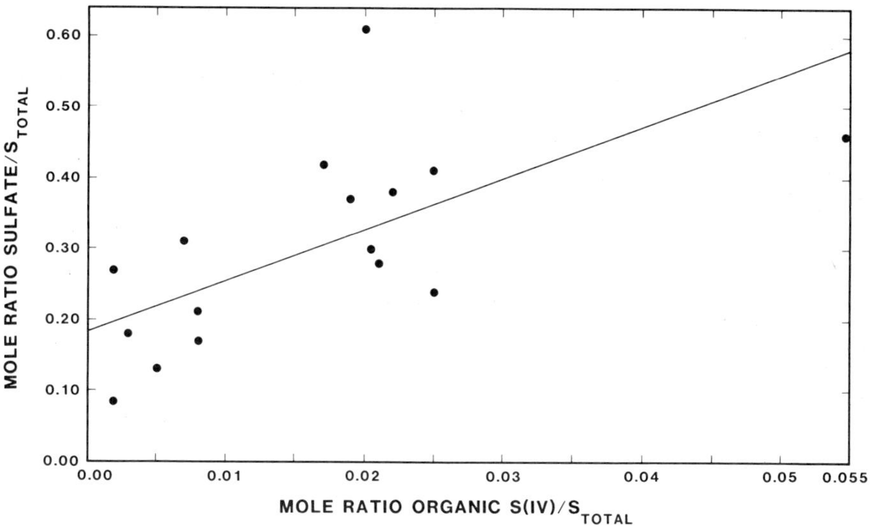

Figure 17. Comparison of fraction of total sulfur present as aerosol sulfate to fraction present as organic S(IV) compound(s) in air masses sampled over the northeastern United States. (Data are from Richter, 1981).

northeastern United States during the summer months is 1–2% SO_2(g) per hour (Eatough et al., 1981e), the data in Figure 17 predict that the daily average rate of formation of organic S(IV) during the same time period will be about 0.2% SO_2(g) per hour. This value is in excellent agreement with rates predicted from the data in Figure 16.

4. SIGNIFICANCE OF AEROSOL S(IV) SPECIES

4.1. Inorganic S(IV)

No studies have been previously reported that would allow an unequivocal evaluation of the toxicologic implications of the inorganic S(IV) chemistry reviewed in this chapter. The inorganic S(IV) species were originally identified by us in an analytical program (Hansen et al., 1975) in support of an epidemiologic study at a copper smelter (Smith et al., 1977). Only limited data were obtained on the concentration of inorganic S(IV) to which workers were exposed. Analyses of the data suggest that the inorganic S(IV) aerosol species may be responsible for part of the observed effects of pollution on the pulmonary system of smelter workers (Smith et al., 1977); however, results were inconclusive because of the limited data and the confounding effects of high SO_2(g) concentrations. Alarie et al. (1973) and McJilton et al. (1973) have reported increased respiratory irritant effects of HSO_3^--containing aerosols compared to SO_2(g). The exposure in these experiments, however, was to simple sulfite salts, which might be quite different chemically from the transition metal S(IV) complexes observed in environmental samples. Laboratory studies by Amdur et al. (1971, 1975), that report an increased irritant effect for animals exposed to several metal salts and $SO_2(g)$ may be more related to possible toxicologic effects from the inorganic S(IV) complexes. A pulmonary response was observed in animals exposed to a mixture of $CuSO_4$- or $FeSO_4$-containing aerosols and $SO_2(g)$ that was greater than expected, assuming that all $SO_2(g)$ incorporated in the aerosol was present as H_2SO_4. In cooperation with Amdur and Underhill, we have analyzed aerosols produced in a manner identical to that used in their earlier experiments (Eatough et al., 1978b; Eatough and Hansen, 1981). The results of these analyses (Table 3, Section 2.2) indicate that the animals were exposed to a mixture of sulfuric acid and inorganic S(IV) salts. If we assume that the effects of the various chemical species to which the animals were exposed are additive, an estimate of comparative toxicity can be made. If the data in Amdur et al. (1975) and Eatough et al. (1978b) are combined by use of this assumption, we calculate that a Cu(?)–S(IV) aerosol in the submicrometer size range is three times more irritating than a corresponding H_2SO_4 aerosol in the animal model used by Amdur. It is interesting to note that the only metal sulfate aerosols that showed a "synergistic" effect with $SO_2(g)$ were those where the metal ion (Fe, Cu, Zn, Mn) is known or believed to form S(IV) complexes [see Section 2.2 and Dyson and Quon (1976) and Barrie and Georgii (1976)]. If these various S(IV) species are responsible at the low concentrations reported here for the "synergestic" effect previously reported, additional toxicologic work is warranted. Techniques for producing Fe(III)–S(IV)-containing aerosols have been reported (Hilton et al., 1979; Schlesinger et al., 1980) that may be suitable for inhalation toxicology studies. Toxicologic studies are currently under way at New York University by use of the aerosol generation techniques described by Schlesinger et al. (1981). The results of these studies should provide a better basis for estimating the toxicity of aerosol inorganic S(IV) species.

4.2. Organic S(IV)

No studies have been reported that allow an estimation of the potential toxicity of the aerosol organic S(IV) species. The most important aspect of organic S(IV) toxicology may be the mutagenic and carcinogenic properties of the compound(s). Many compounds that are believed to be structurally related to the organic S(IV) compound found in aerosols (see Section 3.1) are known to be mutagens and/or carcinogens (e.g., Hoffman, 1980), and these compounds generally act as alkylating agents. Bacterial mutagenicity tests are currently underway on purified samples of the aerosol organic S(IV). The ultimate evaluation of the carcinogenicity threat of the organic S(IV) species probably must await structural identification. The high concentration of this species observed in the environment (Table 11, Section 3.3), especially in urban centers, underscores the importance of completing this research.

REFERENCES

Alarie, Y., Wakisaka, I., and Oka, S. (1973). Sensory irritation by sulfite aerosols. *Environ. Physiol. Biochem.* **3,** 182–184.

Allegrini, I. and Mattogno, G. (1978). Analysis of environmental particulate matter by means of photoelectron spectroscopy. *Sci. Total Environ.* **9,** 227–232.

Amdur, M. O. (1971). Aerosols formed by oxidation of sulfur dioxide. *Arch. Environ. Health* **23,** 459–468.

Amdur, M. O., Bayless, T., Urgo, V., Dubriel, M., and Underhill, D. W. (1975). Respiratory response of guinea pigs to sulfuric acid and sulfate salts. Symposium on *Sulfur Pollution and Research Approaches,* U. S. Environmental Protection Agency, May 27–28.

Barrie, L. and Georgii, H. W. (1976). An experimental investigation of the absorption of sulfur dioxide by water drops containing heavy metal ions. *Atmosph. Environ.* **10,** 743–749.

Carlyle, D. W. (1971). A kinetic study of the aquation of sulfitoiron(III) ion. *Inorg. Chem.* **10,** 761–764.

Charlson, R. J., Covert, D. S., Larson, T. V., and Waggoner, A. P. (1978). Chemical properties of tropospheric sulfur aerosol. *Atmosph. Environ.* **12,** 39–53.

Colucci, A. V. (1976). Sulfur oxides: Current status of knowledge. Electric Power Research Institute Report EA-316.

Danilczuk, E. (1964). Influence of the redox potentials of components on the complex properties with particular respect to $[Fe(SO_3)_n]^{3-2n}$ and $[Cr(SO_3)_n]^{3-2n}$, *Studia Soc. Sci. Torun.,* Sect. B. 5 (4), 88 pp; *Chem. Abstrs.* 62:9858e.

Danilczuk, E. and Swinarski, A. (1961). The complex ion $[Fe^{III}(SO)_{3n}]^{3-2n}$. *Roczniki Chemii* **35,** 1563–1572.

Dasgupta, P. K., Mitchell, P. A., and West, P. W. (1979). Study of transition metal ion–S(IV) systems. *Atmosph. Environ.* **13,** 775–782, 1725–1729.

Dasgupta, P. K., DeCesare, K. B., and Brummer, M. (1982). Determination of S(IV) in particulate matter. *Atmosph. Environ.* **16,** 917–927.

Dyson, W. L. and Quon, J. E. (1976). Reactivity of zinc oxide fume with sulfur dioxide in air. *Environ. Sci. Technol.* **10,** 476–481.

Easter, R. C., Busness, K. M., Hales, J. M., Lee, R. N., Arbuthnot, D. A., Miller, D. F., Sverdrup, G. M., Spicer, C. W., and Howes, J. E., Jr. (1980). Plume conversion rates in the SURE region. EPRI EA-1498, Electric Power Research Institute, Palo Alto, CA.

Eatough, D. J. and Hansen, L. D. (1979). Study of transition metal ion–S(IV) systems. *Atmosph. Environ.* **13,** 1725–1729.

Eatough, D. J. and Hansen, L. D. (1981). S(IV) chemistry in smelter produced particulate matter. *Am. J. Ind. Med.* **1,** 435–448.

Eatough, D. J., Hansen, L. D., Izatt, R. M., and Mangelson, N. F. (1977). Determination of acidic–basic species in particulates by titration calorimetry, in *Methods and Standards for Environmental Analysis,* W. H. Kirchkoff, Ed., NBS Special Publication 404, U.S. National Bureau of Standards, Gaithersburg, MD, pp. 643–649.

Eatough, D. J., Major, T., Ryder, J., Hill, M., Mangelson, N. F., Eatough, N. L., Hansen, L. D., Meisenheimer, R. G., and Fischer, J. W. (1978a). The formation and stability of sulfite species in aerosols. *Atmosph. Environ.* **12,** 263–271.

Eatough, D. J., Hansen, L. D., Hilton, C. M., and Christensen, J. J. (1978b). Aerosol sulfite chemistry and inhalation toxicology. Final Report TPS 76-647, Electric Power Research Institute.

Eatough, D. J., Izatt, S., Ryder, J., and Hansen, L. D. (1978c). Use of benzaldehyde as a selective solvent for sulfuric acid: Interferences by sulfate and sulfite salts. *Environ. Sci. Technol.* **12,** 1276–1279.

Eatough, D. J., Eatough, N. L., Mangelson, N. F., Ryder, J., Hansen, L. D., Meisenheimer, R. G., and Fischer, J. W. (1979). The chemical composition of smelter flue dusts. *Atmosph. Environ.* **13,** 489–506.

Eatough, D. J., Lee, M. L., Later, D. W., Richter, B. E., Eatough, N. L., and Hansen, L. D. (1981a). Dimethyl sulfate in particulate matter from coal- and oil-fired power plants. *Environ. Sci. Technol.* **15,** 1502–1506.

Eatough, D. J., Hansen, L. D., Lee, M. L., and Mangelson, N. F. (1981b). Measurement of sulfur(IV) and methylated sulfate species in aerosols produced by fossil fuel burning steam plants. Final report to EPRI on contract RP1154-1.

Eatough, D. J., Richter, B. E., Eatough, N. L., and Hansen, L. D. (1981c). Sulfur chemistry in smelter and power plant plumes in the western U.S. *Atmosph. Environ.* **15,** 2241–2253.

Eatough, D. J., Christensen, J. J., Eatough, N. L., Hill, M. W., Major, T. D., Mangelson, N. F., Post, M. E., Ryder, J. F., Hansen, L. D., Meisenheimer, R. G., and Fischer, J. W. (1982). Sulfur chemistry in a copper smelter plume. *Atmosph. Environ.* **16,** 1001–1015.

Ellis, E. C. (1982). Personal communication, Southern California Edison Company.

Farber, R. J., Huang, A. A., Bregman, L. D., Mahoney, R. L., Eatough, D. J., Hansen, L. D., Blumenthal, D. L., Keifer, W. S., and Allard, D. W. (1981). The third dimension in the Los Angeles basin. Air Pollution Control Association Annual Conference, Philadelphia, PA, Paper No. 81-60.5.

Freiberg, J. (1974). Effects of relative humidity and temperature on iron catalyzed oxidation of SO_2 in atmospheric aerosols. *Environ. Sci. Technol.* **8,** 731–734.

Freiberg, J. (1978). Diffusion-coupled oxidation of atmospheric sulfur dioxide. *Nature* **274,** 42–44.

Hansen, L. D. (1981). Unpublished results by L. D. Hansen and co-workers, Brigham Young University.

Hansen, L. D., Eatough, D. J., Whiting, L., Bartholomew, C. H., Cluff, C. L., Izatt, R. M., and Christensen, J. J. (1974). Transition metal—SO_3^{2-} complexes: A postulated mechanism for the synergistic effect of aerosols and SO_2 on the respiratory tract. Trace Substances in Environmental Health V. VIII, D. D. Hemphill, Ed., University of Missouri, Columbia, pp. 383–397.

Hansen, L. D., Eatough, D. J., Mangelson, N. F., Jensen, T. E., Cannon, D., Smith, T. J., and Moore, D. E. (1975). Sulfur species and heavy metals in particulates from a copper smelter. *Proceedings, International Conference on Environmental Sensing and Assessment,* Las Vegas, paper 23-2.

Hansen, L. D., Whiting, L., Eatough, D. J., Jensen, T. E., and Izatt, R. M. (1976). Determination of sulfur(IV) and sulfate in aerosols by thermometric methods. *Anal. Chem.* **48,** 634–638.

Hansen, L. D., Richter, B. E., and Eatough, D. J. (1977). Determination of nitrite by direct injection enthalpimetry. *Anal. Chem.* **49,** 1779–1781.

Hansen, L. D., Richter, B. E., Rollins, D. K., Lamb, J. D., and Eatough, D. J. (1979). Determination of arsenic and sulfur species in environmental samples by ion chromatography. *Anal. Chem.* **51,** 633–637.

Hansen, L. D., Ryder, J. F., Mangelson, N. F., Hill, M. W., Faucette, K. J., and Eatough,

D. J. (1980). Inaccuracies encountered in sulfur determination by particle induced x-ray emission. *Anal. Chem.* **52,** 821–824.

Hegg, D. A. and Hobbs, P. V. (1978). Oxidation of sulfur dioxide in aqueous systems with particular reference to the atmosphere. *Atmosph. Environ.* **12,** 241–253.

Hilton, C. M., Christensen, J. J., Eatough, D. J., and Hansen, L. D. (1979). Fe(III)–S(IV) aerosol generation and characterization. *Atmosph. Environ.* **13,** 601–605.

Hoffman, G. R. (1980). Genetic effects of dimethyl sulfate, diethyl sulfate and related compounds. *Mutat. Res.* **75,** 63–129.

Husar, R. B., Lodge, J. P., Jr., and Moore, D. J., Eds. (1978). Sulfur in the atmosphere. *Atmosph. Environ.* **12** (1–3).

Jones, P. W. (1974). Analysis of non-particulate organic compounds in ambient atmospheres. Air Pollution Control Association, Annual Meeting, June 9–13, Denver, CO, paper 74-265.

Lave, L. and Seskin, E. (1977). *Air Pollution and Public Health,* Johns Hopkins University Press, Baltimore, MD.

Lee, M. L., Later, D. W., Rollins, D. K., Eatough, D. J., and Hansen, L. D. (1980). Dimethyl and monomethyl sulfate: Presence in coal fly ash and airborne particulate matter. *Science* **207,** 186–188.

McJilton, C., Frank, R., and Charlson, R. (1973). The role of relative humidity in the synergistic effect of a sulfur dioxide–aerosol mixture on the lung. *Science* **182,** 503–504.

National Academy of Sciences (1979). *Airborne particulates,* NRC Committee on Toxicology, Washington, DC.

Newman, L. (1981). Atmospheric oxidation of sulfur dioxide: A review as viewed from power plant and smelter plume studies. *Atmoph. Environ.* **15,** 2231–2239.

Nielsen, K. K., Hill, M. W., and Mangelson, N. F. (1976). Calibration and correction methods for quantitative proton induced x-ray emission analysis of autopsy tissues. *Adv. X-ray Anal.* **19,** 511–520.

Novakov, T., Mueller, P. K., Alcocer, A. E., and Otvos, J. W. (1972). Chemical composition of Pasadena aerosols by particle size and time of day. III. Chemical states of nitrogen and sulfur by photoelectron spectroscopy. *J. Colloid Interface Sci.* **39,** 225–234.

Panter, R. and Penzhorn, R. D. (1980). Alkyl sulfonic acids in the atmosphere. *Atmosph. Environ.* **14,** 149–151.

Penzhorn, R. D. and Filby, W. G. (1976). A method for the identification of sulfur-containing acids in atmospheric aerosols. *Staub-Reinhalt. Luft* **36,** 205–207.

Richter, B. E. (1981). S(IV) and alkylating agents in airborne particulate matter. Ph.D. dissertation, Brigham Young University, Provo, UT.

Sachner, M. A., Ford, D., Fernandez, R., Cipley, J., Perez, D., Kwoka, M., Reinhart, M., Michaelson, E. D., Schreck, R., and Wanner, R., (1978). Effects of sulfuric acid aerosol on cardiopulmonary function of dogs, sheep and humans. *Am. Rev. Resp. Dis.* **118,** 497–510.

Sadtler (1980). *Sadtler Standard Proton and Carbon-13 NMR Spectra.* Sadtler Research Laboratories, Inc., Philadelphia, PA.

Sato, T., Matani, S., and Okabe, T. (1979). The oxidation of sodium sulfite with nitrogen dioxide with special reference to analytical methods for nitrogen-sulfur compounds produced in the reaction system. *Nippon Kagaku Kaishi* **1979,** 869–878.

Schlesinger, R. B., Gurman, J. L., and Chen, L.-C. (1980). The production and characterization of a transition metal [Fe(III)]–S(IV) aerosol. *Atmosph. Environ.* **14,** 1279–1287.

Schwartz, S. E. and Newman, L. (1978). Processes limiting oxidation of sulfur dioxide in stack plumes. *Environ. Sci. Technol.* **12,** 67–73.

Shy, C. M., and Finklea, J. F. (1973). Air pollution affects community health. *Environ. Sci. Technol.* **7,** 204–208.

Smith, T. J., Peters, J. M., Reading, J. C., and Castle, C. M. (1977). Pulmonary impairment from chronic exposure to SO_2. *Am. Rev. Resp. Dis.* **116,** 31–41.

U.S. Environmental Protection Agency (1975). Position paper on regulation of atmospheric sulfates. Research Triangle Park, NC, EPA-450/2-75-007.

West, P. W. and Gaeke, G. C. (1956). Fixation of sulfur dioxide as disulfitomercurate(II) and subsequent colorimetric estimation. *Anal. Chem.* **28,** 1816–1819.

6

SULFUR, ACIDIC AEROSOLS, AND ACID RAIN IN THE EASTERN UNITED STATES

John W. Winchester

Department of Oceanography
Florida State University
Tallahassee, Florida 32306

1. INTRODUCTION

The discharge of massive amounts of gaseous sulfur dioxide to the atmosphere, principally in the most populated areas of the Northern Hemisphere, by the combustion of coal and oil containing up to several percent sulfur may be causing widespread effects on human health and welfare by the action of SO_2 itself, acidic sulfate aerosol particles, or acidified rainwater (EPA, 1980). Gaseous sulfur emissions globally from anthropogenic combustion sources during 1976 were about 100 million tons, an estimate that may be compared with the annual total from all natural sources of gaseous sulfur compounds, principally biogenic, also about 100 million tons, in addition to 44 million tons as sulfate in sea spray particles (Cullis and Hirschler, 1980). Unlike the widely distributed natural sources over terrestrial and marine areas, the anthropogenic sources tend to be concentrated mainly in North America, Europe, and Asia in areas totaling less than 10% of the Earth's surface. Pollution sulfur generally remains in the atmosphere for up to only a few days to a week or so before removal by wet or dry processes (Junge, 1963; Rodhe, 1978) and should thus be deposited largely within several hundred to a few thousand kilometers from its sources at most. Consequently, the pollution sulfur is not distributed globally; and thus a more drastic modification of air quality is expected in areas where the pollution sources are concentrated than would be implied by a globally uniform distribution of anthropogenic and natural atmospheric sulfur.

This chapter summarizes research carried out at the Florida State University on the basis of measurements of aerosol sulfur concentrations at surface sites in the eastern United States and in remote locations of the Southern Hemisphere, to provide the reader with an appreciation of the extent to which the atmosphere characteristic of the United States has become modified from its natural composition. Elemental analyses of aerosol particles, collected by cascade impactors and time-sequence streaker filter samplers, and of rainwater have been performed by particle-induced x-ray emission (PIXE). Some of the data have been examined to determine relationships with some of the factors that may control the rate of formation of aerosol sulfuric acid and sulfates by oxidation of gaseous sulfur dioxide. By use of the relationships found, an attempt has been made to reconstruct the pattern of expected geographic variability of sulfate aerosol concentrations in the Atlantic coastal region of the southeastern United States for the post-1950 period when air pollution emissions of SO_2 in the United States have been substantial. The results of the reconstruction are compared with the distribution of lung cancer mortality in the same region, and the possible significance of the comparison is discussed.

2. EXPERIMENTAL TECHNIQUES

2.1. Aerosol Sampling by Impactors and Streakers

With the development of the PIXE analysis technique during the 1970s, aerosol particle samplers were designed to utilize the high sensitivity and small dimensions of the sample area analyzed by proton beam bombardment. These samplers require only small air volumes that can be drawn by miniature vacuum pumps having modest electric power requirements, permitting the samplers to be deployed in locations inaccessible to large field equipment. Samples may be collected on towers, remote mountain peaks, ships at sea, and aircraft as well as at multiple sites in ground-based sampling networks. Cascade impactors are used for particle-size fraction sampling and time-sequence filter samplers ("streakers") for time-series concentration measurements.

A schematic diagram of a cascade impactor suitable for PIXE analysis of particle size fractions is given in Figure 1. Each stage consists of a single orifice and an impaction surface spaced optimally from half to one jet di-

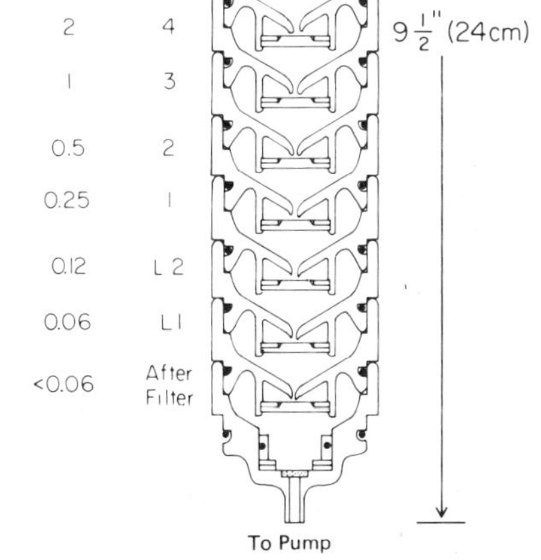

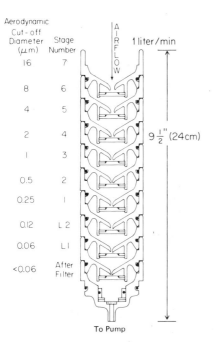

Figure 1. Single-orifice cascade impactor. Particles are collected in aerodynamic-diameter fractions by impaction on greased thin-film surfaces beneath the jets. Stage 1 serves as a critical flow orifice and allows 1 liter min^{-1} sampling rate. Nuclepore after-filter material collects particles smaller than the stage 1 cutoff diameter in a region of low pressure. If optional low-pressure stages L2 and L1 are used, two additional size cuts are realized. (From PIXE International Corporation, Tallahassee, Florida.)

ameter away (Marple and Willete, 1976). A Nuclepore afterfilter collects particles smaller than 0.25 μm aerodynamic diameter (μmad) if the optional low-pressure stages L1 and L2 are not used. The samples are small spots that adhere to greased impaction surfaces, and they may be analyzed by placement in a proton beam of slightly larger dimensions. The impactor airflow rate is about 1.0 liter min^{-1}, and typical sampling times range from 1 hour or less in polluted air to 1 day or longer in very clean air. Under field conditions the impactor is positioned face down so that air enters from below, as protection against rain and falling debris.

The streaker sampler for time sequence filter sampling is shown in Figure 2 (Nelson, 1977). It consists of a smooth Nuclepore filter in contact with a smooth sucking orifice that draws air through the filter as the filter is rotated relative to the orifice so that a continuous time-dependent streak sample is obtained. (A linear streaker is also used where the orifice is drawn

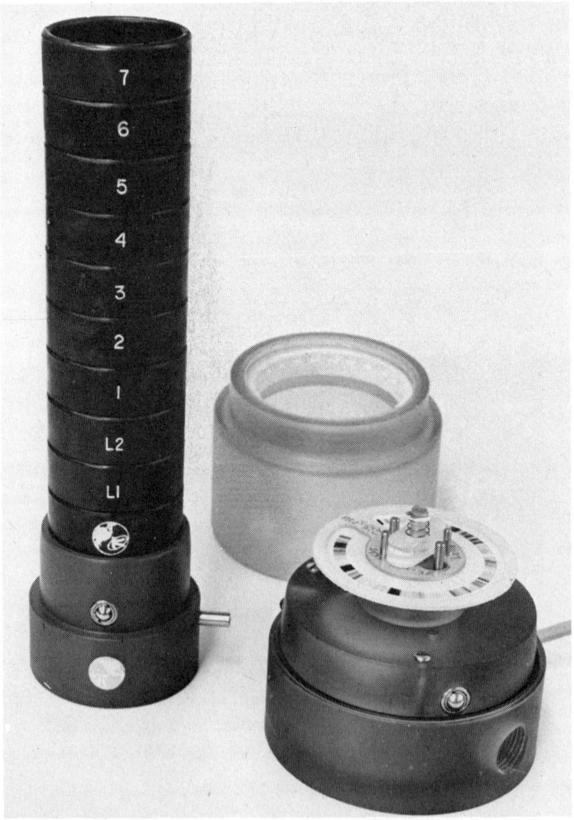

Figure 2. Streaker sampler, right, with previously exposed filter in place. Filter rotates past a sucking orifice which draws air through a small portion of the filter corresponding to 2 hours of sampling. Left, cascade impactor of the type shown in Figure 1. (From PIXE International Corporation, Tallahassee, Florida.)

along the length of a Nuclepore filter strip during sampling.) By using 0.4-μm pore-size filters and a 2 × 5 mm rectangular orifice, a flow rate of about 0.8 liter min^{-1} at 500-torr vacuum is obtained, and particle collection efficiency is nearly quantitative for all particle sizes (Liu and Lee, 1976). A single filter is normally used to obtain a 7-day sample; analyses are performed by stepping the filter in the proton beam for 84 steps of 2-h sampling-time resolution. Operating characteristics of the streaker and impactor samplers are discussed by Winchester (1982).

2.2. Elemental Analysis by PIXE

Particle-induced x-ray emission analysis procedures at the Florida State University employ a proton beam of 3–5 MeV energy that bombards the sample in vacuum for one to a few minutes (Johansson et al., 1975). The PIXE method was first proposed by Johansson et al. (1970) and has subsequently been developed into a routine analytical tool (Cahill, 1980; Johansson and Johansson, 1976) of widespread application (Johansson, 1977; 1981). In principle, the method is akin to x-ray fluorescence (XRF) in that characteristic K, L, or M radiation from the elements present in a sample is excited by incident radiation—energetic protons in PIXE and energetic x-rays in XRF. Unlike XRF, however, the incident energy is focused so that

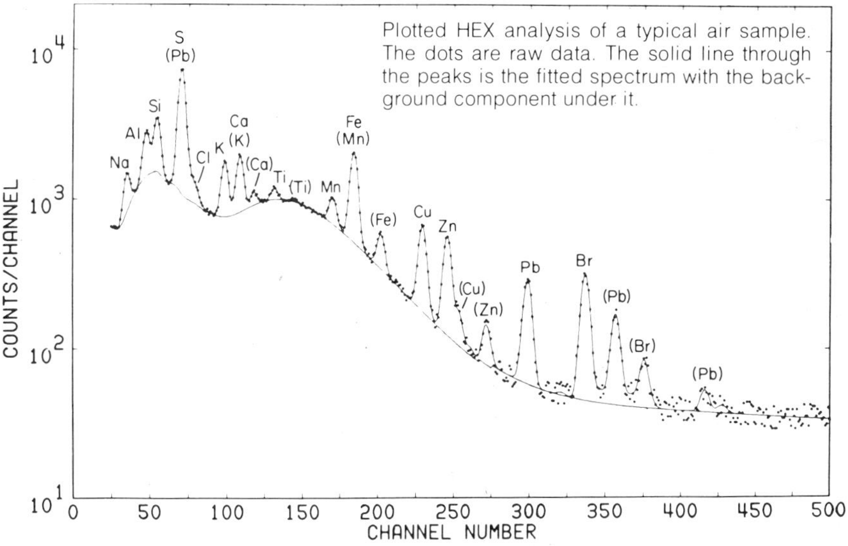

Figure 3. An x-ray spectrum obtained in PIXE analysis. The peaks are of energy, given by channel number, characteristic of the elements indicated, and their magnitudes are proportional to the amounts of the elements present in the sample. The spectrum has been fitted by the computer program HEX to resolve the elements from a background component. (From PIXE International Corporation, Tallahassee, Florida.)

very intense radiation can be delivered to a small sample area. Moreover, the x-ray spectra (obtained by solid-state detectors in both methods) do not contain the high-energy incident radiation component found in XRF, thus improving the ability to resolve heavy elements at low concentrations in the sample. A typical x-ray spectrum is shown in Figure 3 with indications of 15 elements that have been resolved by computer data processing by use of the program HEX, based on an earlier program REX (Kaufmann et al., 1977). The full list of elements that are often determined above detection limits (a few tenths to a few nanograms) in aerosol samples is Mg, Al, Si, P, S, Cl, K, Ca, Ti, V, Cr, Mn, Fe, Ni, Cu, Zn, Ga, Ge, As, Se, Br, Rb, Sr, Zr, Pb, and Bi. In some samples additional elements may be resolved. The method is nondestructive, permitting analysis of a sample more than once for quality control.

3. REMOTE BACKGROUND AEROSOL SULFUR CONCENTRATIONS

3.1 Continental

Considerable amounts of anthropogenic gaseous sulfur dioxide and particulate sulfate occur in the atmosphere of North America, Europe, and Asia where populations are dense and there is industrial activity. The potential for long-range transport of this pollution over a thousand kilometers or more is well known (Ottar, 1980), and even the Arctic has been suspected of being contaminated as a result (Rahn and McCaffrey, 1980). The extent of the contamination may be determined by comparison with measurements at remote locations considered to be free of the transport of pollution into them. Since any Northern Hemisphere site may be impacted to some degree by long-range air pollution transport, measurements in the Southern Hemisphere are needed to provide this baseline atmospheric chemical information. For a baseline applicable to Northern Hemisphere temperate latitudes, measurements should be made at temperate latitude locations in the Southern Hemisphere that are remote from pollution sources. Using PIXE methods, which are especially suitable for such measurements (Winchester, 1981), we have made such measurements on the South American continent at several locations between Manaus, Brazil (3°S) and Punta Arenas, Chile (53°S) in winter and summer seasons (Lawson and Winchester, 1979). Altitudes ranged from sea level (Salvador, Brazil) to 5200 m (Chacaltaya Mountain, Bolivia); this range in altitude permits a meaningful average concentration to be estimated for background aerosol sulfur near the surface of the continent.

Figure 4 presents aerosol sulfur concentrations on the basis of impactor and streaker samples from South America. The impactor data indicate the presence of two particle size groups, or modes: fine particles seldom exceeding 1 μmad and averaging 0.25–0.5 μmad; coarse particles, generally

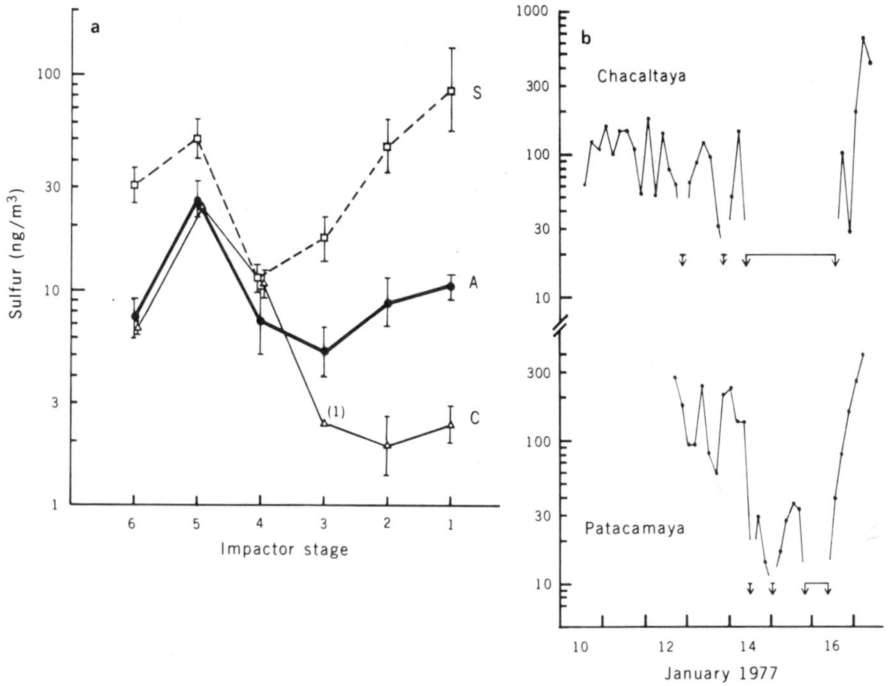

Figure 4. (*a*) Geometric mean, with its standard error, of aerosol sulfur concentrations (austral summer, 1977) for Salvador, Brazil (S, nine samples), Chacaltaya Mountain, Bolivia (C, two simultaneous samples), and all other sites (A, 34 samples from five locations—near Punta Arenas, Chile; Bariloche, Argentina; Patacamaya and Zongo, Bolivia; and Manaus, Brazil). Impactor stage designations are in sense opposite from that in Figure 1 and correspond to the aerodynamic diameter ranges: 6 (after filter), <0.25 μm; 5, 0.25–0.5 μm; 4, 0.5–1 μm; 3, 1–2 μm; 2, 2–4 μm; 1, >4 μm. The maximum at stage 5 (0.25–0.5 μm) indicates a fine mode that is distinct from a coarse mode of several micrometer particles in source location and process. (*b*) Aerosol sulfur concentrations in 4-hour time steps of streaker samplers at two remote Bolivian sites: on Chacaltaya Mountain (altitude 5200 m) and on the Altiplano near Patacamaya (altitude 3800 m). (From Lawson and Winchester, 1979; copyright 1979, American Association for the Advancement of Science.)

exceeding 1 μmad. We understand the coarse-particle sulfur to be derived from the dispersion of condensed material, mainly sea spray but also soil dust. At Salvador, near the seacoast, the coarse-particle sulfur concentrations are high, whereas at Chacaltaya, far from the sea, they are low. We understand the fine-particle sulfur to be derived from the oxidation of trace sulfur gases, SO_2 and/or its reduced sulfur precursors. At all the remote locations in South America about the same concentrations of fine-mode aerosol sulfur are found, averaging about 50 ng m^{-3}. Because this value is reasonably independent of latitude or altitude over the continent, the sources of the background aerosol sulfur must be widespread and may ultimately be either continental or maritime. However, aerosol sulfur concentrations in fine-mode particles sampled from the Southern Ocean south of Australia may be much lower, indicating that the composition of near polar air cannot be considered representative of temperate latitude air (Andreae, 1982). Instead, we may provisionally accept an aerosol sulfur concentration of about 50 ng m^{-3} as a natural level for comparison with Northern hemisphere temperate latitudes.

The streaker data shown in Figure 4 illustrate that the aerosol sulfur concentration may be variable in time periods of a few hours from less than 10 to more than 100 ng m^{-3}. This variability may be due to a number of factors, including physical mixing or transport and chemical formation of aerosol particles from gaseous sulfur. The sudden decrease to low concentrations on January 14 observed at both sites in Bolivia is indicative of factors operating on a large scale, such as incursion of "cleaner" air. Thus when we speak of an average clean air background aerosol sulfur concentration of 50 ng m^{-3}, we should realize that this is the average of concentrations that may vary severalfold in either direction, although the average may not be strongly dependent on specific location within the latitude range investigated here.

3.2. Marine

Aerosol composition in the remote marine atmosphere has been determined at Samoa (14°S) in the southeastern trade winds by sampling with cascade

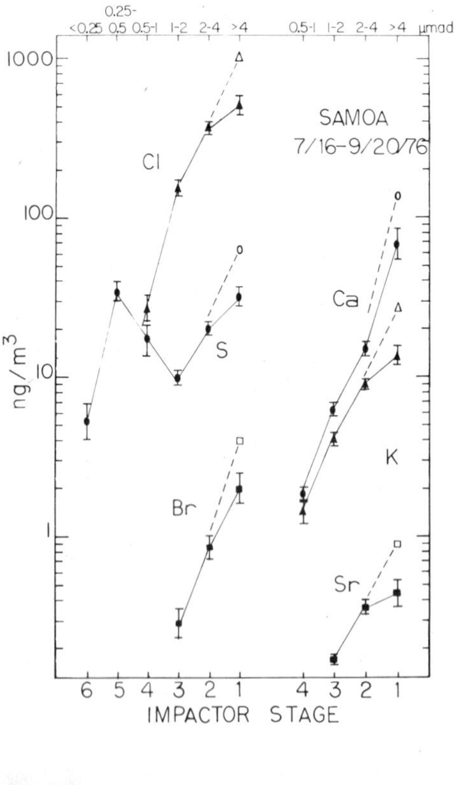

Figure 5. Geometric mean, with its standard error, of element concentrations in 17 cascade impactor sample pairs, plotted against impactor stage as in Figure 4, collected in Samoa. Unfilled points represent concentrations after correction for sampling bias due to wind speed. (From Maenhaut et al., 1981; copyright 1981, American Geophysical Union.)

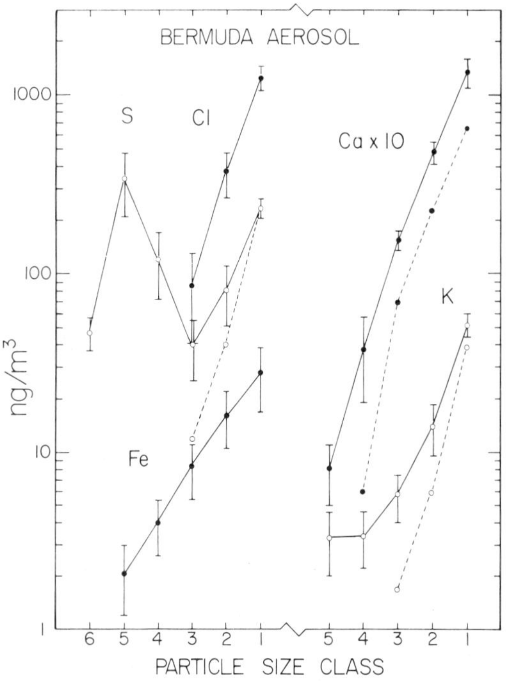

Figure 6. Geometric mean, with its standard error, of element concentrations in 10 cascade impactor samples, plotted against impactor stage as in Figure 4, collected in Bermuda. Dashed lines represent trends for components of sulfur, calcium, and potassium due to sea spray. Iron and part of calcium and potassium are considered to be of terrestrial origin, chlorine is of sea spray origin, and most of fine sulfur is from conversion of air pollution gaseous SO_2 in the atmosphere. (From Berg and Winchester, 1978; after Meinert and Winchester, 1977; copyright 1978, Academic Press Inc., London Ltd.)

impactors mounted on a tower 50 m above sea level (Maenhaut et al., 1981). The distributions with particle size of the concentrations of some of the elements determined by PIXE are shown in Figure 5. Three aerosol groups can be identified on the basis of these distributions: fine-particle sulfur centered in the 0.25–0.5 μmad range; coarse-particle sea spray contained in particles greater than about 1 μmad, with much of the mass consisting of particles larger than 4 μmad; and an additional very coarse particle component consisting of much of the calcium and significant concentrations of several heavy metals in the size fraction greater than 4 μmad, attributed to substances concentrated at the sea–air interface and rendered airborne by the action of wind on water. Processes for marine aerosol generation have been reviewed elsewhere (Berg and Winchester, 1978), including jet and film drop production by bursting air bubbles at the sea surface. For our present purposes, it is important to note that the geometric mean fine sulfur concentration at Samoa was 63 ng m^{-3}, in approximate agreement with the South American result of about 50 ng m^{-3} of aerosol sulfur in the fine mode. This agreement suggests that the temperate latitude concentrations may be maintained by the gradual oxidation of trace sulfur gases with similar concentrations over land and sea areas, rather than maintained by localized processes. Without answering whether the ultimate sources of the trace sulfur gases are terrestrial or oceanic, we use 50 ng m^{-3} as an expected aerosol sulfur concentration near the surface in clean temperate–latitude regions.

At Bermuda, about 1000 km east of Cape Hatteras, North Carolina, the marine aerosol cannot be considered to be free of pollution. Tower-mounted impactors upwind of the island during the autumn indicated the presence of three aerosol components in the size distributions shown in Figure 6 (Berg and Winchester, 1978; Meinert and Winchester, 1977): coarse-particle sea spray containing chlorine, most of the coarse sulfur, and much of the coarse calcium and potassium; a dust component attributed to terrestrial sources containing iron and part of the calcium and potassium; and fine-particle sulfur centered at 0.25–0.5 μmad. Fine aerosol sulfur concentrations average about 500 ng m^{-3}, a factor of 10 greater than found in Samoa or South America. Most of this must be attributed to air pollution transport from the continent.

4. AEROSOL SULFUR, WATER VAPOR, AND TEMPERATURE RELATIONSHIPS

4.1. Evidence from Northern Indiana

Eastern U.S. aerosol sulfur concentrations are often several micrograms per cubic meter, two orders of magnitude greater than concentrations in the Southern Hemisphere, with higher concentrations during summer conditions, whereas gaseous SO_2 concentrations may be higher during winter (Mueller et al., 1980). The chemical form of aerosol sulfur is a mixture of sulfuric acid and ammonium sulfate (Tanner et al., 1981) in particles that may be either solid or liquid depending on relative humidity and the composition of the mixture (Tang et al., 1978). Models for photochemical and other mechanisms for formation of sulfuric acid from sulfur dioxide predict faster rates under summer conditions (e.g., Altshuller, 1979; Middleton et al., 1980).

We have attempted to determine more precisely the dependence of aerosol sulfur concentrations on water vapor partial pressure and on temperature, two atmospheric variables that exhibit seasonal variability and may affect the formation rate of sulfuric acid aerosol. In our approach we compare the short-term variability of each—2-hourly for sulfur by streaker sampler and 3-hourly for temperature and water vapor pressure from data of the U.S. National Weather Service. The overall variability in aerosol sulfur concentration, which may be a factor of 10 or more in a few hours, is undoubtedly due to the combined effects of several physical and chemical factors. However, a correlation between sulfur and either water vapor or temperature would nonetheless imply that the latter variables cause variability in the aerosol formation rate, since parallel variation in the three variables by purely physical or dynamic processes would be an unlikely coincidence.

We have examined 2-hourly concentration measurements for October 1–31, 1977, at the nine Class I sites of the EPRI-SURE network (Mueller et al., 1980) and have found a degree of correspondence in time of high sulfur with high water vapor concentration at all nine of the sites. In contrast, no correspondence with relative humidity was observed. At most of the sites only some of the high sulfur concentrations matched high water vapor concentration, and no overall correlation between sulfur and water concentration was found, so that only a part of the variability in sulfur could be linked to water. However, at Fort Wayne, Indiana (SURE site 7) the time series correspondence was visibly strong and an overall correlation was found. A correlation with temperature was also found. In this section are summarized the results presented in greater detail elsewhere (Winchester and Leslie, 1982).

During the first half of October 1977 six cold fronts passed over the sampling site, marked by dew point maxima but generally without rainfall. On all six occasions short-term maxima in aerosol sulfur concentration were observed, in addition to other sulfur maxima not associated with frontal passage. The pattern of aerosol sulfur concentration variability in time was similar to that of water vapor but unlike that of relative humidity. Some similarity between aerosol sulfur and gaseous sulfur dioxide was observed, although most SO_2 measurements at the rooftop site were below the detection limit of 1 ppb, and much of the SO_2 variability may have been caused by removal at the ground, thus obscuring any correspondence to aerosol sulfur.

The overall correlation between aerosol sulfur and water vapor concentrations is illustrated in Figure 7, a scatter diagram of 2-hourly data. Three features stand out: a trend of correlation is visibly apparent for both night and day subsets of data; their slopes are steep and approximately the same; and the ranges of concentrations of both S and H_2O are approximately the same, with slightly higher means during the day. By inspection, the points fall along a trend line given approximately by $S \propto (H_2O)^3$.

The degree of correlation between aerosol sulfur concentration and water vapor partial pressure, as well as temperature and relative humidity, has been examined thoroughly by computations based on hourly values derived from the 2-hour sulfur measurements and on hourly values of temperature and dew point given by original observer records at the National Weather Service station nearest the aerosol sampling site (10 km distant). The values of R^2 for correlations in regression computations for several combinations of variables are given in Table 1. Since the logarithm of the vapor pressure varies nearly linearly with reciprocal absolute temperature and, by an Arrhenius

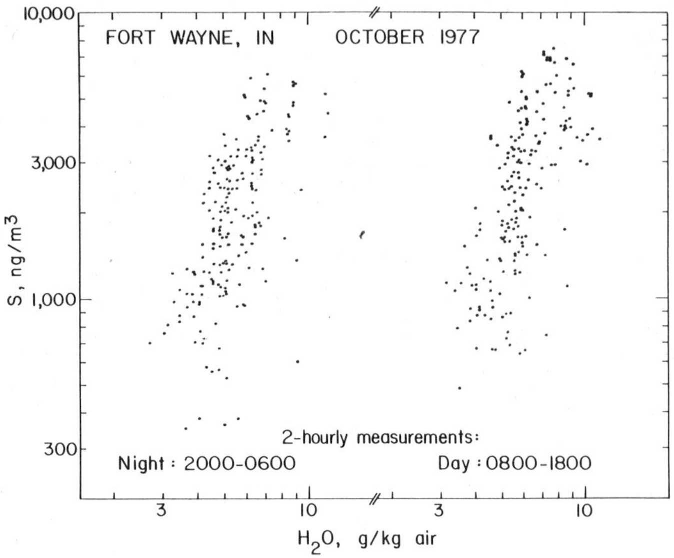

Figure 7. Aerosol sulfur concentrations and specific humidities at EPRI–SURE site 7 for October 1–31, 1977, plotted logarithmically for nighttime and daytime separately. Regression calculations for each 182 data point subset indicate correlation coefficients $R > .6$ ($P \ll .001$) and an approximate relationship $S \propto (H_2O)^3$. (From Winchester and Leslie, 1982; copyright 1982, American Geophysical Union.)

rate equation, the logarithm of the rate of aerosol sulfur formation can be expected to vary linearly with reciprocal absolute temperature, the regressions were performed out of convenience between $\log_{10}$ S, $1/T_{dp}$, $1/T$, and $1/T_{dp} - 1/T$ as measures of aerosol sulfur concentration, water vapor pressure, ambient temperature, and relative humidity, respectively, indicated by S, W, T, and RH in Table 1.

Highly significant correlations are found between sulfur and water vapor and between sulfur and temperature, as well as between water vapor and temperature, but not between relative humidity and sulfur, water vapor, or temperature. The degree of correlation of sulfur with water vapor is somewhat stronger than with temperature, suggesting that water and temperature act independently on sulfur instead of water and sulfur responding separately to temperature. A chemical formation rate of aerosol sulfur from SO_2 by a mechanism that depends on water activity as well as temperature could account for these observations. Higher steady-state concentrations of aerosol sulfur would result from faster formation rates. We thus are led to interpret the variability in the concentration of aerosol sulfur as due in part to variability in its formation rate, a rate that is dependent on water vapor concentration and temperature. The water vapor pressure is a measure of water activity in solution droplets at equilibrium.

The functional dependence of sulfur on water vapor concentration and

Table 1. R^2 Values for Fort Wayne, Indiana, October 1–31, 1977[a]

Set	Range	n	S–W	S–T	S–W,T	S–RH	W–T	W–RH	T–RH
All	01–24 h	743	0.410	0.342	0.434	0.003	0.563	0.018	0.303
Day	09–20 h	372	0.424	0.319	0.461	0.007	0.404	0.178	0.180
Night	21–08 h	371	0.383	0.456	0.454	0.036	0.854	0.011	0.072
Warm	$t > 50°F$	354	0.242	0.124	0.265	0.064	0.185	0.543	0.185
Cool	$t \leq 50°F$	389	0.297	0.204	0.299	0.012	0.584	0.099	0.133

[a] Correlations among hourly aerosol sulfur, water vapor, temperature, and relative humidity. The symbols S, W, T, and RH represent $\log_{10}S$, $1/T_{dp}$, $1/T$, and $1/T_{dp}-1/T$, respectively, as measures of aerosol sulfur concentration, water vapor pressure in terms of dew point temperature, ambient temperature, and ambient relative humidity. In each case the first symbol represents the dependent variable and the second symbol the independent variable in the regression. The column S–W,T represents regressions with both W and T as independent variables. Values of $R^2 > 0.10$ imply probability of no correlation $P < 0.001$ for $n = 100$ independent observations, whereas $R^2 < 0.03$ implies $P > 0.1$; 743 hourly data points were used in the regressions.

on temperature can be estimated either from the regression or by graphical means. Since regressions require one of the variables to be considered as independent, a graphical procedure is preferred in cases such as this where there is random error in both variables. We have found that the distributions of measured aerosol sulfur, water vapor, and SO_2 concentrations were approximately lognormal, except for extreme values, as shown in Figure 8. Reciprocal absolute temperature was also normally distributed. Probability plots permit precise determination of both the median values and the geometric standard deviations. For those variables for which a correlation has been shown to be significant, such as sulfur and water or sulfur and temperature, the functional dependence is obtained by the ratio of the widths of the distributions of the variables. By this procedure we find that aerosol sulfur concentration $\propto (P_{H_2O})^{3.08}$ for its dependence on water vapor pressure. For dependence on temperature, we may express this in terms of the corresponding saturation vapor pressure, i.e., aerosol sulfur concentration $\propto (P_{H_2O}^{saturation})^{1.213}$. Both exponents are best-fit values and are subject to uncertainties in the first decimal place. The water vapor dependence is in agreement with the less precise value estimated from Figure 7. The temperature dependence represents a doubling for a temperature rise of about 9°C, a value close to that found for many chemical reactions in solution.

The functional dependence found would predict higher aerosol sulfur concentrations during summer months when temperatures and absolute humidities are highest and are thus qualitatively in agreement with the observations already mentioned. We interpret this to imply a more rapid formation of sulfate or sulfuric acid aerosol from SO_2 by a process that depends on temperature and water vapor, such as by oxidation in solution controlled by water activity and temperature. The empirical relationships found here may be general although demonstrable only in data sets where variability in aerosol sulfur concentration is due predominantly to the variability in chemical reaction rate.

4.2. Some Properties of Sulfuric Acid

A variation in chemical reaction rate with variation in water vapor pressure, in atmospheres where water is quite abundant, suggests that the apparent overall rate is governed by molecular equilibria in solution. Although this is certainly speculative at the present time, it is useful to recall how water vapor pressure is related to the amount of water in sulfuric acid solution droplets that are in equilibrium with the atmosphere. Figure 9 presents information (International Critical Tables, 1928) that has been known for many years, specifically, that sulfuric acid is a good drying agent, even when it contains a substantial amount of water. Water is strongly bound to H_2SO_4 and may be unavailable to other dissolved substances, such as SO_2, which may also enter into hydration and ionization equilibria. For instance, if a

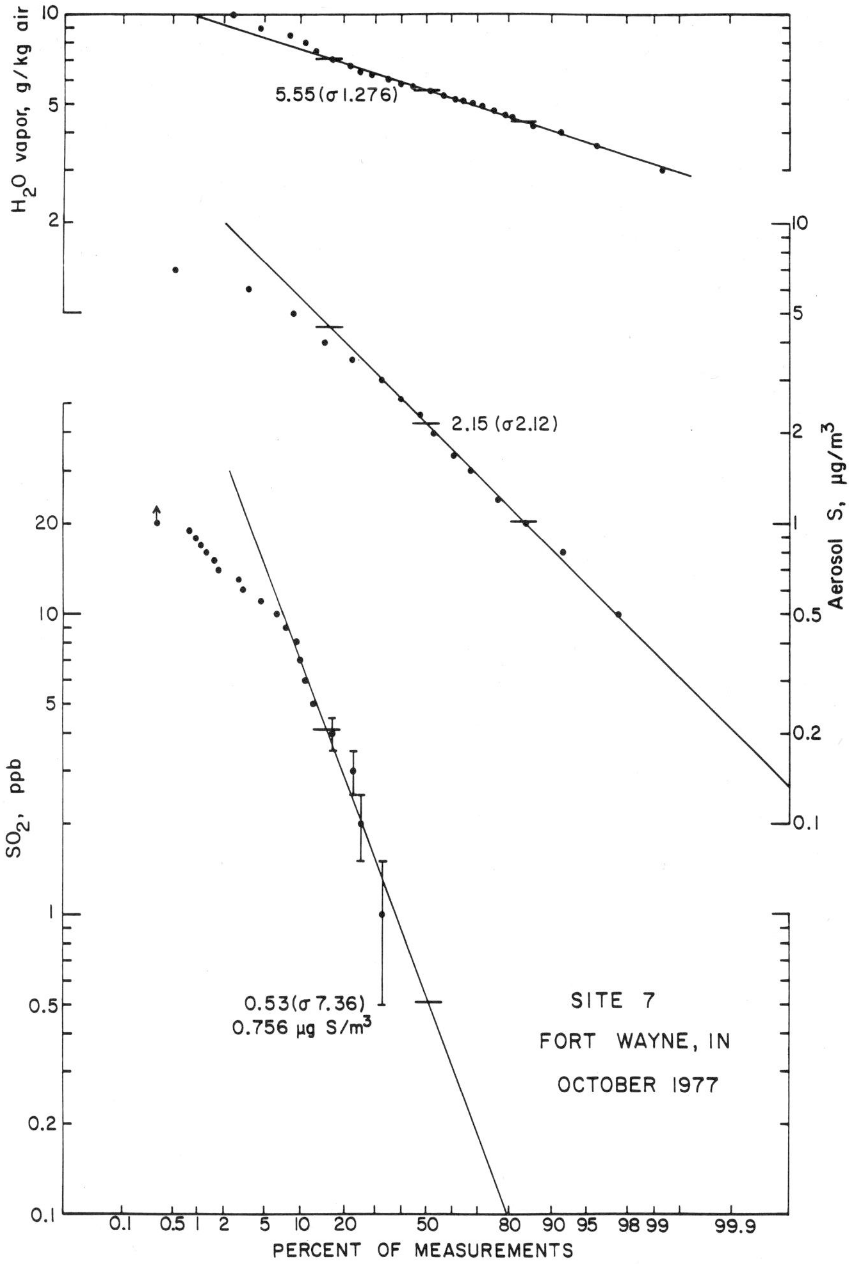

Figure 8. Percentage of measurements in excess of plotted values of 364 2-hourly data for gaseous SO_2, aerosol S, and H_2O vapor concentrations at Fort Wayne site 7, October 1–31, 1977. Lognormality over most of the ranges is indicated by linearity; 50% concentrations and geometric standard deviations (16–84 percentile ranges) indicate a width of the aerosol sulfur distribution intermediate between those of SO_2 and H_2O. Since only integer values of ppb SO_2 (0, 1, 2, . . .) were reported, the assumed ranges of validity at the lowest four values are indicated for comparison with the lognormal trend. Temperatures were found to be normally distributed with a mean and $\pm s$ range of 51 ± 8°F (10.56 ± 4.4°C). (From Winchester and Leslie, 1982; copyright 1982, American Geophysical Union.)

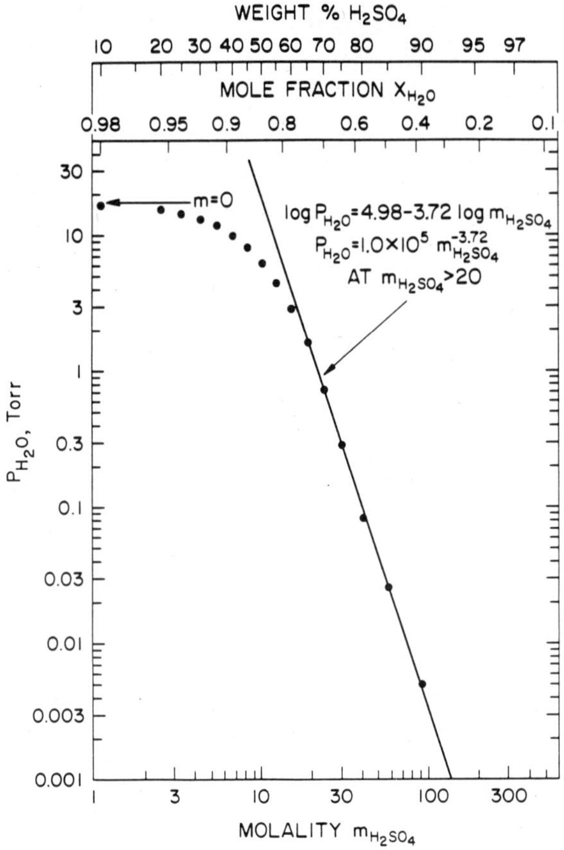

Figure 9. Partial pressures of water vapor in equilibrium with aqueous sulfuric acid solutions at 20°C. The term P_{H_2O} is a measure of the chemical potential or activity of dissolved water for any equilibrium process in solution. For $m_{H_2SO_4} > 20$, P_{H_2O} decreases as the 3.72 power of increasing molality. A value of $P_{H_2O} = 3$ mm Hg corresponds to 2.5 g H_2O per kilogram of air at atmospheric pressure. (From Winchester and Leslie, 1982; after *International Critical Tables*, 1928; copyright 1982, American Geophysical Union.)

dissolved species of S(IV), such as SO_3^{2-}, requires three water molecules in its formation from SO_2, its relative concentration may vary approximately as the third power of water activity. If the inherent rate of oxidation of such a species is sufficiently rapid, an overall rate of oxidation of dissolved S(IV) may appear to vary as water activity to the third power. Although we are not yet able to show that such a mechanism is in fact operative, we consider that the finding of a strong dependence between aerosol sulfur concentration and water vapor pressure, but not relative humidity, may eventually be explained by a process such as this. The finding of a temperature dependence typical of reactions in solution reinforces this expectation.

5. CLIMATIC AND AEROSOL ENVIRONMENT OF THE SOUTHEASTERN UNITED STATES

5.1. Temperature and Water Vapor Distribution Patterns

The finding of a correlation in the eastern United States between aerosol sulfur concentration and water vapor pressure and temperature, on the supposition that this implies a dependence of aerosol sulfur formation rate on water vapor and temperature, leads one to attempt to predict expected steady-state aerosol concentrations in other regions where few measurements have been reported. The southeastern region along the Atlantic coast is especially interesting for such a prediction since it is a region of marked climatological gradients and presumably atmospheric chemical gradients as well. By means of known temperature and humidity profiles with latitude and a reasonable physical model for the distribution of primary air pollution SO_2 in the region, it should be possible to estimate the profile of expected sulfuric acid or sulfate concentration with latitude for different seasons and as an annual average. We need to examine first the known distribution of temperature and water vapor in the region.

The distribution of mean maximum temperature for July in four states, North Carolina, South Carolina, Georgia, and Florida, is presented in Figure 10, constructed from data taken from *Climatography of the United States*, No. 60 (1959; 1976). A significant gradient of rising temperatures from the seacoast inland is due to the somewhat cooler temperatures of the Gulf Stream (Stommel, 1966). However, no gradient from north to south is found, either over land or in the waters of the Gulf Stream.

Winter conditions are summarized in Figure 11 on the basis of data also from *Climatography of the United States*, No. 60 (1976). A marked gradient of colder temperatures in the north to warmer conditions in the south is indicated by the steadily decreasing number of January heating degree days at seven airport weather stations. Although the Gulf Stream moderates the coastal climate because of its somewhat warmer temperature during winter, the Gulf Stream does not exhibit a strong north–south temperature gradient. Therefore, the terrestrial temperature gradient must represent a steadily decreasing average mixing ratio from north to south of cold continental air to warmer maritime air. There is reason to expect that similar trends in this mixing ratio should be found in summer as well as winter, although only in winter does temperature serve as a convenient index because of the contrast between cold continental air and much warmer marine air during winter in this region.

Figure 11 also presents the expected water vapor pressure trends with latitude for January, increasing by a factor of about 2 from north to south with slightly higher values in the afternoon than at dawn. Consequently, for the winter season we find that both temperature and water vapor content

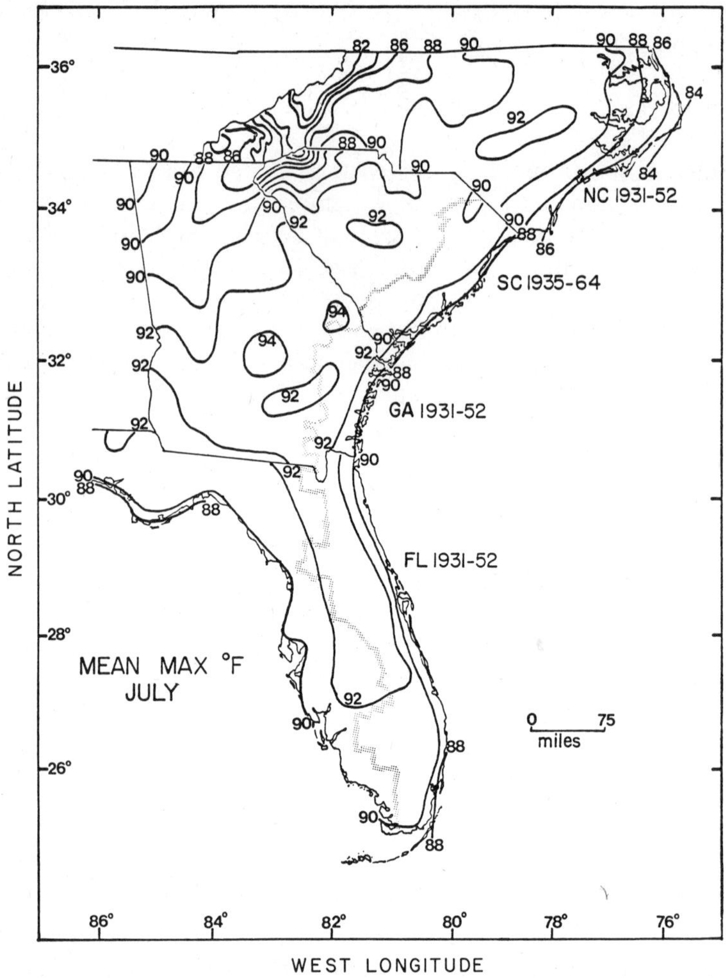

Figure 10. Climatic gradients in the Atlantic coastal zone of southeastern United States are illustrated by mean July maximum temperature contours reported for each of the states. Stippled outline is boundary of coastal and next adjacent inland counties for which epidemiologic data are available. During summer the temperature gradients are strongest from the coast toward inland; in winter they are strongest from north to south along the coast. (From Winchester et al., 1981; after *Climatography of the United States,* No. 60, 1959, 1976.)

of the atmosphere increase as one moves south into peninsular Florida. For the summer season a more uniform distribution of these climatic parameters with latitude is found. If they affect the formation rate of aerosol sulfur, the rate should increase with decreasing latitude during winter but remain uniform (and high) during summer. The average aerosol sulfur concentration distribution should be a combination of these gradients and the distribution of primary SO_2 with latitude.

5.2. Aerosol Sulfur Concentrations and Rain Acidity in Florida

High-quality aerosol sulfur concentration measurements have been made at 10 sites along the length of Florida during summer and late fall seasons (Ahlberg et al., 1978; Leslie et al., 1978). Approximately six cascade impactor samplings of 24 h each were made in each season at near-surface sites in urban and nonurban locations. Nearly all the sulfur was found in a fine mode, mostly in the two size fractions 0.25 to 0.5 and 0.5 to 1 μmad, reflecting its probable formation by the oxidation of gaseous SO_2. Concen-

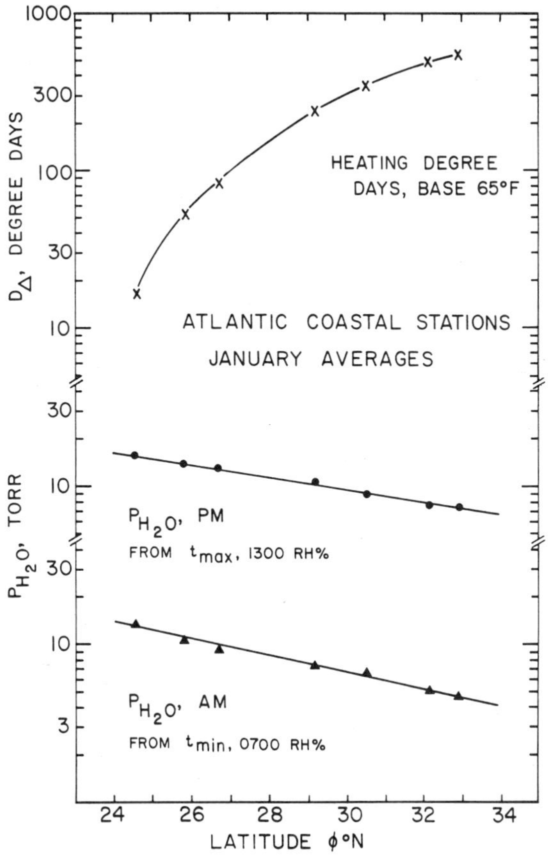

Figure 11. Heating degree days D_Δ and predawn and afternoon values of water vapor pressure P_{H_2O} for January plotted against latitude of seven reporting meteorological stations along the Atlantic coast. Most of the annual D_Δ is due to cool weather of December, January, and February, with an increasing trend with increasing latitude. For both predawn and afternoon hours P_{H_2O} decreases approximately exponentially with increasing latitude, in accordance with conventional Continentality and Oceanicity indexes, on the basis of seasonal temperature changes, and with the higher moisture contents generally found over the ocean. Afternoon P_{H_2O} is somewhat higher than predawn P_{H_2O}. (From Winchester et al., 1981; after *Climatography of the United States*, No. 60, 1976.)

trations of several hundred to more than 1000 ng m^{-3} indicate the probable air pollution origin of the aerosol sulfur. The trends—of the concentrations in six different particle size fractions for the two seasons with location in Florida—are shown in Figure 12. Most of the fine aerosol sulfur, impactor stages 4 and 5, exhibits a nearly monotonic decrease with location from Pensacola in the northwest of Florida to Miami in the southeast. The decrease is a factor of almost 10 in the summer and about 2 in the late fall samples, most clearly seen for stage 4 (0.5–1 μmad). We understand the summer trend to be due mainly to a similar trend of primary air pollution SO_2 concentration, which decreases with decreasing latitude. This decrease reflects the decreasing impact of polluted air masses from large sources located north of Florida and the corresponding increase in maritime composition of the atmosphere, as one moves south into peninsular Florida. The

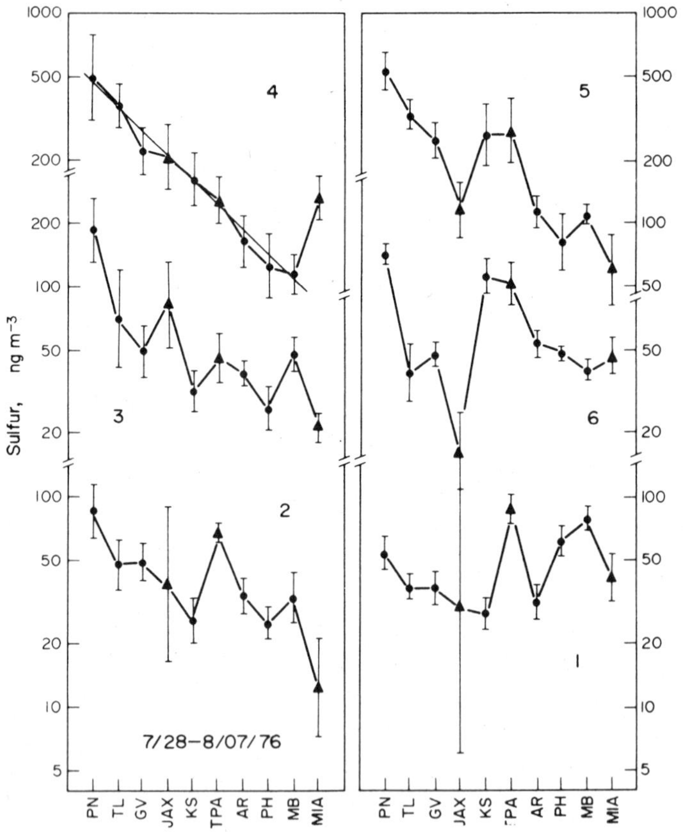

Figure 12. Aerosol sulfur concentrations during summer and late fall seasons, as geometric means with standard errors of approximately six samples taken in each season. Concentrations in each particle size fraction are given separately, with impactor stage designations as in Figure 4. Points are plotted against a sampling site index along the length of Florida from Pensacola (PN) to Miami (MIA). (From Leslie et al., 1978; copyright 1978, Pergamon Press, Ltd.)

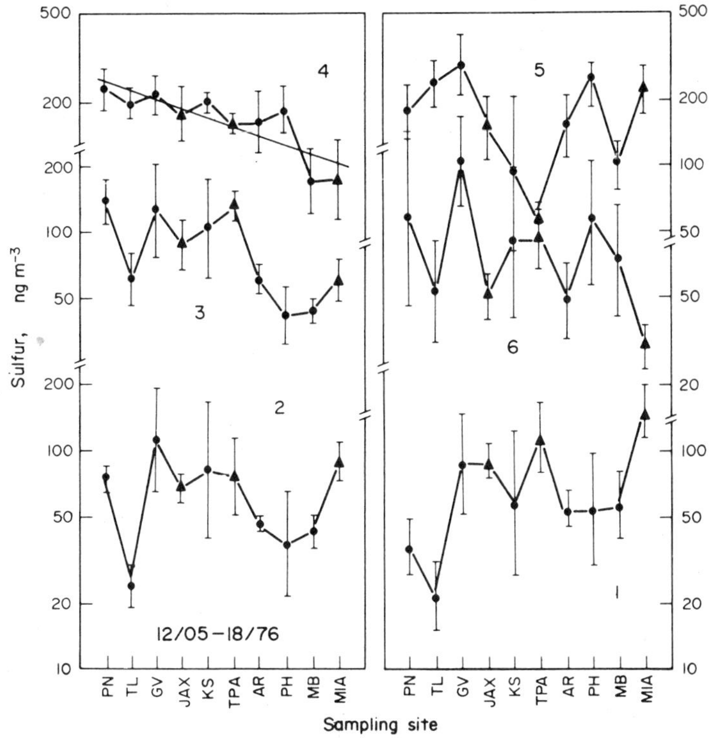

Figure 12. (*Continued*)

winter trend is less steep and may be considered to reflect the compensating effect of a rising rate of aerosol sulfur formation in competition with the decrease in primary SO_2 concentration from north to south.

The aerosol concentration profiles in Florida suggest that most fine-mode sulfur is the result of long-range transport into the state, rather than the result of intrastate air pollution emissions of SO_2. Although Florida sources may be responsible for a portion that is of great air quality importance, e.g., the finest particle sizes or particles that are unusually acidic, the effect of Florida contributions cannot readily be detected by measurement of total aerosol sulfur concentrations without particle size fractionation. Under certain circumstances, however, a local contribution can be discerned by careful evaluation of meteorological as well as chemical evidence (Young and Winchester, 1980).

The acidity of rain, on the other hand, may be affected to a greater degree by local SO_2 emissions than is the case for aerosol sulfur concentrations. Evidence for this viewpoint is presented in Figure 13, which shows relationships between the concentration of dissolved sulfur in rainwater measured by PIXE and rainwater pH measured directly after sample collection. Every rain event in Tallahassee, Florida, between November 7, 1978 and January

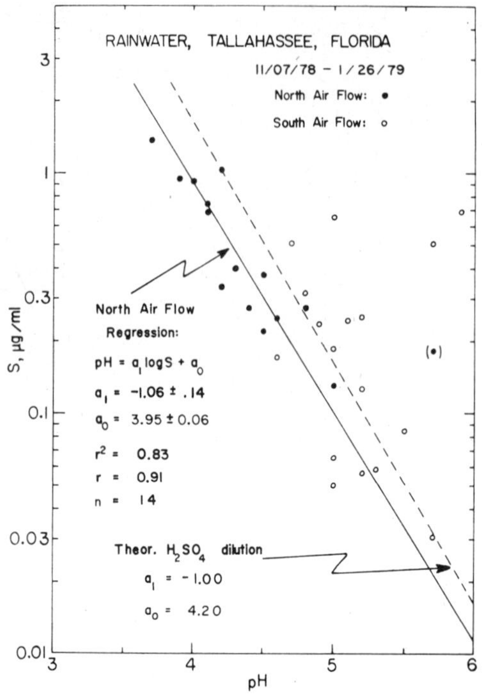

Figure 13. Relationship between pH and dissolved sulfur concentration in all rain events in Tallahassee, FL, from November 7, 1978 to January 28, 1979. Data have been divided into two subsets according to synoptic conditions of the rain events, mainly north airflow (filled points) and mainly south airflow (open points). The dashed line represents the trend expected by dilution of sulfuric acid with water. (From Tanaka et al., 1980a; copyright 1980, American Geophysical Union.)

26, 1979 was sampled for one to a few hours each (Tanaka et al., 1980a). The rain events were divided into two subsets on the basis of local meteorological observations: one that was associated with airflow from the north and expected long-range air pollution transport; and one that was associated with marine airflow from the Gulf of Mexico, expected to bring sea salt into the area. The marine rainfall rate was observed to be much more intense than the rate of north rainfall, a difference that should affect the degree of dilution of dissolved constituents in the rain.

The data in Figure 13 fall into two distinct groups: a north airflow set, where 14 of the 15 points show a strong correlation between the logarithm of the dissolved sulfur concentration and the pH; and a south airflow set, where no such correlation is found. A trend line calculated by regression for the 14 north airflow points shows a slope and position (slightly to the acid side) of the trend expected if all the dissolved sulfur in the rain were sulfuric acid that, through ionization, produced the acidity measured as pH. The lack of a significant correlation for the 17 south airflow points is attributed to marine sulfate, which does not impart acidity to the rainwater. If, in the case of north airflow rains, the acidity were due mainly to precipitation scavenging of fine aerosol sulfur—which, although less efficient than coarse particle scavenging, may remove half the sulfur from the air (Tanaka et al., 1980b)—we must conclude that the aerosol sulfur was primarily in the form of sulfuric acid. Such a conclusion is unwarranted in view of the general evidence that sulfuric acid, although present, probably does not regularly exceed ammonium sulfate in concentration (Ahlberg and Winchester, 1978; Tanner et al., 1981). Consequently, we are led to believe that rainwater acidity may be due more to absorption of gaseous SO_2 into liquid water during the rain event and its oxidation to sulfuric acid in aqueous solution. In atmospheres where gaseous SO_2 concentrations exceed those of aerosol sulfur, the former, which could be due mainly to local sources, may dominate in the formation of acidity in rain and give rise to the observed trend.

6. EXPECTED TRENDS OF SOUTHEASTERN U.S. SULFATE CONCENTRATIONS

6.1. A Simple Physical and Chemical Model

We now attempt to combine the available evidence to estimate the profile of expected concentration of aerosol sulfur, present as sulfuric acid and sulfate, with latitude in the Atlantic coastal region of the southeastern United States. This treatment has been fully described elsewhere (Winchester et al., 1981) and is only briefly summarized here. In essence, the estimate is obtained by multiplying the profile of primary air pollution SO_2 with the profile of the rate of conversion to aerosol sulfur. This is done by use of meteorological data for each month of the year and then summing over the months to obtain an annual average expected profile of aerosol sulfur concentration, assuming that a steady state exists in the horizontal transport of aerosol at any location so that concentration is proportional to formation rate.

For the latitude profile of primary SO_2 at any time of the year, we use the winter season heating degree day index (Figure 11). A simple two-component mixing model along the Atlantic coast is outlined in Figure 14. The model visualizes a relatively uniform continental air mass, characterized by its average temperature and primary pollution SO_2 concentration, which mixes in the coastal zone with a relatively uniform marine air mass, characterized by its average temperature and a negligibly small pollution SO_2 concentration. Mixing may occur in any season of the year, but in winter the temperature contrast between continental and marine air may be used as an index of how the mixing ratio may vary with latitude. Fortuitously, the mixing ratio and SO_2 concentration profiles are given exactly by the

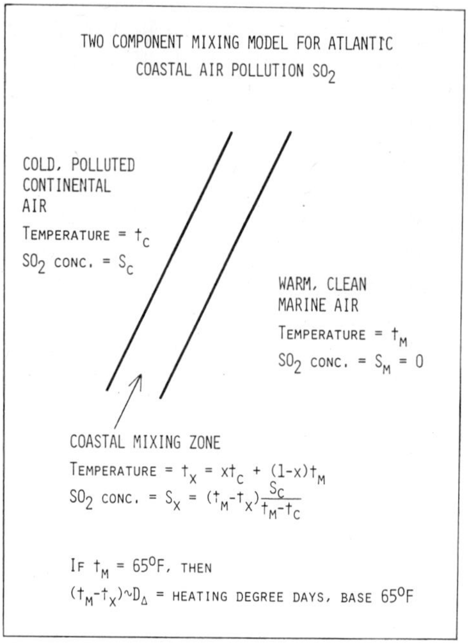

Figure 14. Schematic representation of wintertime mixing of cold and polluted continental air with clean marine air in the region of the Gulf Stream. The temperature contrast provides a convenient measure of mixing ratio between air masses for predicting primary pollution SO_2 concentrations in the coastal zone. (From Winchester et al., 1981.)

heating degree-day index, calculated relative to 65°F, a temperature close to that of the Gulf Stream during winter in the Atlantic coastal region. In other words, as we move south along the Atlantic coast, we expect the concentration of primary pollution SO_2 to decrease in proportion to the heating degree-day index, as a simple consequence of mixing polluted upwind air with much cleaner marine air. We assume the SO_2 profile with latitude to be similar at all times of the year, although we can estimate it only for the winter when temperature contrasts exist.

Next, we must estimate quantitatively how the expected aerosol sulfur formation rate should vary with latitude in any month of the year. The rate constant for the formation is taken to be proportional to the product of water vapor pressure and saturation vapor pressure (a measure of ambient temperature), each raised to an appropriate power. We use the values found for northern Indiana in October 1977, i.e., rate constant $k_S \propto (P_{H_2O})^{3.08}(P_{H_2O}^{saturation})^{1.213}$. We plot the monthly values of k_S evaluated in this way for each of six meteorological stations in the Atlantic coastal region shown in Figure 15. In the winter the expected aerosol sulfur formation rates show a strong north–south gradient, being about 10 times faster in Miami than in Charleston. In the summer the rates are similar throughout the region but 10–100 times faster than the winter rates.

A third parameter needed to complete the analysis according to this simple physical and chemical model is an estimate of the seasonal variability of primary pollution SO_2 concentrations in the region. We have assumed that the profile with latitude is similar in summer and winter (given by heating degree-day index), but we must assume that the concentration itself may exhibit seasonal variability. For example, in winter the penetration of cold northern air into the southeast region may raise primary pollution levels over those typical of summer conditions, when the position of the polar front is further north. We may make a rough estimate of this effect by using the aerosol sulfur measurements in Florida by Leslie et al. (1978). If the summer measurements for those sites that were near the Atlantic coast are divided by the relatively invariant expected rate constants k_S, a profile of expected primary pollution SO_2 is obtained. It is close to that of the heating degree-day index, consistent with our assumptions in the simple model. If the winter measurements are divided by the strongly latitude dependent k_S values, the winter expected profile of primary pollution SO_2 is obtained. Again, it is close to that of the heating degree-day index; this result adds a measure of confidence in the simple model. However, the concentrations of SO_2 differ more than tenfold between the two seasons. In the absence of more detailed understanding, we assume a sinusoidal variation in primary SO_2 concentra-

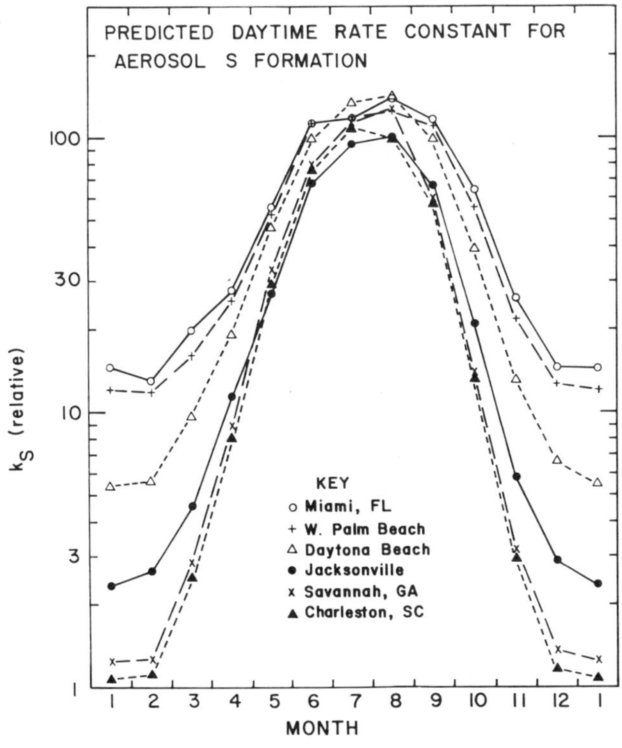

Figure 15. Expected relative values of a rate constant k_S for the conversion of gaseous SO_2 to sulfuric acid aerosol along the southeastern U.S. Atlantic coast, on the basis of temperature and water vapor measurements. (From Winchester et al., 1981.)

tion, calibrate it by means of these measurements, and estimate the appropriate relative SO_2 concentrations for each month of the year, higher in winter than in summer.

6.2. Predicted Trends

The trends with latitude of expected relative aerosol sulfur concentrations are given by the product of the three terms of our simple model: the expected conversion rate constant k_S; the expected latitudinal dependence of SO_2 given by heating degree days; and the expected temporal dependence of SO_2 given by calibration with Florida measurements of aerosol sulfur concentration. The results for each month of the year are given in Figure 16. Two features of these trends stand out. First, during winter the concentrations are higher in the south than in the north, indicating that the more rapid aerosol sulfur formation rate in the south more than offsets the expected

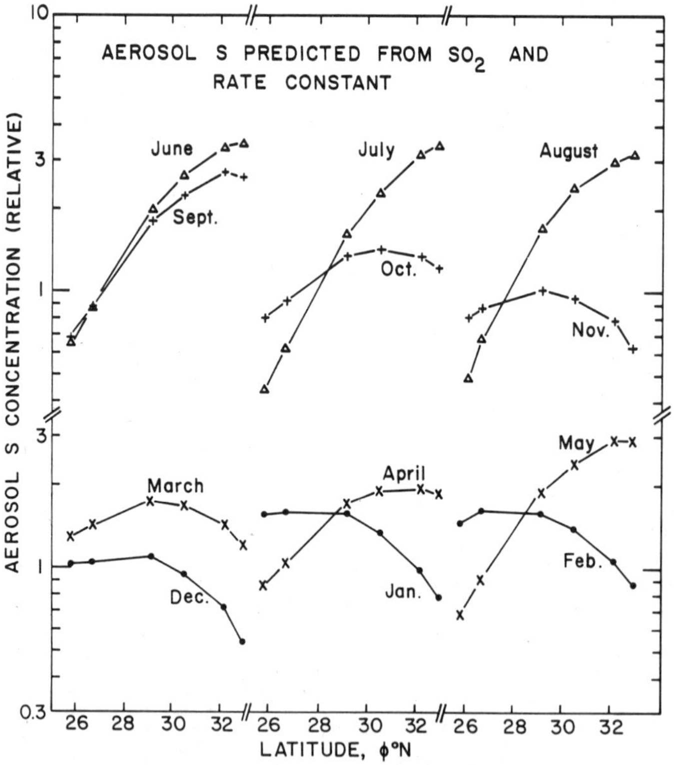

Figure 16. Sulfuric acid aerosol concentration, in relative units, expected for each month as a function of latitude, obtained by multiplying the conversion rate constant k_S, the heating degree day value D_Δ, and a seasonally varying SO_2 concentration factor. These curves summed over the 12 months represent the annual average aerosol sulfur concentration profile with latitude, shown in Figure 17. (From Winchester et al., 1981.)

lower primary pollution SO_2 concentration. During summer the opposite is found; viz., the concentrations are higher in the north because of higher SO_2 concentrations and no great change in aerosol sulfur formation rate with latitude. Second, during summer the concentrations tend to be substantially higher than during winter, except in the extreme south, a reflection of the generally warmer and more humid conditions characteristic of the southeast region during the summer than in winter.

The profile of the average annual expected relative aerosol sulfur concentration with latitude is obtained by summing over the monthly profiles. Such an average may be useful in estimating long-term exposures to aerosol sulfur, such as in human exposure studies. This, of course, would represent a trend on a regional geographic scale, smoothed over local perturbations by urban and industrial centers, but is nonetheless potentially useful for evaluation of other trends in nonurban areas. This average annual trend in relative aerosol sulfur concentration is illustrated in Figure 17.

6.3. Comparison with Lung Cancer Mortality Trends

Implicit in our interest in applying a predictive model for aerosol sulfur concentrations to the southeastern United States are the generally high age-adjusted mortality rates for lung cancer reported in the region since 1950 (Mason and McKay, 1973; Mason et al., 1975). Although the available epidemiological data do not distinguish smokers from nonsmokers, it may be assumed that most lung cancer deaths are among tobacco smokers. Significant geographic trends may be discerned among white males, the group within which the reported numbers of deaths due to lung cancer are largest, thus permitting the most precise mortality rates to be computed. Along the Atlantic seaboard the rates are generally higher in the northeast Florida and southeast Georgia region than in South Carolina or southern Florida. The rates have also steadily risen throughout this coastal region during 1950–1975. Among the 10 metropolitan counties in the United States with the highest rates for 1970–1975, Duval County (Jacksonville), Florida was first (Blot et al., 1982). These trends hint at an air quality factor that may affect the risk of lung cancer.

Epidemiologic interest in the possibility of a link between air pollution and lung cancer is not new (International Symposium, 1978). Most interest in this question has been stimulated by the observation that age-adjusted lung cancer mortality rates are usually higher in urban areas than in nearby nonurban areas and by the suspicion that carcinogens in urban air may add to those from other sources (e.g., tobacco smoke or occupational exposure) to enhance the lung cancer risk in cities. However, there are many urban–nonurban demographic and other differences that could be important, thus complicating the search for a clean-cut answer on the basis of urban–nonurban comparisons. Another possible impediment to finding an answer is that the effect of air pollution exposure to individuals who are also exposed to tobacco smoke or other sources of carcinogens may be multiplicative, so as to enhance lung cancer risk in such individuals. This possibility may not have been given sufficient attention in previous epidemiologic research.

In our research in the southeastern United States, where we have predicted gradients of pollution aerosol sulfur concentration on the basis of a simple model and measured gradients of climatic variables through a predominantly nonurban region, we avoid the complications of comparing urban with nonurban areas. These predicted gradients are unique to aerosol sulfur, which is a secondary atmospheric chemical reaction product from primary sulfur dioxide gaseous pollution, and not to air pollution in general. It is thus

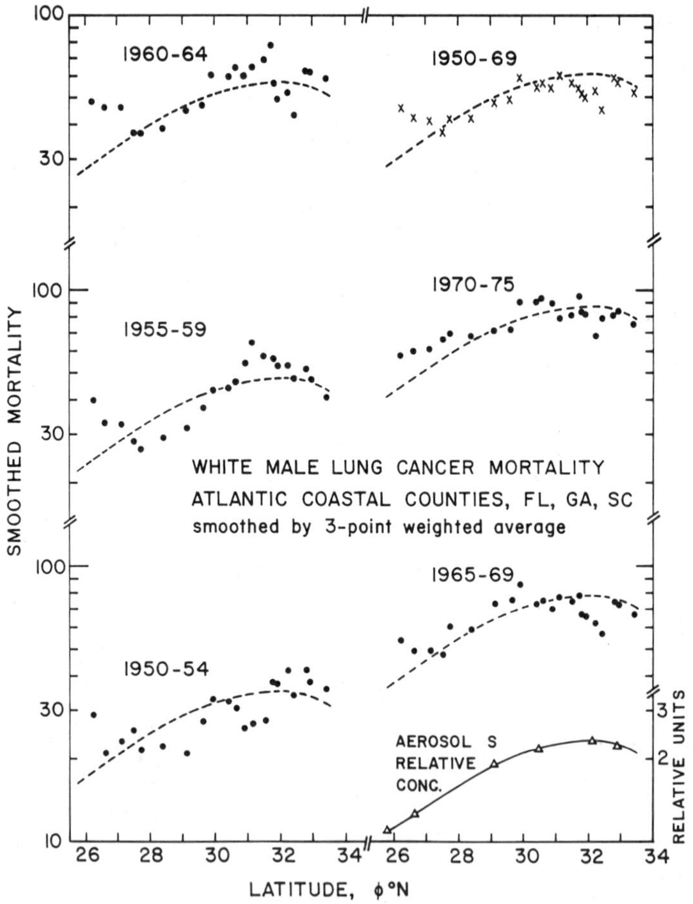

Figure 17. Lower right: average annual sulfuric acid aerosol concentration profile with latitude obtained by summing monthly curves in Figure 16. Remaining six plots: smoothed mortalities for the periods indicated obtained by averaging three adjacent counties weighted according to expected random errors in death count. The lower right curve is overlain on each for comparison. (From Winchester et al., 1981.)

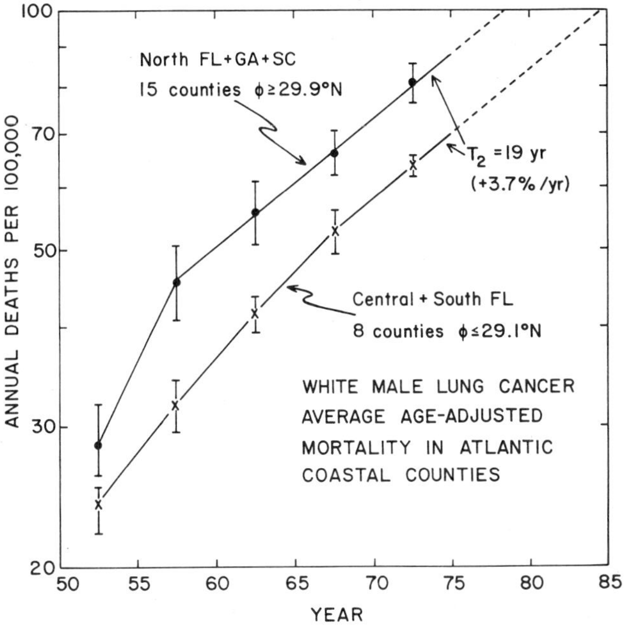

Figure 18. White male age adjusted mortalities for the five pentads 1950–1954 to 1970–1975 (except 1972) plotted against year as the averages with standard errors of two groups of coastal counties, one lying to the north of, and the other to the south of, Flagler, FL (which is not included in either group). Mortalities average 25% higher in the northern group, but both groups have reached a 3.7% annual rate of increase (doubling time T_2 = 19 years) after experiencing a much greater rate of increase in the 1950s. (From Winchester et al., 1981.)

natural to compare these gradients with the patterns of lung cancer mortality. Since sulfur by itself is not known to be carcinogenic, any correspondences between concentration and mortality would suggest a combined effect, possibly multiplicative, of fine-particulate sulfur and carcinogenic agents from other sources.

Smoothed lung cancer mortality data by county for white males, broken down into five successive 5-year intervals from 1950 to 1975, are shown in Figure 17. A full tabulation of the primary data and other data pertaining to the counties in the southeastern Atlantic coastal region is given elsewhere (Winchester et al., 1981). Overlain on each set of smoothed mortality data is the relative aerosol sulfur trend estimated by our simple model. Features common to the sulfur and mortality trends are generally higher values in the north than in the south and the existence of a broad maximum in the northern region, so that lower values are found in South Carolina than in southeastern Georgia or northern Florida, but with lowest values in southern Florida.

It should be noted that the general levels of lung cancer mortality have risen from the vicinity of 30 deaths per 100,000 white males annually in 1950–1954 to nearly 100 deaths in 1970–1975 (Figure 18). This steadily rising death rate corresponds to a doubling time of about 19 years or a rise of about 4% per year on the average. The rate of rise is not unlike the probable rise in SO_2 emissions (and resulting aerosol sulfur concentrations) from tall stack stationary air pollution sources in the eastern United States, as indicated by emissions inventories or, more simply, by some suitable index of industrial activity, such as electric power generation from sulfur-containing fossil fuels (see *World Almanac,* 1982). Our trends of both aerosol sulfur concentration and lung cancer mortality with latitude pertain largely to nonurban areas and thus are not due to urban–nonurban differences in pollution levels, demography, and other factors.

7. CONCLUSIONS

In this summary of research carried out at the Florida State University we have shown that natural background concentrations of aerosol sulfur may be about 100 times lower than concentrations currently found in atmospheres of the eastern United States. The latter levels must be considered to be the result of large-scale addition of gaseous SO_2 to the atmosphere, as a consequence of combustion of fossil fuels and of the conversion of this SO_2 to sulfuric acid and sulfate aerosol. We have summarized evidence that the conversion rate may be correlated with temperature and the content of water vapor in the atmosphere, suggesting a conversion process involving oxidation of SO_2 in aerosol solution droplets. The resulting aerosol may be transported over long distances so that Florida may be a receptor state with an average influx of aerosol sulfur in excess of that formed from intrastate SO_2 emissions. However, acid rain may be more the result of SO_2 oxidation at the time of rain formation and more a function of nearby SO_2 sources.

If one is concerned with potential biological effects from exposure to aerosol sulfur, which is at least partially in the form of sulfuric acid, it is of interest to attempt to estimate the concentrations expected along the Atlantic coast of Florida, Georgia, and South Carolina, in view of the southeastern U.S. climate and the expected dependence of aerosol sulfur formation rate—and steady-state concentration—on temperature and water vapor concentration. The attempt has been made by assuming the general validity of a simple physical and chemical model. Highest concentrations of aerosol sulfur are predicted at times and locations of highest temperatures and water vapor concentrations. The profile with latitude of these concentrations resembles the profiles of lung cancer mortality in white males since 1950, with a temporal rise in mortality approximating the probable rise in SO_2 emissions from air pollution sources in the eastern United States. There is reason, therefore, to examine further the possibility of a link between human exposures to fine-aerosol sulfur and the induction of malignancies in persons also exposed to primary carcinogens. We do not suggest that aerosol sulfur, as sulfuric acid or as sulfate particles, is a primary carcinogen, but rather that it may serve to enhance the risk of cancer in those also exposed to cancer-causing agents. In this respect the geographic variability of aerosol sulfur in the southeastern Atlantic coastal zone could account for much of the variability in lung cancer mortality.

A possible process for enhancing lung cancer risk in tobacco smokers who breathe acid-polluted marine air has been suggested by Keenan (1982). Gaseous formaldehyde and hydrogen chloride are known to react to form bis(chloromethyl) ether, BCME, a demonstrated and highly potent lung carcinogen (Fishbein, 1979; Sax, 1981). Formaldehyde may comprise about 100 ppm of tobacco smoke (Wynder and Hoffmann, 1967). In an aerosol containing chloride, e.g. from sea salt, HCl is released by reaction with sulfuric or nitric acid air pollutants (cf. Berg and Winchester, 1978). The reaction may occur either in the atmosphere or within a cigarette after the external (unequilibrated) aerosol mixture of fine sulfuric acid and coarse alkaline salt particles is drawn in by the smoker. According to this proposed mechanism, reaction of the HCl with formaldehyde may produce BCME in inhaled tobacco smoke at concentrations sufficient to present a lung cancer risk to the smoker and to others who breathe smoke-contaminated air. This proposed mechanism is currently undergoing evaluation.

ACKNOWLEDGMENTS

The research summarized in this review made use of resources of the Florida State University and was assisted in part by grants from the U.S. Environmental Protection Agency, the National Institutes of Health, the Office of Sea Grant, the Florida Sulfur Oxides Study, and the North Atlantic Treaty Organization.

REFERENCES

Ahlberg, M. S., Leslie, A. C. D., and Winchester, J. W. (1978). Characteristics of sulfur aerosol in Florida as determined by PIXE analysis. *Atmosph. Environ.* **12,** 773–777.

Ahlberg, M. S. and Winchester, J. W. (1978). Dependence of aerosol sulfur particle size on relative humidity. *Atmosph. Environ.* **12,** 1631–1632.

Altshuller, A. P. (1979). Model predictions of the rates of homogeneous oxidation of sulfur dioxide to sulfate in the troposphere. *Atmosph. Environ.* **13,** 1653–1661.

Andreae, M. O. (1982). Marine aerosol chemistry at Cape Grim, Tasmania and Townsville, Queensland. *J. Geophys. Res.* **87,** 8875–8885.

Berg, W. W., Jr. and Winchester, J. W. (1978). Aerosol chemistry of the marine atmosphere, in *Chemical Oceanography,* 2nd ed., Vol. 7, J. P. Riley and R. Chester, Eds., Academic Press, New York, pp. 173–231.

Blot, W. J., Davies, J. E., Brown, L. M., Nordwall, C. W., Buiatti, E., Ng, A., and Fraumeni, J. F., Jr. (1982). Occupation and the high risk of lung cancer in northeast Florida. *Cancer* **50,** 364–371.

Cahill, T. A. (1980). Proton microprobes and particle-induced X-ray analytical systems. *Ann. Rev. Nucl. Part. Sci.* **30,** 211–252.

Climatography of the United States, No. 60 (1959). Florida, Georgia, South Carolina, North Carolina. U.S. Weather Bureau, Washington, DC.

Climatography of the United States, No. 60 (1976). Florida, Georgia, South Carolina, North Carolina. National Climatic Center, Asheville, NC.

Cullis, C. F. and Hirschler, M. M. (1980). Atmospheric sulfur: Natural and man-made sources. *Atmosph. Environ.* **14,** 1263–1278.

Environmental Protection Agency (EPA) (1980). *Air Quality Criteria for Particulate Matter and Sulfur Oxides,* Vol. 1, External Review Draft No. 1, Research Triangle Park, NC, Summary and Conclusions, 152 pp.

Fishbein, L. (1979). *Potential Industrial Carcinogens and Mutagens.* Elsevier, New York and Amsterdam, p. 251.

International Critical Tables, (1928). McGraw-Hill, New York, Vol. 3, p. 303.

International Symposium on General Air Pollution and Human Health with Special Reference to Long-Term Effects (1978). Report and working papers, in *Environmental Health Perspectives,* No. 22, pp. 1–123.

Johansson, S. A. E., Ed. (1977). Proceedings of the International Conference on Particle Induced X-ray Emission and its Analytical Applications. *Nucl. Instr. Meth.* **142,** 1–338.

Johansson, S. A. E., Ed. (1981). Particle Induced X-ray Emission and its analytical applications—Proceedings of the 2nd International Conference on PIXE and Its Analytical Applications. *Nucl. Instr. Meth.* **181,** 1–537.

Johansson, S. A. E. and Johansson, T. B. (1976). Analytical application of particle induced X-ray emission. *Nucl. Instr. Meth.* **137,** 473–516.

Johansson, T. B., Akselsson, R., and Johansson, S. A. E. (1970). X-ray analysis: elemental trace analysis at the 10^{-12} g level. *Nucl. Instr. Meth.* **84,** 141–143.

Johansson, T. B., Van Grieken, R. E., Nelson, J. W., and Winchester, J. W. (1975). Elemental trace analysis of small samples by proton induced X-ray emission. *Anal. Chem.* **47,** 855–860.

Junge, C. E. (1963). *Air Chemistry and Radioactivity,* Academic Press, New York, p. 190 et passim.

Keenan, James J. (1982). Information transmitted verbally on August 4 and 27..

Kaufmann, H. C., Akselsson, K. R., and Courtney, W. J. (1977). REX—a computer programme for PIXE analysis. *Nucl. Instr. Meth.* **142,** 251–257.

Lawson, D. R. and Winchester, J. W. (1979). Atmospheric sulfur aerosol concentrations and characteristics from the South American continent. *Science* **205,** 1267–1269.

Leslie, A. C. D., Ahlberg, M. S., Winchester, J. W., and Nelson, J. W. (1978). Aerosol characteristics for sulfur oxide health effects assessment. *Atmosph. Environ.* **12,** 729–733.

Liu, B. Y. H. and Lee, K. W. (1976). Efficiency of membrane and Nuclepore filters for submicrometer aerosols. *Environ. Sci. Technol.* **10,** 345–350.

Maenhaut, W., Darzi, M., and Winchester, J. W. (1981). Seawater and nonseawater aerosol components in the marine atmosphere of Samoa. *J. Geophys. Res.* **86,** 3187–3193.

Marple, V. A. and Willete, K. (1976). Impactor design. *Atmosph. Environ.* **10,** 891–896.

Mason, T. J. and McKay, F. W. (1973). *U. S. Cancer Mortality by County: 1950–1969.* DHEW Publ. No. (NIH)74-615, U.S. Government Printing Office, Washington, DC.

Mason, T. J., McKay, F. W., Hoover, R., Blot, W. J., and Fraumeni, J. F., Jr. (1975). *Atlas of Cancer Mortality for U.S. Counties: 1950–1969.* DHEW Publ. No. (NIH)75-780, Washington, DC.

Meinert, D. L. and Winchester, J. W. (1977). Chemical relationships in the North Atlantic marine aerosol. *J. Geophys. Res.* **82,** 1778–1782.

Middleton, P., Kiang, C. S., and Mohnen, V. A. (1980). Theoretical estimates of the relative importance of various urban sulfate aerosol production mechanisms. *Atmosph. Environ.* **14,** 463–473.

Mueller, P. K., Hidy, G. M., Warren, K., Lavery, T. F., and Baskett, R. L. (1980). The occurrence of atmospheric aerosols in the northeastern United States, in *Aerosols: Anthropogenic and Natural, Sources and Transport,* T. J. Kneip and P. J. Lioy, Eds.; *Ann. N.Y. Acad. Sci.* **338,** 463–482.

Nelson, J. W. (1977). Proton induced aerosol analysis: Methods and samplers, in *X-Ray Fluorescence Analysis of Environmental Samples,* T. G. Dzubay, Ed., Ann Arbor Science Publishers, Ann Arbor, MI, pp. 19–34.

Ottar, B. (1980). The long range transport of sulfurous aerosol to Scandinavia, in *Aerosols: Anthropogenic and Natural, Sources and Transport,* T. J. Kneip and P. J. Lioy, Eds., *Ann. N.Y. Acad. Sci.* **338,** 504–514.

Rahn, K. A. and McCaffrey, R. J. (1980). On the origin and transport of the winter Arctic aerosol, in *Aerosols: Anthropogenic and Natural, Sources and Transport,* T. J. Kneip and P. J. Lioy, Eds., *Ann. N.Y. Acad. Sci.* **338,** 486–503.

Rodhe, H. (1978). Budgets and turn-over times of atmospheric sulfur compounds. *Atmosph. Environ.* **12,** 671–680.

Sax, N. I. (1981). *Cancer Causing Chemicals.* Van Nostrand, New York, p. 6 et passim.

Stommel, H. (1966). *The Gulf Stream: A Physical and Dynamical Description.* 2nd ed., University of California Press, Berkeley.

Tanaka, S., Darzi, M., and Winchester, J. W. (1980a). Sulfur and associated elements and acidity in continental and marine rain from north Florida. *J. Geophys. Res.* **85,** 4519–4526.

Tanaka, S., Darzi, M., and Winchester, J. W. (1980b). Short term effect of rainfall on elemental composition and size distribution of aerosols in north Florida. *Atmosph. Environ.* **14,** 1421–1426.

Tang, I. N., Munkelwitz, H. R., and Davis, J. G. (1978). Aerosol growth studies—IV. Phase transformation of mixed salt aerosols in a moist atmosphere. *J. Aerosol. Sci.* **9,** 505–511.

Tanner, R. L., Leaderer, B. P., and Spengler, J. D. (1981). Acidity of atmospheric aerosols. *Environ. Sci. Technol.* **15,** 1150–1153.

Winchester, J. W. (1981). Particulate matter and sulfur in the natural atmosphere. *Nucl. Instrum. Meth.* **181,** 367–381.

Winchester, J. W. (1982). Atmospheric sampling, in *McGraw-Hill Yearbook of Science and Technology,* McGraw-Hill, New York, pp. 116–118.

Winchester, J. W., and Leslie, A. C. D. (1982). Water vapor and temperature dependence of aerosol sulfur concentrations at Fort Wayne, Indiana, October 1977, in *Heterogeneous Atmospheric Chemistry,* D. R. Schryer, Ed., Geophysical Monograph Series, Vol. 26, American Geophysical Union, Washington, DC pp. 250–256.

Winchester, J. W., Leysieffer, F. W., and Park, Y. C. (1981). *Geographic Distributions of Lung Cancer Mortality and Estimated Sulfuric Acid Aerosol Concentration in Southeastern U.S. Atlantic Coastal Counties.* Technical Report, Departments of Oceanography and Statistics, Florida State University, Tallahassee (unpublished).

World Almanac and Book of Facts (1982). Newspaper Enterprise Association, New York. See also prior annual editions of the *World Almanac.*

Wynder, E. L., and Hoffmann, D. (1967). *Tobacco and Tobacco Smoke.* Academic Press, New York, p. 413.

Young, G. S., and Winchester, J. W. (1980). Association of non-marine sulfate aerosol with sea breeze circulation in Tampa Bay. *J. Appl. Meteorol.* **19,** 419–425.

7

LAGRANGIAN STUDIES OF ATMOSPHERIC POLLUTANT TRANSFORMATIONS

Bernard D. Zak

Sandia National Laboratories
Albuquerque, New Mexico 87185

---

1. INTRODUCTION

It has long been recognized that the Lagrangian (see Appendix A) frame of reference—one that moves with the flow of air—is a natural choice for studies of pollutant behavior in the atmosphere (Seinfeld et al., 1973). In Lagrangian experiments the focus is on a volume of air of limited extent as that volume is transported by the wind. Sensors are systematically deployed in that volume to make whatever measurements may be required to meet experimental objectives.

In comparison to Eulerian studies, in which measurements are made at or over fixed sites, Lagrangian studies offer several advantages. Most importantly, lateral flow into and out of a Lagrangian volume is typically much less than that for a comparable Eulerian volume because the mean wind in the Lagrangian frame of reference is approximately zero. Under some meteorological conditions for limited periods of time, an appropriately defined Lagrangian volume may even be considered to be isolated; that is, exchange of material between the volume and its surroundings may be neglected. This assumption is the basis of the so-called air parcel concept. Under these conditions the Lagrangian volume approximates a "smog chamber without walls" (Moses, 1976). However, it is by no means necessary or even desirable to impose the assumption of isolation in the interpretation of Lagrangian data. Rather, this assumption is typically treated as a zeroth-order approximation to which exchange may be treated as a perturbation.

Another advantage is that in Lagrangian experiments, the size of the study volume is small in comparison to the length of its trajectory. Consequently, with a fixed level of available resources, more data can be obtained on the behavior of the species of interest in this volume than would be obtained if the same resources were uniformly deployed over the entire region of potential interest. Measurement resources are concentrated rather than dispersed. This factor is of particular importance in long-range studies.

A related advantage is that Lagrangian experiments permit relatively easy comparison of experimental data with Lagrangian models (Peters and Carmichael, Chapter 12 of this volume). Thus a fruitful interplay between field measurement and modeling is encouraged. Although Lagrangian models themselves may not be optimal for regulatory purposes, they do serve as invaluable investigative tools in developing understanding of the chemistry and physics of air pollution—an understanding that can be tested with Lagrangian experiments.

Finally, provided an appropriate Lagrangian marker system is used, Lagrangian experiments largely eliminate the impact of uncertainties in the mean wind field on the interpretation of results. The Lagrangian marker gives trajectory information directly. The position of the volume under study as a function of time is well known. Consequently, discrepancies that arise between model predictions and measurements are less likely to be assignable to errors in wind data.

In this chapter the major types of Lagrangian experiment that have been carried out in the lower atmosphere are reviewed. In each case we focus on one or two experiments, describing them in sufficient detail so that the reader may gain some understanding of the scope of the experimental effort involved and what has been learned through the experiments. Finally, we describe some new developments in Lagrangian techniques and discuss their future use.

2. LAGRANGIAN MARKERS

2.1. Development

The history of Lagrangian pollutant transformation studies is closely coupled with the history of scientific ballooning. The point of contact is the Lagrangian marker—the balloon system designed to remain embedded in and mark a Lagrangian volume for study. There is another kind of Lagrangian marker: a release of tracer material (Barr and Gedayloo, 1979). Such releases, which are most often used to study dispersion, are not discussed in detail here.

For the past several decades balloons have been used to carry scientific payloads into the upper atmosphere to study its dynamics, composition, and radiative environment. Although some experiments which have been balloon borne in the past are now carried by satellites, balloons remain the work horses of high-altitude research. The National Center for Atmospheric Research (NCAR); the Department of Energy (DOE); the National Aeronautics and Space Administration (NASA); the U.S. Army, Navy, and Air Force; and many research organizations in other countries continue to rely on balloon technology. Hundreds of high-altitude research balloons are launched every year.

Under these circumstances it was perhaps inevitable that the advantageous properties of constant-volume balloons (CVBs) be first recognized and widely applied by researchers interested in the upper atmosphere (Giles and Angell, 1963; Lally, 1959). The useful characteristic of CVB systems is that they possess an equilibrium altitude to which they will tend to return if displaced. This equilibrium altitude is determined by Archimedes' principle—a body is acted on by a buoyant force equal to the weight of fluid it displaces. Thus a CVB system seeks an altitude at which the product of the density of the air it displaces and its volume is equal to the mass of the system. At that altitude it is neutrally buoyant. At any other altitude there is a force tending to restore it to its equilibrium altitude. However, for CVBs used as Lagrangian markers, the restoring force is sufficiently small that the balloon tends to be swept along with vertical as well as horizontal air motions.

The decade of the sixties saw the application of CVBs on a large scale in the GHOST (global horizontal sounding technique) program and in its French counterpart, the Eole program (Morel and Bandeen, 1973; NCAR, 1969). In these programs it was found that at 200 mbar and above CVBs had lifetimes that routinely permitted multiple circuits of the earth.

While the GHOST program was under development, researchers at the Environmental Science Services Administration, the predecessor agency to the National Oceanic and Atmospheric Administration (NOAA), began experimenting with tetrahedral-shaped CVBs called *tetroons* (Figure 1) at lower altitudes (Angell and Pack, 1960; 1962). As the evidence accumulated that CVBs were acceptable Lagrangian markers, these researchers began to use them to compare Lagrangian and Eulerian properties of turbulence (Angell, 1964, 1974; Angell et al., 1971a) and then to study circulations and other structures in the atmosphere (Angell et al., 1968, 1969; Angell, 1972). Tetroon experiments were also conducted in urban areas to study flow patterns and turbulence as influenced by a city (Angell et al., 1966, 1971a, 1973). In one such study over New York City, a helicopter flew along near the tetroon to provide accurate height data and vertical temperature profiles (Hass et al., 1967).

In 1969 a particularly significant tetroon study was done in the Los Angeles Basin (Angell et al., 1972). Here, as in the New York study, a helicopter accompanied the tetroons. In this case, however, the accompanying helicopter made what may be the first Lagrangian measurements of oxidant concentrations. In the conclusion to the paper describing this study, one can see the plan for the Los Angeles Reactive Pollutant Program (LARPP) beginning to take shape.

2.2. Technology

Periodic position measurement is fundamental to the use of Lagrangian markers. Various position measurement techniques have been developed for this application. One of the first methods used was triangulation with

Figure 1. Tetroon being readied for release for flight over New York City. Pilot balloon attached to tetroon provides for rapid ascent to float altitude. Courtesy of J. K. Angell.

radio direction finders. Later, sun angle sensors on board the balloons were employed. Next, radio-navigation systems such as LORAN and Omega came into prominence. Finally, satellite techniques began to be used.

All these techniques serve reasonably well to track Lagrangian markers on a global scale. However, for pollutant transport and transformation studies, the spatial resolution of these techniques is not always adequate. The NOAA researchers have developed the modified radar technique shown in Figure 2. Each marker balloon carries a lightweight transponder that emits a pulse on 403 MHz when it receives a pulse from an M33 radar. The timing of the return pulse relative to the initial radar pulse gives the distance very accurately. The angular position of the radar antenna when the return pulse is received gives the direction. This and other tetroon techniques are described in an article in the NCAR publication "Balloon Technology" (Angell, 1975).

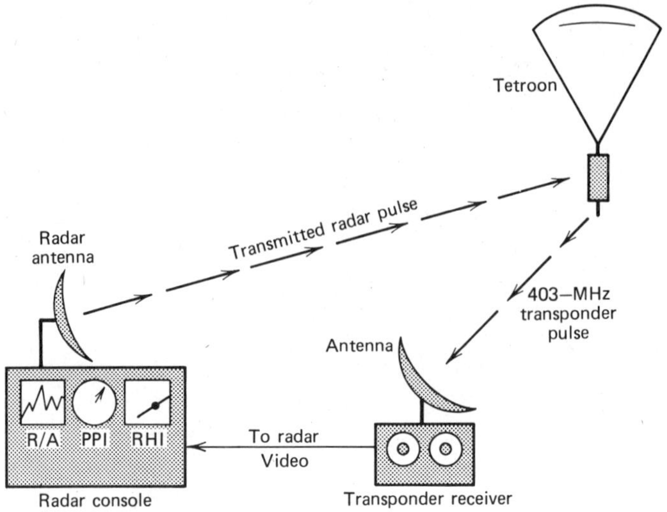

Figure 2. Radar tracking system for transponder-equipped tetroons (Angell, 1975).

As a spinoff from the GHOST program, Lally (1967) published what is essentially a handbook of CVB technology. It remains extraordinarily useful. A more recent review of the capabilities and behavior of CVBs is also noteworthy (Tatom and King, 1977). Hoecker (1975) has described a universal technique for setting the equilibrium altitude for constant-level balloons, and Hanna and Hoecker (1971) have investigated the response of CVBs to vertical winds. In addition, the would-be Lagrangian experimenter will want to be aware of the *Scientific Ballooning Handbook* (Morris, 1975). The earlier literature on the use of constant volume balloons is available through a review article by Angell (1961).

3. LOS ANGELES REACTIVE POLLUTANT PROGRAM (LARPP)

3.1. Operations

The LARPP was the first major study of pollutant transformation in a Lagrangian frame. Sponsored jointly by the Coordinating Research Council and the Environmental Protection Agency, it was carried out in the Los Angeles Basin in September, October, and November 1973. Several groups participated: EPA (instrumented helicopter and ground-level radiation measurements); NOAA (tetroon operations and radar tracking); California Air Resources Board (mobile van and laboratory support for gas chromatograph analysis of bag samples); General Research Corporation (data management); Systems Innovation Corporation (Lidar); and Metronics Associates, Inc. (tracer and overall operation control). Approximately 60 people were involved in the field operation (Perkins, 1974).

Full-scale operations were conducted on 35 days covering a range of smog conditions from light to heavy. Operations were started as early in the day as possible from a site having a significant pollutant source and a long forecast trajectory. Five sites were used, but two sites were used most often (downtown Los Angeles, 17 tests; Downey, 12 tests).

In a typical operation three tetroons ballasted to fly below the inversion base were launched simultaneously from a single location. The position of each was established by radar at 1-min intervals and plotted at the Operations Center. As soon as the centroid of the tetroon pattern was established, fluorescent particle tracer was released by helicopter in a 0.5-mile pattern around the centroid approximately midway between the ground and inversion base. After completion of the tracer release, the first of two instrumented Bell 212 helicopters began a series of constant-altitude rectangular flight patterns around the centroid starting near the inversion base. A second radar continuously tracked the aircraft and reported its position to the Operations Center. The rectangular pattern was flown at four levels, the lowest 200 ft above the ground. Normally, the aircraft would descend, ascend, and descend a second time through each level before returning to base. The second aircraft began at the top level while the first aircraft was completing its final flight pattern at the lowest level. Except for air traffic control constraints, it was possible to maintain the flight pattern close to the tetroon centroid. The Air Resources Board van was directed along the centroid path by the Operations Center. A summary of LARPP experiments is given in Table 1.

Aircraft and van measurements included O_3, NO, NO_x (concurrently), CH_4, nonmethane hydrocarbon, CO, tracer concentration, and air temperature. Bag samples for GC analysis were taken from all vehicles. In addition, the aircraft measured dew point and aerosol light scattering. The aircraft instrumentation (Figure 3) has been described by Evans (1973).

Lidar measurements from a separate mobile van were made along the

Figure 3. EPA instrumented helicopter as used in the LARPP study. Courtesy of R. B. Evans.

Table 1. Listing of LARPP Experiments for Tetroon Triads Tracked for at Least 1 h

Operation	Date	Launch Site	Launch Time (Local)	Tracking Duration (h)	Wind Speed ($m\ s^{-1}$)	Tetroon MSL Heights (m)
1	9–4	Redondo	1200	2.5	4.0	480, 470, 440
5	9–18	Downtown[a]	1030	2.0	1.4	390, 370, 390
8	9–24	Downey	0900	4.5	1.9	480, 450, 460
9	9–25	Long Beach	1000	4.0	2.1	560, 400, 410
10	9–27	Downtown	0900	2.5	2.5	170, 260, 250
11	9–28	Downtown	0830	4.0	2.8	250, 250, 240
12	10–1	Downey	0830	4.5	1.5	370, 410, 490
13	10–2	Pomona	1400	1.5	3.2	720, 850, 790
14	10–4	Downtown	1930	2.5	1.1	300, 310, 340
15	10–5	Downey	1200	2.0	2.6	350, 300, 280
16	10–10	Downey	0830	1.0	1.7	280, 250, 260
17	10–11	Downtown	0800	4.5	2.3	420, 290, 390
18	10–12	Downtown	0930	3.5	2.4	510, 390, 420
19	10–15	Downey	0930	3.5	1.4	390, 350, 380
20A	10–16	Downtown	0930	2.0	3.2	440, 460, 450
20B	10–16	Downtown	1300	1.0	1.9	420, 280, 290
21	10–17	Downtown	0930	4.5	2.6	390, 360, 360
22A	10–18	Downey	0900	2.5	2.9	320, 260, 360
22B	10–18	Downey	1130	2.0	2.0	290, 280, 340
23	10–24	Downey	0900	2.5	0.8	360, 260, 290
24	10–25	Downtown	0830	4.0	1.4	370, 300, 280
25	10–26	Downtown	1000	6.0	1.9	340, 390, 320
26	10–27	Downtown	0900	3.0	2.7	360, 360, 370
27	10–28	Downey	1100	2.5	2.8	550, 380, 400
28	10–29	Downey	0830	2.0	2.1	190, 220, 230
29	10–30	Downey	0830	2.5	1.5	250, 250, 260
30	10–31	Downtown	0830	1.5	3.1	310, 350, 350
31A	11–1	Downtown	0730	2.5	2.8	270, 270, 280
31B	11–1	Downtown	1130	1.5	3.0	400, 390, 530
32	11–2	Downey	1200	1.5	5.1	360, 360, 350
33	11–5	Downey	0700	6.0	1.0	310, 340, 310
34A	11–6	Downtown	0730	3.0	1.9	280, 280, 270
34B	11–6	Downtown	1100	2.0	2.1	390, 450, 410
35	11–7	Downtown	0830	2.5	1.2	280, 290, 330

Source: Angell et al. (1975).
[a] "Downtown" refers to downtown Los Angeles.

projected track during most operations. Continuous daily records of total ultraviolet (UV) radiation were taken throughout the operation period at five stations from the coast to San Bernardino, 80 km inland. In addition, UV in 100-Å bandwidths was measured at two locations, El Monte and Mount Disappointment (altitude 1719 m).

Collateral surface air monitoring data were obtained during a portion of the operational period by two fixed vans operated respectively by GM and Battelle. Both were located in West Covina.

3.2. Data Applications

3.2.1. Transport and Dispersion

The LARPP data base has been used by many researchers both to gain an understanding of the behavior of air pollutants in the Los Angeles Basin and to test general physical and chemical air quality models.

The NOAA group responsible for fielding and tracking the tetroons analyzed their data to elucidate diffusion processes over Los Angeles (Angell et al., 1975) and to investigate the source and trajectory of polluted air reaching the Riverside–San Bernardino area (Angell et al., 1976).

Edinger (1975) carried out a preliminary analysis of the LARPP data that resulted in a qualitative description of the evolution of air parcel shape under varying conditions. Of particular interest is the interaction with sun-heated slopes of the bordering mountains and how this interaction can create isolated elevated layers of polluted air.

On the basis of the LARPP measurements made on November 5, 1973, LARPP Operation 33, Panofsky (1975) developed a model for vertical diffusion coefficients in a growing urban boundary layer. Panofsky used the time-dependent Lagrangian vertical profiles of the essentially inert species carbon monoxide and methane, as well as other available meteorological data from LARPP to evaluate parameters in the model. Panofsky and his co-workers (Crane et al., 1977) have extended the model to other LARPP operations. The results were compared with a second-order closure prediction, with laboratory measurements of particle dispersion, and with the assumption of equality between momentum- and mass-transfer coefficients in the free convective limit. It was concluded that normalized mixing coefficients such as those given by the model could be represented as "universal" functions of the height: mixing depth ratio. These functions are small at both small and large height relative to the mixing depth but reach a maximum at about half the mixing depth. The vertical diffusion coefficients are so large in the middle of the boundary layer that pollutant concentrations there are effectively independent of height.

3.2.2. Photochemistry

Calvert (1976a) has analyzed the data from LARPP Operation 33 to test the theory of ozone generation. He examined the pollutant concentration ratio $[O_3][NO]:[NO_2]$, which should equal the rate constant ratio k_1/k_3, where k_1 is the rate constant for the photolysis of nitrogen dioxide and k_3 the rate constant for the reaction of O_3 with NO to yield NO_2 and O_2. The first rate constant is dependent on the flux of UV light capable of photolyzing NO_2.

Consequently, the concentration ratio is compared with a model involving the time dependence of the UV flux. Actually, two models were used: one that ignores UV scattering and absorption within the boundary layer and one which accounts for these processes. In both cases the agreement was reasonably good. Although there were some discrepancies between the experimental data and the theories, Calvert concluded that the data confirm the essential features of the accepted ozone formation mechanism.

The largest discrepancies were observed in early morning when convective mixing was weak. Calvert further proposed that under these conditions, it is particularly important to take into account the inhomogeneity of reactant concentrations, as discussed by Hilst (1973). Seinfeld (1977) and Feigley (1978) have examined the effects of inhomogeneity on the data in greater detail and agree with this assessment.

Calvert (1976b) has also analyzed LARPP Operation 33 data in connection with hydrocarbon involvement in smog formation. He found, as expected, that the ratio of the concentration of reactive hydrocarbon to that of acetylene decreased regularly and that the ratio $[NO_2]:[NO_x]$ inceased regularly throughout the day. He was also able to make estimates of HO-radical concentrations from the observed rates of removal of several hydrocarbons and from the change in the ratios $[NO_x]:[CH_4]$ and $[NO_x]:[C_2H_2]$ with time. He concluded that the average ambient level of HO radical in the morning hours of November 5, 1973 in the Lagrangian volume sampled was $(1.0 \pm 0.8) \times 10^{-7}$ ppm. He also concluded from the LARPP data that realistic models of photochemical smog must include all classes of hydrocarbons as well as CO and aldehydes.

3.2.3. Smog Chamber Validation

Feigley and Jeffries (1979) have noted that much of the current knowledge of the complex chemical and physical factors that affect the production of ozone and other secondary pollutants in ambient air is based on smog-chamber experimentation. They also point out that it is not known how well reactions in a smog chamber represent reactions in the atmosphere. Simplified reactant mixes, wall interactions, and other factors raise questions that cannot be easily answered or dismissed. Thus there is a need to validate smog-chamber data by comparing them with ambient data. Lagrangian data are most suitable for this purpose. Consequently, Feigley undertook chamber validation experiments on the basis of LARPP data as a thesis project at the University of North Carolina (UNC) (Feigley et al., 1979, 1982).

Again, LARPP Operation 33, the operation during which the largest number of complete helicopter patterns were flown (49), was chosen for detailed study. The triad of tetroons was launched from the Downey site at 0715 PST. The tetroons were ballasted to fly at 310, 340, and 310 m MSL and were tracked for more than 6 hours. This was possible because the wind speed was only about 1 m s^{-1}. The tetroons remained at 30–60 m above the ground for most of the operation but moved to higher altitudes after 1300 PST.

Conditions for simulating the chemistry occurring during LARPP Operation 33 were obtained from the LARPP data base, from a source reconciliation technique similar to that of Mayrsohn and Crabtree (1976) and from a vertical diffusion model (Panofsky, 1975).

During the fall of 1976, some 10 simulation experiments were performed at the UNC outdoor smog chambers under varying conditions of insolation and ambient temperature. The run of October 13, 1976 most closely replicated the data from LARPP Operation 33 (Figure 4).

Considering the potential sources of error, Feigley concluded that the agreement between UNC smog chamber and LARPP data was good. Two factors in particular were highlighted as potentially being responsible for the generally lower ozone concentrations observed in the chamber in comparison to LARPP. The simulations involved a synthetic mix of hydrocarbons designed to behave similarly to what was found in Los Angeles effluents. Formaldehyde was not included even though it was known to be a significant constituent of automobile exhaust. It was not thought to be important. Later experiments at the UNC smog chamber have shown that it is important (Feigley et al., 1979). It has also been found that the presence of aged pollutants in the smog chamber from the previous day can enhance the rate and extent of NO oxidation. Although such aged pollutants were probably present in the LARPP study volume, they were not present in the UNC simulation runs.

3.2.4. Model Development

A major motivation for LARPP was the conviction that the stumbling block in the ability to model the behavior of reactive photochemical air pollutants

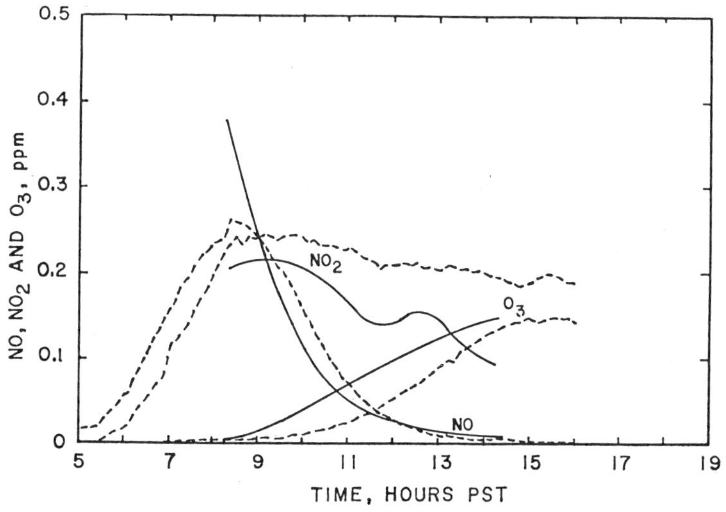

Figure 4. Comparison of NO, NO_2, and O_3 profiles for October 13 simulation (dashed line) and LARPP zero-gradient region (solid line) (Feigley et al., 1979).

in urban areas was the lack of a good data base, rather than lack of modeling capabilities. The LARPP was designed to supply that data base. The Coordinating Research Council also sponsored the development of a Lagrangian model to be based on LARPP data by Environmental Research and Technology, Inc; that model is called ELSTAR (Environmental Lagrangian Simulator of Transport and Atmospheric Reactions; Lloyd et al., 1979a).

In ELSTAR a volume of air (typically 1 × 1 × 0.75 to 5 × 5 × 1.5 km) is advected along a trajectory determined by the local wind. Local winds are derived either by interpolation from nearby meteorological stations or directly from tetroon data. The air volume is divided into a number of layers (normally five) of variable vertical depth. As the air volume is advected along its path, it receives primary pollutants from surface and elevated sources. The pollutants within the moving air volume undergo vertical turbulent diffusion and chemical reaction, with solar UV radiation driving the photochemistry. The model computes the concentrations of 39 chemical species within each vertical layer at discrete points along the trajectory. These include O_3, NO_x, six reactive hydrocarbon groups, NO, nitrous acid, PAN, and free-radical intermediates of known importance in the formation of photochemical smog. The major simplifying assumptions employed are: *K*-theory is appropriate for describing turbulent diffusion; horizontal diffusion is negligible (area source assumed); wind speed and direction are vertically uniform; and the vertical component of wind velocity is negligible. However, lateral diffusion from major point sources is taken into account (Lloyd et al., 1979b).

The chemistry mechanism was developed and first tested against smog chamber data obtained from the University of California, Riverside. The hydrocarbons and oxygenated hydrocarbons are lumped into six general chemical classes: paraffins; ethylene; olefins other than ethylene; aromatic hydrocarbons; formaldehyde; and aldehydes other than formaldehyde.

The LARPP operations were partitioned into a "learning set" and a "test set." The input data was refined, and methodologies developed to obtain best fits between predicted and observed concentrations on the "learning set." The model was then exercised in a relatively "hands off" mode on the "test set." Three LARPP operations (Operations 25, 29 and 33) were taken as the learning set. Eight others (Operations 12, 15, 16, 17, 19, 23, 24, and 25) comprised the test set. The nature of the agreement obtained on the two sets can be inferred from Figures 5 and 6, which compare measurements and predictions for Operations 33 and 23, respectively.

The model results as a whole were sensitive to the values selected for the initial conditions. Concentrations were unavailable for some important species. No data were available for HCHO and the higher aldehydes or for HONO—species that photolyze under ambient conditions to produce radicals. Given the uncertainty in input parameters and model assumptions, the creators of ELSTAR conclude that it performed well on most LARPP operations. Further evaluation of ELSTAR confirmed this initial assessment

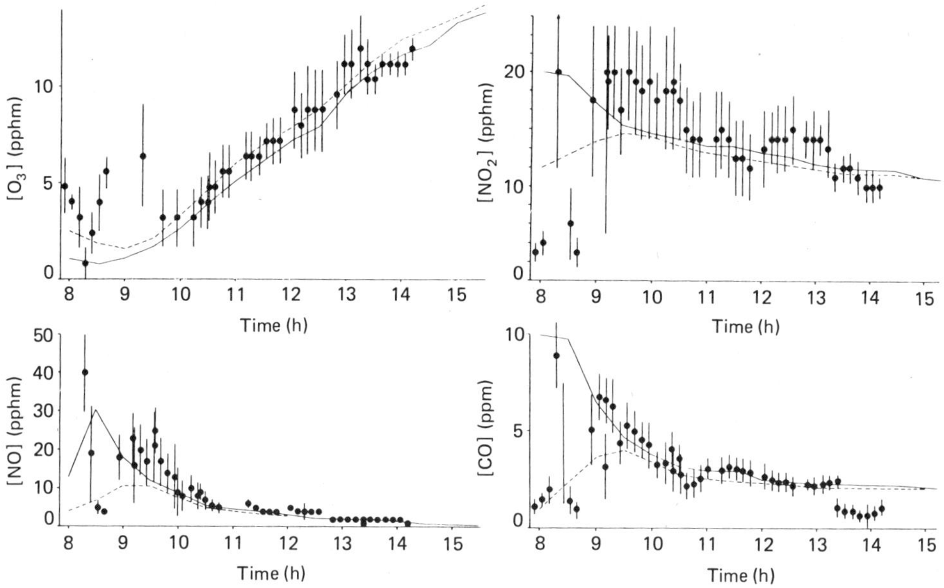

Figure 5. Comparison of ELSTAR predictions with LARPP Operation 33 above-ground-level data. Solid line gives ground-level prediction; dashed line gives center of mixing-layer prediction. The unit pphm represents parts per hundred million mixing ratio. Operation 33 was part of the "learning set" (Lloyd et al., 1979b).

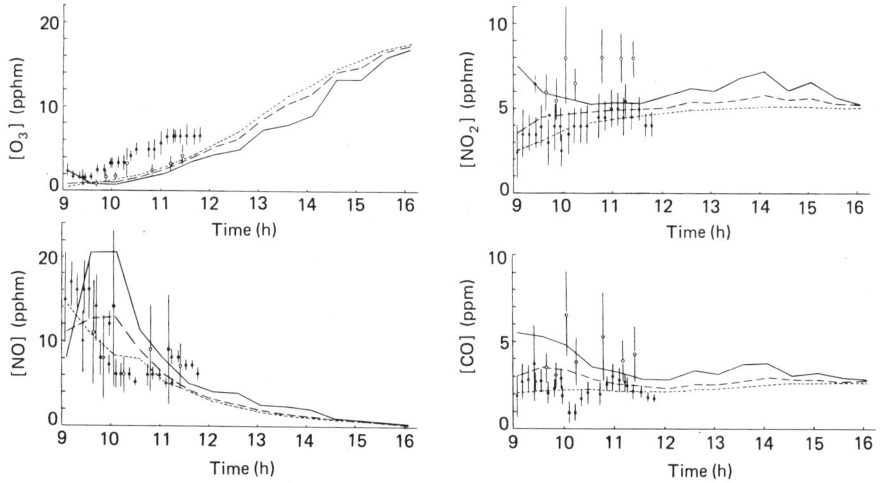

Figure 6. Comparison of ELSTAR predictions with LARPP Operation 23 data. Solid line gives ground-level prediction; short dash, center of mixing-layer prediction; long dash, intermediate altitude prediction. Solid symbols give above-ground-level LARPP data; open symbols, ground-level LARPP data. The unit pphm represents parts per hundred million mixing ratio. Operation 23 was part of the "test set" (Lloyd et al., 1979b).

(Godden et al., 1980). Potential users are referred to the ELSTAR User's Guide (Lurmann, 1979).

3.3. LARPP Data Availability

The LARPP data archive is available on magnetic tape from the National Technical Information Service (NTIS). Descriptions of the available data are given by Eschenroeder (1976) and Martinez (1976).

4. AIRCRAFT LAGRANGIAN PLUME STUDIES

4.1. General Considerations

Instrumented aircraft plume studies fall into two categories: those that focus on urban plumes; and those that focus on power plant or other major point source plumes. Most studies have concerned power plants, especially coal-fired power plants, because of their importance as sources of sulfur in the atmosphere (Meagher, 1980).

Under most meteorological conditions of interest, the plume from a major isolated point source remains identifiable by instrumented aircraft for 50–100 km downwind. Under some conditions plumes are sufficiently cohesive that they may be followed by instrumented aircraft for two to three times as far (Gillani et al., 1978). Under these circumstances the plume itself may be used to perform part of the function of a Lagrangian marker— to define the effluent trajectory. If the information available from plume location is supplemented with good winds-aloft data, it is possible to perform at least approximately Lagrangian plume studies without benefit of a Lagrangian marker balloon system.

In fact, in the majority of aircraft plume studies to date, little effort has been made to take data in a Lagrangian frame of reference, either by using a Lagrangian marker or by appropriately employing winds-aloft data. Rather, it is often implicitly assumed that the plume under study is time invariant.

Especially in areas where major sources are not very well isolated—the industrial northeastern United States—most studies have been constrained to conditions that permitted the plume to be easily followed. Thus many tended to be limited to relatively isolated cohesive plumes early in the day, and for only a few tens of kilometers downwind. Under these conditions, little evidence was found of atmospheric oxidation of sulfur dioxide from power plants beyond the immediate vicinity of the source, that is, between the stack and the first aircraft plume cross section (Forrest and Newman, 1977; Meagher et al., 1978). The conclusion was drawn by some that what was observed in these studies was typical of the behavior of sulfur dioxide in the atmosphere as a whole. On the other hand, earlier measurements reported much higher oxidation rates, but these have come to be regarded in many quarters as suspect (Newman, 1981). Longer-range studies that did make use of good winds-aloft data have shown that afternoon oxidation rates far downwind of the source typically fall in the 1–4%-h^{-1} range (Gillani and Wilson, 1980). Recent measurements in urban plumes, or in mixed urban-power plant plumes have shown even higher oxidation rates. These are summarized by Alkezweeny (1980). Although patterns are beginning to emerge from the now sizable body of measurements that have been made, controversy still thrives in this research area.

There are several sources for the persistent uncertainty surrounding atmospheric sulfur chemistry. Most fundamental is the fact, now more generally appreciated, that the phenomena are complex and influenced by many factors. Second, the concentration measurements themselves, particularly for aerosol sulfate, are difficult to make and subject to a variety of subtle errors. Finally, the inherent variability of the atmosphere, coupled with the fact that the atmospheric chemistry of interest is slow, imposes formidable problems in both the conduct of experiments and the interpretation of results. Lagrangian experiments ameliorate the difficulties from this last source.

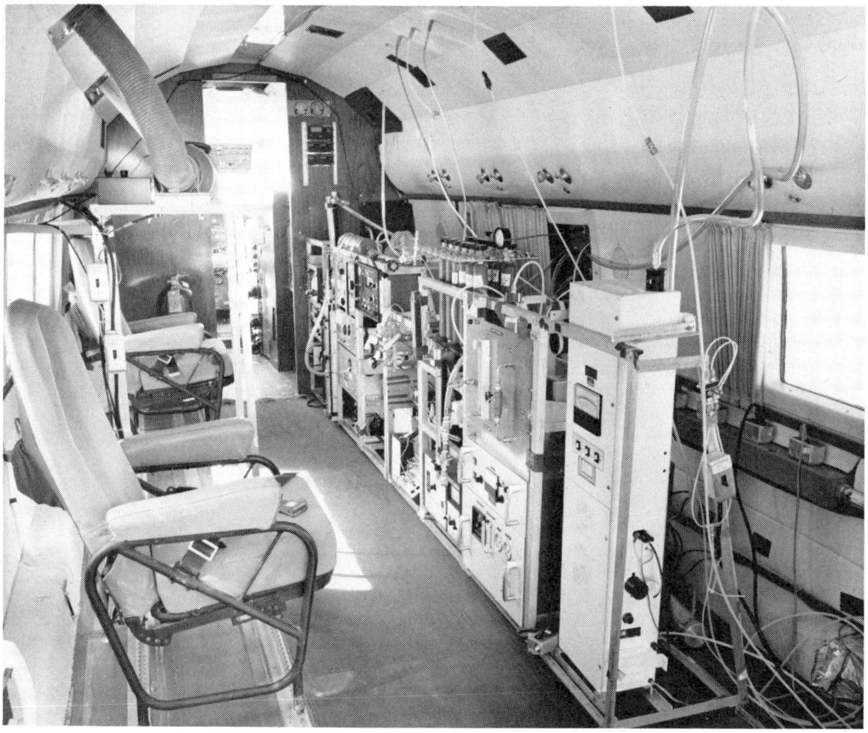

Figure 7. Interior of PNL instrumented DC-3. Courtesy of A. J. Alkezweeny.

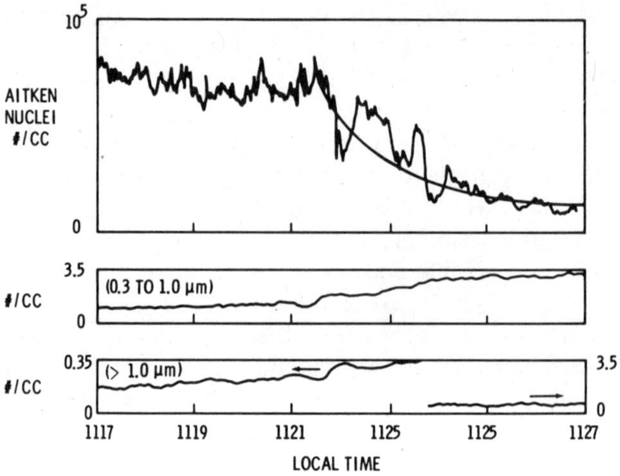

Figure 8. Aitken nuclei and larger particle concentrations downwind of St. Louis at an altitude of 450 m on August 1, 1973. Data taken in a Lagrangian frame (Alkezweeny, 1978). Reprinted with permission of the American Meteorological Society.

4.2. Experiments

The atmospheric chemistry group at Battelle Pacific Northwest Laboratories came early to an appreciation of the advantages of doing plume studies in a Lagrangian frame of reference. Over the last several years Lagrangian experiments have become their standard procedure. The experiments utilize either a tetroon release, winds-aloft data, or both. The winds-aloft data are

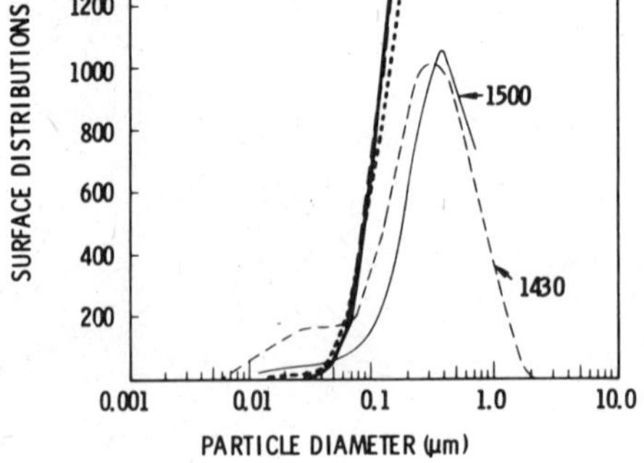

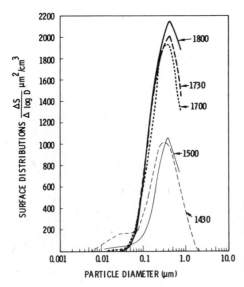

Figure 9. Surface distribution of aerosol particles measured at various times downwind of St. Louis in a Lagrangian frame on August 17, 1974. (Alkezweeny, 1978). Reprinted with permission of the American Meteorological Society.

acquired by a wind sensing system incorporated into the instrument complement of their aircraft, a DC-3 (Figure 7).

Early Lagrangian studies were undertaken as part of project METROMEX (Changnon et al., 1971). The objective of these studies, carried out for the most part during the summers of 1973–1975, was to study sulfur dioxide to sulfate conversion and aerosol dynamics in the St. Louis, Missouri urban plume (Alkezweeny, 1976, 1978). Measurements were taken in the vicinity of a tetroon, which served as the Lagrangian marker. It is of interest to note that electronic tracking of the tetroon was not employed. Over the relatively short distances involved in these experiments, visual tracking was found to be adequate.

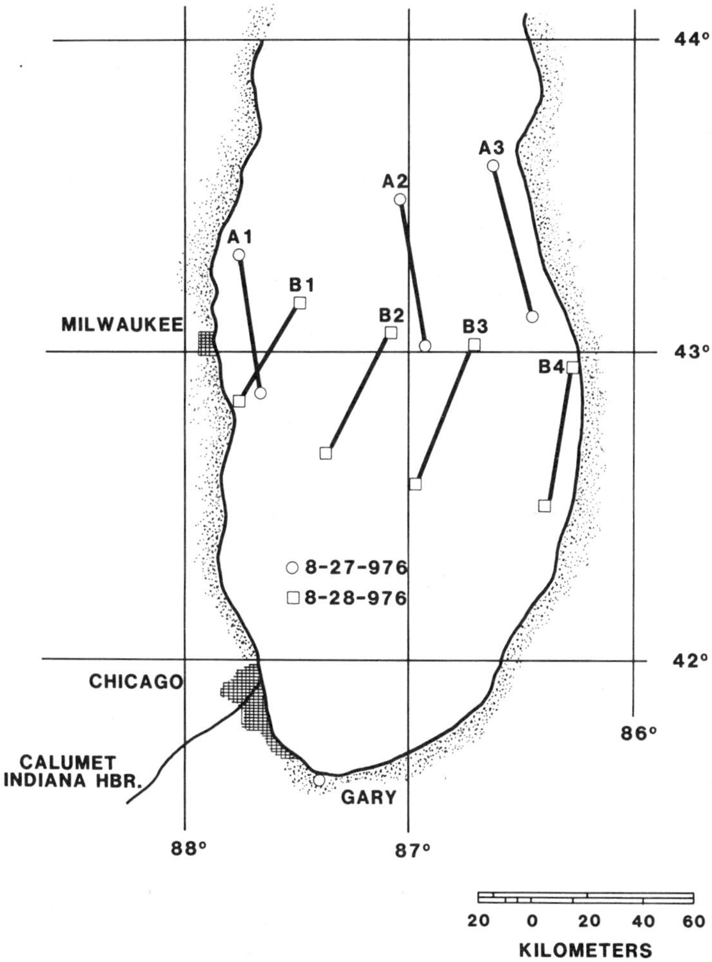

Figure 10. Aircraft sampling routes downwind of Milwaukee during August 27–28, 1976 experiments (Alkezweeny, 1980).

In these experiments it was determined that the rate of SO_2 to SO_4^{2-} conversion in the St. Louis, Missouri, urban plume during summer afternoons was about 11% per hour. Typical observations of particle dynamics in the Lagrangian frame, made with an Aitken nuclei counter, an electrical aerosol analyzer, and a modified optical particle counter, are shown in Figures 8 and 9.

More recently, PNL studies have been carried out in the vicinity of Chicago, Illinois, Gary, Indiana, and Milwaukee, Wisconsin. The experiments take advantage of the high degree of atmospheric stability regularly found over Lake Michigan to study cohesive urban and power plant plumes from shoreline sources without the complication of downwind sources (Alkezweeny et al., 1977).

Of particular interest are the data from August 27 and 28, 1976 on the Milwaukee plume (Figure 10). On these 2 days similar meteorological conditions and SO_2 loadings prevailed. On both days sampling began at about noon. Both days were also characterized by ozone buildup. Despite the similarities, the two days exhibited marked differences in sulfate production.

On August 27 no detectable sulfur conversion was observed. On August 28 a rate of 6.8% per hour was measured. Such wide variations under similar meteorological conditions were again observed during experiments conducted during the summer of 1978. It was hypothesized that the difference between the two days that may account for the difference in conversion rate was that on August 28 the air advecting into the Milwaukee area contained aged pollutants, whereas on August 27 the incoming air was rather clean (Alkezweeny, 1980).

5. LAGRANGIAN MEASUREMENT PLATFORM (LAMP)

5.1. Concept

In the Lagrangian experiments discussed thus far, virtually all the chemical data and most of the meteorological data from the study volume were collected by instrumented aircraft. In most cases the distance and time scales involved were relatively modest—tens of kilometers and a few hours of pollutant transport time. Under these circumstances the instrumented aircraft could survey the study volume frequently. In LARPP Operation 33, for instance, 49 complete patterns were flown in 6 h. This was possible because the travel distance of the volume of interest was short—wind speed was very low.

As the travel time and distance of interest expand, the frequency with which aircraft data can be taken in the study volume falls markedly. The flying time to and from the Lagrangian volume becomes a significant fraction of the total time the aircraft can remain airborne. In addition, longer-range studies require that the study volume itself be proportionately expanded.

Consequently, the cross-sections or other survey patterns cover more distance and consume more flight time. For studies on a scale of hundreds of kilometers, one does well to obtain an aircraft survey every 3–5 h.

The sparsity of data in long range studies can in part be overcome by employing more than one aircraft. However, there is another approach. Onc may expand the load-carrying capability of the Lagrangian marker balloon so that it becomes a measurement platform in its own right. If this is done, continuous real-time data can be collected along the trajectory. In addition, it becomes feasible to take time-integrated particulate and gas samples that are not also spatially integrated over the plume cross section. Thus a LAMP provides a different kind of data set that complements that available from aircraft.

For long-range pollutant transformation studies, a LAMP may offer certain operational advantages over a simple marker as well. To date, Lagrangian markers have been uncontrolled CVBs that have a significant probability of being lost through either malfunction or destruction at the ground surface. With the greater load-carrying capability of a LAMP, provision may be made to control its altitude to avoid premature termination of the experiment. This can be of special value where trajectories pass over mountainous terrain. In addition, a LAMP can carry a sophisticated radio navigation system to ease the problem of long-range tracking.

Of course, a LAMP is just a Lagrangian marker with data acquisition capabilities. The type and extent of those capabilities may be chosen according to the needs of the experiment.

5.2. Project Da Vinci

5.2.1. Operations

The idea of using a large, manned helium balloon as a Lagrangian measurement platform in the boundary layer was proposed by V. Simons and R. Englemann in the early 1970s. In November 1974 the first Da Vinci experiment (DV I) was conducted. It was a proof-of-concept flight conducted in a clean air environment in New Mexico (Englemann et al., 1975; Zak et al., 1981).

In June and July 1976, two more experiments (DV II and DV III) were conducted. The objective of these experiments was to study the long-range behavior of effluents from the St. Louis urban area. At that time, St. Louis was the site of the Regional Air Pollution Study (Schiermeier, 1978) and the Midwest Interstate Sulfur Transformation and Transport (MISTT) project (Wilson, 1978). In collaboration with RAPS and MISTT, the DV II and DV III experiments provided information on the behavior of the mixed urban–industrial plume from St. Louis out to hundreds of kilometers downwind (Zak, 1981a).

The DV II experiment was launched on June 8, 1976 at 0856 CDT from a site about 25 km west of the St. Louis City Center. The experiment began under stagnation conditions. The LAMP spent 12 hours over St. Louis and its suburbs before being carried off to the east by a nocturnal low-level jet (Figure 11). The LAMP was carried by the jet across Illinois through the night. It landed in southern Indiana the following morning, a little more than 24 h after launch.

During the afternoon hours over the St. Louis area the LAMP was embedded in a dilute SO_2 plume, probably from the Meramec power plant south of the city. During the night the LAMP was embedded in an intense SO_2 plume from one or more of the sources in the Alton, IL area.

The DV III experiment was launched from the same site at 0725 CDT on July 23 under clear-sky, brisk-wind conditions. The LAMP was immediately carried across the city, passing just to the north of the Gateway Arch. For the next 16 h the LAMP was carried to the east over Illinois and Indiana (Figure 12). From late morning and through the afternoon, the LAMP was embedded in the edge of the plume from the Labadie power plant. Because of a thunderstorm that threatened the LAMP near the Indiana–Ohio border

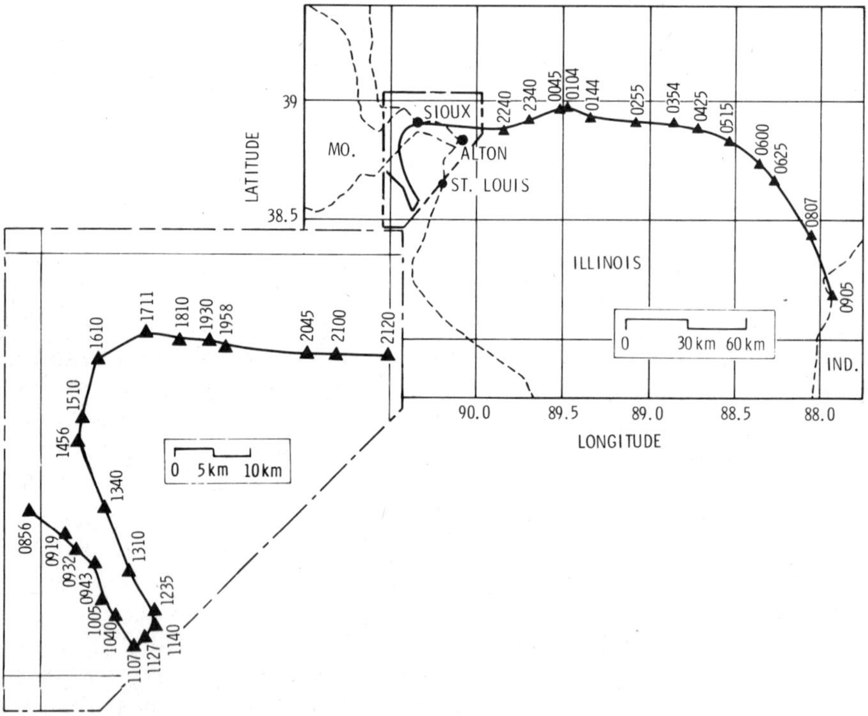

Figure 11. DV II LAMP trajectory. Numbers adjacent to locations are times, CDT. Insert at lower left gives blowup of trajectory over St. Louis metropolitan area.

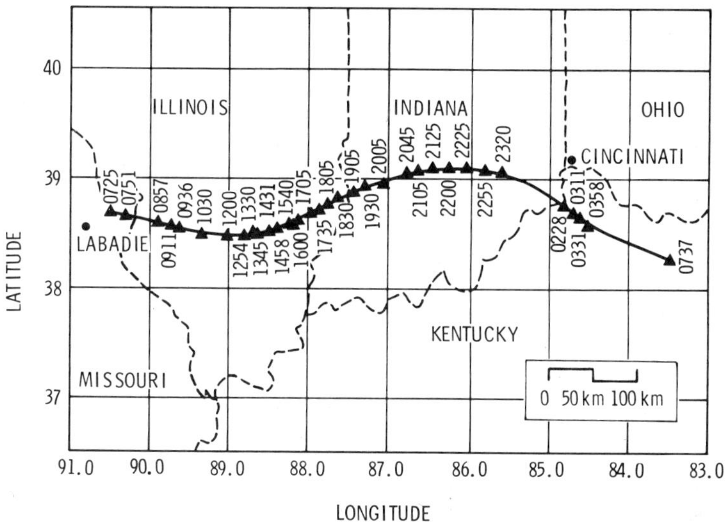

Figure 12. DV III LAMP trajectory.

the altitude was increased several thousand feet to use winds at the higher altitude to put greater distance between the LAMP and the storm. Although the Lagrangian experiment was considered terminated at that time, the LAMP flew on through the night and landed the following morning near Moorehead, Kentucky, in the foothills of the Appalachians, 24 h after launch.

The Da Vinci experiments involved the collaboration of a large number of researchers. Table 2 gives the instrument complement of the LAMP for DV II and III and lists the individuals and organizations involved. In addition, aircraft measurements were made by the MISTT team and by R. Pueschel of NOAA. Pilot-balloon measurements of winds aloft were supplied by a RAPS mobile unit. The Research Triangle Institute instrumented mobile van made ground-level pollutant concentration measurements along the trajectory of DV II. The Argonne National Laboratory mobile acoustic sounders also made atmospheric structure measurements along the paths of both DV II and DV III. The Lagrangian measurements were supplemented by data from the 25 station RAMS (Regional Air Monitoring System) network, which operated in support of the RAPS program.

5.2.2. Results

Several interesting results emerged from the Da Vinci St. Louis Experiments. One of the more noteworthy concerned the effective nocturnal lifetime of ozone under fair-weather conditions. During DV II the ozone concentration built up to over 300 micrograms per cubic meter during the day. At night, as the LAMP was swept along by the nocturnal jet, very little

Table 2. Meteorological and Chemical Measurements for Da Vinci II and III

Measurement	Investigators
Temperature, pressure, and relative humidity	H. Ballard and M. Izquierdo Atmospheric Sciences Laboratory and University of Texas at El Paso
Sling psychrometer	R. Engelmann NOAA/Environmental Research Laboratories
Eddy diffusivity and relative air motion	P. MacCready and J. Mullen AeroVironment, Inc.
Atmospheric diffusion	J. Angel and W. Hoecker NOAA/Air Resources Laboratory
Sulfur dioxide Pulsed-fluorescence-type monitor	R. Schellenbaum and B. Zak Sandia National Laboratories
Sulfur dioxide and sulfate Tandem-filter pack	J. Forrest, L. Newman, and S. Schwartz Brookhaven National Laboratory
Elemental composition and sulfur speciation of aerosols X-ray fluorescence (XRF) Electron spectroscopy for chemical analysis (ESCA)	R. Dod, R. Giauque, C. Hollowell, S. Chang, T. Novakov, and G. Traynor Lawrence Berkeley Laboratory
Elemental composition of aerosols Instrumental neutron activation analysis (INAA)	J. Ondov and R. Ragaini Lawrence Livermore Laboratory
Elemental composition of aerosols Proton-induced x-ray emission (PIXE)	W. Berg, D. Lawson, A. Leslie, J. Nelson, K. Pillotte, and J. Winchester Florida State University
Elemental composition and size distribution of aerosols Scanning electron microscopy (SEM) X-ray energy dispersive analysis (XEDA)	R. Pueschel and C. VanValin NOAA/Environmental Research Laboratories
Size distribution of aerosols Quartz crystal microbalance (QCM) cascade impactor	M. Fowler and W. Sedlacek Los Alamos Scientific Laboratory
b_{scat} Integrating nephelometer	R. Schellenbaum and B. Zak Sandia National Laboratories
Ice nuclei and Aitken nuclei Ice crystal counter Expansion chamber	R. Pueschel and C. VanValin NOAA/Environmental Research Laboratory
Ozone Ultraviolet absorption monitor (Dasibi)	R. Schellenbaum and B. Zak Sandia National Laboratories

Table 2. (Continued)

Measurement	Investigators
Ozone Solid-state fluorescer-type ozonesonde	H. Balland and M. Izquierdo Atmospheric Sciences Laboratory and University of Texas at El Paso
Hydrocarbons, halocarbons, and carbon monoxide Gas chromatography–grab sample	W. Bach, C. Decker, L. Ripperton, J. Sickles, II, F. Vukovich, and J. Worth Research Triangle Institute
Hydrocarbons, halocarbons, and nitrous oxide Gas chromatography–grab sample	D. Cronn and R. Rasmussen Washington State University
Neon, hydrogen, methane, and carbon monoxide Gas chromatography–grab sample	L. Heidt and R. Woods National Center for Atmospheric Research and Sandia National Laboratories

decrease in ozone concentration was observed (Figure 13). The data were fitted to an exponential decay, and the nocturnal "half-life" of the ozone concentration was found to be about 84 h. Other researchers treated the data somewhat differently and arrived at a half-life of 116 h (Decker et al., 1978).

The next morning at the landing site, as the ground-based inversion broke up, the RTI van observed an increase in the ground-level ozone concentrations essentially to the value that had been measured aloft before landing.

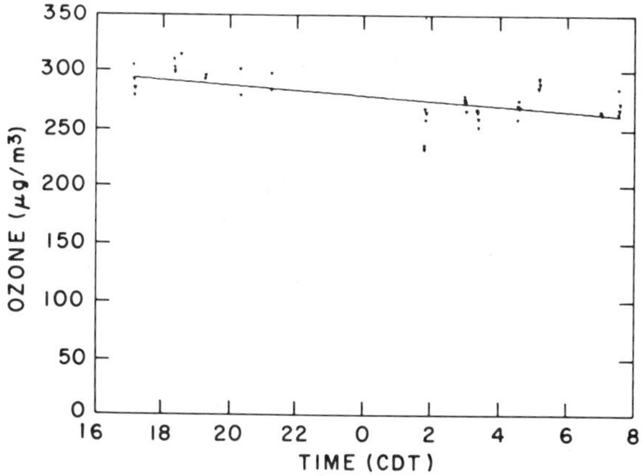

Figure 13. Fit of decaying exponential function to DV II LAMP ozone data taken in a Lagrangian frame downwind of St. Louis. Data correlated with major sulfur dioxide peaks have been suppressed. Best-fit exponential yields an effective lifetime of 121 h, or a half-life of ozone concentration of 84 h.

As part of the Da Vinci analysis effort, Bekowies (1982) has modeled the data from DV II. He applied an existing carbon bond mechanism developed for a smog chamber (Whitten and Hogo, 1977) and incorporated it into a two-box model that simulated the behavior of air above and below an inversion, the height of which was time dependent. For ozone, he obtained very good agreement with both the LAMP data and the RTI van data.

During the day and evening an ozone concentration pattern was observed for DV III similar to that of DV II, but a nocturnal lifetime could not be inferred since the experiment was terminated at around midnight. It is interesting to note that for both DV II and DV III, ozone concentration continued to grow in the downwind plume until late afternoon, when the UV flux began to drop. This suggests that the ozone production may have resumed in the aged plume the following morning. It is unfortunate that the experiments had to be terminated before this could be confirmed.

Another interesting ozone phenomenon was observed by the RTI van. At ground level a subsidiary ozone peak occurred around 0100 CDT during DV II. It correlated with changes in the wind speed and direction at the LAMP, with an increase in turbulence measured on the LAMP and with changes in the apparent depth of the inversion observed by the acoustic sounders. It is hypothesized that we observed in uncommon detail a fairly common summertime phenomenon—transient instability of the inversion under a low-level jet.

The results from the sulfur chemistry were equally interesting. Rates of SO_2-to-SO_4^{2-} conversion were calculated for the periods during which the LAMP was in SO_2 plumes. Over the city of St. Louis on the afternoon of June 8 an oxidation rate of 8.5% per hour was found (Figure 14). In this analysis the pulsed-fluorescence SO_2 data and the LBL XRF particulate sulfur data were used. However, other participating researchers, using their

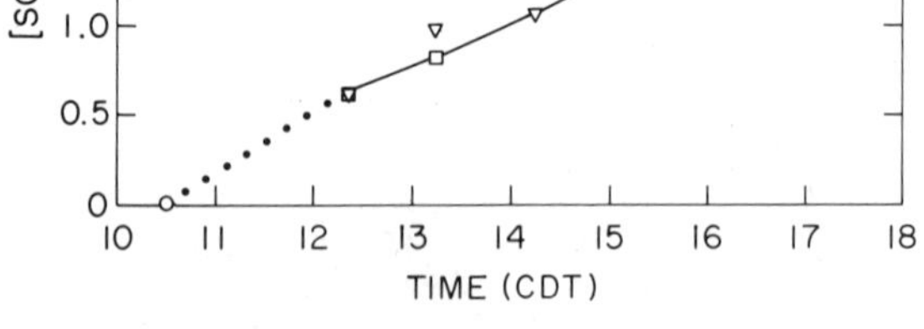

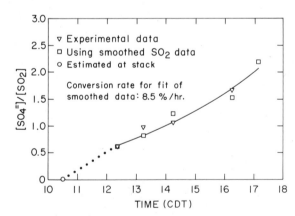

Figure 14. Ratios of particulate sulfur: gaseous SO_2 molar concentrations on LAMP during DV II over St. Louis. Conversion rate for best fit is 8.5% per hour. X-ray fluorescence analysis of filters for particulate sulfur courtesy of R. Giauque, Lawrence Berkeley Laboratory.

own data from the LAMP (BNL tandem filter packs) and their own method of analysis, concluded that the oxidation rate during this period was less than or equal to 4% per hour (Forrest et al., 1979). The quality of the fit of the data shown here lends credence to the higher rate. However, it is estimated that the uncertainty in such rate calculations is about ±50% at the 90% confidence level. Hence the discrepancy is not as large as would appear.

During the early hours of June 9 in the plume of one or more sources from the Alton area, a nocturnal conversion rate of 1.1% per hour was calculated, again with a ±50% uncertainty. On the afternoon of July 23, during the period 1332–1713 CDT, a conversion rate of 4.4% per hour was found; during this period the LAMP was on the fringe of the Labadie plume. This period began 7 h and 175 km downwind from the source. Chang (1979), using a different data set from the LAMP (BNL filter packs) and a different method of analysis, found a conversion rate of 4.2% per hour for this period, in excellent agreement with our calculations.

The sulfur oxidation rate analysis used by the author focuses on the particulate : gaseous sulfur ratio. This method was developed to treat the Da Vinci data but is of quite general applicability. It is less sensitive to assumptions regarding deposition and dilution rates than other methods in current use (Zak, 1981a,d) and preserves the advantage of ratio analysis (Alkezweeny and Powell, 1977).

Dod et al. (1982) analyzed sequential particulate samples taken onboard the LAMP by XRF and ESCA. By coupling the two data sets, these researchers were able to examine the time dependence of the ion ratio of $NH_4^+ : SO_4^{2-}$ (Figure 15). The monotonic increase in this ratio suggests that one may be seeing progressive neutralizaton of acidic aerosol by ambient ammonia. Furthermore, it appears that more rapid neutralization occurs during the afternoon period. This could be due to more rapid mixing rather than photochemical involvement.

5.3. Tennessee Plume Study LAMP Flights

The Tennessee Plume Study (TPS) was the first major field effort of the EPA's STATE (sulfur transformation and transport in the environment) program. This was a large-scale collaborative study focused on the 2600-MW_e coal-fired Cumberland power plant, near Clarksville, Tennessee. The LAMP experiments as part of the TPS were undertaken with the joint support of DOE and the EPA (Gay et al., 1981).

Two experiments took place. The first, on August 20, 1978, was launched at 1024 CDT from a site 0.25 mile from the base of the 1000-ft-tall Cumberland stacks. The LAMP moved off to the southwest, entering the plume within 20 min of liftoff. It remained embedded in the plume for the remainder of the experiment. The LAMP landed near the Tennessee–Mississippi border at 1835 CDT. The trajectory covered about 160 km.

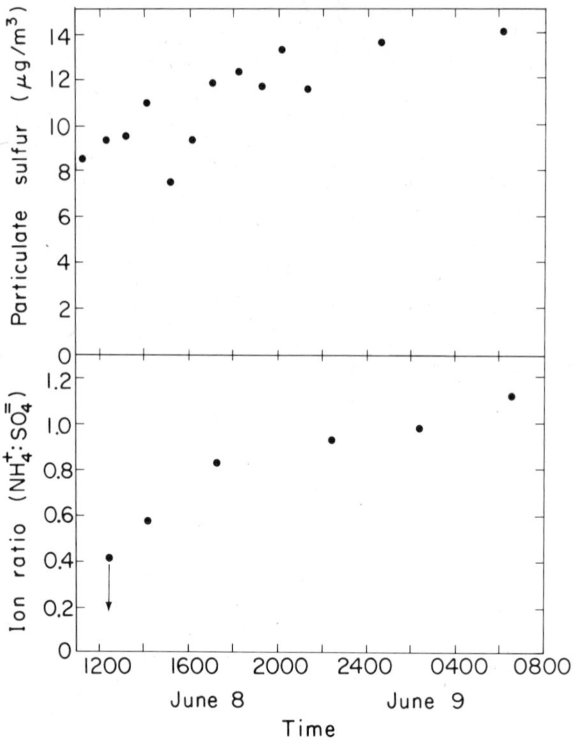

Figure 15. DV II LAMP XRF particulate sulfur data, and ESCA ammonium : sulfate ion ratios. The arrow on the initial ion ratio indicates that the ratio is an upper limit (Dod et al., 1982).

The second flight was launched on August 25, 1978 at 1018 CDT from the same site, but under stagnation conditions. Because of a slow climbout and wind shear, the LAMP was displaced perhaps as much as 1 km from the plume centerline by the time it reached plume altitude. The fringe of the plume engulfed the LAMP after about 1.5 h. For the remainder of the experiment, the LAMP remained on the extreme edge of the Cumberland plume. The LAMP was landed at 1636 CDT, a mere 35 km from the launch site.

The balloon used for the TPS flights was a spherical 10-m-diameter constant-volume helium balloon comprised of a Kevlar fabric–Mylar laminate that had been flown several times before by the Aerospace Corporation (Heinsheimer and Neushul, 1978). It was about the size of a hot-air sport balloon. The instrumentation payload planned for the TPS experiments is given in Table 3. As noted there, two real-time monitors were not flown because the actual lift capability of the balloon did not quite meet its specifications.

Simultaneous aircraft measurements were made during both flights by the Brookhaven National Laboratory Islander aircraft. Winds-aloft and ground-level data were provided by other TPS researchers.

For approximately an hour before to an hour after the LAMP launch, SF_6 tracer was released into the Cumberland stacks. Syringe grab samples for SF_6 analysis were taken every 15–20 min onboard the LAMP. The SF_6 behavior from the August 20 experiment is shown in Figure 16. The data from the August 25 experiment are shown in Figure 17. A comparison of the two illustrations demonstrates the gross difference between the placement of the LAMP in the plume on the two days.

The behavior of the ratio of particulate to gaseous sulfur concentrations observed in the two experiments is also quite different (Figures 18 and 19). On August 20, near the center of the plume, only slow oxidation occurred

Table 3. Instrumentation Payload for Summer 1978 Experiments

		Average Power (W)	Mass (kg)
Scientific Instrumentation			
Pulsed-fluoresence/SO_2 monitor[a]		(25)	(19)
SO_2/SO_4^{2-} tandem filter packs (6)		15	4.1
Five-stage cascade impactors (10)		3	11.3
Hydrocarbon grab samplers (6)		—	5
Dasibi ozone monitor[a]		(30)	(9)
SF_6 tracer sampler syringes			1
Temperature and relative humidity monitor (aspirated)		3	2.3
Radar altimeter			3.2
Pressure altitude monitor		2	3.2
Automatic position indicator (LORAN)[a]		(30)	(3.2)
Minicomputer controller and data logger		—	12.2
Systems Instrumentation			
VHF transceiver	Pilot's package	30	13.6
VOR instrumentation			
Rate-of-climb meter			
Pressure altitude sensor			
Radar transponder			
Superpressure transducer with alarm		—	0.9
Emergency locator transmitter		—	1.3
UHF transceiver		—	4.5
Balloon temperature transducers		—	0.1
Strobe lights		9	1.2
Telemetry transmitter		12	0.4
4-kWh lithium batteries (14)		—	8
Totals for actual flight payload		74	72

[a] Removed from actual flight payload because of weight constraints.

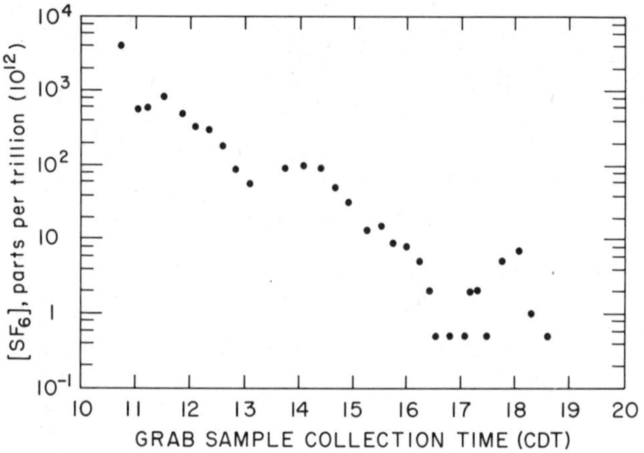

Figure 16. Sulfur hexafluoride concentration measured on board the LAMP versus time during August 20, 1978 experiment. SF_6 gas chromatographic analysis courtesy of E. Sasaki and F. Shair of California Institute of Technology.

for the first hour or so. Over this initial period, a conversion rate of 0.4% per hour was calculated. However, for the remainder of the flight, an average conversion rate of 6.4% per hour was found. On August 25, on the extreme edge of the plume, the calculated conversion rate from the stack to first measurement period was 16.7% per hour, whereas the average conversion rate for the remainder of the flight was 2.7%. Even taking into account the substantial uncertainty associated with the initial measurement, it is clear that a major difference exists between the measurements for the initial period during the two experiments. The difference is consistent with the Gillani–Wilson (1980) hypothesis that plume effluent–background air inter-

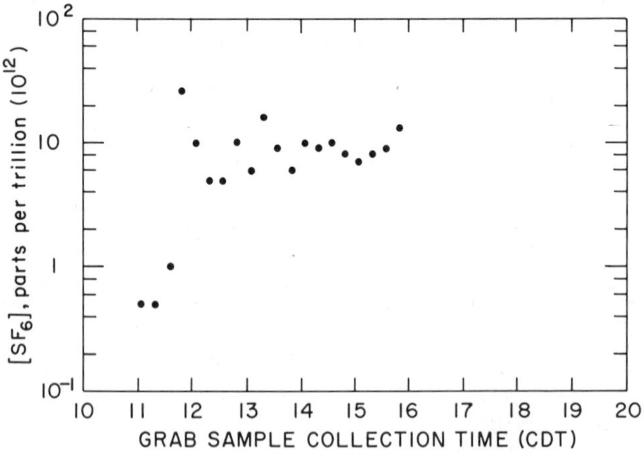

Figure 17. Sulfur hexafluoride concentration measured on board the LAMP versus time during August 25, 1978 experiment. SF_6 gas chromatographic analysis courtesy of E. Sasaki and F. Shair of California Institute of Technology.

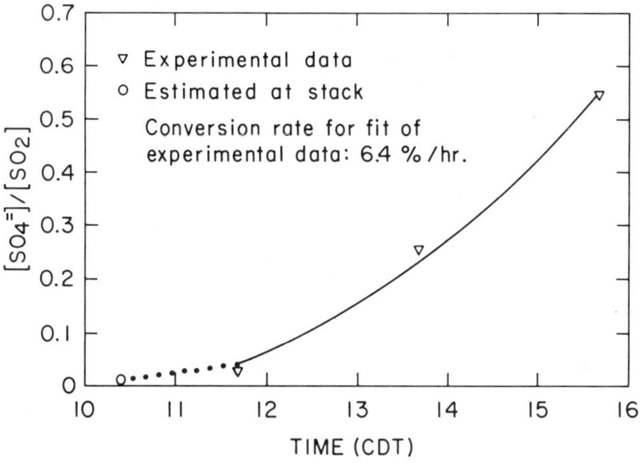

Figure 18. Ratios of molar concentration of particulate sulfur to sulfur in form of sulfur dioxide observed on LAMP during August 20, 1978 experiment. Data from analysis of BNL tandem filter packs.

actions are of primary importance to sulfur dioxide to sulfate conversion in power plant plumes.

5.4. Summary of Results

A summary of the sulfur conversion rate determinations made during both the Da Vinci and Tennessee Plume Study LAMP experiments is given in Table 4, along with a compilation of the conditions under which the measurements were made.

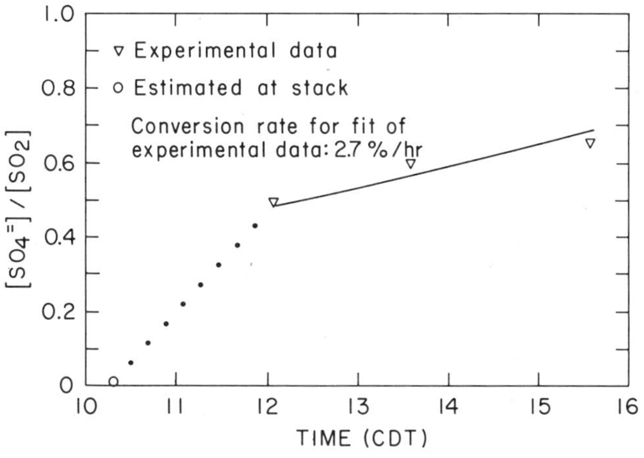

Figure 19. Ratios of molar concentration of particulate sulfur to gaseous sulfur in form of sulfur dioxide observed on LAMP during August 25, 1978 experiment. BNL filter pack data.

Table 4. SO_2-to-Sulfate Conversion Rates and Attendant Conditions Measured by LAMP

Experiment	Rate (%) h^{-1}	Time (CDT)	Plume Age (h)	Excess SO_2 (ppb)	Solar Irradiation[a] (Ly min^{-1})	O_3 Concentration[b] (ppb)	Temperature[c] (°C)	Relative Humidity[c] (%)	NMHC[d] (ppm C)	Comments
8/20/78 TPS I										
Stack to 1st meas.	0.4	1024–1141	0–1.3	Stack-480	1.0	—	20	63	0.9	Mostly sunny
Fitted ratio series	6.4	1141–1541	1.3–5.3	480-30	1.2	135	22	58	0.8	Mostly sunny
8/25/78 TPS II										
Stack to 1st meas.	16.7	1018–1208	0–1.8	Stack-20	.95	—	26	42	0.4	Mostly sunny
Fitted ratio series	2.7	1208–1535	1.8–5.3	20-14	.95	105	21	35	0.6	Mostly sunny
6/8/76 DV II										
Stack to 1st meas.	6.1	1030–1222	0–1.9	Stack-10	1.2	95–120	—	—	—	Sunny with haze

Fitted ratio series	8.5	1222–1708	1.9–6.6	10-2	0.9	120–150	14	70	—	Sunny with haze
6/9/76 DV II										
Stack to 1st meas.	0.7	2100–0213	0–5.2	Stack-50	0	150	20	53	—	Night
Fitted ratio series	1.1	0213–0422	5.2–7.4	50-35	0	150	20	53	—	Night
7/23/76 DV III										
Stack to 1st meas.	3.4	0630–1332	0–7	Stack-8	—	50–80	—	—	—	Clear
Fitted ratio series	4.4	1332–1713	7–10.7	8-2	—	80–95	—	—	—	Clear

[a] For the TPS experiments, the solar irradiation data were provided by the Tennessee Valley Authority (Gillani, 1979b); for the June 8, 1976 Da Vinci flight, these data were abstracted from Decker et al. (1978).

[b] For the TPS experiments, the ozone data were provided by Brookhaven National Laboratory instrumented aircraft (Gillani, 1979a); for the Da Vinci experiments, they were provided by Sandia National laboratories onboard instrumentation (Zak, 1981a).

[c] For the TPS experiments, the LAMP temperature and relative humidity data were provided by Sandia National Laboratories in collaboration with the University of Denver; for the Da Vinci experiments, they were provided by Sandia in collaboration with the Atmospheric Sciences Laboratory and the University of Texas at El Paso.

[d] The nonmethane hydrocarbon data come from analysis of grab samples by the Research Triangle Institute taken on board the LAMP (Gay, 1981).

6. DEVELOPMENTS IN LAGRANGIAN TECHNOLOGY

6.1. New Systems

As the emphasis in air pollution research has shifted from local to long-range effects, the interest in Lagrangian technology has increased. The shift coincided with developments in microelectronics that facilitate miniaturization of tracking systems and other instrumentation. The result is that the Lagrangian markers now under development also incorporate some limited measurement capability, becoming mini-Lagrangian measurement platforms. At the same time, the full-capability LAMP is shrinking in size.

A group at the EPA Meteorology Branch at Research Triangle Park is developing a Lagrangian marker that carries a standard FAA aircraft transponder and an encoding altimeter. The transponder shows up on all FAA radars. Observers stationed at FAA air traffic control centers in the region in which an experiment is being conducted can track the marker and relay position information to the operations center for the experiment. Controllers at the operations center can then vector instrumented aircraft to the marker. The constraint is that the marker must be flown at an altitude high enough to be within line of sight of FAA radars during the experiment—typically a few thousand feet. The system was flown a number of times with good success during the 1980 PEPE (Prolonged Elevated Pollution Episode) field program (Ching, 1981).

Another marker system is under development by the NOAA group at Idaho Falls (Dickson, 1982). It makes use of a LORAN rebroadcast system similar to that employed by LORAN radiosondes. The rebroadcast LORAN signals from the marker are received either at a ground station or on board an instrumented aircraft. A decoding system extracts the marker location information from the LORAN signals. Recent improvements in the LORAN network should make this system usable over much of the United States.

The author's group at Sandia National Laboratories (SNL) is experimenting with a marker system that is similar in principle to a ranging radiosonde (Woods, 1982). A ground or airborne portable interrogator broadcasts a radio signal modulated with an audio tone. The marker receives and rebroadcasts the signal on a different radiofrequency. The interrogator receives the rebroadcast signal and compares the phase of the received audio with the audio signal being transmitted. The phase shift gives the distance to the marker. When coupled with a directional antenna, the system gives the location of the marker relative to the interrogator. A similar system is in use for tracking stratospheric balloons by a group at Utah State University (Pound, 1980). This system will permit instrumented aircraft to home on the marker and execute circular survey patterns at a predetermined radius relative to the marker at selected altitudes. This procedure will make aircraft surveys far more reproducible from survey to survey.

Each marker tracking system described above has advantages and disadvantages. However, microelectronics offers the possibility of combining the features of two or even all three systems without exceeding the weight constraints (6 lb or 2.7 kg) imposed by Federal Aviation Regulation 101.

The author's group is also completing the development and testing of an automated, remotely piloted LAMP (Figure 20). Whereas previous versions of the LAMP required an on-board pilot, the automated LAMP will not. It will be flown, as stratospheric research balloons are, by remote control. Because it incorporates a CVB, the altitude will be self-regulating. However, intervention by a remotely located pilot in its vicinity will be necessary for launch and landing and perhaps at dawn and dusk, when the temperature of the system changes markedly.

The automated LAMP includes a radio navigation system that automatically scans all DME (radio navigation) channels and acquires up to five DME stations within range including portable units that may be deployed.

The instrumentation consists of the usual complement of sensors and samplers carried on well-equipped instrumented aircraft. The present payload is 200 kg with the possibility for expansion to 450 kg. The instrumentation is controlled by an on-board microcomputer that also formats the data, records it on magnetic tape, and telemeters it to ground or airborne portable units. The system is described in more detail elsewhere (Zak, 1981b,c).

Figure 20. LAMP payload being prepared for field experiment. Instrument suitcase at the lower left is the portable data reception and remote control unit. The payload can also be used to measure and sample close in to elevated stacks if suspended 30 m below a helicopter.

6.2. Future Uses

Three related long-range air pollution problems are strong candidates for research involving the application of Lagrangian markers and measurement platforms: acid rain (sulfur and nitrogen chemistry); regional visibility degradation; and long-range oxidant transport phenomena. In each case the use of the Lagrangian systems will enable researchers to extend their understanding of the transport, transformation, and removal processes occurring far downwind of source areas. The goal will be to encompass the entire multiple-day "life cycle" of the species of interest.

In the near term a particularly attractive use for the Lagrangian measurement platform is in the acid rain program. It has been pointed out by Gillani (Gillani et al., 1981; Gillani and Wilson, 1982) that very little information exists on the interaction of effluent plumes with clouds. He also points out that a Lagrangian measurement platform embedded in a cloud and suitably equipped to take sequential cloud water samples would make an excellent tool with which to investigate cloud droplet chemistry. Such Lagrangian experiments could not only study the interaction of industrial or urban plumes with clouds, but could also involve intentional in-cloud reactant releases in the vicinity of a LAMP. This latter possibility offers an exciting prospect. The gap between chamber experiments and ambient measurements would finally be bridged.

APPENDIX A: WHAT DOES THE TERM "LAGRANGIAN" MEAN?

The term "Lagrangian" is used in at least two distinct but related senses in the air pollution literature. A third meaning concerning Lagrange's formulation of Newtonian mechanics need not concern us here (Goldstein, 1950), nor need a fourth from planetary astronomy that refers to libration points of satellites (Smith et al., 1982). In the air pollution literature, perhaps the more common and broad meaning is "moves with the flow of air." In this usage the term is descriptive, and not intended to be precise. This is the usage in the title of this chapter.

In turbulence theory and fluid mechanics, however, a somewhat more precise and rigorous meaning is intended. See, for example, the contrast between the Lagrangian and Eulerian description of the turbulent transport of material given by Peters and Carmichael in Chapter 12 of this volume. A "Lagrangian" approach focuses on "fluid particles," or "fluid elements," the motion of which can be followed in detail (e.g., Batchelor, 1967). Not a great deal is said about such "fluid particles" by most authors, beyond how their motion is described. However, Seinfeld (1975), for one, goes into more detail:

> By a "fluid particle" we mean a volume of fluid large compared with molecular dimensions but small enough to act as a point which exactly

> follows the fluid. The "particle" may contain fluid of a different composition than the carrier fluid, in which case the particle is referred to as a "marked particle."

With our present-day understanding of fluids as composed of discrete molecules rather than continuous matter, this may be as close as one can get to the idealized "fluid element" originally introduced by Lagrange (1787) and explained by Dugas (1955).

Other authors describe a Lagrangian frame of reference with a little different slant. Drake and Barrager (1979) state:

> Reference frames may be fixed at the earth's surface, at the source of the pollutant (for either fixed or moving sources), or on a puff of pollutant as it moves downwind from the source. Reference frames fixed at the earth's surface or on the source are called Eulerian (because of their relation to the advecting and diffusing pollutant), while frames fixed on a puff of pollutant are called Lagrangian.

Here a "puff" is equated to the Lagrangian fluid particle. In some ways this point of view is of more heuristic value. Intuitively, one recognizes that a puff of gas disperses in time, and loses its identity. The same would be true, of course, of the entity described by Seinfeld, but that is not quite so immediately apparent. In fact, the Lagrangian "fluid particle" of fluid mechanics and turbulence theory is an idealization, an entity that does not lose its identity in time but that can, in principle, be followed indefinitely.

Pasquill (1974) makes passing reference to this idealization:

> Because of the continuity of diffusion processes, down the scale of motion to the ultimate molecular agitation, the view is sometimes expressed that the term "Lagrangian velocity" has no obvious meaning in the case of a fluid, except presumably when referred to a single molecule. In practice this difficulty is evaded by thinking either in terms of an element of fluid so small that its own diffusive spread is negligible compared with its translation under the action of larger-scale turbulence, or in terms of a solid particle of negligible buoyancy.

Pasquill's comment is helpful in that it implicitly introduces the idea that what one may consider to be a Lagrangian fluid element depends on the duration and scale of the phenomenon of interest. The concept of a Lagrangian fluid element may vary with the type of application. Clearly, it may not be identical in the analysis of turbulent flow through a 25-μm aperture, long-range air pollution on a scale of 1000 km, and the formation of stars from the intergalactic medium.

Pasquill's comment brings another question to mind. Why not simply consider Lagrangian fluid elements to be individual molecules? Then no question of idealization would arise. A strong reason for not making this choice is that in the structure of the theory, fluid elements are treated by the mechanics of continuous media rather than by the mechanics of discrete particles. The motions of individual molecules cannot be so treated. Thus, to make this choice would be to surrender much of the value of the concept.

To summarize, the concept of a Lagrangian fluid particle or fluid element has been of enduring value, particularly in the area of atmospheric diffusion (Taylor, 1921). It is, however, an idealization. This idealization is intimately related to the treatment of fluids as continuous media rather than discrete molecules. Because fluid particles or elements are idealizations that have no physical reality, one may choose to fit the concept to the problem at hand, optimizing its usefulness. In this tailoring the duration and scale of motion of interest must be taken into account.

So we have come full circle. The meaning of the term "Lagrangian" intended in fluid mechanics and turbulence theory is perhaps not quite so exact as might at first appear. One may even say that this second meaning of the term overlaps to some extent the more descriptive and less precise usage.

It is important to have this background in mind when interpreting data taken in a Lagrangian frame of reference. If one does not, one might either make inappropriate mathematical assumptions or fall into unprofitable discussions about what is or is not "truly" Lagrangian.

APPENDIX B: ACRONYMS

BNL	Brookhaven National Laboratory
CARB	California Air Resources Board
CRC	Coordinating Research Council
CVB	Constant-volume balloon
DME	Distance measuring equipment
DOE	Department of Energy
DV I, II, III	Da Vinci Experiments 1, 2, and 3, respectively
ELSTAR	Environmental Lagrangian Simulator of Transport and Atmospheric Reactions (model)
EPA	Environmental Protection Agency
ESCA	Electron spectroscopy for chemical analysis
FAA	Federal Aviation Administration
GC	Gas chromatograph
GHOST	Global horizontal sounding technique
GM	General Motors
LARPP	Los Angeles Reactive Pollutant Program
LAMP	Lagrangian Measurement Platform
LBL	Lawrence Berkeley Laboratory

LORAN	Long-range (radio) navigation
METROMEX	Metropolitan Meteorological Experiment
MISTT	Midwest Interstate Sulfur Transformation and Transport (project)
NASA	National Aeronautics and Space Administration
NCAR	National Center for Atmospheric Research
NOAA	National Oceanic and Atmospheric Administration
NTIS	National Technical Information Service
PAN	Peroxyacyl nitrate
PEPE	Prolonged Elevated Pollution Episode (experiment)
PNL	(Battelle) Pacific Northwest Laboratories
RAMS	Regional Air Monitoring System
RAPS	Regional Air Pollution Study
RTI	Research Triangle Institute
SNL	Sandia National Laboratories
STATE	Sulfur Transformation and Transport in the Environment (program)
TPS	Tennessee Plume Study
UHF	Ultra-high frequency
UNC	University of North Carolina
UV	Ultraviolet (light)
VHF	Very high frequency
VOR	VHF omni-directional range
WU	Washington University
XRF	X-ray fluorescence

ACKNOWLEDGMENTS

A special note of thanks goes to Roy Evans of the USEPA for giving me a verbal guided tour of the Los Angeles Reactive Pollutant Program. I take pleasure in acknowledging the following individuals for promptly responding to my requests for reports, reprints, photographs, and unpublished documents: A. J. Alkezweeny (PNL); J. G. Edinger (UCLA); B. Gay (EPA); N. Gillani (WU); and H. E. Jeffries (UNC). I am also grateful to G. Rowe for conducting the literature search, to K. Jackson for coordinating the work on illustrations, and to S. Pike for her remarkable alchemy in turning poor handwritten text into beautifully typed manuscript. Finally, I would like to thank H. W. Church, R. O. Woods, and R. E. Akins for reviewing the manuscript and offering many helpful suggestions.

REFERENCES

Alkezweeny, A. J. (1976). Growth of aerosol in an urban plume, in *Atmospheric Pollution*, M. M. Benarie, Ed., Elsevier, Amsterdam, pp. 233–242.

Alkezweeny, A. J. (1978). Measurement of aerosol particles and trace gases in METROMEX. *J. Appl. Meteorol.* **17,** 609–614.

Alkezweeny, A. J. (1980). Gas to particle conversion in urban plumes. Presented at the 73rd Annual Meeting of the Air Pollution Control Association, Montreal, June 22–27, 1980.

Alkezweeny, A. J. and Powell, D. C. (1977). Estimation of transformation rate of SO_2 to SO_4 from atmospheric concentration data. *Atmosph. Environ.* **11,** 179–182.

Alkezweeny, A. J., Young, J. A., Lee, R. N., Busness, K. M., and Hales, J. M. (1977). Transport and Transformation of Pollutants in the Lake Michigan Area, in *Proceedings of the 4th Joint Conference on the Sensing of Environmental Pollutants, New Orleans, LA,* American Chemical Society, pp. 853–855.

Angell, J. K. (1961). Use of constant level balloons in meteorology. *Adv. Geophys.* **8,** 137–219.

Angell, J. K. (1964). Measurements of Lagrangian and Eulerian properties of turbulence at a height of 2,500 ft. *Quart. J. R. Meterol. Soc.* **90,** 57–71.

Angell, J. K. (1972). A comparison of circulations in transverse and longitudinal planes in an unstable planetary boundary layer. *J. Atmosph. Sci.* **29,** 1252–1261.

Angell, J. K. (1974). Lagrangian–Eulerian time-scale relationship estimated from constant volume balloon flights past a tall tower. *Adv. Geophys.* **18A,** 419–431.

Angell, J. K. (1975). The use of tetroons for probing the atmospheric boundary layer. *Atmosph. Technol.* **7,** 38–43.

Angell, J. K. and Pack, D. H. (1960). An analysis of some preliminary low-level constant level balloon (tetroon) flights. *Mon. Weather Rev.* **88,** 235–248.

Angell, J. K. and Pack, D. H. (1962). Analysis of low-level constant volume balloon (tetroon) flights from Wallops Island. *J. Atmosph. Sci.* **19,** 87–98.

Angell, J. K., Pack, D. H., Holzworth, G. C., and Dickson, C. R. (1966). Tetroon trajectories in an urban atmosphere. *J. Appl. Meteor.* **5,** 565–572.

Angell, J. K., Pack, D. H., and Dickson, C. R. (1968). A Lagrangian study of helical circulations in the atmosphere. *J. Atmosph. Sci.* **25,** 707–717.

Angell, J. K., Pack, D. H., and Delver, N. (1969). Brunt–Vaisala oscillations in the planetary boundary layer. *J. Atmosph. Sci.* **26,** 1245–1252.

Angell, J. K., Pack, D. H., Hoecker, W. H., and Delver, N. (1971a). Lagrangian–Eulerian time-scale ratios estimated from constant volume balloon flights past a tall tower. *Quart. J. R. Meteorol. Soc.* **97,** 87–92.

Angell, J. K., Pack, D. H., Dickson, C. R., and Hoecker, W. H. (1971b). Urban influence on nighttime airflow estimated from tetroon flights. *J. Appl. Meteor.* **10,** 194–204.

Angell, J. K., Hoecker, W. H., Dickson, C. R., and Pack, D. H. (1973). Urban influence on a strong daytime air flow as determined from tetroon flights. *J. Appl. Meteorol.* **12,** 924–936.

Angell, J. K., Dickson, C. R., and Hoecker, W. H. (1975). Relative diffusion within the Los Angeles Basin as estimated from tetroon triads. *J. Appl. Meteorol.* **14,** 1490–1498.

Angell, J. K., Dickson, C. R., and Hoecker, W. H. (1976). Tetroon trajectories in the Los Angeles Basin defining the source of air reaching the San Bernardino–Riverside area in late afternoon. *J. Appl. Meteorol.* **15,** 197–204.

Barr, S. and Gedayloo, T. (1979). *Proceedings of the Atmospheric Tracers and Tracer Application Workshop, Los Alamos, New Mexico, May 22–24, 1979.* Los Alamos Scientific Laboratory Report LA-8144-C.

Batchelor, G. K. (1967). *An Introduction to Fluid Dynamics,* Cambridge University Press, Cambridge, p. 71.

Bekowies, P. J. (1982). Lagrangian measurement platforms in the validation of smog chamber chemistry, in *Final Report on Project Da Vinci,* Vol. 3, *Reports of Participants in Da Vinci II and III,* B. D. Zak, R. M. Holland, P. S. Homann, and K. M. Jackson, Eds., Sandra National Laboratories Report SAND 78-0403/3.

Calvert, J. G. (1976a). Test of the theory of ozone generation in Los Angeles atmosphere. *Environ. Sci. Technol.* **10,** 248–256.

Calvert, J. G. (1976b). Hydrocarbon involvement in photochemical smog formation. *Environ. Sci. Technol.* **10,** 256–262.

Chang, T. Y. (1979). Estimate of the conversion rate of SO_2 to SO_4 from the Da Vinci flight data. *Atmosph. Environ.* **13,** 1663–1664.

Changnon, S. A., Huff, F. A., and Semonin, R. G. (1971). METROMEX: An Investigation of inadvertant weather modification. *Bull. Am. Meteorol. Soc.* **52,** 958–967.

Ching, J. (1981). NOAA, Research Triangle Park, NC. Private communication.

Crane, G., Panofsky, H. A., and Zeman, O. (1977). A model for dispersion from area sources in convective turbulence. *Atmosph. Environ.* **11,** 893–900.

Decker, C. E., Sickles, J. E., Bach, W. D., Vukovich, F. M., and Worth, J. J. B. (1978). *Project Da Vinci II: Data Analysis and Interpretation,* Research Triangle Institute Report on Contract 68-02-2568.

Dickson, C. R. (1982). NOAA, Idaho Falls, ID. Private communication.

Dod, R. L., Giauque, R., Hollowell, C., Taynor, G., and Novakov, T. (1982). Da Vinci: Analysis of particulate samples by ESCA and x-ray fluorescence spectroscopy, in *Final Report on Project Da Vinci,* Vol. 3: *Reports of Participants in Da Vinci II and III,* Sandia National Laboratories Report SAND 78-0403/3.

Drake, R. L. and Barrager, S. M. (1979). *Mathematical Models for Atmospheric Pollutants,* Electric Power Research Institute Report EPRI EA-1131.

Dugas, R. (1955). *A History of Mechanics,* Edition du Griffon, Neuchatel, Switzerland.

Edinger, J. G. (1975). Preliminary analysis of LARPP data. Presented at the Scientific Seminar on Automotive Pollutants, Washington, DC, February 12.

Engelmann, R. J. et al. (1975). *Project Da Vinci Program Plan,* U.S. Energy Research and Development Administration Report ERDA-62.

Eschenroeder, A. (1976). *Los Angeles Reactive Pollutant Program (LARPP) Summary of Data Management Activities,* Environmental Research and Technology Document P-1656/3.

Evans, R. B. (1973). Aerial air pollution sensing techniques, in *Proceedings of the Second Conference on Environmental Quality Sensors, Las Vegas, Nevada, October 10–11, 1973.*

Feigley, C. E. (1978). Correspondence. *Environ. Sci. Technol.* **12,** 843–845.

Feigley, C. E. and Jeffries, H. E. (1979). Analysis of processes affecting oxidant and precursors in the Los Angeles Reactive Pollutant Program (LARPP) Operation 33. *Atmosph. Environ.* **13,** 1369–1384.

Feigley, C. E. Jeffries, H. E., and Carpenter, M. A. (1979). Smog chamber validation using Lagrangian atmospheric data, U.S. Environmental Protection Agency Report EPA-600/3-79-050.

Feigley, C. E., Jeffries, H. E., and Kamens, R. M. (1982). An experimental simulation of Los Angeles reactive pollutant program (LARPP) operation 33—Part I. Experimental simulation in an outdoor smog chamber. *Atmosph. Environ.* **16,** 1989–1996.

Forrest, J. and Newman, L. (1977). Further studies on the oxidation of sulfur dioxide in coal-fired power plant plumes. *Atmosph. Environ.* **11,** 465–474.

Forrest, J., Schwartz, S. E., and Newman, L. (1979). Conversion of sulfur dioxide to sulfate during the Da Vinci flights. *Atmosph. Environ.* **13,** 157–167.

Gay, G. T., Zak, B. D., Barker, B., Holland, R. M., and Homann, P. S. (1981). *Lagrangian Measurement Platform Flights in Support of the Tennessee Plume Study: Field Effort and Data,* Sandia National Laboratories Report SAND 79-1336.

Giles, K. C. and Angell, J. K. (1963). A Southern Hemisphere horizontal sounding system. *Bull. Am. Meteorol. Soc.* **44,** 687–696.

Gillani, N. V. (1979a). *Project STATE, Tennessee Plume Study, Data Volume, BNL Aircraft Measurements.* STATE Data Center Report, Dept. of Mech. Eng., Washington University, St. Louis MO.

Gillani, N. V. (1979b). *The LAMP Days (August 20 and 25, 1978). A Compilation of Selected Meterological and Chemical Aircraft Data of the Mission which Included Participation of the Lagrangian Air Monitoring Platform.* STATE Data Center Report, Dept. of Mech. Eng. Washington University, St. Louis MO.

Gillani, N. V. and Wilson, W. E. (1980). Formation and transport of ozone and aerosols in power plant plumes, in *Aerosols: Anthropogenic and Natural, Sources and Transport,* T. J. Kniep, and P. J. Lioy, Eds., New York Academy of Sciences, New York, pp. 276–296.

Gillani, N. V. and Wilson, W. E. (1982). Gas-to-particle conversion of sulfur in power plant plumes. II. Observations and parameterization of liquid phase conversions. *Atmosph. Environ.* (in press).

Gillani, N. V., Husar, R. B., Husar, J. D., Patterson, D. E., and Wilson, W. E. (1978). Project MISTT: Kinetics of particulate sulfur formation in a power plant plume out to 300 km. *Atmosph. Environ.* **12,** 589–598.

Gillani, N. V., Kohli, S., and Wilson, W. E. (1981). Gas-to-particle conversion of sulfur in power plant plumes. I. Parameterization of the conversion rate for dry, moderately polluted ambient conditions. *Atmosph. Environ.* **15,** 2293–2313.

Godden, D., Lurmann, F., Lloyd, A., Nittu, B., and Kwan, C. (1980). *Further Evaluation of ELSTAR (Environmental Lagrangian Simulator of Transport and Atmospheric Reactions) in Los Angeles under Different Meteorological Conditions,* Environmental Research and Technology Document P-A037.

Goldstein, H. (1950). *Classical Mechanics,* Addison-Wesley, Reading, MA.

Hanna, S. R. and Hoecker, W. H. (1971). The response of constant density balloons to sinusoidal variations of vertical wind speeds. *J. Appl. Meteorol.* **10,** 601–604.

Hass, W. A., Hoecker, W. H., Pack, D. H., and Angell, J. K. (1967). Analysis of low level constant volume balloon (tetroon) flights over New York City. *Quart. J. R. Meteorol. Soc.* **93,** 483–493.

Heinsheimer, T. F., and Neushul, P. C. (1978). *ATMOSAT-7 Flight Report.* Aerospace Corporation El Segundo, CA, Report ATR-78 (7708-01)-1

Hilst, G. (1973). *A Coupled Two Dimensional Diffusion and Chemistry Model for Turbulent and Inhomogeneously Mixed Reaction Systems,* Aeronautical Research Associates of Princeton, Inc., Princeton, NJ.

Hoecker, W. H. (1975). A universal procedure for deploying constant volume balloons and for deriving vertical air speed for them. *J. Appl. Meteorol.* **14,** 1118–1124.

Lagrange, J. L. (1787). *Mechanique Analytique,* chez la veuve Desaint, Paris.

Lally, V. E. (1959). Satellite satellites: A conjecture on future atmospheric sounding systems. *Bull. Am. Meteorol. Soc.* **41,** 428–432.

Lally, V. E. (1967). *Superpressure Balloons for Horizontal Soundings of the Atmosphere,* National Center for Atmospheric Research Technical Note NCAR-TN-28.

Lloyd, A. C., Lurmann, F. W., Godden, D. G., Hutchins, J. F., Eschenroeder, A. Q., and Nordseick, R. A. (1979a). *Development of the ELSTAR Photochemical Air Quality Simulation Model and Its Evaluation Relative to the LARPP Data Base,* Environmental Research and Technology Document P-5287-500.

Lloyd, A. C., Lurmann, F., Godden, D., and Hutchins, J. (1979b). Development, testing, and possible applications of ELSTAR (Environmental Lagrangian Simulator of Transport and Atmospheric Reactions), in *Proceedings of the Specialty Conference on Ozone/Oxidants: Interactions with the Total Environment, October 14–17, Houston, Texas,* Air Pollution Control Association, pp. 396–406.

Lurmann, F. (1979). *User's Guide to the ELSTAR Photochemical Air Quality Simulation Model,* Environmental Research and Technology, PB 80-109184; NTIS/DF-79/001A.

Martinez, J. R. (1976). *Los Angeles Reactive Pollutant Program (LARPP) Guide to Modelers' Archive,* Environmental Research and Technology Document P-1656.

Mayrsohn, H., and Crabtree, J. H., (1976). Source reconciliation of atmospheric hydrocarbons. *Atmosph. Environ.* **10,** 137–143.

Meagher, J. F. (1980). Overview (of natural and anthropogenic sources), in *Atmospheric Sulfur Deposition,* D. S. Shriner, C. R. Richmond, and S. E. Lindberg, Eds., Ann Arbor Science, Ann Arbor, MI, pp. 33–34.

Meagher, J. F., Stockberger, L., Bailey, E. M., and Huff, O. (1978). The oxidation of sulfur dioxide to sulfate aerosols in the plume of a coal-fired power plant. *Atmosph. Environ.* **12,** 2197–2203.

Morel, P. and Bandeen, W. (1973). The Eole experiment: Early results and current objectives. *Bull. Am. Meteorol. Soc.* **51,** 298–306.

Morris, A. L. (1975). *Scientific Ballooning Handbook,* National Center for Atmospheric Research Technical Note NCAR-TN/1A-99.

Moses, H. (1976). U.S. Department of Energy, Washington DC. Comments made at Project Da Vinci press conference, St. Louis, Missouri.

NCAR (1969). *GHOST, a Technical Summary,* National Center for Atmospheric Research, Boulder, CO.

Newman, L. (1981). Atmospheric oxidation of sulfur dioxide: A review as viewed from power plant and smelter studies. *Atmosph. Environ.* **15,** 2231–2239.

Panofsky, H. A. (1975). A model for vertical diffusion coefficients in a growing urban boundary layer. *Boundary Layer Meteorol.* **9,** 235–244.

Pasquill, F. (1974). *Atmospheric Diffusion,* Wiley, New York, p. 87.

Perkins, W. A. (1974). The Los Angeles Reactive Pollutant Program, paper presented at the LARPP Symposium, November 12–14, 1974.

Pound, E. (1980). Private communication, Utah State University, Logan.

Roberts, P. T. and Friedlander, S. K. (1975). Conversion of SO_2 to sulfur particulate in the Los Angeles atmosphere. *Environ. Health Perspectives* **10,** 103–108.

Schiermeier, F. A. (1978). RAPS field measurements are in. *Environ. Sci. Technol.* **12,** 644–648.

Seinfeld, J. H., Hecht, T. A., and Roth P. M. (1973) *Existing Needs in the Observational Study of Atmospheric Chemical Reactions,* U.S. E.P.A. Report EPA-R4-73-031.

Seinfeld, J. H. (1975). *Air Pollution: Physical and Chemical Fundamentals,* McGraw-Hill, New York.

Seinfeld, J. H. (1977). Correspondence. *Environ. Sci. Technol.* **11,** 1218–1219.

Smith, B. A., et al. (1982). A new look at the Saturn system: The Voyager 2 images. *Science* **215,** 504–537.

Tatom, F. B. and King, R. L. (1977). *Constant Volume Balloon Capabilities for Aeronautical Research,* NASA Contractor Report NASA CR-2805.

Taylor, G. I. (1921). Diffusion by continuous movements. *Proc. Lond. Math. Soc.* **20,** 196–212.

Whitten, G. Z. and Hogo, H. (1977). *Mathematical Modeling of Simulated Photochemical Smog,* U.S. Environmental Protection Agency Report EPA-600/3-77-011.

Wilson, W. E. (1978). Sulfates in the atmosphere: A progress report on project MISTT. *Atmosph. Environ.* **12,** 537–547.

Woods, R. O. (1982). Sandia National Laboratories. Private communication.

Zak, B. D. (1981a). *Final Report on Project Da Vinci: A Study of Long Range Air Pollution Using a Balloon-Borne Lagrangian Measurement Platform, Vol. 1: Overview and Data Analysis.* Sandia National Laboratories Report SAND 78-0403/1.

Zak, B. D. (1981b). *Sandia National Laboratories' "Skyhook" Measurement Platform,* unpublished technical note.

Zak, B. D. (1981c). *SNL Air Pollution Research Systems,* unpublished technical note.

Zak, B. D. (1981d). Lagrangian measurements of sulfur dioxide to sulfate conversion rates. *Atmosph. Environ.* **15,** 2583–2591.

Zak, B. D., Holland, R. M., Jr., Homann, P. S., and Jackson, K. M. (1981). *Final Report on Project Da Vinci: A Study of Long-Range Air Pollution Using a Balloon-Borne Lagrangian Measurement Platform,* Vol. 2, *Reports of Participants in Da Vinci I,* Sandia National Laboratories Report SAND 78-0403/2.

8

TURBULENT TRANSPORT OF OZONE TO SURFACES COMMON IN THE EASTERN HALF OF THE UNITED STATES

Marvin L. Wesely

Radiological and Environmental Research Division
Argonne National Laboratory
Argonne, Illinois 60439

1. INTRODUCTION

Ozone is a photochemical oxidant found in the troposphere in concentrations sufficiently large to be quite important in photochemical processes over most regions of the world. These processes are intimately coupled with the reactions of nitrogen oxides on both urban and global scales, as described in Chapters 9–12 of this volume. Also, it is well established that ozone can be directly harmful to surface materials and living organisms (e.g., National Research Council, 1977). For such reasons, considerable research effort has been directed toward computing budgets of atmospheric ozone, a task that requires knowledge of the sources and sinks of ozone and of other factors that influence ozone concentrations in the lower atmosphere. For example, Aldaz (1969), Fabian and Junge (1970), and Galbally and Roy (1980) summarize estimates of surface deposition in attempts to calculate the total amount of ozone removed on a global scale. The present chapter likewise examines ozone removal rates over many types of surface, but on temporal and spatial scales that are smaller. Specifically, surfaces common in the eastern half of the United States are studied, in support of the U.S. Environmental Protection Agency's Northeast Regional Oxidant Study (NEROS) (e.g., Clarke et al., 1982).

Estimation of surface removal rates over continental distances requires considerable simplification or extrapolation from knowledge of processes that occur over individual types of surfaces. Our first step is to gain that knowledge of detailed information by short-term intensive field experiments (lasting typically 2 weeks) at carefully chosen sites. These sites or combinations of sites fit into broad land-use categories that can be mapped by use of satellite imagery. For example, maize and soybean fields obviously belong in the category of agricultural land or cropland. The spatial and temporal resolutions eventually required are partially dictated by the capabilities of the numerical models employed to compute the vertical fluxes. For NEROS, although we assume that great spatial detail is needed, the spatial resolution undoubtedly will be limited by the broadness of the land-use categories chosen. Averaging times in field experiments are typically 0.5 h, compared to 1–6 h time steps commonly employed in numerical simulations.

At this time, not all important surface types in the eastern half of the United States have been studied thoroughly. Notably lacking are complete studies over deciduous forest and urban areas. However, deposition velocities over these surfaces might be inferred indirectly from studies over similar surfaces [e.g., see Wesely and Hicks (1977), for the case of vegetation].

The purpose here is to present data on ozone turbulence properties in the atmospheric surface layer, especially those that aid the study of removal rates, and to present a summary of recent eddy-correlation measurements of ozone fluxes. To facilitate identification of the factors that control the removal of ozone at the surface, information provided by others on ozone fluxes is utilized also. Other investigators have seldom used eddy correlation to determine ozone flux; rather, another micrometeorological technique, the profile method, has most often been used (e.g., Garland and Derwent, 1979; Regener, 1957; Rich et al., 1970). In the profile method, measured vertical profiles of mean ozone concentration, wind speed, and temperature are applied in micrometeorological flux-gradient equations to compute the flux. In the related ozone Bowen-ratio method, the ozone flux is computed as the product of heat flux times an ozone concentration difference divided by the heat content (via temperature) difference across the same height interval. This has been employed successfully by Leuning et al. (1979), with simultaneous direct measurement of canopy stomatal resistances to facilitate determination of the partitioning of flux between the vegetation and the soil surface. Another technique, sometimes referred to as the "box" method, estimates ozone uptake by measuring the decay of ozone concentration in an enclosure placed on the field surface (e.g., Aldaz, 1969; Galbally and Roy, 1980). Approaches employing enclosures in the laboratory also have provided important data on the ability of individual types of surfaces to destroy ozone (e.g., Rich et al., 1970; Thorne and Hanson, 1972; Turner et al., 1973).

2. EDDY-CORRELATION METHOD

2.1. Flux Measurement

The eddy-correlation technique is a fairly common micrometeorological method to measure directly the vertical flux densities of heat, mass, and momentum within the atmospheric surface boundary layer. These fluxes, although measured above the surface, are the same as the average exchange rates occurring at a broad expanse of the air–surface interface upwind of the detection point, provided there are no significant local sources or sinks of the quantity of interest in the intervening layer of air. Moreover, the surface upwind for a distance of at least 100 times the measurement height should be flat and have uniform surface characteristics, to avoid internal boundary layers that can cause a change of flux with height. Thus it is desirable to place the sensors quite close to the surface when the surface of interest is contained within a small field. As is discussed in Section 2.2, the proximity of the sensors to the surface is limited by the frequency response of the sensors.

For eddy correlation, fast-response sensors are placed as close together as possible and data are acquired to enable computation of the covariance between fluctuations of the vertical wind speed and the quantity of interest. The flux of ozone is given by

$$F_\chi \equiv \overline{w\chi} = \overline{w'\chi'} + \overline{w}\,\overline{\chi} , \qquad (1)$$

where w is the vertical wind speed, χ is the ozone concentration, overbars indicate time averages, and primes denote deviations from the means. The right-hand portion of Eq. (1) should actually consist of four terms that result from the expansion of $\overline{w\chi} = \overline{(\overline{w} + w')(\overline{\chi} + \chi')}$, but two terms are not shown because they are zero since the factors they contain, $\overline{w'}$ and $\overline{\chi'}$, are by definition equal to zero. Furthermore, $\overline{w}$ is set to zero by electronic filtering of the signal, so that $\overline{w}\,\overline{\chi} = 0$ is obtained. Then F_χ is simply equal to the measured covariance $\overline{w'\chi'}$. However, a real $\overline{w}$ usually does exist because of the effects of evaporative Stefan flow at the surface and air density fluctuations at the point where w is measured. When these effects are large above a surface that removes ozone very slowly, a small correction to measured F_χ is needed. All the present eddy-correlation data have been adjusted for nonzero $\overline{w}$ by the procedures explained by Wesely et al. (1981).

In addition to the ozone flux, the fluxes of momentum, sensible heat, and water vapor are often measured because these quantities are important in interpreting the results in the present type of study. In all measurements reported here, the vertical wind is sensed with a propeller anemometer, temperature with a microbead thermistor, and water vapor content with a Lyman alpha hygrometer. The reader is referred to past publications on eddy fluxes of ozone frequently cited in this chapter for greater detail on the sensors and techniques used. [See also Kanemasu et al. (1979) for a summary of some of the relevant aspects of the eddy-correlation technique.]

2.2. Frequency Response Requirements

One of the major difficulties in the eddy-correlation technique seldom discussed fully is the compensation for the effects of poor frequency response. Many cospectral studies have shown that most of the vertical transport of scalar quantities such as heat content is associated with cyclical frequencies n near 0.1 $\overline{u}/z$ (units: Hz), where $\overline{u}$ is the mean horizontal wind speed and z is the height above the surface; eddy-correlation sensors should be capable of responding to fluctuations at frequencies of nearly $n = 2\overline{u}/z$ (Kaimal, 1973; Kaimal et al., 1972). Since profiles of $\overline{u}$ tend to be logarithmic functions of height within the lower 10 m of the atmosphere, the frequencies of the eddies responsible for the vertical transport of quantities such as heat and ozone increase as the surface is approached. Clearly, frequency responses of at least 1 Hz are needed in normal wind conditions when sensors are deployed at a height of a few meters.

The ozone sensors used here are chemiluminescent devices consisting of a reaction chamber through which air is forced after passing through a rather long sample tube. The exponential response time t_c evaluated for a small step change in ozone concentration in the air entering the chamber is an adequate description of response characteristics within the reaction chamber, which are governed largely by the process of purging the chamber of changing ozone amounts. Both the response time and the delay (transit) time Δt_c, the time needed for a change in concentration at the sample inlet located near the vertical wind sensor to reach the reaction chamber, are minimized but are still a few tenths of a second.

When t_c and Δt_c as well as the sample height z, wind speed $\overline{u}$, and atmospheric stability conditions are known, the fluxes can be adjusted to values that would have been obtained if the sensors had ideal frequency response. Corrections of up to 30% are quite possible, but confidence in large-percentage corrections depends a great deal on knowing the response and delay times very accurately. Furthermore, unless cospectra are calculated for all flux data, the cospectral properties of the fluxes for a given data collection period are rarely known with sufficient precision for very large correction factors to be calculated with sufficient accuracy.

For example, consider the case of a propeller anemometer and an ozone sensor used in neutral conditions above a moderately rough surface. We assume that the propeller has a response length of 2 m, so that the time response is $t_w = 2/\overline{u}$, with negligible delay time (Hicks, 1972). Correction factors evaluated for various assumed values of the response and delay times of the ozone sensor are shown in Table 1.

The correction factors were numerically computed on the basis of techniques described by Hicks (1972) by use of cospectra presented by Kaimal et al. (1972). Table 1 shows that an increase in Δt_c has about the same effect as an increase in t_c. As a rough guide, we apparently can assume that with a propeller anemometer the percentage change in the correction factor is proportional to $\overline{u}(t_c + \Delta t_c)/z$; for a value of 0.45, this treatment yields a correction of approximately 30%. Use of a vertical wind sensor with faster response does not greatly improve the uncorrected estimates in this case in which a substantial delay time is assumed in the other sensor, because of the effects of phase shifts induced by the lagged signal. It is expected that in slightly unstable conditions the percentage error given by the correction factors can be reduced by a factor of about 0.8 and in moderately unstable conditions, about 0.6. In stable conditions, the correction factors, computed with the cospectra provided by Kaimal (1973), can become extremely large.

The present sensors are usually deployed at a height near 6 m. Thus, with the values of $(t_c + \Delta t_c)$ that range from 0.3 to 0.6 s for the present ozone sensors, correction factors in neutral conditions can vary from 1.04 to 1.4

Table 1. Examples of Underestimates of Pollutants Eddy Fluxes Due to Poor Time Response Characteristics of Eddy-Correlation Sensors, in Neutral Atmospheric Stability

z (m)	$\bar{u}$ (m s^{-1})	t_w (s)	t_c (s)	Δt_c (s)	Flux Correction Factor
2	3	0.67	0.3	0.3	1.58
2	3	0.67	0.3	0.6	1.73
2	3	0.67	0.6	0.3	1.74
2	3	0.67	0.6	0.6	1.94
2	6	0.33	0.3	0.3	1.95
2	6	0.33	0.3	0.6	2.60
2	6	0.33	0.6	0.3	2.39
2	6	0.33	0.6	0.6	3.21
4	3	0.67	0.3	0.3	1.31
4	3	0.67	0.3	0.6	1.39
4	3	0.67	0.6	0.3	1.40
4	3	0.67	0.6	0.6	1.50
4	6	0.33	0.3	0.3	1.50
4	6	0.33	0.3	0.6	1.81
4	6	0.33	0.6	0.3	1.75
4	6	0.33	0.6	0.6	2.11
8	3	0.67	0.3	0.3	1.16
8	3	0.67	0.3	0.6	1.19
8	3	0.67	0.6	0.3	1.21
8	3	0.67	0.6	0.6	1.26
8	6	0.33	0.3	0.3	1.26
8	6	0.33	0.3	0.6	1.42
8	6	0.33	0.6	0.3	1.40
8	6	0.33	0.6	0.6	1.58

in winds of 1–6 m s^{-1}. Fluxes corrected by more than 30% are discarded because of uncertainties in the form of cospectra, especially for nonneutral conditions, and in the difficulty in determining (t_c + Δt_c) with precision greater than 0.1 s. This procedure allows the final flux estimate to be accurate to within ±10% if no other sources of error are introduced.

With complete digital processing of the data, some of the errors associated with finite t_c and Δt_c can be lessened. The delay times can be set to zero by simple lagging of the w signal. If cospectra are computed for each data collection period, values of cospectral density can be adjusted for the effects of poor response and then the cospectra integrated to produce corrected estimates of the fluxes. The cospectra were not adjusted for the data on fluxes presented here, but the delay times were often effectively eliminated by digital techniques.

3. TURBULENCE PROPERTIES

3.1. Inference of Spectral Properties

The above analysis of the effects of inadequate frequency response contains an implicit assumption that ozone turbulent fluctuations have properties similar to those of temperature fluctuations since the Kaimal et al. (1972) normalized cospectra applied are actually for w and temperature T rather than w and χ. In fact, we are often forced to assume some degree of similarity between temperature and trace substance behavior in the atmospheric surface layer, for lack of good alternative assumptions. For example, corrections for the effects of atmospheric stability on flux-gradient relationships (such as discussed below in Section 4.1) are often presumed to be the same for sensible heat flux and fluxes of other scalar quantities. This has been shown in numerous studies to be a good approximation at least for water vapor content; field experiments have also confirmed this for carbon dioxide (Allen et al., 1974; Sinclair et al., 1975).

It has been shown that temperature and water vapor content fluctuations are highly correlated in the lower planetary boundary layer (Friehe et al., 1975; McBean and Miyake, 1972; Phelps and Pond, 1971; Swinbank and Dyer, 1967; Thorpe et al., 1973; Wesely and Hicks, 1978; Wyngaard et al., 1978). Correlation coefficients between T and specific humidity q fluctuations at frequencies associated with vertical transport are nearly unity over warm, wet surfaces. This is probably because the surface can appear as a nearly uniform source of both heat and water vapor that are swept up from the surface by practically the same turbulent wind motions. If the high degree of correlation extends to other scalar quantities, the large amount of work on T spectra, w and T cospectra, and heat content flux-gradient relationships, as well as other investigations on temperature fluctuations that treat flux-carrying eddies, can be applied more confidently to scalar quantities such as ozone.

Figure 1 presents a typical example of spectral correlation coefficients r found above a field of soybeans in the afternoon for pairs of T, q, and χ (ozone concentration), computed as the cospectral densities divided by the square root of the product of the spectral densities. Additional information is given in Table 2. Near $n = 0.08$ Hz, where the peaks of the cospectra of w with T, q, and χ occur, T and q are highly correlated as expected, whereas χ is strongly negatively correlated with both T and q. The latter correlation is negative because the flux of ozone is directed downward, opposite to the fluxes of heat and water vapor. For the case of negative (downward) heat flux, the signs of the correlation coefficients with T are reversed (Figure 2), but the shapes of the curves are about the same. These large magnitudes of the spectral correlation coefficients indicate that the same turbulent motions carry all these scalar quantities vertically. Above $n = 0.2$ Hz, the

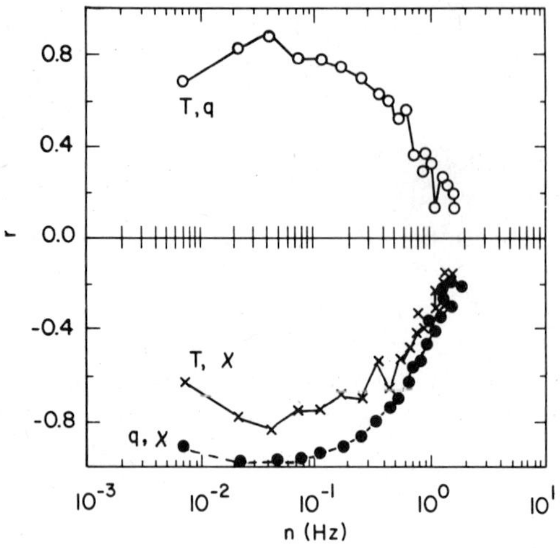

Figure 1. Spectral correlation coefficients obtained for a 25-min period centered at 1345 CST on August 16, 1979. Other information is given in Table 2.

correlation r lessens in magnitude because of noise, phase lags introduced by the sensors, and spatial separations between the sensors (5–40 cm). At the low frequencies, $|r|$ decreases because of the introduction of turbulent fluctuations not associated with vertical transport, especially for the cases of greater atmospheric stability. As a result of these measurements, assumptions that vertical transfer and the associated spectra of various scalar quantities are similar are strongly supported.

3.2. Variances

Variances in the atmospheric surface layer are dominated by fluctuations associated with vertical transport: because of the large magnitudes of the spectral correlation coefficient between fluctuations of different scalar quantities at the frequencies where vertical fluxes occur, we expect the variances, when properly normalized, to exhibit a high degree of similarity. The usual

Table 2. Measurements Taken During Periods When Data for Figures 1 and 2 Were Obtained[a]

Time (EST)	u_* (cm s^{-1})	$\bar{u}$ (cm s^{-1})	H (W m^{-2})	L_wE (W m^{-2})	z/L	$\bar{\chi}$ (ppb)	$-F_\chi/\bar{\chi}$ (cm s^{-1})
1345	28	277	54	393	−0.200	45	0.72
0715	36	397	−11	125	0.003	24	0.51

[a] Sensors were placed about 5.2 m above a soybean field in Lancaster County, PA, during the middle of August 1979. In addition to the variables defined in the text, u_* is the friction velocity, H the sensible heat flux, L_wE the latent heat flux, L the Obukhov stability length, and $-F_\chi/\bar{\chi}$ the ozone deposition velocity.

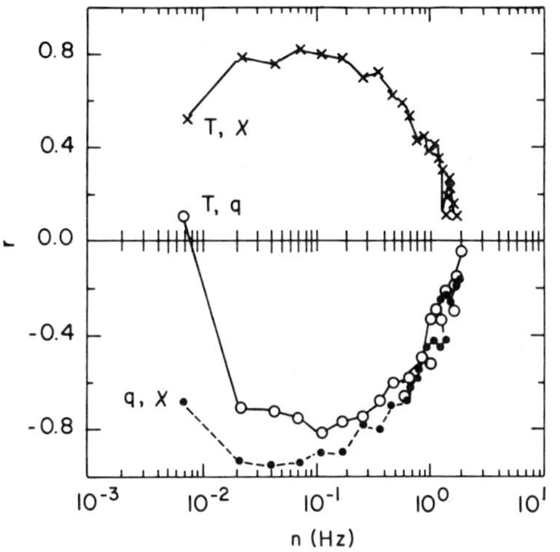

Figure 2. Spectral correlation coefficients as in Figure 1 except for 0715 CST on August 15, 1979.

procedure of normalization is to divide the standard deviation σ by the appropriate "starred" scale χ_*, which is the vertical flux divided by the friction velocity (u_*), as discussed in Section 4.1. These are plotted versus the atmospheric stability parameter z/L [Eq. (5)], where L is the Obukhov stability scale length.

Figure 3 presents values of σ/χ_* derived from 25-min data collection periods during a 2-week experiment above soybeans in Licking County, Ohio in August 1980. To remove trends, a running mean removal with an effective time constant of 200 s was applied. As indicated in Figure 3, the values of σ/χ_* for water vapor, ozone, and carbon dioxide are practically the same but they are slightly greater than a standard of comparison for unstable conditions, σ_T/T_* provided by the "Kansas" data of Wyngaard et

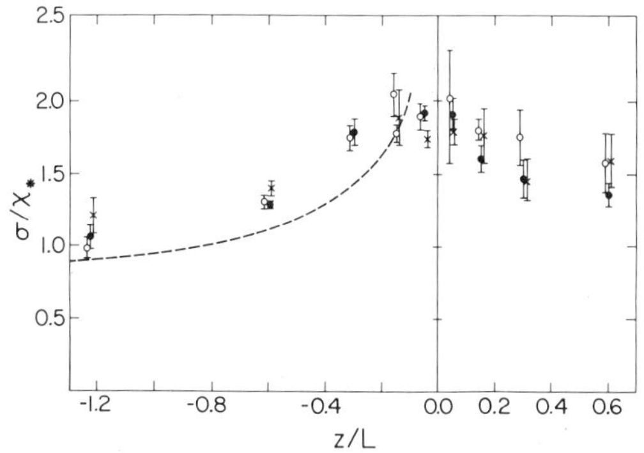

Figure 3. Normalized standard deviations of concentrations of water vapor (●), ozone (○), and carbon dioxide (×). Standard error bars are shown. The dashed curve for unstable conditions is given by $\sigma_T/T_* = 0.95(-z/L)^{-1/3}$.

al. (1971). No explanation for this discrepancy has been found. Values of σ_T/T_* from the Ohio site are not shown because of the small amount of data except at very small sensible heat fluxes, when σ_T/T_* tends to become excessively large. The latter phenomenon also occurs for the three scalar quantities employed in Figure 3, but data with extremely small vertical fluxes were selectively removed from consideration. The available good measurements of σ_T/T_* tend to agree with the values of σ/χ_* shown. Hence values of σ/χ_* for different scalars appear to be the same for a given atmospheric stability.

Since the normalized standard deviation σ_1/χ_{1*} of one scalar quantity is approximately equal to that, σ_2/χ_{2*}, for another scalar quantity measured simultaneously at the same height, the relationship

$$F_1 = \frac{\sigma_1 F_2}{\sigma_2} \tag{2}$$

can be applied to determine F_1. Variances are relatively easy to measure compared to eddy-correlation estimates of fluxes. Thus if F_2 can be determined by any means, a dependable measure of F_1 might be obtained more easily than the procedures used in this chapter. For example, F_2 might be derived from sensible or latent heat flux obtained by the well known energy-balance Bowen-ratio technique (e.g., Kanemasu et al., 1979). Furthermore, the effects of a large delay time in a pollutant sensor could be avoided since it would not necessarily reduce the variance measured. For pollutant sensors with a response time slightly too large for use for eddy correlation but small enough to allow detection of a substantial amount of the variance associated with the flux-carrying eddies, selected electronic filtering of high frequencies from a signal obtained with a temperature sensor, for example, would allow effective use of Eq. (2). Such techniques have not been used for the ozone fluxes in this chapter, but examination of the statistics presented in Figure 3 at least provided a check on the veracity of the ozone flux estimates.

4. DEPOSITION EQUATIONS

4.1. Profile Equations

Flux-profile relationships in the turbulent atmospheric surface layer are rather well established for the case of heat and, to a slightly lesser extent, water vapor. As discussed in Section 3.1, there is strong evidence that the behavior of these and other scalar quantities (but not necessarily momentum, a vector quantity) are very similar over surfaces with rather simple geometry that allow wind to transfer the different quantities to and from the surface with practically the same turbulent motions. Hence we assume that the expression for profiles of temperature in the turbulent air can be readily modified to the case of any scalar quantity. For a mean concentration $\overline{\chi}_z$ at height z, the appropriate expression is

$$\bar{\chi}_s - \bar{\chi}_z = \chi_* k^{-1}\left[\ln\left(\frac{z}{z_{0\chi}}\right) - \Psi_\chi\right], \tag{3}$$

where $\bar{\chi}_s$ is the concentration of the air when in contact with the surface, $\chi_* = \overline{w'\chi'}/u_*$ is a frequently used χ scale, $k = 0.4$ is the von Karman constant, $z_{0\chi}$ is the roughness scale length for χ, and Ψ_χ is a term for taking into account the diabatic effects caused by vertical density gradients. Both Ψ_χ and $z_{0\chi}$ need further explanation. We assume that Ψ_χ is the same as that for heat Ψ_H, for which there are readily available empirical expressions. The expression by Dyer and Hicks (1970) serves the purpose here:

$$\Psi_\chi = \exp\left\{0.598 + 0.39\ln\left(\frac{-z}{L}\right) - 0.09\left[\ln\left(\frac{-z}{L}\right)\right]^2\right\} \tag{4a}$$

for $z/L < 0$ and

$$\Psi_\chi = -5\frac{z}{L} \tag{4b}$$

for $z/L \geqslant 0$. The stability parameter z/L is given as

$$\frac{z}{L} = -zkg\frac{H + L_w E/14}{\rho c_p u_*^3 \theta}, \tag{5}$$

where g is the acceleration due to gravity, H is the sensible heat flux, $L_w E$ is the latent heat flux, ρ is the density of air, c_p is the specific heat of air at constant pressure, u_* is the friction velocity, and θ is the absolute temperature.

The term $z_{0\chi}$ represents a somewhat artificial length; it is actually the height at which $\bar{\chi}$ extrapolates to the value $\bar{\chi}_s$. Garratt and Hicks (1973) provide a means for determination of $z_{0\chi}$ when the aerodynamic roughness scale length z_0 is known. In practice, z_0 can be determined from measurements of wind speed profiles and the friction velocity or roughly estimated from knowledge of surface characteristics. Over relatively smooth surfaces such as water, snow, or bare soil, $z_{0\chi}$ is estimated as

$$z_{0\chi} \approx \frac{D_\chi}{ku_*}, \tag{6}$$

where D_χ is the molecular diffusivity of χ in air. Over relatively rough surfaces such as vegetation, $z_{0\chi}$ can be obtained from

$$\ln\left(\frac{z_0}{z_{0\chi}}\right) \approx a, \tag{7}$$

where a is a numerical constant equal to about 2 when χ represents temperature. Equation (7) has been verified by Heilman and Kanemasu (1976). The value of a should be adjusted when D_χ has a numerical value different from the molecular diffusivity κ for heat, usually by the procedure of multiplying by $(\kappa/D_\chi)^{2/3}$ (Monteith, 1973). As discussed by Wesely and Hicks (1977), the procedure to determine a used by Galbally and Roy (1980), for example, that provides for larger values of a, should be used only over surfaces with large bluff elements. The type of vegetation investigated here is made of fibrous elements and provides a rather uniform surface.

Combination of Eq. (3) with (7) results in the relationship

$$\bar{\chi}_s - \bar{\chi}_z = \chi_* k^{-1} \left[\ln\left(\frac{z}{z_0}\right) + 2\left(\frac{\kappa}{D_\chi}\right)^{2/3} - \Psi_\chi \right] \qquad (8)$$

for vegetation. Clearly, the factor $2(\kappa/D_\chi)^{2/3}$ for rough surfaces (or the corresponding $\ln(kz_0u_*/D_\chi)$ for smoother surfaces) represents the increase of the ratio $(\bar{\chi}_s - \bar{\chi}_z)/\overline{w'\chi'}$ due to the effects of the sublayer of air in contact with the surface.

4.2. Resistances

Equation (8) does not quite complete the needed characterization of deposition processes. Although it tells us how to relate $F_\chi = \overline{w'\chi'}$ to $(\bar{\chi}_s - \bar{\chi}_z)$ when u_*, z_0, and z/L are known, a calculation of F_χ would require estimation of $\bar{\chi}_s$ as well as $\bar{\chi}_z$. This surface concentration is not easily measured nor readily derived from measured parameters or basic principles. In fact, estimates of deposition or flux for practical application usually require derivation of F_χ only from $\bar{\chi}_z$ in combination with estimates of parameters related to u_*, z_0, and atmospheric stability. This desired relationship between F_χ and $\bar{\chi}_z$ is formally expressed as the well-known deposition velocity $v_d = -F_\chi/\bar{\chi}_z$.

To deal with the problem of not knowing $\bar{\chi}_s$, an understanding of the ability of the surface to remove substance χ from the atmosphere must be known. Then we can set $\bar{\chi} = 0$ somewhere in the nongaseous substrate and insert an additional term in Eq. (8) representing a surface resistance:

$$\bar{\chi}_z = -\chi_* k^{-1} \left[\ln\left(\frac{z}{z_0}\right) + 2\left(\frac{\kappa}{D_\chi}\right)^{2/3} + r_c k u_* - \Psi_* \right]. \qquad (9)$$

Now we must determine the nature of r_c for various surfaces.

Equation (9) can be rearranged to become

$$-\frac{\bar{\chi}_z}{F_\chi} = (u_* k)^{-1} \left[\ln\left(\frac{z}{z_0}\right) - \Psi_\chi \right] + 2(u_* k)^{-1} \left(\frac{\kappa}{D_\chi}\right)^{2/3} + r_c, \qquad (10)$$

where each term on the right-hand side has units of seconds per centimeter and represents a concentration difference divided by a vertical flux. By analogy to Ohm's law for electrical circuits, each of these represents a resistance to the vertical flux (current) driven by the concentration gradient (potential difference). The term $r_a \equiv (u_* k)^{-1}[\ln(z/z_0) - \Psi_\chi]$ is an aerodynamic resistance in the layer of air from height z to z_0, $r_s \equiv$

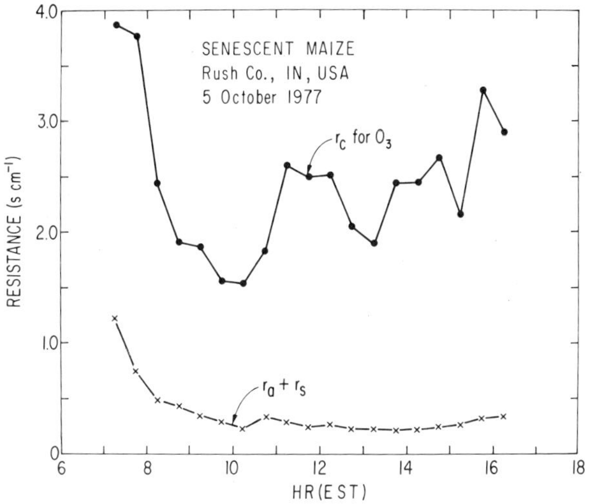

Figure 4. Example of resistances to ozone flux during the daytime above senescent maize (after Wesely et al., 1978).

$2(u_*k)^{-1}(\kappa/D_\chi)^{2/3}$ is the bulk resistance offered by the quasilaminar layer of air closest to the surface ($r_s \equiv u_*\kappa^{-1}\ln(kz_0u_*/D_\chi)$ is the form for smooth surfaces), and r_c is the surface resistance remaining. It is readily seen that if $(r_a + r_s)$ is small compared to r_c, it follows that $\overline{\chi}_s$ is near the values of $\overline{\chi}_z$ above and r_c mostly controls the value of $-\overline{\chi}_z/F_\chi$. The resistance r_c is the quantity we seek to characterize for ozone. For example, Figure 4 shows a sample of data on resistance to ozone flux where the surface resistance clearly is the dominating resistance below a height of a few meters (Wesely et al., 1978).

Rather than Eq. (10), the corresponding expression involving deposition velocity and resistances is frequently used:

$$|v_d| = (r_a + r_s + r_c)^{-1}. \tag{11}$$

In the present study v_d is measured, r_a and r_s are calculated, and r_c is determined as the residual quantity. It should be remembered that Eq. (11) can provide an estimate of r_c only as a property of the surface somehow averaged; very little information is provided directly on the resistance of different types of surface element located in the same field.

5. ESTIMATES OF SURFACE RESISTANCES

5.1. Row Crops

The physiological activity of the type of live vegetation considered here typically has a strong diurnal variation, primarily in response to solar radiation. The small pores, or stomata, on the leaves of vegetation open during the daytime to permit the exchange of water vapor, carbon dioxide, and oxygen between leaves and the atmosphere. This provides a pathway for ozone to sites at inner leaf surfaces, where rapid reactions can take place. Thus, in the unlikely situation that no ozone is destroyed at the soil beneath the plant canopy and at outer leaf surfaces, surface resistance to ozone uptake would be proportional to the bulk stomatal resistance of the plant canopy. Since work in enclosure chambers has shown that ozone uptake by leaves is strong through leaf stomata but very weak at the waxy outer covering of leaves (Rich et al., 1970; Thorne and Hanson, 1972), it is likely that most of the ozone is removed through leaf stomata and at the soil surface. This was found to be the case for maize in the field (Leuning et al., 1979; Wesely et al., 1978), with soil uptake accounting for 20–50% of the loss during the daytime in midsummer.

Figure 5 shows the average resistance r_c at two maize fields during the daytime. For the earlier case, described by Wesely et al. (1978), a minimum value near 1.0 s cm^{-1} occurs in the morning, which closely parallels bulk stomatal resistance inferred from values of r_c for water vapor. This minimum occurs rather early in the day, probably as a result of stomata closing later because of increasing water stress on the plants. In parallel with the canopy resistance, an effective resistance near 4 s cm^{-1} for the soil has been found (Wesely et al., 1978). For more well-watered maize, data from the DeWitt County site show a minimum near 1.0 s cm^{-1} slightly after 1200 h and are nearly coincident with the maximum in solar radiation.

In comparison to Figure 5, Figure 6 for a field of senescent maize ready for harvest shows a 50% increase in minimum r_c and less of a diurnal cycle (Wesely et al., 1978). The existence of a small but noticeable diurnal cycle

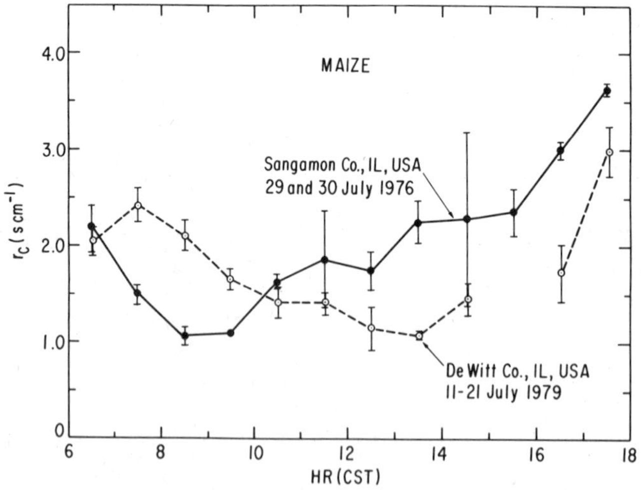

Figure 5. Residual surface resistances derived from measurements during the daytime over lush maize, mostly under cloudless skies. Standard error bars are shown.

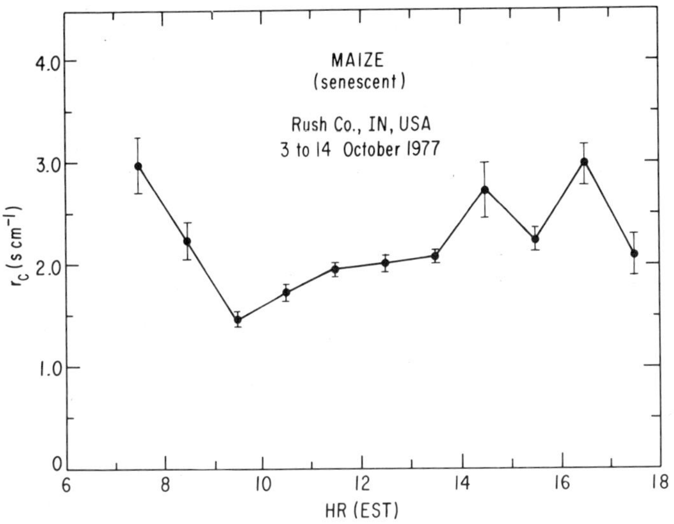

Figure 6. As in Figure 5, except over senescent maize.

in Figure 6 suggests that there are sites in the brown canopy, soil surface, and perhaps small patches of grassy weeds where the capacity to destroy ozone varies in response to changes in solar radiation, temperature, humidity, or other environmental factors that change diurnally. Further indication is given by the value of $r_c = 5.7$ s cm^{-1} obtained at night, a considerably higher value than during the daytime.

The lowest values encountered of surface resistances for ozone have occurred over soybean fields, especially in strong winds or when ground cover was incomplete. Figure 7 shows data that were obtained during two NEROS experiments. At the earlier study in Lancaster County (Wesely et

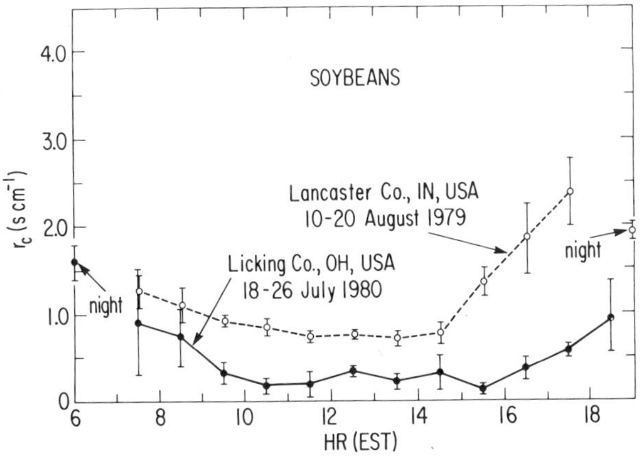

Figure 7. As in Figure 5, except over soybeans.

al., 1982), the maximum deposition velocity at 5.2 m over a full soybean plan canopy was typically about 0.85 cm s^{-1} near midday, but values as high as 1.2 cm s^{-1} were recorded during very windy conditions. In contrast, deposition velocities at the Licking County site were typically 1.3 cm s^{-1}, reaching nearly 2.0 cm s^{-1} during strong winds. The soybeans were shorter and did not provide as much ground cover as at the Lancaster site; approximately 25% of the soil surface was exposed between the rows as seen from above the 75-cm-tall canopy. Since the plants were vigorously growing, the midday value of r_c for water vapor were nearly as small as they were at Lancaster despite the greater plant canopy there. Evaporation from the soil at Licking was probably somewhat retarded by the dry soil crust usually present, but the dryness probably aided ozone destruction (Turner et al., 1973). The surface soil was wetter at Lancaster because of more shading by the plant canopy.

At night the mean surface resistance at Licking County was only slightly smaller than at Lancaster. Figure 7 shows that nighttime values of r_c was 1.4–2.0 s cm^{-1} at both sites. Buoyancy-driven convection during the daytime may have greatly aided transport to the warm soil and the nearby soybean leaves at Licking. If so, and if the canopy provided an effective minimum r_c of 1.0 s cm^{-1}, near that likely at Lancaster, the effective soil resistance that acts in parallel with canopy resistance would have to be as small as 0.4 s cm^{-1} to achieve the minimum values of 0.3 s cm^{-1} shown for Licking in Figure 7. This is rather small but is possible for sufficiently dry soil [e.g., see the range of values given by Galbally and Roy (1980)]. Such a situation implies that approximately 70% of the ozone destruction occurs at the soil surface during the daytime.

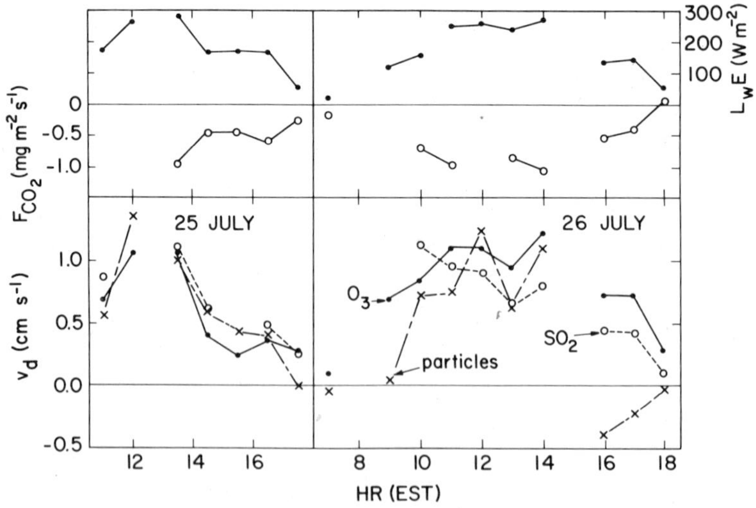

Figure 8. Fluxes and deposition velocities measured during 2 days at the Licking County site.

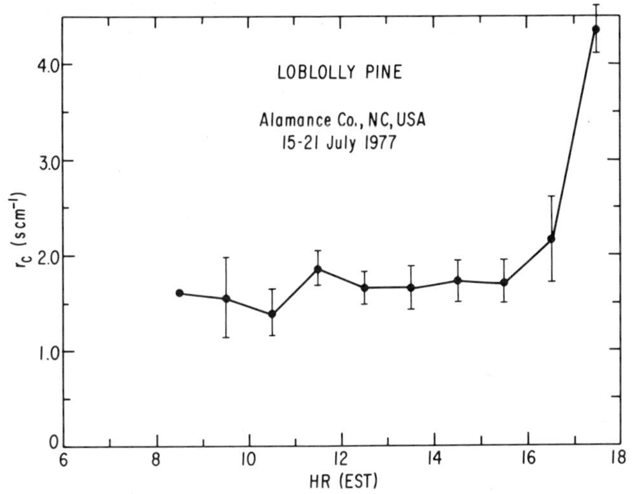

Figure 9. As in Figure 5, except over a pine forest.

Figure 8 shows fluxes and deposition velocities for other quantities during two days at Licking. The fluxes of carbon dioxide and water roughly chart the trends of air–plant exchange, and these are closely followed by the deposition velocities for both ozone and sulfur dioxide, as expected. Similar to that of ozone, deposition of sulfur dioxide to vegetation is controlled by stomatal resistance; this material also can be rapidly removed by soil (Garland, 1978). The flux of particles, in this case for those primarily 0.03–0.1 μm in diameter, is associated with a minimum resistance ($r_s + r_c$) of about 1.0–1.5 s cm^{-1} to grass (Wesely et al., 1977) and pine forest (Hicks et al., 1982) in the afternoon but is about one-third as large at Licking. Apparently, the results for both sulfur dioxide and small particle fluxes support the contention that unusually vigorous exchange of some substances can occur at a field with a partial covering of soybean row crop under certain soil conditions.

5.2. Forests

Figure 9 gives the average resistances found at a pine forest during a field experiment described by Hicks et al. (1982). A rather large minimum value of about 1.4 s cm^{-1} is found for r_c, whereas the nighttime value is about 12 s cm^{-1}. Since these data were taken during quite dry conditions, slightly smaller daytime values might be found in other cases. Galbally and Roy (1980) suggest that a value of 1.2 s cm^{-1} is a typical minimum, and it is well known that deciduous forest can have considerably smaller bulk stomatal resistances, perhaps as small as 0.5 s cm^{-1} for ozone during ideal conditions. Lenschow et al. (1982) find surface resistances as low as 0.5 s cm^{-1} determined from aircraft measurements 20–30 m above a forest consisting of about 80% coniferous and 20% deciduous trees.

In 1981 an experiment was performed above deciduous trees in the winter when leaves were absent; the results are shown in Table 3. Measurements were taken at a height of about 10 m above the treetops, which extended about 30 m above the ground. Because of the difficulty of determining the effective height of measurement, resistances are not calculated, but the average deposition velocities of 0.19–0.37 cm s^{-1} indicate that an appropriate surface resistance r_c is 2.5–4.0 s cm^{-1} during the daytime. At night, deposition velocities are near 0.05 cm s^{-1}, which points to surface resistances of approximately 20 s cm^{-1}. It is difficult to explain the diurnal variation evident, because of the lack of sufficiently active vegetation in the vicinity. Perhaps convective conditions during the daytime allowed rather deep penetration of turbulent air into the porous canopy where the outer surfaces of twigs, barks, etc. could provide sites effective in ozone destruction. At night, stable stratification of air might inhibit such deep mixing.

The last three entries in Table 3 show a case when about 10 cm of snow completely covered the leaf litter on the ground. Since fresh snow greatly retards ozone removal (Galbally and Roy, 1980; Wesely et al., 1981), a difference would be seen between the last three entries and the others if significant deposition ever took place at the litter. It is not surprising that such deposition appears unimportant, because significant wind and turbulent mixing within a few meters of the ground were not observed during the experiment. These results imply that aerodynamic resistances for transport all the way to the forest floor were very large.

5.3. Grass

Figure 10 shows that the daytime surface resistance of short pasture grasses has a minimum value near 2 s cm^{-1}; additionally, some 15 half-hour data

Table 3. Ozone Deposition Measurements Above a Leafless Deciduous Forest in North Carolina in 1981

Time	Snow?	v_d (cm s^{-1})	Number of 0.5-h Samples	$\bar{u}$ (m s^{-1})
January 25, day	No	0.28 ± 0.06	4	1.5
January 25–26, night	No	0.08 ± 0.01	21	2.7
January 26–27, night	No	0.06 ± 0.01	17	2.6
January 28, day	No	0.26 ± 0.03	8	2.3
January 28–29, night	No	0.03 ± 0.01	14	2.8
January 29, day	No	0.19 ± 0.04	16	4.1
January 31, day	Yes	0.37 ± 0.10	5	1.6
January 31–February 1, night	Yes	0.01 ± 0.01	11	1.8
February 1, day	Yes	0.26 ± 0.05	10	1.6

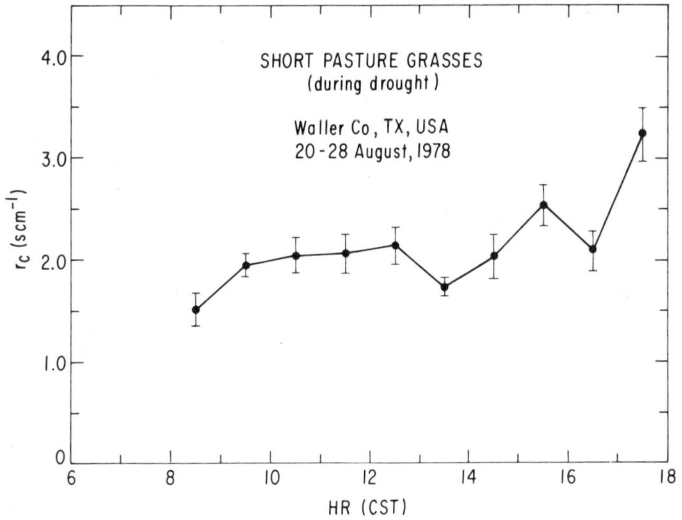

Figure 10. As in Figure 5, except over grass.

collection periods at night (not shown) produced values averaging $r_c = 2.9$ s cm^{-1}. These results and the deposition velocities associated are consistent with those presented by Regener (1957), Garland and Derwent (1979), and Galbally and Roy (1980), all of whom used methods other than eddy correlation. Garland and Derwent did not find a significant variation of r_c from night to day for the same values of u_*, but Galbally and Roy do find such a variation that is determined with methods that should not depend on u_*. For the case of poorly watered thick grass, Galbally and Roy find values very similar to those presented here (see their Figure 5 for Site 1 during January).

5.4. Surfaces Without Open Leaf Stomata

5.4.1. Vegetation and Soil

Many of the cases with inactive leaf stomata have already been discussed and are summarized in Table 4. Soils, except when waterlogged (Wesely et al., 1981), seem to have a surface resistance near 1 s cm^{-1} and provide a significant sink for ozone even under rather short plant canopies at night; the values are $r_c \approx 3$ s cm^{-1} for grass and $r_c = 1.4$–2.0 s cm^{-1} for soybeans. For the taller canopies of senescent maize, loblolly pine, and leafless deciduous forest, the nighttime values of r_c at night increase with height of canopy to about 6, 12, and 20 s cm^{-1}, respectively. Such high values suggest that the soil surface is not a very effective sink under very tall canopies. However, penetration of turbulent air into the canopy and to the soil surface beneath shorter canopies seems to be greatly aided by heating of the surface and the resulting convective mixing, as evidenced by the lower values of

Table 4. Surface Resistances Found in the Present Study (Except Where Noted) for Surfaces That Do Not Have Vegetation with Open Leaf Stomata

Site	r_c (s cm^{-1})
Freshwater, moderate wind speeds	94 ± 3
Seawater [from summary by Galbally and Roy, (1980)]	6–44
Snow, with no blowing snow	34 ± 3
Bare soil, waterlogged	10 ± 1
Bare soil (Turner et al., 1973)	0.2–2.0
Short mixed grass during drought (at night)	2.9 ± 0.3
Senescent maize (at night)	5.7 ± 0.4
Senescent maize (daytime)	~2
Soybeans, full cover (at night)	1.9 ± 0.1
Soybeans, 75% cover (at night)	1.6 ± 0.2
Loblolly pine (at night)	12 ± 1
Deciduous forest in winter (at night)	~20
Deciduous forest in winter (daytime)	2.5–4.0

r_c during the daytime for leafless deciduous forest, sensecent maize, and the soil beneath the partial soybean canopy (as discussed in Section 5.1). An alternative explanation suggested by Galbally and Roy (1980) is that photoactivation of sites for ozone destruction occurs during the daytime. Further experiments are needed to determine whether this is the case.

5.4.2. *Water and Snow*

The surface resistance of the freshwater of Lake Michigan has been found to be quite large, near 100 s cm^{-1}, but even this is much less than expected if chemical reactions are not important (Wesely et al., 1981). By comparison, Aldaz (1969) reports data corresponding to a value of $r_c \approx 50$ s cm^{-1} for distilled water in the laboratory and $r_c \approx 10$ s cm^{-1} for an open freshwater reservior. Garland et al. (1980) also report a value of 50 s cm^{-1} for the liquid-phase resistance r_c of distilled water and suggest that water impurities cause a lowering of resistance. Apparently the Lake Michigan surface water has unusually low amounts of chemicals that react with ozone. Galbally and Roy (1980) give a range of r_c of 10–83 s cm^{-1} for tap water and for lake water *in situ* and state that the extent of mixing or wave motion produces the lower values. No significant variation of r_c for changes in wind speed and wave development was detected at the Lake Michigan site, but measurements were made in only light to moderate winds; water spray would very likely increase the transfer rates.

The entry for seawater in Table 4 is obtained from various results summarized by Galbally and Roy (1980). Laboratory studies by Garland et al. (1980) indicate that r_c should be 19–32 s cm^{-1}, with chemical reactions with iodide and other (unidentified) substances causing most of the ozone destruction. A set of eddy-correlation measurements from aircraft by Lenschow et al. (1982) points to values of r_c = 15–20 s cm^{-1} in fairly strong winds. As yet, a definite trend of r_c with wind speed has not been proved; on the basis of data available, it appear most likely that r_c is 15–30 s cm^{-1}, with a weak dependence on wind speed.

For snow, Galbally and Roy (1980) report a median resistance of 16 s cm^{-1} but find a great deal of variation depending on the freshness of snow. Such a strong variation was not detected in the eddy-correlation results reported by Wesely et al. (1981), whose estimate of $r_c \approx$ 34 s cm^{-1} is given in Table 4.

6. SURFACE RESISTANCES FOR BROAD LAND-USE CATEGORIES

Table 5 gives estimates of r_c for land-use categories similar to those that have been suggested for NEROS. The procedures used to derive the values closely follow the methods employed by Sheih et al. (1979). Pasquill stability classifications A, B, and C are for daytime conditions, D for near-neutral conditions or when solar radiation is suppressed greatly by cloud cover, and E and F for nocturnal cases. It should be remembered that in very stable conditions beyond category F the deposition velocity is reduced greatly by

Table 5. Surface Resistances to Ozone Uptake and Surface Roughnesses for Selected Land-Use Categories as a Function of Pasquill Stability Class [a]

Land-Use Category	z_0 (cm)	r_c (s cm^{-1}) A, B, C	D	E, F
1. Agricultural land	25	0.7	2.0	4.0
2. Rangeland	5	1.5	2.0	3.0
3. Mixed 1 and 2	10	1.0	2.0	3.5
4. Deciduous forest	100	0.6	3.0	15
5. Coniferous forest	100	1.5	4.0	15
6. Mixed forest including wetland	100	0.7	3.0	15
7. Nonforested wetland	20	1.0	2.5	3.0
8. Freshwater lakes	~0.01	100	100	100
9. Ocean[a]	~0.01	20	20	20
10. Urban areas[a]	100	3.0	4.0	4.0
1. Bare soil	0.2	1.0	1.0	1.0
1. In autumn before harvest	20	2.0	4.0	4.0
1–3,7,8. With snow	0.05	30	30	30
2. Brown grass in winter	3	2.0	3.0	3.0
4–6. In cold weather	100	3.0	20	20
7. Brown vegetation	20	2.0	4.0	4.0

[a] Values apply during summer except where noted.
[b] Values for this type apply to all seasons.

aerodynamic resistances, probably to values near 0.05 cm s^{-1}. This is a fairly common occurrence at night over continental areas that have smooth to moderately rough surfaces.

In most cases the values of r_c shown in Table 5 reflect the findings presented in this chapter. Exceptions include nonforested wetland, for which a behavior quite similar to agricultural land is assumed. Category 10, urban areas, is interpreted to be only the highly developed areas that have little vegetation and as such do not constitute a very important category since the percentage of the total landscape covered is quite small. The values given for category 10 are meant to show only moderately high resistances with little diurnal trend.

Section 5 shows that different agricultural crops can have substantially different surface resistances and that these differences may be increased when some crops provide only partial cover of the soil surface. Early in the growing season, a large amount of ozone destruction occurs at the soil surface. Late in the season, the total exchange with vegetation decreases because of plant senescence. Such considerations clearly indicate that Table 5 and similar summaries are of limited value. Estimates of ozone removal at specific locations and times are best made by a detailed examination of the characteristics of the surfaces in the vicinity.

7. CONCLUSIONS

Section 5 presents resistances that can be used to calculate deposition velocities by application of the equations given in Section 4 or some other scheme for estimation of $(r_a + r_s)$. Put in the proper framework, such as that presented by Sheih et al. (1979), the values of r_c can be applied to provide estimates of deposition velocity over rather large regions. The removal rates are greatest to lush vegetation, less to senescent or water-stressed vegetation, and smallest to water surfaces. Pine forest, maize, and soybeans exhibit strong diurnal trends in the ability to remove ozone. Residual resistances calculated for the surfaces typically are at a minimum during morning or midday, when they are 0.7–1.5 s cm^{-1}, depending on availability of soil moisture. Since the gas-phase resistance $(r_a + r_s)$ to vertical transport is typically 0.5 s cm^{-1} below heights near 5 m, the corresponding maximum deposition velocities at such heights are typically 0.9–0.5 cm s^{-1}, calculated according to Eq. (11). The largest deposition velocities yet detected approach 2 cm s^{-1} above a partial soybean cover, presumably because of the very efficient removal of ozone by both the plants and the large amount of soil exposed. Senescent maize and very dry grass provide weak diurnal variations, and the relative importance of soil uptake usually appears greater than for lush vegetation. Also, taller, denser plant canopies seem to retard transport of ozone to the soil surface beneath more effectively than do less substantial canopies. There is some evidence that ozone destruction at the lower parts of the canopy and at the soil surface is enhanced during the daytime, perhaps by convective mixing to active sites or by photoactivation of sites.

Very small ozone destruction rates are found at surfaces of snow, very wet bare soil, and lake water. For snow, the surface resistance is about 34 s cm^{-1}, fairly constant over a wide range of conditions, indicating that the maximum deposition velocity is about 0.03 s cm^{-1}. For cold waterlogged soil, the surface resistance is about 10 s cm^{-1}, which corresponds to a maximum deposition velocity of about 0.1 cm s^{-1}. This is about an order of magnitude less than for drier soil. The greatest resistances measured are for the surface of Lake Michigan, near 95 s cm^{-1}, which corresponds to about 0.01 cm s^{-1} for deposition velocity.

The behavior of the turbulent fluctuations of ozone concentration in the atmospheric surface layer is quite similar to that of other scalar quantities such as temperature, water vapor content, and sulfur dioxide and carbon dioxide concentrations. Evidence for this is provided by the large magnitude of the correlation coefficient among pairs of temperature, humidity, and ozone variations and by the nearly identical behavior of normalized standard deviations of fluctuations of ozone, water vapor content, and carbon dioxide.

ACKNOWLEDGMENTS

This work was supported primarily by the U.S. Environmental Protection Agency through consecutive interagency agreements IAG-78-DX0193, AD-89-F0123-0, and AD-89-F1686-0 to provide information for the Northeast Regional Oxidant Study (NEROS). Environmental Protection Agency support also came from the Multistate Atmospheric Power Production Pollution Study (MAP3S), and substantial funding was received from the U.S. Department of Energy for work concerning the transport and removal of pollutants in coastal areas. The spectral analyses were performed by R. L. Hart of Argonne National Laboratory (ANL), and field work was assisted by D. R. Cook of ANL and J. A. Eastman, formerly of ANL. Thanks are due to J. L. Durham of the EPA and B. B. Hicks, formerly of ANL, for their advice and guidance.

8. NOMENCLATURE

a Numerical value of $\ln(z_0/z_{0\chi})$ for rough surfaces

c_p Specific heat of air at constant pressure

D_χ Molecular diffusivity of χ in air

E Water vapor vertical flux density, $\rho\overline{w'q'}$

F Vertical flux density

F_χ	Vertical flux density of χ (usually ozone), $\overline{w'\chi'}$
g	Acceleration due to gravity
H	Vertical sensible heat flux density, $\rho c_p \overline{w'T'}$
k	Von Karman constant
L	Obukhov stability length
L_w	Latent heat of water vaporization
n	Frequency (Hz)
q	Specific humidity of air
r	Spectral correlation coefficient
r_a	Aerodynamic resistance from z to z_0
r_c	Surface (residual) resistance
r_s	Resistance of quasilaminar layer from z_0 to $z_{0\chi}$
T	Dry-bulb air temperature
T_*	Turbulence temperature scale, $H/\rho c_p u_*$
t_c	Exponential response time of chemical sensor
t_w	Exponential response time of w sensor
u_*	Horizontal wind speed
u	Surface friction velocity
v_d	Deposition velocity at specific height, $-F_\chi/\overline{\chi}$
w	Vertical wind velocity
z	Height above the surface
z_0	Aerodynamic roughness scale length
$z_{0\chi}$	Surface scale length for χ
Δt_c	Delay time of chemical sensor
θ	Potential air temperature
κ	Thermal diffusivity in air
ρ	Air density
σ	Standard deviation
σ_T	Standard deviation of temperature fluctuations
χ	Concentration (usually ozone)
χ_*	Starred χ scale, F_χ/u_*
$\overline{\chi}_s$	Mean concentration of χ at the surface
$\overline{\chi}_z$	Mean concentration of χ at height z
Ψ_χ	Diabatic influence function, equal to that for heat

REFERENCES

Aldaz, L. (1969). Flux measurements of atmospheric ozone over land and water. *J. Geophys. Res.* **75**, 6943–6946.

Allen, L. H., Jr., Hanks, R. J., Aase, J. K., and Gardner, H. R. (1974). Carbon dioxide uptake by wide-row grain sorghum computed by the profile Bowen ratio. *Agron. J.* **8,** 35–41.

Clarke, J. F., Ching, J. K. S., Brown, R. M., Westburg, H., and White, J. H. (1982). Regional transport of ozone, *Preprint Volume AMS/APCA Third Joint Conference on Applications of Air Pollution Meteorology,* American Meteorological Society, San Antonio, TX, January 12–15, 1982.

Dyer, A. J. and Hicks, B. B. (1970). Flux-gradient relationships in the constant flux layer. *Quart. J. R. Meteorol. Soc.* **96,** 715–721.

Fabian, P. and Junge, C. E. (1970). Global rate of ozone destruction at the earth's surface. *Arch. Meteorol. Geoph. Biokl., Ser. A.* **19,** 161–172.

Friehe, C. A., LaRue, J. C., Champagne, R. H., Gibson, C. H., and Dreyer, G. F. (1975). Effects of temperature and humidity fluctuations on the optical refractive index in the marine boundary layer. *J. Opt. Soc. Am.* **65,** 1502–1511.

Galbally, I. E. and Roy, C. R. (1980). Destruction of ozone at the earth's surface. *Quart. J. R. Meteorol. Soc.* **106,** 599–620.

Garland, J. A. (1978). Dry and wet removal of sulphur from the atmosphere. *Atmosph. Environ.* **12,** 349–362.

Garland, J. A. and Derwent, R. G. (1979). Destruction at the ground and the diurnal cycle of concentration of ozone and other gases. *Quart. J. R. Meteor. Soc.* **105,** 169–183.

Garland, J. A., Elzerman, A. W., and Penkett, S. A. (1980). The mechanism for dry deposition of ozone to water surfaces. *J. Geophys. Res.* **85,** 7488–7492.

Garratt, J. R. and Hicks, B. B. (1973). Momentum, heat and water vapour transfer to and from natural and artificial surfaces. *Quart. J. R. Meteorol. Soc.* **99,** 680–687.

Heilman, J. L. and Kanemasu, E. T. (1976). An evaluation of a resistance form of the energy balance to estimate evapotranspiration. *Agron. J.* **68,** 607–617.

Hicks, B. B. (1972). Propeller anemometers as sensors of atmospheric turbulence. *Boundary-Layer Meteorol.* **3,** 214–228.

Hicks, B. B., Wesely, M. L., Durham, J. L., and Brown, M. A. (1982). Some direct measurements of atmospheric sulfur fluxes over a pine plantation. *Atmosph. Environ.* (in press).

Kanemasu, E. T., Wesely, M. L., Hicks, B. B., and Heilman, J. L. (1979). Techniques for calculating energy and mass fluxes, in *Modification of the Aerial Environment of Crops,* B. J. Barfield and J. F. Gerber, Eds., ASAE Monograph No. 2, pp. 156–182, American Society of Agricultural Engineers, St. Joseph, MO.

Kaimal, J. C. (1973). Turbulence spectra, length scales and structure parameters in the stable surface layer. *Boundary-Layer Meteorol.* **4,** 289–309.

Kaimal, J. C., Wyngaard, J. C., Izumi, Y., and Coté, O. R. (1972). Spectral characteristics of surface-layer turbulence. *Quart. J. R. Meteorol. Soc.* **98,** 563–589.

Lenschow, D. H., Pearson, R., Jr., and Stankov, B. B. (1982). Measurements of ozone vertical flux to ocean and forest. *J. Geophys. Res.* **87,** 8833–8837.

Leuning, R., Neumann, H. H., and Thurtell, G. W. (1979). Ozone uptake by corn (*Zea Mays* L.): A general approach. *Agric. Meteorol.* **20,** 115–135.

McBean, G. A. and Miyake, M. (1972). Turbulent transfer mechanisms in the atmospheric surface layer. *Quart. J. R. Meteorol. Soc.* **98,** 383–398.

Monteith, J. L. (1973). *Principles of Environmental Physics,* American Elsevier, New York.

National Research Council (1977). *Ozone and Other Photochemical Oxidants,* Committee on Medical and Biological Effects of Environmental Pollutants, National Academy of Sciences, Washington, DC.

Phelps, G. T. and Pond, S. (1971). Spectra of the temperature and humidity fluctuations of fluxes of moisture and sensible heat in the marine boundary layer. *J. Atmosph. Sci.* **28,** 918–928.

Regener, V. H. (1957). The vertical flux of atmospheric ozone. *J. Geophys. Res.* **62,** 221–228.

Rich, S., Waggoner, P. E., and Tomlinson, H. (1970). Ozone uptake by bean leaves. *Science* **169,** 79–80.

Sheih, C. M., Wesely, M. L., and Hicks, B. B. (1979). Estimated dry deposition velocities of sulfur over the eastern United States and surrounding regions. *Atmosph. Environ.* **13,** 1361–1368.

Sinclair, T. R., Allen, L. H., Jr., and Lemon, E. R. (1975). An analysis of errors in the calculation of energy flux densities above vegetation by a Bowen-ratio profile method. *Boundary-Layer Meteorol.* **8,** 129–139.

Swinbank, W. C. and Dyer, A. J. (1967). An experimental study in micrometeorology. *Quart. J. R. Meteorol. Soc.* **93,** 494–500.

Thorne, L. and Hanson, G. P. (1972). Species differences in rates of vegetal ozone absorption. *Environ. Pollut.* **3,** 303–312.

Thorpe, M. R., Banke, E. G., and Smith, S. D. (1973). Eddy correlation measurements of evaporation and sensible heat flux over Arctic sea ice. *J. Geophys. Res.* **78,** 3573–3584.

Turner, N. C., Rich, S., and Waggoner, P. E. (1973). Removal of ozone by soil. *J. Environ. Qual.* **2,** 259–264.

Wesely, M. L. and Hicks, B. B. (1977). Some factors that affect the deposition rates of sulfur dioxide and similar gases on vegetation. *J. Air Pollut. Control Assoc.* **22,** 260–263.

Wesely, M. L. and Hicks, B. B. (1978). High-Frequency temperature and humidity correlation above a warm wet surface. *J. Appl. Meteorol.* **17,** 123–128.

Wesely, M. L., Hicks, B. B., Dannevik, W. P., Frisella, S., and Husar, R. B. (1977). An eddy-correlation measurement of particulate deposition from the atmosphere. *Atmosph. Environ.* **11,** 561–563.

Wesely, M. L., Eastman, J. A., Cook, D. R., and Hicks, B. B. (1978). Daytime variations of ozone eddy fluxes to maize. *Boundary-Layer Meteorol.* **15,** 361–373.

Wesely, M. L., Cook, D. R., and Williams, R. M. (1981). Field measurement of small ozone fluxes to snow, bare wet soil, and lake water. *Boundary-Layer Meteorol.* **20,** 459–471.

Wesely, M. L., Eastman, J. A., Stedman, D. H., and Yalvac, E. D. (1982). An eddy-correlation measurement of NO_2 flux to vegetation and comparison to ozone flux. *Atmosph. Environ.* **16,** 815–820.

Wyngaard, J. C., Coté, O. R., and Izumi, Y. (1971). Local free convection, similarity, and budgets of shear stress and heat flux. *J. Atmosph. Sci.* **28,** 1171–1182.

Wyngaard, J. C., Pennell, W. T., Lenschow, D. H., and LeMone, M. A. (1978). The temperature–humidity covariance budget in the convective boundary layer. *J. Atmosph. Sci.* **35,** 47–58.

9

FATE OF NITROGEN OXIDES IN URBAN ATMOSPHERES

*Larry G. Anderson**

Environmental Science Department
General Motors Research Laboratories
Warren, Michigan 48090

* Present Address: Department of Chemistry University of Colorado at Denver, Denver, CO 80202.

1. INTRODUCTION

Nitrogen oxides (NO_x = NO + NO_2) play an important role in the urban atmosphere. They are intimately involved in the complex chain of chemical reactions that lead to the formation of photochemical smog. In addition, nitrogen oxides have been implicated in a variety of health and environmental effects.

In considering the fate of nitrogen oxides in urban atmospheres, the major emphasis of this paper is to describe the chemical and physical processes undergone by nitrogen oxides as they are transported downwind. Some of these processes transform nitrogen oxides to other nitrogenous species, hence interrupting the chain of reactions leading to smog formation. Other processes lead to the ultimate removal of the nitrogenous species from the atmosphere. In addition, an attempt is made to quantify the importance of these processes by examining ambient atmospheric data from several urban atmospheres.

To place the role of urban nitrogen oxides in perspective, the discussion begins with a consideration of the global nitrogen cycle, identifying both the natural components of the cycle and human impact on this cycle. The discussion then turns to the identification of the anthropogenic sources of nitrogen oxides, which are found to be heavily concentrated in urban areas. The chemical and physical processes that lead to the formation of other gaseous and particulate nitrogenous species and to the ultimate removal of these nitrogenous species from the atmosphere are then considered. The techniques used in the measurement of the nitrogenous species are described next, followed by a brief overview of the measurements that have been made in urban areas. An attempt is then made to evaluate the NO_x removal from several urban atmospheres. Finally, conclusions about the fate of nitrogen oxides in the urban atmosphere are discussed, along with an identification of some research needs.

2. GLOBAL NITROGEN CYCLE

The global nitrogen cycle is briefly outlined here; a more extensive discussion of this cycle can be found in Chapter 10 of this volume by Stedman and Shetter. The global nitrogen cycle is the description of the flow of nitrogenous species between the various reservoirs in the hydrosphere, lithosphere, atmosphere, and biosphere. The distribution of nitrogenous species in the major reservoirs is shown in Table 1. This table indicates that the main reservoirs for nitrogen in the environment are the atmosphere and sediments. Nitrogen in the atmosphere is present primarily as molecular nitrogen, which is quite unreactive and not readily usable in other parts of the environment. The stability of molecular nitrogen is responsible for the atmosphere being the largest reservoir for nitrogen. The sedimentary reservoir for nitrogen has become large through the slow but continuous loss of nitrogen to the sediments, where it is relatively inaccessible to the rest of the environment. This loss of nitrogen to the sediments may be approximately balanced by the return of nitrogen from the weathering of igneous rocks and from volcanic action (Delwiche, 1970).

Nitrogen is an essential nutrient for all living organisms. The primary source of this nitrogen is the atmosphere. However, nitrogen is not useful to most organisms until it is "fixed" or converted to a form that can be chemically utilized by the organisms. The "natural" fixation of nitrogen occurs by two types of process. One is the action of a comparatively few microorganisms that are capable of converting molecular nitrogen taken from the atmosphere to ammonia (NH_3), ammonium ion (NH_4^+), and organic nitrogen compounds that can be utilized by other organisms. The other "natural" nitrogen fixation process occurs in the atmosphere by the action of some ionizing phenomena, such as cosmic radiation or lightning, on molecular nitrogen. This leads to the formation of nitrogen oxides in the atmosphere, which are ultimately deposited on the Earth's surface as nitrates (NO_3^-) and are also biologically useful.

In addition to these "natural" nitrogen fixation processes, human activities have led to additional nitrogen fixing processes. These fall into three basic categories: biological fixation; industrial fixation; and combustion. Humans have increased the cultivation of legumes, which have a symbiotic relationship with certain microorganisms capable of nitrogen fixation. This increasing cultivation of legumes provides both an increase in the soil nitrogen and a valuable food crop. Industrial nitrogen fixation consists primarily of the production of ammonia for fertilizer use. Combustion can also

Table 1. Distribution of Nitrogenous Species in the Major Environmental Reservoirs (Delwiche, 1977)

Reservoir	Tg N[a]
Atmosphere	3.9×10^9
Plants and animals	1.0×10^4
Sea	9.9×10^5
Soil	3.3×10^5
Sediments	2.0×10^8

[a] Tg N = 10^{12} g of nitrogen.

Table 2. Comparison of Rates of Natural and Anthropogenic Nitrogen Fixation (Delwiche, 1977)

Source	Process	Rate ($Tg\ N\ yr^{-1}$)
Natural	Biological	60
	Atmospheric	7.4
Anthropogenic	Biological	69
	Industrial	40
	Combustion	20

lead to the fixation of nitrogen as nitrogen oxides (NO or NO_2) and, to a much lesser extent, nitrous oxide (N_2O) and ammonia (NH_3). Table 2 presents Delwiche's (1977) estimates of the magnitudes of the rates of these nitrogen fixation processes. On the basis of these estimates, anthropogenic routes to nitrogen fixation are about twice the size of those of "natural" nitrogen fixation.

Since 1950 industrial fixation of nitrogen for fertilizer use has increased from about 6 Tg yr^{-1} to about 40 Tg yr^{-1} (Delwiche, 1970; 1977), and this rapid growth is expected to continue in order to boost agricultural productivity. This continued shifting of the balance in the nitrogen cycle could lead to potentially serious environmental problems, such as eutrophication of aquatic ecosystems (National Research Council, 1978) or stratospheric ozone depletion due to increases in N_2O emissions from fertilized soils (Logan et al., 1978; National Research Council, 1978).

A global budget for nitrogen oxides in the atmosphere has been constructed by Söderlund and Svennson (1976). Table 3 shows their summary of global source strengths for NO_x. Their budget was based on the assumption that in view of the short turnover time for nitrogen oxides and particulate

Table 3. Global Source Strengths for Nitrogen Oxides in the Atmosphere (Söderlund and Svensson, 1976)

Source	Rate ($Tg\ N\ yr^{-1}$)
Combustion	19
From the stratosphere	0.3
Atmospheric conversion of NH_3	3–8
Losses from soils	1–14
Atmospheric (lightning, etc.)	?
Total (excluding atmospheric)	23–41

nitrate in the atmosphere, the sources for nitrogen oxides in the atmosphere must balance their removal.

The processes that remove nitrogen oxides from the atmosphere are deposition, both wet and dry. The second column in Table 4 summarizes the estimates by Söderlund and Svensson (1976) for the deposition of nitrogen oxides and particulate nitrate. The basis for the estimates of the wet deposition of NO_3^- was discussed previously (Söderlund and Svensson, 1976). The dry deposition estimates were based on deposition velocities in the range from 0.3 to 0.8 cm s^{-1} for the gas-phase nitrogenous species and 0.05 to 0.15 cm s^{-1} for the particulate phase, NO_3^-. These deposition velocities were used with estimated global mean concentrations of NO_x and NO_3^- to determine the deposition rates shown in Table 4.

Recent data for the atmospheric concentrations of NO, HNO_3, and NO_3^- in the background atmosphere suggest much lower concentrations than were employed in the calculations by Söderlund and Svennson (1976). Tropospheric concentrations of NO have been reported for equatorial regions of the Pacific peaking near noon at about 4 ppt (parts per trillion, 10^{12}) (McFarland et al., 1979). Kley et al. (1981) have taken the value of 10 ppt as representative of the near-surface NO + NO_2 mixing ratio in clean remote tropical air. The average gas-phase HNO_3 concentration in the remote atmosphere has been reported to range from 0.10 to 0.16 ppb over continental regions and 0.07–0.11 ppb in the marine atmosphere (Huebert and Lazrus, 1980). The corresponding particulate NO_3^- concentrations range from 0.07 to 0.08 ppb (mole fraction) in continental regions and 0.05 to 0.11 ppb (mole fraction) in the marine atmosphere (Huebert and Lazrus, 1980). [The unit ppb (mole fraction) for an aerosol constituent directly corresponds to ppb (volume) for a gaseous constituent. For NO_3^-, 1 ppb (mole fraction) is equivalent to 2.14 ppb (mass fraction) or to 2.58 μg m^{-3} (20°C, 1 atm).] These concentrations were used to obtain revised dry deposition estimates for nitrogen oxides, and the results are shown in column 3 of Table 4.

Table 4. Comparisons of Estimated Deposition Rates for Nitrogen Oxides and Particulate Nitrate

Process	Rate (Tg N yr^{-1}) (Söderlund and Svensson, 1976)	Rate (Tg N yr^{-1}) (This Work)
Wet deposition, continental	13–30	13–30[a]
Wet deposition, oceanic	5–16	5–16[a]
Dry deposition, continental	19–53	1–4
Dry deposition, oceanic	6–17	2–13
Total	43–116	21–63

[a] from Söderlund and Svensson (1976).

The estimated total deposition rate resulting from the use of the more recent values for the concentrations of nitrogenous species in the remote atmosphere is about half that estimated by Söderlund and Svensson (1976). If the atmospheric NO_x source of 7.4 Tg N yr^{-1}, as shown in Table 2 (Delwiche, 1977), is added to the remaining sources listed in Table 3, the resulting estimated source strength for NO_x emissions is in the range of 31–49 Tg N yr^{-1}. This is well within the the range of total removal rates for nitrogenous species, shown in column 3 of Table 4, 21–63 Tg N yr^{-1}. On the basis of the estimated source strengths and concentrations, the tropospheric turnover time for gaseous NO_x, including HNO_3, is estimated to be in the range of 1–4 days and that for particulate NO_3^-, in the range of 3–9 days. Because of these short atmospheric lifetimes, the major effects of emissions of nitrogen oxides are expected to be local or regional rather than global in nature.

3. SOURCES OF NITROGEN OXIDES

The sources of nitrogen oxides in the atmosphere are numerous and varied. These sources include anaerobic bacterial action on nitrogenous compounds in soils and waters, formation in the electrical discharges of lightning, and formation from atmospheric oxidation of ammonia, as well as anthropogenic emissions. As was seen in Section 2, all these sources are likely to contribute significantly to nitrogen oxides on a global scale. However, on an urban scale, which is the focus of this work, anthropogenic emissions are likely to be the most significant source of nitrogen oxides.

The major anthropogenic sources of nitrogen oxides in the atmosphere are emissions from combustion processes. In recent years the use of fossil fuel combustion in transportation and electrical power generation has increased substantially, leading to an increase in the emissions of nitrogen oxides. Figure 1 shows a plot of the trend in estimated nitrogen oxide emissions in the United States between 1940 and 1970 (Cavender et al., 1973). During this 30-yr period nitrogen oxide emissions increased to almost three times the 1940 value, corresponding to an average annual increase of about 3.6% per year. Figure 2 shows an estimate of the yearly trend in NO_x emissions between 1970 and 1977 [U.S. Environmental Protection Agency (USEPA), 1978]. The data in Figure 1 should not be compared with those in Figure 2 since there were changes in the methodology and emission factors used in these two emission trend estimates (USEPA, 1978). The data in Figure 2 suggest that the growth in NO_x emissions has slowed, particularly since 1973. The 1977 NO_x emissions are less than 4% greater than the 1973 emissions. This slowing of the growth of NO_x emissions was probably due to the fuel shortages and energy conservation activities following the "1973 Arab oil crisis."

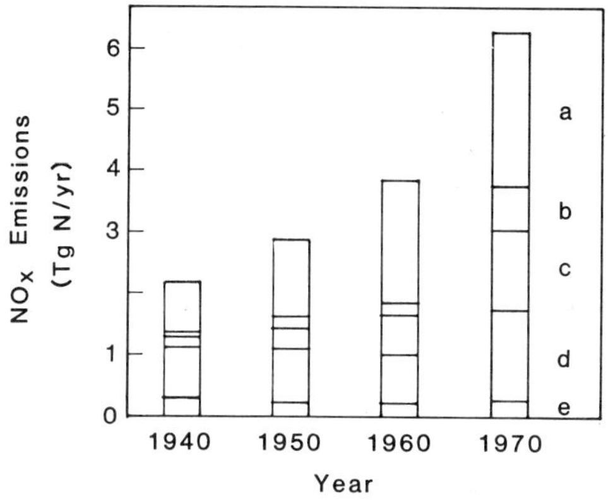

Figure 1. Trend in U.S. emissions of nitrogen oxides between 1940 and 1970 (Cavender et al., 1973). The segments of the bars represent the portions of the emissions from the following sources, listed from top to bottom: (a) highway vehicles; (b) nonhighway vehicles; (c) electric power generation; (d) industrial, commercial, and residential combustion; (e) miscellaneous.

As shown in Figure 2, during 1970–1977 the miscellaneous NO_x emission sources (industrial processes, solid wastes, and forest fires) and the stationary fuel combustion for industrial, commercial, and residential uses have decreased slightly. The fraction of the total NO_x emissions from nonhighway vehicles remained about the same, with the emissions increasing slightly. The bulk of the growth of NO_x emissions has been due to the increase in emissions from stationary combustion for electrical utilities, which have grown from 27% of the total NO_x emissions in 1970 to 31% in 1977, and to the increase in emissions from highway vehicles resulting from increased vehicle miles traveled, which have grown from 27% of the total NO_x emissions in 1970 to 29% in 1977.

Since NO_x emissions are due primarily to transportation, power gener-

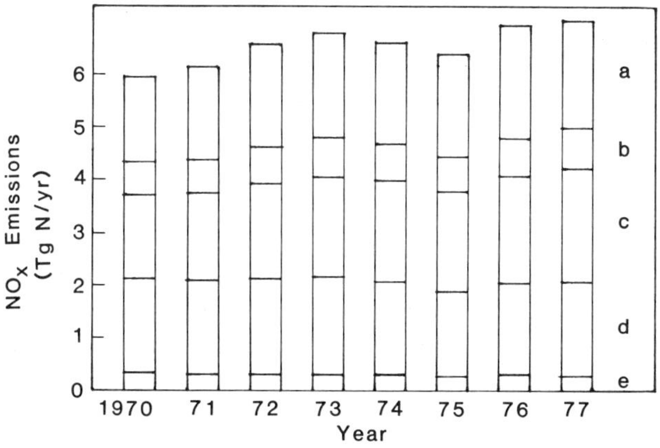

Figure 2. Trend in U.S. emissions of nitrogen oxides between 1970 and 1977 (USEPA, 1978). The segments of the bars represent the portions of the emissions from the following sources, listed from top to bottom: (a) highway vehicles; (b) nonhighway vehicles; (c) electric power generation; (d) industrial, commercial, and residential combustion; (e) miscellaneous.

Table 5. Population and Land Area Corresponding to Various 1975 NO_x Emission Density Classes (USEPA, 1978)

NO_x Emission Density (metric tons km^{-2} yr^{-1})	Percentage of Total Population	Percentage of Total Land Area
>39	29	1
12–39	25	4
3.9–12	18	8
1.2–3.9	18	23
<1.2	10	64

ation, and industrial activity, one would expect the greatest NO_x emissions in the heavily populated and highly industrialized areas of the country. This expectation is supported by the data in Table 5, which shows the percentage of the total U.S. population and the percentage of the total land area in various NO_x emission density classes for 1975 (USEPA, 1978). The regions of the United States included in the two highest NO_x emission density classes contain more than half the total population yet occupy only about 5% of the total land area. Clearly, the major anthropogenic sources of nitrogen oxides are located in urban areas.

4. CHEMISTRY OF NITROGEN OXIDES

Nitrogen oxides are involved in a very complicated sequence of reactions in the atmosphere. The urban atmospheric chemistry is only outlined since it has been discussed previously in much greater detail (Demerjian et al., 1974). The particular emphasis of this section is placed on the transformations of the nitrogen oxides to other nitrogenous species.

4.1. Gas-Phase Processes

The processes to be described are the ones that are generally believed to be of greatest importance under urban atmospheric conditions. Many other reactions involving nitrogen oxides can occur, but these are not discussed. Much of the kinetics data for the reactions to be described has been compiled and evaluated by Hampson and Garvin (1978).

Nitric oxide, NO, can be converted to NO_2 by the thermal oxidation

$$2NO + O_2 \rightarrow 2\,NO_2, \qquad [1]$$

although the rate of this reaction is too slow to be of great significance in the atmosphere. The NO_2 can be readily photolyzed producing a ground state oxygen atom, $O(^3P)$, and NO

$$NO_2 + h\nu\ (\lambda < 410\ nm) \rightarrow NO + O(^3P). \quad [2]$$

The resulting oxygen atom then reacts with molecular oxygen to form ozone (O_3):

$$O(^3P) + O_2 + M \rightarrow O_3 + M. \quad [3]$$

The O_3 reacts readily with NO to reform NO_2:

$$O_3 + NO \rightarrow NO_2 + O_2, \quad [4]$$

or it can react more slowly with NO_2 to form nitrogen trioxide (NO_3):

$$O_3 + NO_2 \rightarrow NO_3 + O_2. \quad [5]$$

The product NO_3 reacts most rapidly with NO:

$$NO_3 + NO \rightarrow 2NO_2, \quad [6]$$

or it can react with NO_2 to form dinitrogen pentoxide (N_2O_5):

$$NO_3 + NO_2 + M \rightarrow N_2O_5 + M. \quad [7]$$

This N_2O_5 may thermally decompose, reforming NO_2 and NO_3:

$$N_2O_5 + M \rightarrow NO_2 + NO_3 + M, \quad [8]$$

or it may react with water to form nitric acid ($HONO_2$):

$$N_2O_5 + H_2O \rightarrow 2HONO_2. \quad [9]$$

This reaction is believed to occur quite slowly in the gas phase but may also occur heterogeneously. Nitrous acid (HONO) can be formed by the reaction

$$NO + NO_2 + H_2O \rightarrow 2HONO. \quad [10]$$

Much of the important chemistry of the nitrogen oxides involves their interaction with various free radicals. Before these reactions can be discussed, the processes that generate these free radicals must be identified. One source of radicals is the photolysis of HONO:

$$HONO + h\nu\ (\lambda < 400\ nm) \rightarrow OH + NO. \quad [11]$$

The hydroxyl radical (OH) is one of the most important transient species in atmospheric chemistry. Reaction [11] is believed to be one of the more important sources of hydroxyl radicals in the urban atmosphere during the early morning hours. Another source of hydroxyl radicals may result from the photolysis of O_3,

$$O_3 + h\nu(\lambda < 320\ nm) \rightarrow O_2 + O(^1D), \quad [12]$$

$$O_3 + h\nu(450 < \lambda < 750\ nm) \rightarrow O_2 + O(^3P). \quad [13]$$

The $O(^1D)$ atom produced in reaction [12] is an excited oxygen atom and is very reactive. The $O(^1D)$ atom may react with water to form the OH radical,

$$O(^1D) + H_2O \rightarrow 2OH. \quad [14]$$

However, the probability that this reaction occurs is quite low, since the $O(^1D)$ atom can be collisionally deactivated to an $O(^3P)$ atom with great efficiency. The $O(^3P)$ atom is not capable of reacting with water to produce OH, but rather undergoes reaction [3], reforming O_3.

The hydroxyl radicals can react with both NO and NO_2,

$$OH + NO + M \rightarrow HONO + M, \quad [15]$$

$$OH + NO_2 + M \rightarrow HONO_2 + M, \quad [16]$$

again producing nitrous acid and nitric acid. As was suggested above, nitrous acid is not very stable during the day, since it is readily photolyzed (reaction [11]). Hence, its concentration is expected to be quite low during the day. On the other hand, nitric acid does not photolyze significantly in the urban atmosphere and is quite stable thermally; thus it is expected to be present in much larger quantities and to be a significant reservoir for nitrogenous species in the urban atmosphere.

Reactions [1]–[16] when combined can lead to the formation of small quantities of ozone during the day. However, for generation of the large quantities of O_3 characteristic of urban photochemical smog, some other process must oxidize NO to NO_2 without destroying an O_3. The key to this process is the OH radical and its reactions with hydrocarbons, e.g.,

$$OH + RH \rightarrow H_2O + R \quad [17]$$

to form alkyl radicals. This radical then reacts with O_2,

$$R + O_2 \rightarrow RO_2 \quad [18]$$

to form an alkylperoxy radical. The RO_2 radical is one of the species capable of oxidizing NO to NO_2,

$$RO_2 + NO \rightarrow RO + NO_2, \quad [19]$$

and in doing this, produces an alkoxy radical. Oxygen can abstract a hydrogen atom from this radical,

$$RO + O_2 \rightarrow R'CHO + HO_2, \quad [20]$$

forming the hydroperoxyl radical (HO_2) and a carbonyl compound. This HO_2 can also oxidize NO to NO_2:

$$HO_2 + NO \rightarrow OH + NO_2, \quad [21]$$

regenerating the OH radical, the species that initiated the attack on the hydrocarbon. This simple hydrocarbon oxidation sequence illustrates the type of chemistry that goes on during smog formation.

The hydrocarbon oxidation chemistry (reactions [17]–[21]) leads to the formation of many more radical species and hence makes possible the formation of other nitrogenous compounds. Thus the alkylperoxy radical produced in reaction [18] can also react with NO_2

$$RO_2 + NO_2 \rightarrow RO_2NO_2 \qquad [22]$$

to form an alkylperoxynitrate. The alkoxy radical formed in reaction [19] can react with either NO or NO_2

$$RO + NO \rightarrow RONO \qquad [23]$$

$$RO + NO_2 \rightarrow RONO_2 \qquad [24]$$

to form alkyl nitrites or alkyl nitrates. The HO_2 radical produced in reaction [20] can react with NO_2

$$HO_2 + NO_2 \rightarrow HO_2NO_2 \qquad [25]$$

to form peroxynitric acid.

One other important class of nitrogen compound remains to be formed. The OH radical can react with the aldehyde produced in [20]

$$OH + RCHO \rightarrow H_2O + RCO \qquad [26]$$

to produce an acyl radical, which reacts with O_2

$$RCO + O_2 \rightarrow RC(O)O_2 \qquad [27]$$

to form a peroxyacyl radical. This radical in turn can react with NO_2

$$RC(O)O_2 + NO_2 \rightarrow RC(O)O_2NO_2 \qquad [28]$$

to form a peroxyacyl nitrate. One of these peroxyacyl nitrates, namely, peroxyacetyl nitrate (PAN) (where the R in the product of reaction [28] is $CH_3\cdot$) formed in the reaction of acetaldehyde has been identified as an important oxidant, phytotoxicant, and lachrymator in photochemical smog (National Research Council, 1978).

Most of the nitrogenous compounds formed in reactions [22] through [25] and [28] are relatively unstable and decompose. The main exception to this statement are the alkyl nitrates produced in reaction [24]. Alkyl nitrates are not easily photolyzed and are reasonably stable thermally. Therefore, the concentration of alkyl nitrates would be expected to build up during the day. However, since the alkyl nitrate formation (reaction [24]) must compete with the reaction of the alkoxy radical with oxygen (reaction [20]), the formation of alkyl nitrates is greatly limited. Alkyl nitrites also have a limited formation rate for similar reasons. In addition, alkyl nitrites are more easily photolyzed and are not expected to accumulate in the urban atmosphere. Peroxyalkyl nitrates, peroxynitric acid, and peroxyacyl nitrates all undergo thermal decomposition reactions,

$$RO_2NO_2 \rightarrow RO_2 + NO_2 \quad [29]$$

$$HO_2NO_2 \rightarrow HO_2 + NO_2 \quad [30]$$

$$RC(O)O_2NO_2 \rightarrow RC(O)O_2 + NO_2. \quad [31]$$

Hendry and Kenley (1979) have presented estimates of the atmospheric half-lives of peroxymethyl nitrate, peroxynitric acid, and PAN for different conditions of temperature and NO_2: NO ratio. These estimates are shown in Table 6. The effect of temperature is obviously related to the activation energy for the decomposition. The NO_2:NO ratio influences the competition between the nitrate forming processes [22], [25], and [28] and the oxidation of NO to NO_2 by reactions [19], [21], and

$$RC(O)O_2 + NO \rightarrow RC(O)O + NO_2. \quad [32]$$

Lower temperatures and higher NO_2:NO ratios lead to increased atmospheric half-lives for the peroxynitrates, whereas higher temperatures and lower NO_2:NO ratios lead to shorter atmospheric half-lives. As can be seen from Table 6, even under the most favorable conditions, peroxymethyl nitrate and peroxynitric acid will have half-lives of only a few minutes and hence would be expected to be only minor reservoirs for nitrogenous species. Peroxyacetyl nitrate is sufficiently stable to be of significance under most conditions.

Hendry and Kenley (1979) have suggested that PAN can serve as an OH source as the NO concentration builds up. The following sequence of reactions demonstrates this process:

$$CH_3C(O)O_2NO_2 \rightarrow CH_3C(O)O_2 + NO_2 \quad [33]$$

$$CH_3C(O)O_2 + NO \rightarrow CH_3C(O)O + NO_2 \quad [34]$$

$$CH_3C(O)O \rightarrow CH_3 + CO_2 \quad [35]$$

$$CH_3 + O_2 \rightarrow CH_3O_2 \quad [36]$$

$$CH_3O_2 + NO \rightarrow CH_3O + NO_2 \quad [37]$$

$$CH_3O + O_2 \rightarrow CH_2O + HO_2 \quad [38]$$

$$HO_2 + NO \rightarrow OH + NO_2 \quad [39]$$

net reaction

$$CH_3C(O)O_2NO_2 + 3\,NO + 2O_2 \rightarrow 4\,NO_2 + CO_2 + CH_2O + OH. \quad [40]$$

The net result is that one PAN molecule oxidizes three molecules of NO, forms one molecule of formaldehyde, and forms one OH radical. This sequence of reactions is potentially quite important since it allows OH radical chemistry to continue at night, in the absence of the normally required

Table 6. Estimated Atmospheric Half-Lives for Peroxynitrates (hours) (Hendry and Kenley, 1979)

Compound	Temperature (K)	NO_2:NO Ratio 0.1	1.0
$CH_3O_2NO_2$	275	0.012	0.050
	295	0.001	0.003
HO_2NO_2	275	0.058	0.106
	295	0.003	0.006
$CH_3C(O)O_2NO_2$	275	24.7	103
	295	0.88	3.6

photolytic driving force. Considerable additional work is required for evaluation of the potential importance of this sequence of reactions as an OH source at night for urban atmospheric conditions.

4.2. Particulate Formation

Orel and Seinfeld (1977) have outlined the possible paths for the formation of particulate nitrate. These include (1) the reaction of HNO_3 with NH_3 to produce NH_4NO_3, with its incorporation into existing aerosol, (2) the direct absorption of HNO_3 into an aerosol droplet, (3) the direct absorption of NO and NO_2 into the aerosol, followed by chemical reaction within the droplet, and (4) the absorption of organic nitrates into the aerosol. Each of these possible routes to particulate NO_3^- formation is considered individually.

The relationship between gas-phase concentrations of NH_3, HNO_3, and ammonium nitrate containing aerosols has been demonstrated extensively (Cadle et al., 1982; Doyle et al., 1979; Stelson et al., 1979). If ammonium nitrate is present in the aerosol phase, gaseous NH_3 and HNO_3 would be expected to be present as a result of the volatility of NH_4NO_3. Stelson et al. (1979) have compared measured ambient concentrations of NH_3 and HNO_3 with the NH_3–HNO_3–NH_4NO_3 equilibrium. They found reasonable agreement between the measured concentrations and those dictated by the equilibrium expression, particularly at relative humidities below the NH_4NO_3 deliquescence, where the solid–gas equilibrium might be expected to apply. Similar results have been reported by Doyle et al. (1979) and Cadle et al. (1982). Cadle et al. (1982) also presented results from studies conducted during the winter, at lower temperatures, where the equilibrium NH_3 and HNO_3 concentrations are expected to be considerably lower. They found that the product of the measured NH_3 and HNO_3 concentrations consistently exceeded the equilibrium values. This could be due to a kinetic limitation in the rate at which NH_3 and HNO_3 can react and be incorporated into the aerosol at the lower temperatures.

Tang (1980) has reported the results of equilibrium calculations for the more realistic NH_3–HNO_3–H_2SO_4–H_2O system at 25°C. These calculations suggest that HNO_3 partial pressure increases with decreasing pH of the solution droplet and with decreasing relative humidity. On the other hand, the NH_3 partial pressure should decrease with decreasing solution pH and decreasing relative humidity. Insufficient atmospheric data are available to test these predictions.

The next process to be considered is the direct absorption of HNO_3 into an aqueous aerosol droplet. Because of the large solubility of HNO_3 in water, one might expect a substantial amount of HNO_3 to be incorporated into an aqueous aerosol. However, this is not the case. Levine and Schwartz (1982) have described the equilibrium between gaseous HNO_3 and aqueous-phase HNO_3. For an assumed atmospheric aerosol liquid water content of 10 $\mu g\ m^{-3}$ and an atmospheric HNO_3 concentration of 10 ppb, the fraction of HNO_3 in the aqueous phase is predicted to be only about 5×10^{-3}. On the other hand, if the HNO_3 reacted in the aerosol after it had been absorbed, possibly with NH_3, this sequence of processes might lead to a significant incorporation of NO_3^- into the aerosol. Another process that could lead to particulate NO_3^- formation is the reaction of gaseous N_2O_5 with an aqueous aerosol to form aqueous HNO_3. This process is expected to occur, but the quantity of aqueous HNO_3 in the aerosol would still be limited by the gaseous HNO_3–aqueous HNO_3 equilibrium discussed above.

The third route to particulate nitrate formation is the direct absorption of NO and NO_2 into the aerosol and the subsequent chemical reaction within the droplet. The reactions

$$NO(g) + NO_2(g) + H_2O(l) \leftrightarrows 2H^+ + NO_2^- \qquad [41]$$

$$2\,NO_2(g) + H_2O(l) \rightleftharpoons 2H^+ + NO_2^- + NO_3^- \qquad [42]$$

have been suggested to be of importance in describing particulate NO_3^- formation in the atmosphere, on the basis of equilibrium chemistry considerations (Middleton and Kiang, 1979; Orel and Seinfeld, 1977; Peterson and Seinfeld, 1979). However, on the basis of kinetic considerations for reactions [41] and [42], Lee and Schwartz (1981) have concluded that both reactions are too slow for them to reach equilibrium even in clouds (see also Schwartz and White, this volume, chapter 1), and certainly not in aerosols. Lee and Schwartz also evaluated the kinetics of an iron-catalyzed reaction sequence that parallels reaction [42]:

$$NO_2(aq) + Fe^{2+} \rightleftharpoons NO_2^- + Fe^{3+}, \qquad [43]$$

$$NO_2(aq) + Fe^{3+} + H_2O(l) \rightleftharpoons NO_3^- + 2\,H^+ + Fe^{2+}. \qquad [44]$$

Again this route is too slow to be of significance, even in cloud water. The potential for interaction between aqueous NO_2 and S(IV) species was also considered by Lee and Schwartz (1981). This reaction was found to be of potential importance but could not be evaluated further because of lack of the appropriate understanding of the chemistry and kinetics of this interaction. Additional consideration should also be given to the role that particulate carbon may play in NO_3^- formation in the atmosphere.

The final route to particulate NO_3^- formation to be considered is the absorption of organic nitrates into the aerosol. Difunctional organic acid nitrates have been identified in aerosol samples obtained in the Los Angeles area (Appel et al., 1980a). These acid nitrates were detectable but had concentrations generally more than a factor of 10 lower than the corresponding dicarboxylic acids. The formation of difunctional organic acid nitrates has been investigated in a smog chamber study (Grosjean and Friedlander, 1980) of the oxidation of cyclic and long-chain diolefins. For the oxidation of cyclohexene, a mechanism was proposed that explained the formation of the observed products. Although this route to particulate NO_3^- formation can occur, it is expected to be of relatively minor importance.

On the basis of the preceding discussions, it would appear that the most significant route to particulate NO_3^- formation is through the NH_3–HNO_3–NH_4NO_3 equilibrium. However, additional investigations of possible routes of NO_3^- formation following the absorption of NO and NO_2 into the aerosol should be conducted.

5. REMOVAL PROCESSES FOR NITROGENOUS SPECIES

All nitrogenous species are ultimately removed from the atmosphere by deposition processes, either dry deposition or wet deposition. Dry deposition is simply the process of a gas or particle encountering the surface and sticking to it. Wet deposition requires that the gas or particle be incorporated into a cloud or precipitation before it encounters the surface. In general, dry deposition is a rather inefficient process for removing pollutants, but it is a process that occurs continuously. On the other hand, wet deposition is much more efficient, but it occurs only as isolated events during a small fraction of the time.

5.1. Dry Deposition

Dry deposition is usually discussed in terms of a dry deposition velocity v_d for gases or particles, which is simply the deposition flux F divided by the atmospheric concentration C:

$$v_d = \frac{-F}{C}. \tag{1}$$

The factors that influence dry deposition include meteorological parameters, the properties of the pollutant, and the nature of the surface. The meteorological parameters describe the transport of the pollutant toward the deposition surface. For particles, gravitational settling and Brownian diffusion are determined by pollutant properties and play a major role in quantifying deposition. For gases, the more important pollutant properties are the solubility in water and the chemical reactivity in the aqueous phase. The important deposition surface parameters include the surface moisture and the physiological state of vegetation, i.e., whether the stomata are open or closed; for further discussion see Wesely, this volume, Chapter 8.

The dry deposition velocities for particles depend strongly on the particle size, the friction velocity, and the aerodynamic surface roughness (Sehmel, 1980). The deposition velocity is generally a minimum in the 0.1–1-μm-diameter size range. Sehmel (1980) presents calculations of particle deposition velocities for different particle sizes, surface roughnesses, and friction velocities. Typical values for the deposition velocities for particles in the 0.1–1-μm size range are in the range 0.02–0.1 cm s^{-1}. Deposition velocities in this range are likely to apply to particulate NO_3^- that is in the 0.1–1-μm-diameter range.

Sehmel (1980) has reviewed the literature for field studies of deposition of gaseous NO, NO_2, NO_x, and PAN over grass and crops. The values for NO range from negative to 0.9 cm s^{-1}, and the single value for NO_2 is 1.9 cm s^{-1}; for NO_x, the range is negative to 0.5 cm s^{-1}; and for PAN, a single value of 0.8 cm s^{-1} has been reported. Judeikis et al. (1979) have reported the results of laboratory deposition studies on selected soils and cement and have found deposition velocities for NO_2 in the range of 0.3–0.8 cm s^{-1} and for NO in the range 0.1–0.2 cm s^{-1}.

Deposition velocities for NO_2 on surface waters have been estimated by Lee and Schwartz (1981). They found that on the basis of the limited solubility of NO_2 and its inefficient oxidation in water, the deposition velocity was not controlled by gas-phase mass transport; hence the deposition velocity must be less than about 1 cm s^{-1}. The deposition velocity was found to be aqueous-phase mass transport limited with some possible enhancement because of the slow reaction of NO_2 with water, which bounds the deposition velocity, $0.1 \geq v_d \geq 5 \times 10^{-4}$ cm s^{-1}. Lee and Schwartz (1981) also conclude that if the deposition velocities for NO_2 are as high as have been reported, the NO_2 must react with the plant material, not just be taken up in the water in the plant tissue. Because of its high solubility, the deposition of HNO_3 to surface waters and wet surfaces is expected to be gas-phase mass transport limited, or of the order of 1 cm s^{-1}.

Clearly, there is a need for continued studies of the dry deposition rates of NO, NO_2, HNO_3, PAN, and particulate NO_3^-. Experimental studies need to be performed to test the theoretical understanding of the process and to provide a basis for the extension of the theory.

5.2. Wet Deposition

Wet deposition is again a process that differs between particles and gases. For aerosols, there are two processes that must be considered: the aerosol attaching to the condensed water and the condensed water falling out. The slower of these two processes determines the rate of wet deposition of an aerosol. The removal rates for aerosols depend strongly on the solubility, size, and concentration of the aerosol (Scott, 1979). The cloud growth mechanism will also affect the efficiency of the removal by changing the "washout ratio," which is the mass per unit volume of the pollutant in the rain divided by the mass per unit volume of the pollutant in the air. For rain from warm clouds, the washout ratio would be in the range of 10^5–10^6, whereas for snow or cold clouds, the washout ratio would be nearer at 10^5.

Unlike the wet deposition of aerosols, which is a collection or concentrating process, the wet deposition of gases is an equilibrium process. It is necessary to consider not only the gas solubility in water, but any subsequent reactions that may occur after dissolving. The in-cloud scavenging of gases is of real significance only if a chemical transformation occurs within the cloud. As was suggested in Section 4.2, no significant transformations are expected for NO_2 in cloud water (Lee and Schwartz, 1981). In the case of HNO_3, which is a highly soluble gas, the in-cloud scavenging has been shown to be a relatively unimportant process, since most of the HNO_3 incorporated into the cloud is released to the gas phase on cloud evaporation (Levine and Schwartz, 1982). The below-cloud washout of HNO_3 by falling rain has also been estimated by Levine and Schwartz (1982). They report values in the range 5–50% removal of HNO_3 per millimeter of rainfall, depending on assumptions about the drop size distribution and rainfall rate. Durham et al. (1981) presented the results of calculations for the below-cloud scavenging of HNO_3 that resulted in an initial removal of HNO_3 of 13% per millimeter of of rainfall for a particular drop size distribution and rainfall rate. Clearly, HNO_3 is predicted to be removed from the atmosphere very efficiently by rainfall. Wet deposition of NO, NO_2, and PAN is not likely to be of great significance unless chemical transformations could occur in clouds at a significant rate. Levine and Schwartz (1982) have concluded that the wet deposition and dry deposition of HNO_3 are of comparable importance in the removal of HNO_3 from the atmosphere on an annual-average basis.

6. MEASUREMENT TECHNIQUES FOR NITROGENOUS SPECIES

Some techniques that are currently being used for measurement of nitrogenous species are discussed in this section. Particular emphasis is placed on the limitations and interferences in these techniques. A broader treatment of the measurement techniques for nitrogen oxides has been presented by the National Research Council (1977).

6.1. Nitric Oxide

The current method used for the measurement of NO is a chemiluminescent technique. This technique utilizes the gas-phase chemiluminescent reaction between NO and ozone (O_3) to produce an electronically excited NO_2

$$O_3 + NO \rightarrow NO_2^* + O_2. \quad [45]$$

The excited NO_2 can then emit light in the range from about 600 to 3000 nm, with a maximum near 1200 nm, or it can be collisionally quenched:

$$NO_2^* \rightarrow NO_2 + h\nu, \quad [46]$$

$$NO_2^* + M \rightarrow NO_2 + M. \quad [47]$$

The light emitted in reaction [48] by the excited NO_2 is then monitored as a quantitative measure of the NO concentration. This technique has numerous advantages, including high sensitivity and specific response to NO with no known interferences (Matthews et al., 1977; Winer et al., 1974).

6.2. Nitrogen Dioxide

The current method for the measurement of NO_2 uses the NO–O_3 chemiluminescent technique after quantitatively converting NO_2 to NO. The NO_2 concentration is given by the difference in the chemiluminescence signal when the sample stream passes through the converter and then the chemiluminescence cell (NO + NO_2) and the signal when the sample stream bypasses the converter and directly enters the chemiluminescence cell (NO). The complications with this NO_2 measurement technique arise in the converter stage. One needs a converter that reduces NO_2 to NO with high efficiency but does not convert any other species that may be in the atmosphere into NO. These two criteria have not been met in any of the converters that are currently in use.

One of the earliest converters, which is still in common use, employs a high-temperature, stainless steel thermal unit. This system thermally decomposes NO_2 by the reaction

$$NO_2 \rightleftharpoons NO + \tfrac{1}{2}O_2. \quad [48]$$

Temperatures in excess of 650°C are required to effect an efficient conversion (Breitenbach and Shelef, 1973). In these high-temperature converters, most nitrogenous species, including NH_3 and methylamine, are expected to be converted to NO (Breitenbach and Shelef, 1973; Matthews et al., 1977). The conversion of molecular nitrogen (N_2) and nitrous oxide (N_2O) to NO has been shown to be negligible in these converters (Matthews et al., 1977).

The temperature required for the conversion can be reduced in converters containing metals, such as molybdenum, which can undergo surface oxidation during the reduction of NO_2:

$$2NO_2 + Mo \rightarrow MoO_2 + 2NO. \quad [49]$$

Temperatures in excess of 300°C are required for efficient NO_2 conversion to NO. Temperatures in this region are insufficient for the conversion of NH_3 to NO. Molybdenum converters exhibit good efficiencies for conversion to NO of various organic nitrates and nitrites, such as PAN, ethyl nitrate, and ethyl nitrite, but low efficiency for nitroethane conversion to NO (Winer et al., 1974). Good conversion efficiency for HNO_3 to NO has also been shown by use of this converter (Joseph and Spicer, 1978).

A third type of NO_2-to-NO converter uses heated carbon. Like the molybdenum, the carbon is oxidized while the NO_2 is reduced to NO

$$NO_2 + C \rightarrow CO + NO. \quad [50]$$

This converter operates at even lower temperatures, near 250°C. As with the molybdenum converters, organic nitrates are converted with high efficiencies to NO (Winer et al., 1974), as is HNO_3. Commercial instruments are available that use each of these three types of converter and others based on the same principles.

All these converters lack specificity in the conversion of NO_2, and only NO_2, to NO, with high efficiency. Catalytic-chemical converters that use either ferrous sulfate ($FeSO_4$) or ferrous ammonium sulfate [$Fe(NH_4)_2(SO_4)_2$], have also been tested. Ferrous sulfate has been found to eliminate interference when it is used as a converter for NO_2 to NO in smog chamber studies of chlorine-containing compounds (Joshi and Bufalini, 1978). It has also been reported that $FeSO_4$ does not convert HNO_3 to NO (Bowermaster and Shaw, 1981; Kelly and Stedman, 1979), nor does it convert PAN to NO (Kelly and Stedman, 1979). Apparently, this converter has overcome the interferences arising from the conversion of other nitrogenous compounds, besides NO_2, to NO. However, this converter is not without problems. It has been reported that high humidities reduce the NO_2 to NO conversion efficiency in the $FeSO_4$ converter (Joshi and Bufalini, 1978). Macoy et al. (1980) have tested both $FeSO_4$ and $Fe(NH_4)_2(SO_4)_2$ converters and concluded that great care is necessary in the design of the catalyst bed to assure high conversion efficiencies at the necessary sample flow rates.

Kley et al. (1981) have recently reported the results of NO_2 measurements made by use of a photolytic converter to convert NO_2 to NO. In this system the sample air is passed through a photolysis cell where the NO_2 is partially photolyzed, yielding NO. This NO is then detected by a very sensitive NO chemiluminescence detector (McFarland et al., 1979). This converter is expected to be specific, i.e., it is not expected to convert PAN, HNO_3, or NH_3 to NO. However, it is likely that certain other trace nitrogenous species such as nitrous acid (HNO_2), nitrogen trioxide (NO_3), and some organic nitrates and nitrites might be converted to NO. The photolytic converter does not work at the high efficiencies of the thermal and catalytic converters. Kley et al. (1981) report an efficiency of only 60%. An accurate measurement of NO_2 requires an acurate knowledge of this conversion efficiency, which depends strongly on the flow rate of the sample air through the photolysis cell, the pressure in the cell, and the NO_2 photolysis rate by the lamp. These parameters require careful control and monitoring, and the system requires frequent calibrations for both NO_2 and NO.

Other techniques for the direct spectroscopic measurement of NO_2 are under development. Poizat and Atkinson (1982) have devised an intracavity photoacoustic absorption technique for the measurement of NO_2 by monitoring selected absorption bands in the visible portion of the spectrum. Other groups are constructing infrared diode laser absorption systems with sufficient sensitivity to measure NO_2 in the parts per billion (ppb) concentration range (Reid et al., 1980). If direct, accurate, and reliable techniques can be developed for measuring NO_2, chemiluminescence instruments can be compared on a side-by-side basis with the direct methods to evaluate the importance of the interferences in the chemiluminescence NO_2 measurements.

6.3. Organic Nitrates

Organic nitrates constitute a large class of compounds of which only a few, however, have been detected in the atmosphere (Graedel, 1978). The most important of these compounds is peroxyacetyl nitrate (PAN). The analysis for PAN in the ambient atmosphere is generally performed by the use of a gas chromatograph equipped with an electron capture detector (Stephens, 1969). This technique has been and continues to be used extensively in laboratory and atmospheric measurements of PAN. The technique is also suitable for the analysis of other compounds within this class.

Peroxyacetyl nitrate is a relatively unstable compound, which complicates the calibration of PAN measurements. All the calibration techniques that are used for PAN are referenced to its infrared absorption. Field calibrations of PAN analyzers are performed by several different techniques. One of these techniques involves the preparation of a high concentration of PAN in diluent in the laboratory and the determination of its concentration by infrared measurements. This mixture can then be taken into the field and diluted for gas chromatographic (GC) calibrations (Stephens, 1969). Alternatively, the gas chromatograph can be calibrated in the laboratory for PAN response relative to a stable nitrate, such as *n*-propyl nitrate. Then field calibrations can be performed by noting the gas chromatographic response to the injection of known quantities of *n*-propyl nitrate (Stephens and Price, 1973). A third calibration technique that has been used involves the generation of PAN in the field by the chlorine initiated photooxidation of acetaldehyde in the presence of NO_2 in the air. The generator used must be characterized in the laboratory for the concentrations of PAN produced for various generator parameters, such as flow rates and concentrations (Grosjean, 1981).

6.4. Nitric Acid and Particulate Nitrate

The measurement techniques used for HNO_3 and NO_3^- are so interrelated that their sampling and analysis are discussed in the same section. Considerable difficulty often exists in distinguishing between HNO_3 and NO_3^-. This is due to artifacts that occur in certain sampling procedures. Gaseous HNO_3 and NO_2 can be inadvertently collected on certain particulate filters and, hence, measured as particulate NO_3^-. On the other hand, particulate NO_3^- can sometimes be lost from particulate prefilters either because of reactions of acidic aerosol (H_2SO_4) with particulate nitrate forming particulate sulfate and gaseous HNO_3 or simply because of evaporation. This additional HNO_3 would then be collected and measured as gaseous HNO_3. Another difficulty that exists in the measurements of HNO_3 is quantitatively transferring the HNO_3 from the atmosphere to the collection medium or analyzer. Bowermaster and Shaw (1981) have studied the transmission efficiencies for HNO_3 through various materials. They recommend that HNO_3 sampling lines be made of Teflon and be kept scrupulously clean. The HNO_3 transmission efficiency for most atmospheric sampling systems is unknown.

The chemiluminescence technique for measuring NO has been applied to the measurement of HNO_3. Two approaches to this measurement have been reported. Joseph and Spicer (1978) reported one technique using a Teflon prefilter to eliminate particulate matter from the analyzer. They measure the total NO_x signal after passing the air through a heated molybdenum converter. This signal is a measure of NO + NO_2 + PAN (and other organic nitrates) + HNO_3. A second measurement is made after passing the air through a nylon filter, followed by the molybdenum converter. Nylon filters have been shown to quantitatively remove HNO_3 from a gas stream (Miller and Spicer, 1975). Thus the second signal provides a measure of NO + NO_2 + PAN. The HNO_3 concentration is determined by the difference between these two measurements. There is some potential for artifacts in these measurements. Appel et al. (1980b) have shown that clean Teflon prefilters collect very little HNO_3, but that as the loading of these prefilters increases, the retention of HNO_3 also increases. This suggests that frequent changes of the prefilter are necessary to minimize HNO_3 retention. Laboratory studies have shown that HNO_3 may also be liberated from the filter because of reaction of H_2SO_4 with nitrate salts (Harker et al., 1977). More recent laboratory and field studies suggest that NH_4NO_3 reacts with strong gaseous acids, such as HCl, as well as particulate strong acids, such as H_2SO_4, to release HNO_3 (Appel and Tokiwa, 1981). These reactions can result in a significant loss of NO_3^- from Teflon filters. Particulate NO_3^- can also be lost from filters because of dissociation of NH_4NO_3 on the filter to NH_3 and HNO_3 (Appel et al., 1980b).

The second approach to the measurement of HNO_3, that of using the chemiluminescent NO detector, employs two different converters (Kelly and Stedman, 1979). In this approach four separate measurements are made: (1) passing the air directly through the NO detector provides a measurement of NO; (2) passing the air through the $FeSO_4$ converter followed by the NO detector provides a measurement of NO + NO_2; (3) passing the air through a nylon fiber trap, which quantitatively removes HNO_3, and then a thermal converter operating at 350°C, followed by the $FeSO_4$ converter and the NO detector provides a measurement of NO + NO_2 + organic nitrates (largely PAN); and (4) passing the air through the thermal converter, the $FeSO_4$ converter, and then the NO detector provides a measurement of NO + NO_2 + PAN + HNO_3. The concentrations of the individual components can then be determined by taking differences between the appropriate measurements. As was suggested in Section 6.2, the $FeSO_4$ converter could have some conversion efficiency problems that could render it unsuitable for use; in any event further testing and development appear to be necessary.

A wide variety of filter techniques have been used in the measurement of NO_3^- and HNO_3. Recently, a nitrate artifact has been discovered that severely affects the particulate nitrate data collected by the National Air Surveillance Network. This artifact has been extensively documented in the literature (e.g., Appel et al., 1979; Spicer and Schumacher, 1977, 1979). Two of the common arrangements currently used employ a Teflon prefilter for collection of the particulate nitrate with a nylon filter for collection of HNO_3 (Appel et al., 1980b; Spicer, 1979) or a Teflon prefilter followed by a NaCl impregnated cellulose filter (Appel et al., 1980b; Okita et al., 1976). In a direct comparison of these two techniques, a good correlation was found between the HNO_3 collected on both the nylon and the NaCl-impregnated filters (Appel et al., 1980b). The chemical analysis for nitrate from the filters is done by a variety of different techniques, including ion chromatography and wet chemical techniques.

In view of the complexities in measuring NO_3^- and HNO_3 resulting from the positive and negative artifacts, a denuder difference experiment was designed by Shaw et al. (1979). Two sampling assemblies are used in this experiment. The first assembly (A) consists of a Teflon particulate filter followed by a HNO_3 collection tube containing nylon fibers. The second assembly (B) consists of a diffusion denuder for gaseous acid followed by a Teflon particulate filter and an HNO_3 collection tube. The denuder is simply an array of glass tubes that are coated with a strong base. This denuder will remove acidic gases, in this case HNO_3, and pass aerosol particles. It is necessary to analyze the particulate filters and the HNO_3 collection tubes in each assembly for NO_3^-. These data allows determination of total atmospheric nitrate (NO_3^- + HNO_3), particulate NO_3^-, and gaseous HNO_3. The total atmospheric nitrate is the sum of the NO_3^- measured on the filter and collection tube of assembly A. The filter will collect NO_3^- and some fraction of the HNO_3. The HNO_3 collection tube will collect the HNO_3 that passes through the filter without being retained and any HNO_3 that is volatilized from the filter. The total particulate nitrate is determined from the sum of the NO_3^- measured on the filter and the HNO_3 collection tube of assembly B. The filter will collect the particulate NO_3^-, and the collection tube will collect any HNO_3 that is volatilized from the filter. The gaseous HNO_3 is simply the difference between the total nitrate and the particulate NO_3^-. It should be noted, however, that both the total nitrate and the particulate NO_3^- determined by this technique could be high if any NO_2 were retained on the filters and converted to NO_3^-.

Braman and Shelley (1980) have recently reported work using a tungstic acid preconcentrating technique for collecting ammonia, nitric acid, and particulate ammonium and nitrate. The preconcentration is achieved by using tungstic acid, which can react with both NH_3 and HNO_3, retaining these species. The sampling system uses a quartz tube that is coated with tungsten(VI) oxide by vacuum deposition, followed by a tube packed with tungsten(VI) oxide–coated sand. The air sample is drawn through the hollow tube and then the packed tube. Gaseous NH_3 and HNO_3 are retained in the hollow tube, whereas aerosol particles pass through this tube. Particulate NH_4^+ and NO_3^- are trapped in the packed tube. After sufficient sample has been collected, the tubes are ready for analysis. The analysis of both the hollow and packed tubes follows the same procedure. The tube to be analyzed is first heated to about 400°C, thus causing release of the NH_3 or NH_4^+ as NH_3, by reversing the reaction with tungstic acid. The HNO_3 or NO_3^- is also released but is converted to NO_2. The NH_3 and NO_2 released from the sample tube then pass through a hollow tungsten(VI) oxide–coated tube. This tube traps the NH_3. The NO_2 goes on to a high-temperature converter and a chemiluminescent NO detector. The resulting signal is a measurement of the amount of HNO_3 or NO_3^- present in the sample. The tungsten(VI) oxide tube in the analysis system is then heated, again releasing the NH_3. The NH_3 is converted to NO and measured. This signal represents the NH_3 or NH_4^+ present in the sample. Additional investigations are necessary, but apparently NO_2 does not interfere. However, aliphatic amines respond as NH_3.

6.5. Other Nitrogenous Species

Platt et al. (1979) have recently developed a long path differential optical absorption spectroscopic technique for the measurement of UV or visible absorbing species. This technique has been applied to NO_2, looking at a very narrow portion of the absorption spectrum at 323–335 nm. The optical path lengths used in these studies were typically in the range 1–5 km, and the measured absorptions were no more than a few tenths of a percent. For determination of concentrations of the various species, reference spectra in the range of observation for the absorbing species are needed. Then appropriate fractions of the reference spectra are subtracted from the atmospheric spectra, beginning with the strongest features in the spectrum. These fractions are then related to the concentration of the absorbing species in the path. This technique has also been applied to the measurement of nitrous acid (HONO) (Perner and Platt, 1979; Platt et al., 1980a; Platt and Perner, 1980) and nitrogen trioxide (NO_3) (Platt et al., 1980b, 1981). To increase the sensitivity of the measurements, it is desirable to extend the path length, but this would restrict the measurements of periods when the visibility is sufficient to allow any measurements to be made.

7. URBAN MEASUREMENTS OF NITROGENOUS SPECIES

The most extensive measurements of urban air have been made in the Los Angeles, California area. The results of a few of these studies are discussed briefly to provide examples of the concentrations of nitrogenous species that have been measured. In the 1973 study in West Covina, extensive measurements of NO, NO_2, PAN, and HNO_3 were reported (Spicer, 1977). It was found that PAN and HNO_3 constituted a significant fraction of the nitrogen oxides (NO + NO_2 + PAN + HNO_3), ranging from 2 to 54% and averaging about 11% (Spicer et al., 1976). During the midafternoon the PAN + HNO_3 fraction averaged about 25%. The maximum hourly average PAN concentration reported was 46 ppb on a day when the maximum hourly average O_3 was 0.213 ppm; the maximum hourly average HNO_3 concentration reported was 40 ppb on a day when the maximum hourly average O_3 was 0.271 ppm (Spicer, 1977). The results of a more recent study in Claremont during September–October 1980 have been reported (Grosjean, 1981). In this study the PAN + HNO_3 fraction of the total nitrogen oxides ranged from 1 to 39% and averaged about 18% on the basis of 4-h averaged data. The highest PAN value reported was 47 ppb as a single measurement, and the highest HNO_3 was 14 ppb as a 4-h average. These data suggest that PAN and HNO_3 can contribute significantly to the signal from chemiluminescent analyzers.

Spicer (1977) showed that PAN concentrations in West Covina peak at essentially the same time as do O_3 concentrations, but the results were less clear for HNO_3. Grosjean (1981) reported that O_3, PAN, and HNO_3 all reach their maximum in Claremont in the late afternoon. The 4-h time resolution of the HNO_3 measurements does not establish the coincidence in the maximum concentrations for HNO_3 and O_3 with much confidence. Stedman et al. (1981) have also reported the results from measurements of HNO_3 made in Claremont during September 1980. These measurements, which were made by use of the chemiluminescent technique, suggest that HNO_3 and O_3 concentrations reach their maximum values at very nearly the same time, midafternoon.

Particulate NO_3^- concentrations have also been reported for the Claremont study (Grosjean, 1981). Concentrations of NO_3^- ranged between 0.2 and 17 ppb (mole fraction). Particulate NO_3^- was found to have very different diurnal behavior from that reported for HNO_3. The NO_3^- concentration was found to peak at night, rather than in the afternoon hours when the HNO_3 concentration reached its maximum. The NO_3^- was found to range from 0.2% of the total nitrogen (NO + NO_2 + PAN + HNO_3 + NO_3^-) during the day up to nearly 17% at night, with an average over the entire sampling period of about 4%.

The measurement of other nitrogenous species have also been reported in the Los Angeles Basin. Nitrogen trioxide (NO_3) was detected in differential optical absorption experiments conducted in both Claremont and Riverside (Platt et al., 1980b). Nitrogen trioxide was detected only after sunset and during time periods when both NO_2 and O_3 are relatively high. This is expected for the following two reasons. First, the reaction of O_3 with NO_2 (reaction [5]) is the primary source of NO_3. Second, the presence of high O_3 at night suppresses the concentration of NO, which is primarily responsible for NO_3 loss by reaction [6]. The maximum NO_3 reported was 355 ppt, which was measured when NO_2 was 82 ppb and O_3 was 196 ppb. The maximum NO_3 concentration was generally reached within 1–2 h after sunset; the NO_3 decayed rapidly, often to near the detection limit by midnight. Platt et al. (1981) have reported NO_3 concentrations consistent with the equilibrium with N_2O_5 (reactions [7] and [8]). They have suggested that this rapid NO_3 decay at night might be due to the interaction of NO_3 and/or N_2O_5 with liquid water in the atmosphere since the decay seems to be more important during periods of high relative humidity and fog. The potential importance for this removal process for nitrogenous species requires further investigation.

Measurements of HNO_2 by use of the differential optical absorption technique have also been reported (Platt et al., 1980a). Nitrous acid was found to build up slowly throughout the night, reaching its maximum concentration near sunrise. The HNO_2 maxima reported for measurements in both Claremont and Riverside ranged from 1.1 to 4.1 ppb. After sunrise the HNO_2 decayed rapidly because of the photolysis reaction [11]. This provides an important source of OH radicals in the early morning hours to initiate photochemical processes.

The precipitation chemistry in the Los Angeles Basin has been studied by Liljestrand and Morgan (1981). They concluded that the annual net acidity flux from precipitation was only about 15% of that in the northeastern United States because of the lower annual precipitation and the higher contribution from alkaline sources. Their data showed that the ratio of nitrate to non-sea-salt sulfate increases from coastal to inland sites and that sulfuric acid is the dominant acid near the coast, whereas HNO_3 becomes the dominant acid as one moves further inland. In addition, they concluded that about 2% of the annual emissions of sulfur oxides and about 1% of the annual emissions of nitrogen oxides are scavenged by precipitation in the Los Angeles Basin.

8. EVALUATION OF URBAN NO_x REMOVAL

The discussion in this section concentrates on various attempts to quantify NO_x removal from four different urban areas: Los Angeles, St. Louis, Houston, and Denver. Spicer (1977) has proposed a method for calculating NO_x loss by using CO as an inert tracer of urban atmospheric motions. Increases in the measured CO : NO_x ratio $(CO/NO_x)_m$ over the CO : NO_x ratio from the emissions inventory $(CO/NO_x)_e$ are interpreted as NO_x losses according to the relationship

$$NO_x \text{ loss} = \left[\frac{(CO/NO_x)_m}{(CO/NO_x)_e} - 1\right](NO_x)_m. \tag{2}$$

This approach to the evaluation of NO_x removal has numerous limitations, which are pointed out in the application of this approach to data collected from each city. Other approaches to the evaluation of NO_x removal are also discussed, when applicable, for these urban areas.

8.1. Los Angeles

Spicer (1977) applied his NO_x loss calculation approach to data collected during a 29-day period from August through September 1973 by the Los Angeles Air Pollution Control District in Azusa. The slope for the CO–NO_x linear regression was 18.4 for the data, and the correlation coefficient was 0.73. The emissions inventory used in this work gave the CO : NO_x emissions ratio as 14.3, which resulted in a NO_x loss of 35 ppb. These NO_x measurements were made by use of a colorimetric technique that responds to both PAN and HNO_3. If the response of the colorimetric technique and the sampling efficiency for PAN and HNO_3 is quantitative, the formation of PAN and HNO_3 cannot account for any of this 35 ppb NO_x loss.

Spicer et al. (1976) performed NO_x loss calculations on data collected during a 15-day period in September 1973 in West Covina. The resulting NO_x loss was only 18 ppb, when a CO : NO_x emissions ratio of 15.0 was used. The NO_x data used in this analysis were corrected for the PAN and HNO_3 interferences by subtracting the independently measured concentrations of PAN and HNO_3. It was suggested that the difference between the results for West Covina and for Azusa may have been due to inaccuracies in the CO data from Azusa, which used nondispersive infrared detection of CO rather than the gas chromatographic technique used in the West Covina study. The nitrogenous product species measured in West Covina during this study were PAN, 12 ppb; HNO_3, 3 ppb; and NO_3^-, 1 ppb (mole fraction). These species account for a total of 16 ppb of the NO_x loss.

A more recent and extended emissions inventory has been reported for the South Coast Air Basin for 1973 (Chock et al., 1981). This inventory gives a CO : NO_x emissions ratio of 15.3, which would result in an NO_x loss for Azusa of 25 ppb and for West Covina of 15 ppb. This emissions inventory states that 98% of the CO emissions are from vehicular sources, whereas 65% of the NO_x emissions are from vehicular sources. Another 25% of the NO_x emissions are from other area sources, and the remaining 10% are from point sources. If this emissions inventory is applicable, the PAN, HNO_3, and NO_3^- measurements in West Covina can account for all the NO_x loss calculated.

It is clear that the results of these NO_x loss calculations depend strongly on the emissions inventory used. Spicer et al. (1976) reanalyzed data collected during a 3-day period in West Covina by using emissions along the trajectory followed by the air parcel prior to reaching West Covina. They found that the $CO:NO_x$ emissions ratio depended strongly on the trajectory followed. They used these data to calculate hourly values of the NO_x loss for the three afternoons. It was found that periods of high NO_x loss could occur when product (PAN + HNO_3) concentrations were low, and periods of low NO_x loss could occur when product concentrations were high. It was concluded that these types of calculations contain too many uncertainties and unknowns to be useful on a short time scale.

Calvert (1976) estimated the NO_x removal rate from data collected during Operation 33 of the Los Angeles Reactive Pollutant Program (LARPP) on November 5, 1973. The NO_x removal rate was determined from the time dependence of the NO_x concentration relative to a tracer for atmospheric dilution processes. In these calculations both methane (CH_4), and acetylene (C_2H_2), were used as tracers. The resulting NO_x removal rate was about 8% per hour during the 0800–1300 period. This NO_x removal was attributed to HNO_3 and NO_3^- formation, and not to PAN. But HNO_3, like PAN, is expected to be measured as NO_x with some unknown efficiency. Therefore, only a portion of the HNO_3 can be considered an NO_x loss.

Spicer et al. (1976) also performed an NO_x loss analysis by using a portion of the LARPP data. For Operation 33, it was found that the measured $CO:NO_x$ ratio remained approximately constant at all altitude levels along the trajectory. This suggests that the rate of NO_x loss is not significantly faster than that of CO, the tracer. However, the $NO_x:CH_4$ and $NO_x:C_2H_2$ data suggest that there is a large NO_x loss, as concluded by Calvert (1976). Spicer et al. (1976) found that the the measured $CO:C_2H_2$ ratio decreases along the trajectory. But since both CO and C_2H_2 are emitted primarily in automobile exhaust, this ratio should be relatively independent of time or the trajectory. This result suggests that there may be some problem in the measurement of CO and/or C_2H_2.

Chang et al. (1979) have estimated the rate of NO_x removal by using CO as an inert tracer and the technique described by Calvert (1976). The analysis was performed on averaged sets of data collected at three sites, West Los Angeles, downtown Los Angeles, and Azusa over a 6-year period (1970–1975). The average removal rate of NO_x in the Los Angeles area was found to be at least 4% per hour for the 0800–1600 time period.

8.2. St. Louis

Spicer (1977) performed an NO_x loss calculation for morning and early afternoon sampling periods on 5 days in July 1973. He reported a slope of 14.6 for the $CO–NO_x$ linear regression and a correlation coefficient of 0.73.

The CO:NO_x molar emissions ratio for the St. Louis Air Quality Control Region (AQCR) was reported to be 14.6. The ratio based on the St. Louis city emission inventory was 13.8. Use of the first of these inventories resulted in no NO_x loss at all, whereas the second gave an NO_x loss of 3 ppb. A much different emissions inventory prepared for 1975–1976 in the St. Louis AQCR (Littman, 1979) gave a CO:NO_x molar emissions ratio of 5.46. This inventory indicated that 89% of the CO was from area sources, whereas 11% was from point sources. For NO_x, the emissions inventory has only 28% area sources and 72% from point sources. If these emissions data were appropriate for the July 1973 sampling period, the resulting NO_x loss would be 93 ppb. This again demonstrates the sensitivity of the NO_x loss to the selection of the CO:NO_x emissions ratio.

The Regional Air Pollution Study (RAPS) conducted by the USEPA in St. Louis provides a very extensive data set for the evaluation of NO_x loss. The analysis of a small portion of these data is presented here. These data are averages over the 25-station Regional Air Monitoring System (RAMS) network for June 6–9, 1976, as presented by Decker et al. (1978). Figure 3 shows a linear regression plot of the measured CO concentration versus the NO_x concentration. The resulting slope was 15.1 and the correlation coefficient was 0.92. Data from June 8, 1976 that had been averaged for each of three classes of sites—urban, suburban, and rural—were also analyzed. The slope for the CO–NO_x linear regression was 14.8 for the urban data, 10.4 for the suburban data, and 7.2 for the rural data. The corresponding correlation coefficients were 0.92 for the urban data, 0.70 for the suburban data, and 0.55 for the rural data. This decrease in slopes in going from an urban to rural site is the opposite of that expected as an urban air mass aged. In an aging air mass the NO_x concentration would be decreasing more rapidly than the CO concentration; hence the CO:NO_x ratio should increase. This decrease in the slopes in going from urban to rural data is probably due to

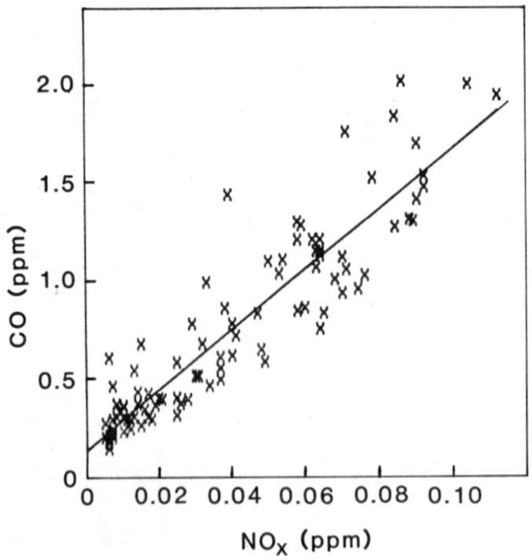

Figure 3. Linear regression of CO versus NO_x for St. Louis RAMS data, June 6–9, 1976.

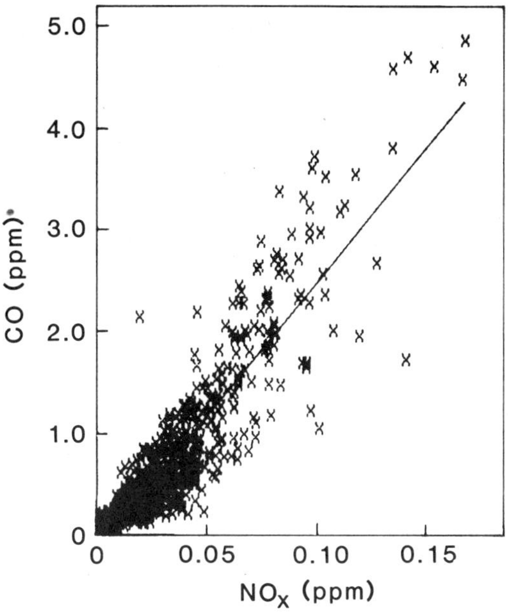

Figure 4. Linear regression of CO versus NO_x for Houston GMR data, August–September 1977.

differences in the emissions that are important in these areas. In the urban area the measurements would be more heavily influenced by vehicular emissions, where the CO : NO_x emissions ratio is 13.5 (Decker et al., 1978), whereas in the rural areas, the measurements would be more heavily influenced by the point-source emissions, which have a CO : NO_x emissions ratio of 1.8. This greatly complicates any NO_x-loss analysis.

8.3. Houston

During the summer of 1977, General Motors Research Laboratories (GMR) participated in the Houston Area Oxidants Study (Monson et al., 1978). This study provided CO and NO_x data for August 7–September 16, 1977 for an urban Houston site. Figure 4 shows the results of the linear regression of CO versus NO_x. The resulting slope is 26.1, and the correlation coefficient is 0.91. The emissions sources for NO_x and CO in Houston are significantly different from those in Los Angeles. On the basis of a 1973 emissions inventory from Harris County, Texas, only about 35% of the NO_x emissions and 50% of the CO emissions are from area sources, primarily transportation related (Pennington, 1976). The major point-source emissions for NO_x are from fuel combustion in industrial and electric power generation applications. The major point sources for CO are different, arising from industrial process losses in petroleum refining and chemical manufacturing. The ratio of the CO:NO_x emission resulting from this inventory is about 6.2. Use of the value of the slope and emissions ratio above and the average NO_x concentration of 27 ppb suggests that the average NO_x loss is about 87 ppb.

This large calculated NO_x loss seems rather high and is probably not real. The use of emissions inventory data that are 4 years older than the atmospheric measurement data is likely to have a significant effect on the calculated NO_x loss. But probably of greater importance is the lack of an adequate tracer for the NO_x. The CO emissions in Houston have important point sources that are different from the major NO_x point sources. If the GMR monitoring site in Houston were influenced only by area-source emissions of CO and NO_x, the CO:NO_x emissions ratio would be 8.9. The calculated NO_x loss would still be high, about 52 ppb. In the most extreme case, assuming that the GMR site was influenced by all the CO emissions and only the area-source NO_x emissions, the CO:NO_x emissions ratio is 17.9. This still results in an NO_x loss of 12 ppb. Even under this extreme attempt to minimize the calculated NO_x loss, there is still a significant calculated NO_x loss.

None of the product compounds PAN, HNO_3, or NO_3^- were measured at the GMR monitoring site during this study. However, PAN and at least a portion of the HNO_3 are expected to have been measured as NO_x. Independent measurements of PAN have been reported for three sites in the Houston area during the time period of this study (Jorgen, 1978). Throughout the study the average daytime PAN concentrations were less than 1 ppb at all three sites. This is not unexpected since the high temperatures in Houston should lead to thermal decomposition of PAN. The temperatures averaged about 28°C during the study and were even higher during the day when PAN formation should be significant. It is not known whether HNO_3 or particulate NO_3^- could have accounted for a significant NO_x loss, which remains highly uncertain.

8.4. Denver

During November and December 1978, GMR and the Motor Vehicle Manufacturers Association conducted a study to characterize the physicochemical nature of Denver's brown cloud. The data from this study have been used in an attempt to quantify the chemical and physical removal processes for nitrogen oxides (Anderson, 1982). In that work a model was constructed to describe the average diurnal variation observed for CO, NO_x, and NO_3^- over the 40-day sampling period. The model including diurnally varying emissions, advection, and entrainment terms was fit to the observed CO concentration profiles by adjusting the mixing height used in the model. Then the appropriate NO_x emissions and chemical reaction and deposition terms were included for NO_x conversion and removal. The model suggested that during the day, the reaction of OH with NO_2 to form HNO_3 occurred at a rate averaging nearly 8% per hour. The HNO_3 could then react with NH_3 to form NH_4NO_3. The rate of the HNO_3 reaction with NH_3 kinetically limits the conversion of HNO_3 to NO_3^-. This conversion rate of HNO_3 to NO_3^- was estimated to be less than 15% per hour. At night, the major particulate NO_3^- formation process appears to be first order in NO_x and occurs at the rate of about 0.5% per hour. However, the chemical nature of this conversion has not been identified. The deposition velocities used in the model were 0.3 cm s^{-1} for NO_x, 1.0 cm s^{-1} for HNO_3, and 0.1 cm s^{-1} for NO_3^-.

A complete description of the gaseous measurements made at the GMR site northeast of downtown Denver has been given by Ferman et al. (1981). Figure 5 shows the results of the linear regression of the measured CO–NO_x concentration for the GMR site. The slope from this plot was 18.5 and the correlation coefficient was 0.93. The CO:NO_x emissions ratio for Denver is about 16.8 (Colorado Department of Health, 1979). If this slope and emissions ratio is used, with the average NO_x concentration of 0.125 ppm, the NO_x loss is calculated to be 13 ppb. As has been discussed before, the PAN and some portion of the HNO_3 were probably included in the NO_x measurements and would not contribute to NO_x loss. The particulate NO_3^- concentration averaged about 1.6 ppb (mole fraction) (Countess et al., 1980). Thus particulate NO_3^- can account for only a small portion of the apparent NO_x loss.

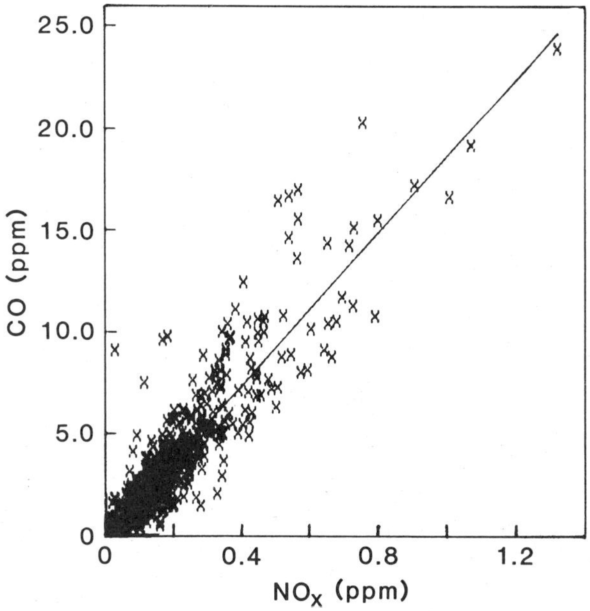

Figure 5. Linear regression of CO versus NO_x for Denver GMR data, November–December, 1978.

The technique described by Calvert (1976) was used to calculate the NO_x loss rate relative to CO. This technique was applied to selected daytime periods on 33 of the sampling days using both 15 minute and hourly averaged data for CO and NO_x. The NO_x removal rate relative to CO averaged 0.11 ± 0.31 h^{-1} (range −0.26 to +1.31 h^{-1}) for the 15-min data and 0.06 ± 0.17 h^{-1} (range −0.15 to +0.81 h^{-1}) for the hourly averaged data. The decay was also calculated using diurnally averaged data for the time period 0700 to 1700 with the resulting NO_x removal rate 0.019 h^{-1}. These NO_x removal rates range from 1.9 to 11% per hour and are much more reasonable values.

The analysis of NO_x loss with the use of CO as a tracer is complicated by the diurnal variation in the emissions of CO and NO_x. Figure 6 shows the diurnal variability in the CO : NO_x emissions ratio in Denver. The average emission rates for CO and NO_x are from a report by the Colorado Department of Health (1979). Heisler et al. (1980) apportioned the NO_x emissions between the sources: motor vehicles, 40%; natural gas combustion, 28%; and electric power generation, 32%. Wolff et al. (1981) presented the normalized diurnal emissions profiles for each of these sources. For convenience, the CO emissions were taken as 100% from motor vehicles, instead of the 93% stated in the emissions inventory. As shown in Figure 6, there are very large variations in the relative magnitude of CO and NO_x emissions. This variability limits one's ability to investigate NO_x losses relative to CO on an hour-by-hour basis.

The OH concentration has been estimated for several days during the study by using the individual hydrocarbon data. These estimates are based upon the decay of propene, butene, or ethene with respect to acetylene, which was chosen as the tracer. Sufficient data were available for nine days to estimate the OH concentration by using one or more of these olefins. The estimated OH concentrations determined for each of the olefins are summarized in Table 7. The overall average of the estimated OH concentrations is 6.9 ± 5.7 × 10^{-8} ppm, and the range of values estimated is 1.1 × 10^{-8} to 2.3 × 10^{-7} ppm. If the average OH concentration and the rate constant for the OH reaction with NO_2 of 7.8 × 10^5 ppm^{-1} h^{-1}—which is appropriate for Denver's wintertime conditions (Anderson, 1980)—are used, the rate of

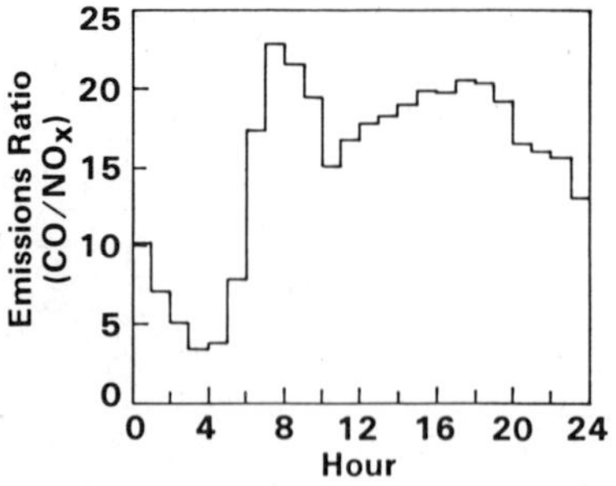

Figure 6. Diurnal variation of the CO : NO_x emissions ratio for Denver.

Table 7. Hydroxyl Concentrations Estimated from the Rate of Olefin Removal Relative to a Tracer

Olefin–Tracer	OH_{min} (ppm)	OH_{max} (ppm)	OH_{ave} (ppm)[a]
C_2H_4–C_2H_2	2.3×10^{-8}	2.3×10^{-7}	$9.8 \pm 11.1 \times 10^{-8}$
C_3H_6–C_2H_2	1.1×10^{-8}	1.4×10^{-7}	$5.7 \pm 4.4 \times 10^{-8}$
C_4H_8–C_2H_2	2.4×10^{-8}	1.7×10^{-7}	$7.3 \pm 5.3 \times 10^{-8}$
Average			$6.9 \pm 5.7 \times 10^{-8}$

[a] The error limits shown are the standard deviations of the estimated OH concentrations from the mean.

NO_2 conversion to HNO_3 is estimated to average about 5.4% per hour. This process ultimately represents a rather substantial sink for NO_x since the HNO_3 thus produced is likely to be incorporated into the aerosol phase as particulate NO_3^- or removed from the atmosphere by deposition.

9. CONCLUSIONS AND RESEARCH NEEDS

Much has been learned about the fate of nitrogen oxides in urban atmospheres in recent years, but much remains to be learned. Problems exist in all areas pertinent to our understanding, including: emissions inventories, ambient concentration measurements, chemical transformation processes, and the physical removal processes. In this section a brief summary of some areas where our understanding has improved is presented, along with the identification of certain areas that require further investigation.

Based upon the ambient atmospheric data from the Los Angeles area, one must conclude that there can be significant concentrations of PAN and HNO_3 present in the atmosphere. The measurements suggest that PAN and HNO_3 average 10 to 20% of the total nitrogen oxides present, ranging considerably higher during the afternoon when their formation is a maximum. This is of particular importance to the understanding of the fate of nitrogen oxides in the urban atmosphere since both PAN and HNO_3 are expected to be dry deposited more efficiently than the other nitrogenous species, NO, NO_2, and NO_3^-.

The particulate NO_3^- concentration in the urban atmosphere has been found to be relatively small, averaging about 4% of the total nitrogenous species present for the Los Angeles data and a little more than 1% for the Denver data. On the basis of these low concentrations for particulate NO_3^- and the inefficient dry deposition for the fine particulate matter, one does not expect particulate NO_3^- to play a major role in the ultimate removal of nitrogen oxides from the urban atmosphere.

The precipitation chemistry studies conducted by Liljestrand and Morgan (1981) concluded that about 1% of the annual emissions of nitrogen oxides within the Los Angeles Basin are locally scavenged by precipitation. Because of the limited precipitation in this area of the country, the dry deposition of the nitrogenous species would presumably be more important than the wet deposition processes. Therefore, it is likely that several percent of the nitrogen oxides emissions will be deposited during the brief time that an air mass remains in the Los Angeles Basin.

Substantial uncertainties were identified in the attempts to evaluate the removal of NO_x from the selected urban atmospheres. However, the data suggest that the NO_x losses are several percent for these measurements made in the urban area, where the nitrogen oxides have had little time to undergo transformations. This is consistent with the estimates of NO_x removal rates, which were in the range 1–8% per hour.

For the discussions presented in this chapter it is clear that the understanding of the fate of nitrogen oxides in the urban atmosphere is incomplete. Numerous problem areas requiring further research have been identified throughout the text of this chapter. The following discussion briefly outlines only a few of these areas where additional research is needed.

The techniques used in the ambient measurements of nitrogenous species require further improvement and testing. Further work is required in the development of techniques for measuring NO_2 that are specific and sensitive and have good time resolution. Continued uncertainty exists regarding the measurements of HNO_3 and particulate NO_3^-. These techniques require continued testing and development, to free the measurements from both positive and negative artifacts, and to improve the time response with which the measurements can be made.

One area of considerable uncertainty is related to the chemistry occurring at night. The recent measurements by Platt et al. (1980a,b) of NO_3 and HONO have demonstrated that there is some very interesting and possibly important chemistry that occurs at night. The suggestion by Hendry and Kenley (1979) that the thermal decomposition of PAN might lead to a significant OH radical source at night also requires further investigation.

Considerable uncertainty remains about the nature of the particulate NO_3^- formation processes in the atmosphere. Many of the processes previously considered to be of importance on the basis of equilibrium considerations now appear to be unimportant on the basis of kinetics arguments. The only process that still appears to be significant is related to the NH_3–HNO_3–NH_4NO_3 equilibrium. Additional research efforts should be directed toward the search for other particulate NO_3^- formation processes.

Simple theories describing both dry and wet deposition processes are being formulated. Unfortunately, there is a lack of experimental data that can test the theoretical understanding of the processes. Additional relevant experimental data would also be useful in providing a basis for extending and improving the theoretical understanding of the deposition processes.

One major area of continuing research related to the problem of understanding the fate of nitrogen oxides in the atmosphere will be the estimation of NO_x removal rates from the atmosphere. As was shown by the discussions in Section 8, there are numerous problems. These problems should be seriously considered prior to future attempts to characterize NO_x removal rates from the urban atmosphere. Some of these problems include the need for (1) accurate measurements of the nitrogenous species NO, NO_2, PAN, HNO_3, and NO_3^- with good time resolution, (2) a tracer having most of the same sources as NO_x, a reliable emissions inventory including both NO_x and the tracer, and the diurnal variations of their emissions, (3) a fairly large data base, to permit averaging of some of the departures from the normal situation described by the emissions inventory, and (4) improved techniques for analyzing NO_x removal data.

The ultimate goal of our efforts to understand the fate of nitrogen oxides in urban atmospheres is the creation of a detailed model that completely describes the system. As has been shown throughout this chapter, the limitations in our knowledge of the pertinent emissions, chemical, and physical processes involved have precluded the development of such models. However, as our knowledge improves, the data necessary for input to the model and testing of the model will become available, thus allowing the development of detailed models of the fate of nitrogen oxides in urban atmospheres.

REFERENCES

Anderson, L. G. (1980). Absolute rate constants for the reaction of OH with NO_2 in N_2 and He from 225 to 389 K. *J. Phys. Chem.* **84,** 2152–2155.

Anderson, L. G. (1982). Chemical and physical processes for removing nitrogen oxides from Denver's atmosphere. General Motors Research Publication GMR-3717, submitted for publication.

Appel, B. R. and Tokiwa, Y. (1981). Atmospheric particulate nitrate sampling errors due to reactions with particulate and gaseous strong acids. *Atmosph. Environ.* **15,** 1087–1089.

Appel, B. R., Wall, S. M., Tokiwa, Y., and Haik, M. (1979). Interference effects in sampling particulate nitrate in ambient air. *Atmosph. Environ.* **13,** 319–325.

Appel, B. R., Wall, S. M., and Knights, R. L. (1980a). Characterization of carbonaceous materials in atmospheric aerosols by high-resolution mass spectrometric thermal analysis. *Adv. Environ. Sci. Technol.* **9,** 353–365.

Appel, B. R., Wall, S. M., Tokiwa, Y., and Haik, M. (1980b). Simultaneous nitric acid, particulate nitrate and acidity measurements in ambient air. *Atmosph. Environ.* **14,** 549–554.

Bowermaster, J. and Shaw, R. W. (1981). A source of gaseous HNO_3 and its transmission efficiency through various materials. *J. Air Pollut. Control Assoc.* **31,** 787–820.

Braman, R. S. and Shelley, T. J. (1980). Gaseous and particulate ammonia and nitric acid concentrations—Columbus, Ohio area—Summer 1980. EPA Report 600/7-80-179.

Breitenbach, L. P. and Shelef, M. (1973). Development of a method for the analysis of NO_2 and NH_3 by NO-measuring instruments. *J. Air Pollut. Control Assoc.* **23,** 128–131.

Cadle, S. H., Countess, R. J., and Kelly, N. A. (1982). Nitric acid and ammonia in urban and rural locations. *Atmosph. Environ.* (in press).

Calvert, J. G. (1976). Hydrocarbon involvement in photochemical smog formation in Los Angeles atmosphere. *Environ. Sci. Technol.* **10,** 256–262.

Cavender, J. H., Kircher, D. S., and Hoffman, A. J. (1973). Nationwide air pollutant emission trends 1940–1970. EPA Publication No. AP-115.

Chang, T. Y., Norbeck, J. M., and Weinstock, B. (1979). An estimate of the NO_x removal rate in an urban atmosphere. *Environ. Sci. Technol.* **13,** 1534–1537.

Chock, D. P., Dunker, A. M., Kumar, S., and Sloane, C. S. (1981). Effect of NO_x emissions rates on smog formation in the California South Coast Air Basin. *Environ. Sci. Technol.* **15,** 933–939.

Colorado Department of Health (1979). Report to the public—1979, Vol. 1, *Activity Summary*, Denver, CO.

Countess, R. J., Wolff, G. T., and Cadle, S. H. (1980). The Denver winter aerosol: A comprehensive chemical characterization. *J. Air Pollut. Control Assoc.* **30,** 1194–1200.

Decker, C. E., Sickles, J. E., Bach, W. D., Vukovich, F. M., and Worth, J. J. B. (1978). Project Da Vinci II: Data analysis and interpretation. EPA Report 450/3-78-028.

Delwiche, C. C. (1970). The nitrogen cycle. *Sci. Am.* **223** (3), 136–146.

Delwiche, C. C. (1977). Energy relations in the global nitrogen cycle. *Ambio* **6,** 106–111.

Demerjian, K. L., Kerr, J. A., and Calvert, J. G. (1974). The mechanism of photochemical smog formation. *Adv. Environ. Sci. Technol.* **4,** 1–262.

Doyle, G. J., Tuazon, E. A., Graham, R. A., Mischke, T. M., Winer, A. M., and Pitts, J. N. (1979). Simultaneous concentrations of ammonia and nitric acid in a polluted atmosphere and their equilibrium relationships to particulate ammonium nitrate. *Environ. Sci. Technol.* **13,** 1416–1419.

Durham, J. L., Overton, J. H., and Aneja, V. P. (1981). Influence of gaseous nitric acid on sulfate production and acidity in rain. *Atmosph. Environ.* **15,** 1059–1068.

Ferman, M. A., Wolff, G. T., and Kelly, N. A. (1981). An assessment of the gaseous pollutants and meteorological conditions associated with Denver's brown cloud. *J. Environ. Sci. Health* **A16,** 315–339.

Gradel, T. E. (1978). *Chemical Compounds in the Atmosphere*, Academic Press, New York.

Grosjean, D. (1981). Critical evaluation and comparison of measurement methods for nitrogeneous compounds in the atmosphere. Final Report. Coordinating Research Council, CAPA-19-81.

Grosjean, D. and Friedlander, S. K. (1980). Formation of organic aerosols from cyclic olefins and diolefins. *Adv. Environ. Sci. Technol.* **9,** 435–473.

Hampson, R. F. and Garvin, D. (1978). Reaction rate and photochemical data for atmospheric chemistry-1977. NBS Special Publication 513.

Harker, A. B., Richards, L. W., and Clark, W. E. (1977). The effect of atmospheric SO_2 photochemistry upon observed nitrate concentrations in aerosols. *Atmosph. Environ.* **11,** 87–91.

Heisler, S. L., Henry, R. C., Watson, J. G., and Hidy, G. M. (1980). The 1978 Denver winter haze study. ERT Document No. P-5417-1.

Hendry, D. G. and Kenley, R. A. (1979). Atmospheric chemistry of peroxynitrates, in *Nitrogenous Air Pollutants: Chemical and Biological Implications*, D. Grosjean, Ed., Ann Arbor Science, Ann Arbor, MI, pp. 137–148.

Huebert, B. J. and Lazrus, A. L. (1980). Tropospheric gas-phase and particulate nitrate measurements. *J. Geophys. Res.* **85C,** 7322–7328.

Jorgen, R. T. (1978). Ambient peroxyacetyl nitrate (PAN) measurements in the Houston area. Rockwell International Report No. HAOS-08.

Joseph, D. W. and Spicer, C. W. (1978). Chemiluminescence method for atmospheric monitoring of nitric acid and nitrogen oxides. *Anal. Chem.* **50,** 1400–1403.

Joshi, S. B. and Bufalini, J. J. (1978). Halocarbon interferences in chemiluminescent measurements of NO_x. *Environ. Sci. Technol.* **12,** 597–599.

Judeikis, H. S., Seymour, S., Stewart, T. B., Hedgpeth, H. R., and Wren, A. G. (1979). Laboratory studies of heterogeneous reactions of oxides of nitrogen, in *Nitrogenous Air Pollutants: Chemical and Biological Implications*, D. Grosjean, Ed., Ann Arbor Science, Ann Arbor, MI, pp. 83–109.

Kelly, T. J. and Stedman, D. H. (1979). Chemiluminescence measurements of HNO_3 in air, in *Current Methods to Measure Atmospheric Nitric Acid and Nitrate Artifacts*, R. K. Stevens, Ed., EPA Report 600/2-79-051.

Kley, D., Drummond, J. W., McFarland, M., and Liu, S. C. (1981). Tropospheric profiles of NO_x. *J. Geophys. Res.* **86,** 3153–3161.

Lee, Y. N. and Schwartz, S. E. (1981). Evaluation of the rate of uptake of nitrogen dioxide by atmospheric and surface liquid water. *J. Geophys. Res.* **86,** 11971–11983.

Levine, S. Z. and Schwartz, S. E. (1982), In-cloud and below-cloud scavenging of nitric acid vapor. *Atmosph. Environ.* **16,** 1725–1734.

Liljestrand, H. M. and Morgan, J. J. (1981). Spatial variations of acid precipitation in Southern California. *Environ. Sci. Technol.* **15,** 333–339.

Littman, F. E. (1979). Regional air pollution study: Emission inventory summarization. EPA Report 600/4-79-004.

Logan, J. A., Prather, M. J., Wofsy, S. C., and McElroy, M. B. (1978). Atmospheric chemistry: Response to human influence. *Transact. R. Soc.* **290A,** 187–234.

Macoy, N. H., Weingarten, R., Pires, A., and Poultney S. (1980). High altitude pollution program, stratospheric measurement system, laboratory performance capability report, chemical conversions techniques. Federal Aviation Administration Report FAA-EE-80-11.

Matthews, R. D., Sawyer, R. F., and Schefer, R. W. (1977). Interferences in chemiluminescent measurements of NO and NO_2 emission from combustion systems. *Environ. Sci. Technol.* **11,** 1092–1096.

McFarland, M., Kley, D., Drummond, J. W., Schmeltekopf, A. L., and Winkler, R. H. (1979). Nitric oxide measurements in the equatorial Pacific region. *Geophys. Res. Lett.* **6,** 605–608.

Middleton, P. and Kiang, C. S. (1979). Relative importance of nitrate and sulfate aerosol production mechanisms in urban atmospheres, in *Nitrogenous Air Pollutants: Chemical and Biological Implications*, D. Grosjean, Ed., Ann Arbor Science, Ann Arbor, MI, pp. 269–288.

Miller, D. F. and Spicer, C. W. (1975). Measurement of nitric acid in smog. *J. Air Pollut. Control Assoc.* **25,** 904–942.

Monson, P. R., Ferman, M. A., and Kelly, N. A. (1978). Houston oxidant field study: Summer 1977. Paper 50.4, presented at 71st Annual Meeting Air Pollution Control Association, Houston, TX, June 1978; General Motors Research Publication GMR-2736.

National Research Council (1977). *Nitrogen Oxides*, National Academy of Sciences, Washington, DC.

National Research Council (1978). *Nitrates: An Environmental Assessment.* National Academy of Sciences, Washington, DC.

Okita, T., Morimoto, S., Izawa, M., and Konno, S. (1976). Measurement of gaseous and particulate nitrates in the atmosphere. *Atmosph. Environ.* **10,** 1085–1089.

Orel, A. E. and Seinfeld, J. E. (1977). Nitrate formation in atmospheric aerosols. *Environ. Sci. Technol.* **11,** 1000–1007.

Pennington, J. E. (1976). A summary of air pollutant emissions in Texas 1973. Texas Air Control Board, Austin, TX.

Perner, D. and Platt, U. (1979). Detection of nitrous acid in the atmosphere by differential optical absorption. *Geophys. Res. Lett.* **6,** 917–920.

Peterson, T. W. and Seinfeld, J. H. (1979). Calculation of sulfate and nitrate levels in a growing, reacting aerosol. *Am. Inst. Chem. Eng. J.* **25,** 831–838.

Platt, U. and Perner, D. (1980). Direct measurement of atmospheric CH_2O, HNO_2, O_3, NO_2 and SO_2 by differential optical absorption in the near UV. *J. Geophys. Res.* **85,** 7453–7458.

Platt, U., Perner, D., and Patz, H. W. (1979). Simultaneous measurement of atmospheric CH_2O, O_3 and NO_2 by differential optical absorption. *J. Geophys. Res.* **84,** 6329–6335.

Platt, U., Perner, D., Harris, G. W., Winer, A. M., and Pitts, J. N. (1980a). Observations of nitrous acid in an urban atmosphere by differential optical absorption. *Nature* **285,** 312–314.

Platt, U., Perner, D., Winer, A. M., Harris, G. W., and Pitts, J. N. (1980b). Detection of NO_3 in the polluted troposphere by differential optical absorption. *Geophys. Res. Lett.* **7,** 89–92.

Platt, U., Perner, D., Schroder, J., Kessler, C., and Toennissen, A. (1981). The diurnal variation of NO_3. *J. Geophys. Res.* **86,** 11965–11970.

Poizat, O. and Atkinson, G. H. (1982). Quantitative detection of NO_2 at atmospheric pressures by visible photoacoustic spectroscopy. *Anal. Chem.* **54,** 1485–1489.

Reid, J., El-Sherbiny, M., Garside, B. K., and Balik, E. A. (1980). Sensitivity limits to a tunable diode laser spectrometer, with application to the detection of NO_2 at the 100 ppt level. *Appl. Opt.* **19,** 3349–3354.

Scott, B. C. (1979). Gas and aerosol scavenging. *Proceedings of the Workshop on Toxic Substances in Precipitation*, Environmental Protection Agency, Jekyll Island, GA.

Sehmel, G. A. (1980). Particle and gas dry deposition: A review. *Atmosph. Environ.* **14,** 983–1011.

Shaw, R. W., Dzubay, T. G., and Stevens, R. K. (1979). The denuder difference experiment, in *Current Methods to Measure Atmospheric Nitric Acid and Nitrate Artifacts*, R. K. Stevens, Ed., EPA Report 600/2-79-051.

Söderlund, R. and Svensson, B. H. (1976). The global nitrogen cycle, in *Nitrogen, Phosphorus and Sulfur—Global Cycles*, B. H. Svensson and R. Söderlund, Eds., SCOPE Report 7, pp. 23–74.

Spicer, C. W. (1977). The fate of nitrogen oxides in the atmosphere. *Adv. Environ. Sci. Technol.* **7,** 163–261.

Spicer, C. W. (1979), Measurement of gaseous HNO_3 by electrochemistry and chemiluminescence, in *Current Methods to Measure Atmospheric Nitric Acid and Nitrate Artifacts*, R. K. Stevens, Ed., EPA Report 600/2-79-051, pp. 27–35.

Spicer, C. W. and Schumacher, P. M. (1977). Interferences in sampling atmospheric particulate nitrate. *Atmosph. Environ.* **11,** 873–876.

Spicer, C. W. and Schumacher, P. M. (1979). Particulate nitrate: Laboratory and field studies of major sampling interferences. *Atmosph. Environ.* **13,** 543–552.

Spicer, C. W., Gemma, J. L., Schumacher, P. M., and Ward, G. F. (1976). The fate of nitrogen oxides in the atmosphere. Second year report. Coordinating Research Council, CAPA-9-71.

Stedman, D. H., West, D. H., and Schetter, R. E. (1981). Measurements related to atmospheric oxides of nitrogen. Final Report. Coordinating Research Council, CAPA-19-81.

Stelson, A. W., Friedlander, S. K., and Seinfeld, J. H. (1979). A note on the equilibrium relationship between ammonia and nitric acid and particulate ammonium nitrate. *Atmosph. Environ.* **13,** 369–371.

Stephens, E. R. (1969). The formation, reactions and properties of peroxyacyl nitrates (PAN's) in photochemical air pollution. *Adv. Environ. Sci. Technol.* **1,** 119–146.

Stephens, E. R. and Price, M. A. (1973). Analysis of an important air pollutant: Peroxyacetyl nitrate. *J. Chem. Educ.* **50,** 351–354.

Tang, I. N. (1980). On the equilibrium partial pressures of nitric acid and ammonia in the atmosphere. *Atmosph. Environ.* **14,** 819–828.

U.S. Environmental Protection Agency (1978). National air quality, monitoring, and emissions trends report, 1977. EPA Report 450/2-78-052.

Winer, A. M., Peters, J. W., Smith, J. P., and Pitts, J. N. (1974). Response of commercial chemiluminescent NO–NO_2 analyzers to other nitrogen-containing compounds. *Environ. Sci. Technol.* **8,** 1118–1121.

Wolff, G. T. Countess, R. J., Groblicki, P. J., Ferman, M. A., Cadle, S. H., and Muhlbaier, J. L. (1981). Visibility-reducing species in the Denver 'brown cloud.' Part II. Sources and temporal patterns. *Atmosph. Environ.* **15,** 2485–2503.

10

THE GLOBAL BUDGET OF ATMOSPHERIC NITROGEN SPECIES

Donald H. Stedman

Departments of Chemistry and
Atmospheric and Oceanic Science
University of Michigan
Ann Arbor, Michigan 48109

Richard E. Shetter

Space Physics Research Laboratory
University of Michigan
Ann Arbor, Michigan 48109

1. INTRODUCTION

Tropospheric ozone was discovered and extensively studied in the 1800s (Fox, 1873) then forgotten in the early 1900s with the discovery of the stratospheric ozone layer (Chappuis, 1880; Hartley, 1881). It was rediscovered in studies of Los Angeles smog in the early 1950s, at which time the important role of gaseous oxides of nitrogen (NO_x = NO + NO_2) was recognized (Leighton, 1961). By 1971 the role of nitrogen oxides in stratospheric ozone chemistry had been described by Johnston (1971) and Crutzen (1971), and simultaneously the importance of NO_x and its interaction with the cycle of tropospheric odd hydrogen radicals was recognized by Levy (1971). By contrast, nitrogen compounds in precipitation [mostly in the form of ammonium (NH_4^+) and nitrate (NO_3^-)] have been studied more or less continuously since the middle 1800s because of their then perceived agricultural significance (Brimblecombe and Pitman, 1980; Cowling, 1982; Smith, 1872).

In 1923 Lewis and Randall (1923) pointed out that even atmospheric nitrogen is thermodynamically unstable: "Even starting with water and air we see by our equations that nitric acid should form until it reaches a concentration of about 0.1 *M* where the calculated equilibrium exists. It is to be hoped that nature will not discover a catalyst for this reaction, which would permit all of the oxygen and part of the nitrogen of the air to turn the oceans into dilute nitric acid." This fact was not commonly recognized until Delwiche (1970) and Lovelock (1971) emphasized that atmospheric N_2 is a steady-state component of a geochemical cycle. These geochemical cycles, which determine the concentrations of the atmospheric nitrogen-containing species, are the subject of this chapter.

Nitrogen compounds known or expected in the troposphere are shown in Table 1. Also shown are the standard enthalpies and free energies of

Table 1. Atmospheric Nitrogenous Species

Compound	Formal Oxidation State	ΔH_f° (298)[a]	ΔG_f° (298)[a]	Global Atmospheric Mixing Ratio (ppm)[b]	Atmospheric Burden Tg N[b,e]
NH_3	−3	−11	−4	1 (−3)[c]	0.3
NH_4^+	−3	−86	−63	2 (−3)[d]	0.6
N_2	0	0	0	7.9 (5)	3.9 (9)
N_2O	1	19	25	0.302	1.4 (3)
NO	2	22	21	5 (−6)	1 (−2)
HONO	3	−19	−11	—	—
NO_2	4	8	12	5 (−5)	0.2
HNO_3	5	−32	−18	7 (−5)	0.14
NO_3^-	5	−50	−27	1 (−4)[d]	0.1
NO_3	6	17	28	—	—
HNO_4	7	−54	—	—	—

[a] NBS (1968); rounded to nearest kcal mol^{-1}.

[b] Estimates derived in this chapter.

[c] The notation 1 (−3) represents 1×10^{-3}.

[d] Mole fraction. The unit ppb (mole fraction) for an aerosol constituent directly corresponds to ppb (volume) for a gaseous constituent. For NO_3^-, 1 ppb (mole fraction) is equivalent to 2.58 μg m^{-3} (20°C, 1 atm).

[e] Tg N = 10^{12} g of nitrogen.

formation and estimates of average tropospheric mixing ratio and total atmospheric burden. We believe that preindustrial atmospheric cycles of all the species in Table 1 were balanced and that a steady state existed between formation and removal terms. This chapter attempts to formulate balanced budgets, and add, only where necessary, discussions of anthropogenic perturbation.

Our ability to understand atmospheric chemistry is based partly on the advancement of technology that enables us to make measurements. This advancement, in turn, is but a ripple carried on a rising tide of per capita energy consumption, particularly in the developed countries. Pollution concomitant with this energy use so dominates locally over the natural cycles that measurement of the atmospheric system free of anthropogenic influence has become remarkably difficult. Even Pt. Barrow, Alaska, once thought to be a pristine location, turns out to be severely influenced by human activity, as described recently in a series of papers edited by Rahn (1981).

2. THE GEOSPHERIC CYCLE

Geochemical cycles of many tropospheric nitrogen species have been discussed by a number of authors, including Junge (1963), Delwiche (1970),

Munn and Bolin (1971), Robinson and Robbins (1970), Burns and Hardy (1975), McConnell and McElroy (1973), Walker (1977), Söderlund and Svensson (1976), and recently Logan et al. (1981). All concur on the need for more information, although not necessarily on which information gaps are the most important. For instance, Logan et al. (1981) state that "Development of a comprehensive model for tropospheric OH is hampered by a lack of data for NO and NO_2." By contrast, Delwiche (1970) is most concerned with fixed nitrogen in the ecosystem and writes that "much work has to be done to develop successful management techniques" for bacterial denitrification to control the buildup of fixed nitrogen.

The most important processes thought to be affecting the atmospheric cycles of the species listed in Table 1 are represented schematically in Figure 1. The treatment of "soil and marine fixed nitrogen" as a single box serves to conceal our lack of knowledge concerning the rates and processes of ammonia oxidation and nitrate reduction in the soil. It also conceals input of nitrogen from weathering of sedimentary rocks and loss of fixed nitrogen by sedimentary deposition.

The processes neglected in Figure 1 are important if one wants to answer the question "is the total budget of mobilized nitrogen (i.e., atmosphere and

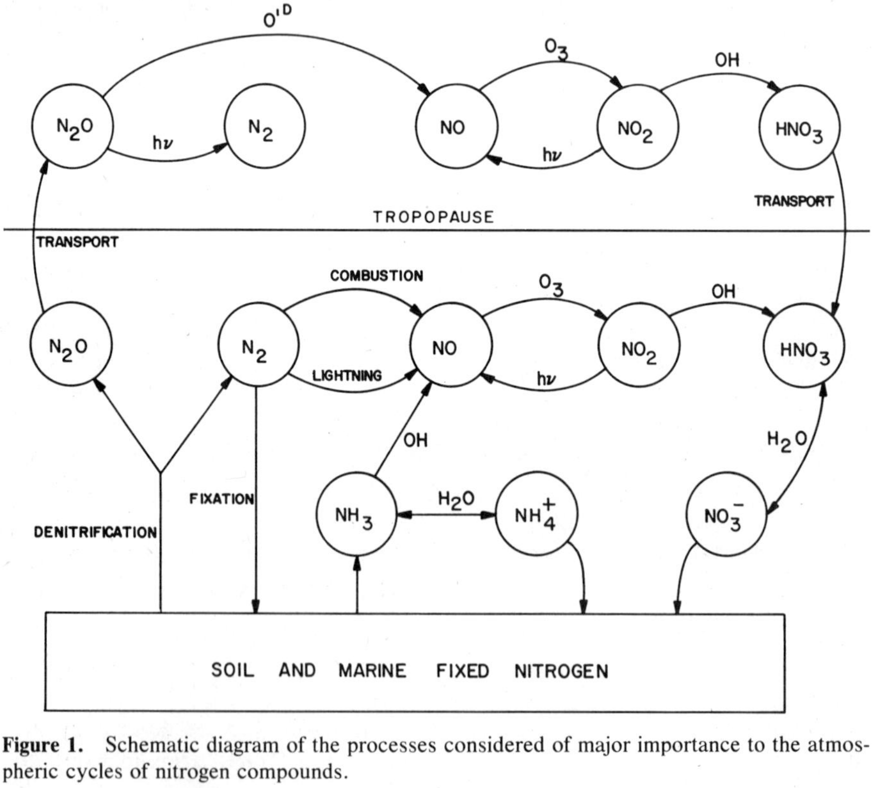

Figure 1. Schematic diagram of the processes considered of major importance to the atmospheric cycles of nitrogen compounds.

Table 2. Terrestrial Nitrogen Inventories According to Several Authors[a]

Reference	Atmosphere	Sediments	Crust
Delwiche (1970)	3.8(9)[b]	4(9)	1.4(10)
Söderlund and Svensson (1976)	3.9(9)	4(8)	1.1(11)
Walker (1977)	3.9(9)	1.4(9)	1.6(7)

[a] Here and throughout this chapter units for terrestrial inventories of nitrogen compounds are Tg N unless otherwise specified.

[b] The notation 3.8(9) represents 3.8×10^9.

sediments) constant, increasing, or decreasing?'' Table 2 shows various estimates for the major reservoirs of terrestrial nitrogen. If we use the largest value for the sedimentary nitrogen reservoir (Delwiche), the 8×10^9 Tg of mobilized nitrogen along with the age of the earth (4×10^9 yr) implies an average outgassing rate of 2 Tg yr^{-1}. (Here and throughout this chapter terrestrial inventories and fluxes of nitrogen compounds are expressed in units of Tg N and Tg N yr^{-1}, respectively; 1 Tg $= 10^{12}$ g.) If we use Walker's value for crustal nitrogen then clearly outgassing has effectively terminated, and if anything, our extra nitrogen fixing capacity may be reducing the total nitrogen budget. If, however, one takes Söderlund and Svensson's or Delwiche's value for the crustal abundance, it is possible that mobilization of nitrogen is still taking place and may continue for several billion years. Alternatively, one may examine volcanic activity to estimate outgassing. Unfortunately, nitrogen analysis in volcanic (and fumarolic) gases is frequently used in such studies to estimate an upper limit for atmospheric contamination of the sample and thus cannot be simultaneously used as a reliable indicator of volcanic nitrogen. However, if we use a $CO_2:N_2$ volume ratio of 9:1 (Holland, 1964) and a present-day volcanic release rate of CO_2 of 2 Tg yr^{-1} (Walker, 1977), we obtain a nitrogen release rate of 0.14 Tg yr^{-1}. Although this value is similar to that obtained using Delwiche's estimate for the sedimentary nitrogen reservoir, it is nevertheless much less than that required to provide the observed mobilized nitrogen. These observations imply significantly enhanced outgassing in early geologic eras but leave open the question of whether the total available nitrogen is increasing or steady. This arises from uncertainty in the estimates of crustal reservoir capacity, which vary over four orders of magnitude.

Table 2 shows reasonable agreement regarding the size of the sedimentary nitrogen reservoir. Sedimentary rocks are cycled by geologic processes independent of their trace chemistry. The turnover time for the sedimentation due to weathering processes can be estimated from carbon, oxygen, or phosphorus cycles. Walker's carbon and oxygen cycle gives a turnover time of 10^8 yr. Estimates from phosphorus cycles described by Pierrou (1975) vary from 7.5×10^7 to 4×10^8 yr. All these are short in comparison to the age of the Earth, and therefore one might expect the sedimentary reservoir to be in steady state, with a lifetime of the order of 10^8 yr.

Söderlund and Svensson (1976) neglect any volcanic juvenile addition but include a significant sedimentation rate of 38 Tg N yr^{-1}. Such a rate is difficult to reconcile with their total sedimentary nitrogen inventory of 4×10^8 Tg, since it implies an average lifetime of sedimentary nitrogen in rocks of only 10^7 yr, which appears too small compared to that derived from the cycles. Delwiche, on the other hand, gives a sedimentation rate of 0.2 Tg yr^{-1}. This, along with his sedimentary nitrogen inventory of 4 Tg, implies a lifetime of 10^{10} yr (large in comparison to the age of the Earth), which is clearly too large to be a significant factor in keeping the geologic inventory near steady state. Walker's estimate for a sedimentary burial rate of 14 Tg yr^{-1} seems more appropriate. This, together with Delwiche's intermediate estimate of 1.4×10^9 Tg for the sedimentary reservoir, gives a turnover time of 10^8 yr for the sedimentary reservoir and implies an input to the biosphere from weathering of the order of 14 Tg yr^{-1}, certainly significantly larger than any plausible input of juvenile nitrogen, which can thus be neglected. The disagreement as to the actual sedimentation rate is not difficult to understand considering the difficulties in determining the net rate of oceanic organic nitrogen sediment accretion caused by the very efficient reuse of fixed nitrogen by organisms in the upper sedimentary layers. Figure 2 shows these slow cycles and the status of the uncertainty surrounding them.

If the concentration of nitrogen in the atmosphere were to vary, the reservoir of either sedimentary or igneous rocks must absorb the variation.

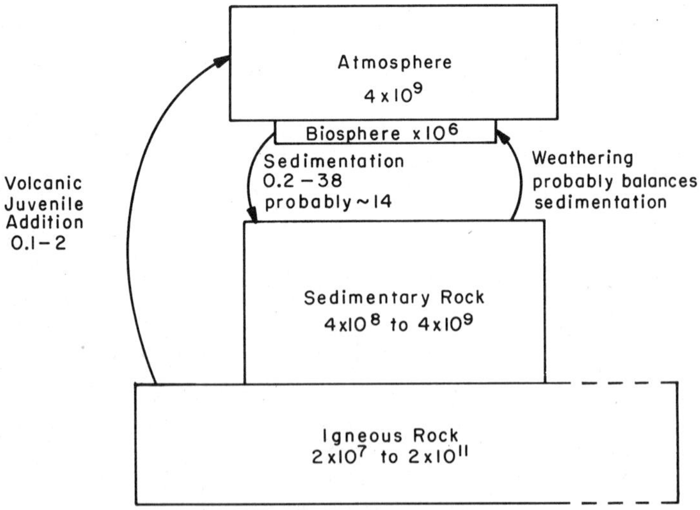

Figure 2. Schematic representation of lithospheric cycles of nitrogen and associated uncertainties. Areas of boxes are suggestive of respective burdens, which are in units of Tg N. Fluxes are in units of Tg N $year^{-1}$.

However, since the time constant for variation is the inventory divided by the flux, we can see from Figure 2 that the time constant for such changes is probably not smaller than 1×10^8 or 2×10^9 yr, respectively. Not something to worry our grandchildren.

Walker (1977) speculates that some 30% more nitrogen may have been in the atmosphere after the first 10^4 yr of outgassing than is now present. He bases this speculation on his small crustal reservoir and the argument that before the development of life, organic nitrogen compounds now contributing to the sedimentary reservoir were not available. However, in the absence of biology, lightning-fixed nitrogen, as oxides of nitrogen, would accumulate in the oceans and on the land, without returning to the N_2 reservoir. Also, N_2 may still be increasing slowly, courtesy of juvenile sources from a crustal reservoir much larger than the one Walker uses. On the basis of the findings discussed above, we believe that there is little justification even for a speculation of other than steady state, or steadily increasing, N_2.

3. THE BIOSPHERIC CYCLE

From Figure 2 it is clear that the time constant for the biospheric component of fixed nitrogen is geologically short. Therefore, before human intervention the cycles coupling the biosphere to the atmosphere and lithosphere should have been in balance. Furthermore, both oceanic and terrestrial components of the biospheric budget must have been balanced separately. Figure 3 shows an attempt to form a preindustrial (and by implication, preagricultural) balanced budget for these components. The balance of total fixed nitrogen requires that fluxes to and from N_2 be equal. Table 3 shows a possible balance sheet.

3.1. Nitrogen Fixation Rate

Some questions immediately arise concerning the preindustrial rate of nitrogen fixation. Most authors agree that the prehistoric standing terrestrial biomass was greater by virtue of the more extensive tropical and boreal forests. However, that does not necessarily imply a greater biological nitrogen fixation rate. For instance, Evdokimova et al. (1975) suggest a 25% decrease in fixation rate when fields become climax forests. Burns and Hardy (1975) suggest that 25% of nitrification is supplied by leguminous crops alone. Furthermore, application of newly mined phosphorus to the soil can be expected to increase nitrogen fixation. If, as a rough approximation, we use a very conservative biospheric ratio of 10:1 N:P atoms then the application of 12.6 Tg P yr^{-1} (Pierrou, 1975) would imply 57 Tg N yr^{-1} fixation. Thirty-six Tg N yr^{-1} is fixed industrially, much of which is applied

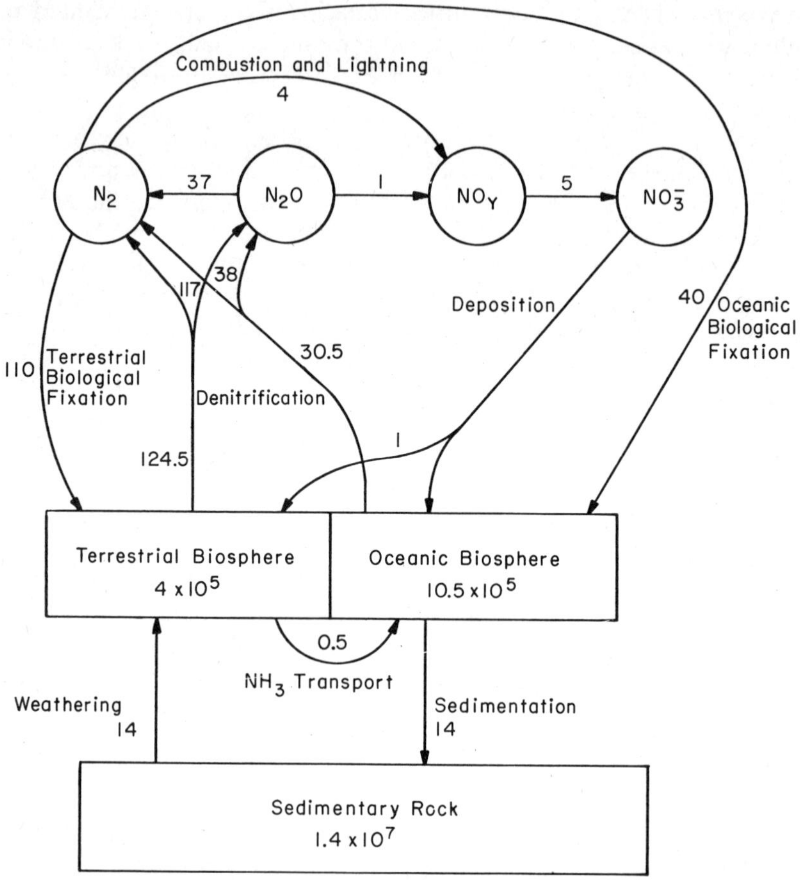

Figure 3. Important fluxes and reservoirs affecting the biospheric budget of fixed nitrogen estimated for prehistoric Earth. Burdens are indicated in Tg N and fluxes in Tg N year^{-1}. NO_y includes all gaseous nitrogen oxides, mainly $NO + NO_2 + HNO_3$. The ammonia cycle is included only as the net transport of nitrogen from land to ocean.

Table 3. Balance of Total Preindustrial Fixed Nitrogen[a]

Nitrogen Fixation Process	Flux (Tg yr^{-1})	Denitrification	Flux (Tg yr^{-1})
Terrestrial biological	110[1.2][b]	As N_2	140[1.5]
Oceanic biological	40[2]	As N_2O	14[3]
Combustion	4[2]		
Total	154		154

[a] Here an throughout this chapter units for terrestrial fluxes of nitrogen compounds are Tg N yr^{-1} unless otherwise specified.

[b] The number in brackets represents the factor by which the value is estimated to be uncertain. Thus, the combustion term 4[2] implies that this term is uncertain between 2 and 8 Tg N yr^{-1}.

with the phosphorus, but 21 Tg N yr^{-1} remains. All the above imply a greater present-day rate of terrestrial biological nitrogen fixation than in the past, perhaps by 25%.

Oceanic pelagic nitrogen fixation is very poorly estimated at the present time. We can apply the same argument above concerning P:N ratios. The present phosphorus intake of the oceans is estimated by Pierrou (1975) to be 20 Tg P yr^{-1} distributed 17:3 between river runoff and airborne input. The P:N atomic ratio in plankton is 1:16 (Walker, 1977); thus 20 Tg P yr^{-1} implies 145 Tg N yr^{-1}. From this we subtract the present-day external inputs of nitrogen to the oceans, estimated at 60 Tg N yr^{-1} (Söderlund and Svensson, 1976), which leaves 85 Tg yr^{-1} as an estimate of present-day oceanic nitrogen fixation derived from a consideration of the phosphorus cycle. This is comparable to Söderlund and Svensson's independent estimate of 30–130 Tg yr^{-1}. If we are correct, and oceanic nitrogen fixation is limited by phosphorus input (not accompanied by fixed nitrogen), any reduced mobilization of phosphorus before agriculture would have severely limited oceanic nitrogen fixation. According to Evdokimova et al. (1975), forests turned to agricultural land mobilize approximately 100 times more phosphorus. If one estimates a global contribution for an area of this process, then preindustrial phosphorus mobilization must have been lower by a factor of 2–4. Since the oceanic phosphorus turnover time is only 10^4 yr, then, unless oceanic phosphorus cycling was more efficient in earlier times (Broecker, 1974), there must have been two to four times less nitrogen fixation in earlier times. We thus reduce the estimate of the oceanic nitrogen fixation rate by a factor of 2 from that given by Söderlund and Svensson to 15–65 Tg yr^{-1}, with an average of 40 Tg yr^{-1}. Note that we have already similarly reduced the estimated rate of input to sedimentary rocks from their value of 38 Tg yr^{-1} to 14 Tg yr^{-1}.

The preindustrial combustion input was derived from the present estimated wildfire input (Crutzen et al., 1979) but doubled to account for the increased forest. It is nevertheless a small term. The denitrification terms in Table 3 are derived from the necessary balance of the preindustrial biosphere.

3.2. Denitrification

For both the terrestrial and oceanic reservoirs to be separately balanced, the transfers by means of sedimentation and weathering must be made up by runoff, ammonia transfer, and differences in denitrification rates. River runoff is at present estimated to introduce 8–13 Tg N yr^{-1} to the oceans, more than balanced by a sea spray return of 10–20 Tg yr^{-1}. We chose to neglect the net difference between these terms, and assume that it was negligible also in the pre-industrial system. The present ammonia flux to the oceans is estimated at 35 Tg N yr^{-1} (Söderlund and Svensson, 1976). As is discussed more fully in Section 4.1, most ammonia flux is the result of agricultural activities, and the estimate by Söderlund and Svensson is probably very much too great. We estimate this flux to be less than 1 Tg N yr^{-1} at present and, allowing that this flux has doubled since prehistoric times, therefore ascribe a flux of 0.5 Tg N yr^{-1} to the NH_3 transport term.

An oceanic denitrification flux of 31 Tg N yr^{-1} is required to balance the oceanic reservoir. This estimate differs little from that given by Delwiche, who suggested a minimum of 30 Tg yr^{-1} on the basis of somewhat different reasoning. A terrestrial denitrification flux of 124.5 Tg yr^{-1} makes up the balance. According to Table 3, the total denitrification flux is designated 10:1 N_2:N_2O. This corresponds to other estimates but is highly uncertain and may well be different between average oceanic and average terrestrial denitrification.

3.3. Human Input

To remind us that Figure 3 is preindustrial, the terrestrial biomass has been estimated to be 20% larger, although for reasons discussed earlier it is thought to fix less nitrogen. Many authors point out that preindustrial climax biomes were significantly more conservative of trace nutrients than is the case in present agricultural practice. Because of this conservative terrestrial biomass, we estimate that there was a factor of 2 less nutrient (particularly phosphorus) flux to the oceans and hence presumably considerably less oceanic biomass and nitrogen fixation than at present.

Delwiche (1970) is justifiably concerned about the impact of human agricultural practices on this balance. Potential impacts arise from:

1. Combustion products of burned biomass.
2. Dramatically increased soil erosion transferring trace nutrients (including N + P) through freshwater to the oceanic reservoir.
3. Intentional nitrogen fixation by the fertilizer industry.
4. Intentional cultivation of nitrogen fixing legumes.
5. Unintended nitrogen fixation by energy- and transport-related combustion.

Discussing these impacts one at a time, Crutzen et al. (1979) estimate that tropical biomass burning may generate as much as 4.5 Tg N yr^{-1} as N_2O and 19 Tg N yr^{-1} as NO, according to Wong's (1978) estimate of CO_2 production. The short-term effect of CO_2 emission may be offset partly by the generation and subsequent burial of carbon that has been charred to a refractory graphitic material, most likely to be ultimately buried rather than returned to biological cycling. There is no apparent direct compensatory effect for the nitrogen fixation.

Increased soil erosion caused in part by the biomass burning (and other agricultural practices) may be an indirect compensation. According to

Granat et al. (1975), the rate of soil erosion is estimated to have been doubled, courtesy of human activity. If we focus on phosphorus, the rate of runoff to the oceans will have at least doubled, to arrive at the present input rate of 17.4 Tg P yr^{-1} (Pierrou, 1975). The present oceanic phosphorus inventory is about 1.2×10^5 Tg, giving a time constant of 7000 yr if we consider the whole ocean. If, however, we consider that the increased phosphate goes mainly into the surface layers, the time constant would reduce to only 100 yrs. The atomic ratio of P:N:C in plankton is approximately 1:16:106 (Walker, 1977). If, as we suspect, the increased phosphorus input to the oceans resulting from human activity is not accompanied by this much nitrogen and carbon, an increase in the oceanic nitrogen and carbon fixation must result. The increase in CO_2 predicted on the basis of fossil-fuel combustion is 5000 Tg CO_2 yr^{-1}. The observed increase is 2000 Tg yr^{-1}, implying an extra uptake of 3000 Tg yr^{-1} in the oceans (Broecker et al., 1979). According to the above ratio, this uptake would require 19 Tg P yr^{-1}. It seems unlikely that the similarity between this number and the present-day phosphorus input (estimated at 17.4 Tg P yr^{-1}) is coincidental. This implies that our profligate use of coal and oil is at least partly compensated by our profligacy in mining and applying phosphorus as a fertilizer.

The concept of "phosphorus matching" has been applied to the anthropogenic carbon budget by Mackenzie (1975) and in some detail by Broecker et al. (1979). Broecker discounts phosphorus matching on the basis of the observation by Stumm (1973) that only 7% of Europe's phosphorus applied as fertilizer and dumped as municipal waste reaches the natural waters. He uses an average rate of phosphorus mining of 2.5×10^{11} mol yr^{-1} (8 Tg P yr^{-1}) and a C:P atomic ratio of 250 in river runoff to show that phosphorus transport can only account for 2% of the carbon budget.

The current rate of phosphorus mining is 12.6 Tg P yr^{-1} (Pierrou, 1975; Stumm, 1973). However, Pierrou gives a rate of river runoff to the oceans almost an order of magnitude larger than Stumm. We used this river runoff rate for our calculation of possible phosphorus matching. This larger rate, and the fact that direct input to the oceans eliminates the 7% efficiency term used by Broecker, gives the larger numbers we obtain. All the above is based on simplistic assumptions of phosphate transfer and the carbon cycle in both the soil and the oceans, which are matters of intense research certainly beyond the scope of this chapter. Perhaps these comments will generate some more thought regarding the "phosphorus matching" concept.

Increased transport of phosphorus by erosion and synthetic phosphorus fertilizer does not compensate the fixed nitrogen cycle. Phosphorus input unaccompanied by extra nitrogen input requires further nitrogen fixation of about 7 g N/g P; thus the total runoff of 17.4 Tg P yr^{-1} would imply 120 Tg N yr^{-1}. This is an upper limit because of the neglect of a number of factors, but it illustrates the suggestion that increased erosional runoff contributes to increased oceanic nitrogen fixation perhaps by as much as 40 Tg yr^{-1}. Fossil-fuel combustion generates 19 Tg N yr^{-1} as NO. As is discussed later, most of this is locally returned to the biosphere, and only a fraction (perhaps 10%) is returned to the oceans.

Runoff of nitrogen-containing organic matter (mostly humic) appears to be almost compensated by return of sea spray droplets that are highly enriched in the surface active organic matter (Blanchard and Syzdek, 1972). Thus all potential impacts increase nitrogen fixation. Contributions are estimated in Table 4.

Table 3 gives a preindustrial nitrogen fixation rate of 154 Tg yr^{-1}. Thus we believe that human activities have already almost doubled the rate of nitrogen fixation in the last 1000 years and, therefore, that the biospheric cycle is at present unbalanced.

The foregoing arguments are all based on estimates of fluxes. There are, however, independent estimates of inventories. Estimates of total terrestrial fixed nitrogen (Tg N) include those by Söderlund and Svensson (1976) (3.3×10^5), Liu et al. (1977), (2.5×10^5), and Delwiche (1970) (9×10^5). If we extrapolate Bolin's (1977) suggestion that there has been perhaps an 8% reduction in the biospheric carbon reservoir in the last 180 yr, there may have been a reduction of as much as 20% in the terrestrial biosphere over the last 1000 years. If we apply this rate to Söderlund and Svensson's (1976) inventory, a net loss of about 70 Tg N yr^{-1} has occurred. This is significantly larger than Crutzen's estimate of 20 Tg N yr^{-1} from biomass burning; possibly the rate of turning over of forests to agricultural land has decreased as suggested by Broecker et al. (1979) on the basis of ^{13}C evidence from Stuiver (1978), although that is not clear. However, Stuiver (1978) derives a 7% decrease in the terrestrial carbon biomass from 1850 to 1950, much in agreement with Bolin (1977). If this is caused by forest conversion to pasture, the nitrogen biomass decrease would be approximately 3% since pasture has a higher N:C ratio (0.01) than forest (0.004). This would imply a flux of fixed nitrogen away from the terrestrial biosphere of 100 Tg yr^{-1}. Again, it is difficult to see where such a flux could fit in with our estimates, but if the reduced terrestrial biomass estimates turn out to be correct, the shift in 100 yr of 10^4 Tg N to some other reservoir will need to be accounted for.

Table 4. Estimates of Human Contributions to Increased Nitrogen Fixation

Activity	Fixation Rate (Tg N yr^{-1})	Reference
Biomass burning	19[3][a]	Crutzen et al. (1979)
Soil erosion	40[2]	This work
Fertilizer production	12.6[1.1]	Söderlund and Svensson (1976)
Legume cultivation	40[1.5]	Burns and Hardy (1975)
Combustion	20[1.1]	Stedman et al. (1978)
Total	132	

[a] Estimated uncertainty factor as in Table 3.

Presumably, the burning of brush and trees does not return fixed nitrogen to N_2 but rather fixes more and delivers it to the atmosphere as NO_x. This then can be transferred to the ocean fixed nitrogen inventory as runoff and as rained-out NO_3^-. An inconsistency is apparent in this process. No budget includes a large net transfer of fixed nitrogen to the oceans other than through ammonia. A further problem is that Crutzen's (1979) estimate for the biomass burning contribution of 20 Tg N yr^{-1} appears to be too low on the basis of the above arguments; however, it in turn is based on Wong's (1978) estimate of CO_2 input, which has been criticized as being a factor of 4 too high (Fahnestock, 1979)!

All the observations discussed above emphasize that we are not fully informed about the historic nitrogen cycle. Our inability to adequately measure nitrogen fixation rates is a major source of our uncertainty as to whether the present fixed nitrogen budget is balanced. This balance could be achieved only if denitrification had dramatically increased to compensate for the increased fixation estimated in Table 3. If we estimate [as Liu et al. (1977) do in considerably more detail] that the denitrification rate can change only as fast as the total biomass changes, the time constant for a change in denitrification rate is roughly $15 \times 10^5/300$ or 5000 yr. Thus we estimate that the present denitrification rate could be increasing at about 0.02% per year. Or, if our interpretation of Stuiver's data (1978) just discussed is correct, this increase could be as much as 0.3% per year.

All the gaseous nitrogen species except N_2 have geologically short atmospheric lifetimes, and thus their present-day budgets must be balanced. Because of the difficulty of estimating prehistoric budgets, present-day versions will be attempted, with only some speculation based on earlier discussions, as to how these gaseous cycles have been affected by human activities. A side effect of these short lifetimes is that the concentrations of these species are in general highly variable in space and time; thus averages of concentrations and fluxes, even when they can be obtained are fraught with errors. The most familiar example of a short-lived atmospheric species is water vapor. We are all acquainted with its heterogeneity in space and time and the spasmodic nature of the precipitation that removes it. In a similar way, the value (or lack of it) to be associated with average atmospheric water vapor concentrations and the errors associated with obtaining such an average will be apparent to the reader. The budgets of trace nitrogen species suffer from all the above caveats plus a lack of data, with some of the available data conflicting.

4. AMMONIA AND AMMONIUM

Ammonia, ammonium, and organic amino compounds are at the lowest oxidation state of nitrogen (−3) and thus are all unstable with respect to oxidation by O_2 to water and N_2 or higher nitrogen oxides. Despite this thermodynamic instability in an oxygen atmosphere, all biological fixation of nitrogen occurs to ammonia, which then enters the soil cycles of nitrification–denitrification; this phenomenon is described in more detail by Rosswall (1976) for instance, and by various authors in the collection edited by Nriagu (1976). Most soil nitrogen is stored as soil organic matter (generally humic substances). This organic nitrogen is mineralized to ammonium by microorganisms. Plants either use this NH_4^+ directly, or it is oxidized microbially to NO_3^- and then used. In most ecosystems the cycling process is very efficient, although some volatilization of NH_3 can occur. Anyone who has lived in a house with a cat or a baby has noticed ammonia released by microbial ammonification and is aware that this activity is both localized and sporadic and thus very difficult to quantify.

Rosswall (1976) suggests that the most active ecosystems with respect to atmospheric emissions are desert and highly fertilized grasslands, but both emit sporadically depending on water availability and fertilizer application. One of the problems facing an attempt to quantify the NH_3 flux from the soil to the atmosphere is the apparently trivial matter of defining the boundary. Denmead et al. (1976) show that large ammonia production rates from the ground can be almost completely absorbed by the plant cover. The soil flux is a strong function of soil temperature and moisture availability. The plant's ability to absorb NH_3 is a function of the plant's growth status and diurnally varying stomatal opening. Finally, the ability of the ammonia in the canopy to be released to the atmosphere is a function of atmospheric turbulence and wind and its ability to penetrate into the canopy. This array of unrelated random variables all controlling a local phenomenon makes any attempt to estimate global flux average over "appropriate" conditions apparently impossible. Dawson (1977) seems to have done a reasonable job of attempting this task. Either the flux must be estimated by balancing a global budget such as shown in Figure 4, or some flux measurement must be made that intrinsically averages over these variables.

4.1. Sources of Atmospheric NH_3 and NH_4^+

A first step in understanding the ammonia flux into the atmosphere is a general examination of likely sources. This discussion starts with the Northern Hemisphere and attempts to point out where the Southern Hemisphere will differ. Since NH_3 has an average atmospheric residence time of only a few days, interhemispheric transport can safely be neglected.

The Northern Hemispheric sources shown in Table 5 are derived from Söderlund and Svensson (1975) with some other sources included and alterations described below. The burning of coal in well-designed combustion chambers releases no ammonia; that which is released in the initial heating of the coal grains is subsequently efficiently burned to NO. Therefore, only poorly burned coal is likely to contribute NH_3 at all. Domestic animals

Table 5. Northern Hemisphere Sources of Atmospheric NH_3 and NH_4^+

Source	Flux ($Tg\ N\ yr^{-1}$)
Wild animal excrement	3[1.8][a]
Human excrement	1.5[2]
Domestic animals	23[1.4]
Industrial losses and fertilizer evaporation	3.5[1.3]
Burning of coal	<2
Soil emissions estimated by balancing the annual budget	51 (estimated by difference)
Total	82

[a] Estimated uncertainty factor as in Table 3.

herded intensively are mostly cattle and swine. These predominate in the Northern Hemisphere over the Southern Hemisphere even more than the land-area ratio. Loss of fertilizer by evaporation is estimated at 17–20 Tg N yr^{-1} of NH_3. This corresponds to other authors' estimates [e.g., the discussion in Liu et al. (1977)].

4.2. Concentration Measurements

There is no set of data on atmospheric concentrations of NH_4^+ or NH_3 adequate to permit performance of an accurate hemispheric (let alone global) average. Table 6 shows some observations from various authors and locations. From this we estimate, with considerable scope for uncertainty, an annual average hemispheric inventory of 0.3 Tg N as NH_3 and 0.6 Tg N as NH_4^+, using an estimated average scale height of 2.5 km (Stedman et al., 1975). This gives sea-level average concentrations of 1 ppb of NH_3 and 2 ppb (mole fraction) of NH_4^+. [The unit ppb (mole fraction) for an aerosol constituent directly corresponds to ppb (volume) for a gaseous constituent. For NH_4^+, 1 ppb (mole fraction) is equivalent to 0.75 $\mu g\ m^{-3}$ (20°C, 1 atm)]

4.3. Sinks for NH_3 and NH_4^+

The next problem is to estimate the sink term, which in this case is mainly rainout of NH_4^+ with additional contributions from dry deposition and reaction of NH_3 with OH.

Miller (1905) shows a remarkable set of data from 1860 to 1890 on rainout

Table 6. Observations of NH_4^+ and NH_3 Concentrations in the Atmosphere[a]

Location	Altitude	Season	Species	Mixing Ratio (ppb)[b]	Reference
Tasmania	Ground level		NH_3	0.12	Ayers and Gras (1980)
West Germany	Ground level	Summer	NH_3	6.3	Georgii and Lenhard (1978)
West Germany	Ground level	Winter	NH_3	2.5	Georgii and Lenhard (1978)
West Germany	2 km	Summer	NH_3	2.1	Georgii and Lenhard (1978)
West Germany	2 km	Winter	NH_3	2.1	Georgii and Lenhard (1978)
Tropical Ocean	Ground level		NH_3	2.5	Georgii and Gravenhorst (1977)
Philippines	8 km		NH_4^+	0.006–0.035	Cadle (1973)
West Germany	Ground level	Summer	NH_4^+	12.0	Georgii and Lenhard (1978)
West Germany	Ground level	Winter	NH_4^+	0.7–4	Georgii and Lenhard (1978)
Palestine, TX	18 km		NH_4^+	0.006	Lazrus et al. (1971)
Barbados	Ground		NH_4^+	0.008	Savoie and Prospero (1982)
Coastal Florida, USA	Ground		NH_4^+	0.12–0.18	Savoie and Prospero (1982)
Pt. Barrow, Canada	Clean air		NH_4^+	0.037	Barrie et al. (1981)
	Polluted air			0.08	Barrie et al. (1981)
Allegheny Mountains, USA			NH_4^+	5.6	Pierson et al. (1980)
Toronto, Canada	Clean air		NH_3	0.84	Vijan and Wood (1981)
Northwestern Pacific Ocean			NH_3	0.14–1.4	Tsunogai (1971)

[a] Most of these data are from rural or remote locations.

[b] Mixing ratios are ppb (volume) for NH_3 and the equivalent ppb (mole fraction) for NH_4^+.

of NH_4^+ and NO_3^-. His compilation shows a very consistent molar ratio of $NH_4^+:NO_3^-$ between 0.84 and 0.94 with a median of 0.88. The only noticeably outlying stations, out of about 30 sites, are New Zealand, Barbados, and British Guiana. Miller's data on average annual NH_4^+ deposition are not so consistent; they range from 0.12 to 1.3 g N m^{-2} yr^{-1} with an average of 0.5 g m^{-2} yr^{-1}. If this deposition rate is assumed for all the Northern Hemispheric land area and if the ratio of land: ocean NH_4^+ precipitation (2.4 globally) from Söderlund and Svensson is applied for evaluation of the ocean term, we obtain a Northern Hemispheric sink of 62 Tg N yr^{-1} (50 on land). If anything, this estimate may be somewhat high because rain collection in the 1800s included dry deposition and occasional avian excrement.

If we now turn to the most recent U.S. data, Wilson et al. (1980) measure annual NH_4^+ deposition with a range from 0.13 to 0.26 g N m^{-2} yr^{-1}, not very different from Miller is data (1905), although 100 yr later. Similarly, in Sweden Ångstrom and Högberg (1952) observe an average of 0.2 in 1947–1950, and the deposition rate at Rothampsted (UK) has been constant from 1860 to 1960 at 0.2–0.35 g N m^{-2} yr^{-1} (Brimblecome and Pitman, 1980). It appears that ammonium deposition has been constant over all the Northern Hemisphere for over 100 yr. Thus, despite the intensive herding of domestic animals, human activity has not recently dramatically increased the NH_4^+ deposition flux.

An alternative technique for estimation of flux terms is by means of ratios to other species. Wilson et al. (1980) measure SO_4^{2-}, NO_3^-, and NH_4^+ in rainfall at four stations in an area fairly representative of the northeastern United States. Their measured $SO_4^{2-}:NO_3^-$ deposition has a sulfur: nitrogen (S:N) ratio by mass of 3.2:1. The Electric Power Research Institute (EPRI, 1981) has gone to some trouble to determine the source inventory in this region and obtain an S:N ratio of 4.3:1 by mass. If we regard the similarity between 4.3 and 3.2 as evidence that rainout at least roughly reflects the source ratio for these short-lived species, the Wilson et al. (1980) $NO_3^-:NH_4^+$ mole ratio of 1.4 (contrasted with Miller's 1880 data, where this ratio was consistent at 0.14) can be used in conjunction with EPRI's NO_x source term (32×10^9 g NO_2 day^{-1}) to arrive at an estimated NH_3 source term of 1 g N cm^{-2} yr^{-1}. Considering that this is a source estimate over the northeastern United States, a land area where fertilizer is applied extensively and cattle are intensively raised, the agreement of all these estimates within a factor of 2 reinforces the suggestion that NH_4^+ has a steady source, little changed over 120 years. The unchanging long-term deposition rate at Rothampsted, UK (Brimblecombe and Pitman, 1980) is further evidence for this contention. These observations led to the suggestion that the Northern Hemispheric rainout rate of NH_4^+ is about 60 Tg N yr^{-1}, of which 50 Tg N yr^{-1} rains on the land. This value is shown in Figure 4.

The other losses to be considered are dry deposition and reaction with the OH free radical (McConnell and McElroy, 1973). Dry deposition is important only for gaseous ammonia. One can neglect dry deposition of

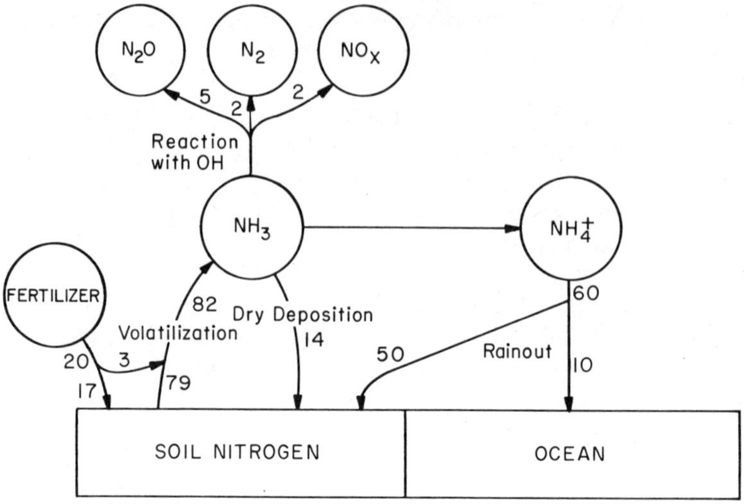

Figure 4. Budget of ammonia and ammonium in the Northern Hemisphere. Fluxes are in Tg N year^{-1}.

NH_4^+ since particulate diffusion coefficients are much lower than those of gases and since most NH_4^+ is in particles too small to show appreciable sedimentation. Dawson (1977) neglects dry deposition, but it is difficult to imagine how this can be correct in view of the canopy losses of NH_3 described by Denmead (1976); however, it is possible, in view of the correlation methods used by Dawson, that he actually calculates net emission rather than gross. We note also the good agreement between his latitudinally averaged emissions and Eriksson's (1952) estimated deposition by latitude. Dawson's prediction of a summer : winter deposition rate ratio of 3 : 1 in the latitude band 30–40°N and 7 : 1 in the band 40–50°N also correlates well with the observations by Wilson et al. (1980), who show a very consistent 4 : 1 summer : winter ratio in NH_4^+ deposition at an array of stations from 38–44°N and with recent studies of soil emission rates by Georgii and Lenhard (1978).

If, ammonia is readily absorbed by plant surfaces as observed, there will be a significant rate of dry deposition, at least in the growing season (the time of maximum emission). Extensive estimates of dry deposition of SO_2 and ozone, respectively, have been made by Sheih et al. (1979) and Galbally and Roy (1978). We believe that ammonia will be at least as efficiently removed by plant surfaces as SO_2, although contact with soil is necessarily a two-way process. If, as several authors suspect (Ayers and Gras, 1980; Georgii and Lenhard, 1978), the ocean surfaces are in equilibrium with atmospheric NH_3, there is no net flux except dry deposition of the excess NH_3 blowing off the land. We use an average saturated [NH_3] over the oceans of 0.5 ppb and an average over land of 1.5 ppb with [NH_4^+] equivalent to about three times [NH_3], and an NH_3 scale height of ~2 km, which has been observed by Georgii and Lenhard (1978) and corresponds to predictions concerning water-soluble gases by Stedman et al. (1975). In summer there is a fairly steady net high pressure over the oceans and on-shore flow. In winter, when the ammonia concentrations are significantly less (we estimate a factor of 4), there is average off-shore flow, but the net annual average dry deposition of <1 Tg N yr^{-1} is negligible. Over the land we use a deposition velocity of 1 cm s^{-1} to the canopy in summer and 0 in winter; thus on average we have 0.5 cm s^{-1} for a total flux of 14 Tg N yr^{-1}.

The reaction of NH_3 with OH has been regarded as a source of oxides of nitrogen (Crutzen, 1971; Logan et al., 1981; McConnel and McElroy, 1973). On the basis of recent data of Stuhl (1978) and rate constant data of Kurasawa and Lesclaux (1979), Logan et al. (1981) show that this reaction is a source if $[NO_x]$ is less than 60 ppt (parts per trillion, 10^{12}) and a sink if $[NO_x]$ is greater than 60 ppt.

The reaction mechanism for the gas-phase oxidation of NH_3 by OH radicals starts with the reaction

$$NH_3 + OH \rightarrow NH_2 + H_2O$$

having a rate constant (Perry et al., 1976)

$$k = 2.9 \times 10^{-12} \exp(-860/T) \text{ cm}^3 \text{ s}^{-1}$$

$$k_{298} = 1.4 \times 10^{-13} \text{ cm}^3 \text{ s}^{-1}.$$

The fate of NH_2 used to be unknown; however, because of the work of Lesclaux and DeMissy (1977), we know that the reaction with O_2 is negligible and that an interesting competition for NH_2 is set up between slow oxidation with relatively abundant ozone to products unknown (but that probably represent a net source of NO_x) and fast reactions with NO_2 and NO that are a net sink for NO_x (although a source of N_2O):

$$NH_2 + O_3 \rightarrow \text{products} \qquad k = 1.6 \times 10^{-14} \text{ cm}^3 \text{ s}^{-1}$$

$$NH_2 + NO_2 \rightarrow N_2O + H_2O \qquad k = 2.0 \times 10^{-11} \text{ cm}^3 \text{ s}^{-1}$$

$$NH_2 + NO \rightarrow N_2 + H_2O \qquad k = 2.3 \times 10^{-11} \text{ cm}^3 \text{ s}^{-1}$$

There is a factor of roughly 10^3 between the rate constants with O_3 and with NO_x; thus at 60 ppb O_3, 60 ppt NO_x is the break-even point between a NO_x source and a NO_x sink. To fully evaluate the effects of these processes, one needs a global correlation of O_3, NH_3, OH, and NO_x. In the absence of such data, we estimate a correlation that gives rise to a large continental NO_x sink and a small clean tropospheric NO_x source.

Hydroxyl reactions are most important in summer when NH_3 is at its highest concentration. The major source of NH_3 is emission from fertile soils into the lower boundary layer. A major component of anthropogenic NO_x input goes into the lower boundary layer, and there is also a non-negligible soil source of NO_x (Section 6.4). This situation conspires to make continental NH_3 a major sink for NO_x, whereas marine NH_3 is a minor source. This "tropospheric NO_x leveling" effect of NH_3 has not apparently been considered previously, and if the numbers given below are correct, it is a significant process in the budgets of both N_2O and NO_x as described in Sections 5 and 6. Thus, over the Northern Hemispheric summer landmass we have an average $[NH_3]$ of 3 ppb in the lowest 2.5 km and an average NO_x of not less than 300 ppt (Section 6.2).

Because of the diurnally varying nature of NH_3 emissions and OH concentration (Logan et al., 1981) an appropriate estimate for summer average [OH] is probably 4×10^6 cm^{-3}. With a rate constant of 1.6×10^{-13} $cm^3\ s^{-1}$, the time constant for NH_3 loss is 1.56×10^6 s (18 days) for a net loss of approximately 8 Tg N yr^{-1} NH_3 and 6 Tg N yr^{-1} of NO_x, making a net source of 10 Tg N yr^{-1} N_2O and 4 Tg N yr^{-1} as N_2. Since the reaction of OH with NO_2 is about 60 times faster than with NH_3, a considerable conversion of NO_2 to HNO_3 takes place while this conversion to N_2O occurs. Again, the situation is a competition, this time for the available OH radicals between NH_3 and NO_2.

In locations where $NO_x < 60$ ppt, any NH_3 in the gas phase will be an NO_x source. Thus Logan et al. (1981) estimate a remote tropospheric NO_x source of 1 Tg N yr^{-1}.

4.4. Atmospheric Lifetimes

As emphasized by Prospero et al. (1982), short-lived species such as aerosols and ammonia necessarily have variable atmospheric lifetimes because of the variability of precipitation events. However, the average lifetime of a species whose principal sink is rainout (NH_4^+) must be at least as long as the average period between precipitation events. This latter term is a strong function of latitude, topography, and the proximity of local water sources. On average, the estimated lifetime of water vapor is 3 days, so the NH_4^+ lifetime must be somewhat longer than that. Our estimated fluxes and inventories give an average lifetime for NH_4^+ of 3.6 days and a lifetime of gaseous NH_3 of 2.8 days. This last number is most uncertain because of our lack of certainty concerning the average ammonia inventory (see Table 6). The average and the statistical distributions of the atmospheric lifetimes of species subject to effective removal by precipitation are discussed in detail by Rodhe and Grandell (1972).

4.5. Other Reduced Nitrogen Compounds

Rainout data for "albumenoid nitrogen" is available for many of the agricultural stations reporting NH_4^+ and NO_3^- data. This material is presumably pollens, organic debris, and humic material associated with soil erosion. It does not enter the cycles of NH_3 or NO_x of concern here; however, it does redistribute fixed nitrogen around the biosphere and pass some into the oceans. The flux is considered small in comparison to river runoff (Söderlund and Svensson, 1975). Aliphatic and aromatic amines were considered by the National Research Council (1977). They decided that there was no evidence for other than occupational exposure to any of the compounds considered or for significant sources to the biosphere.

4.6. The Southern Hemisphere

There are very little data on Southern Hemispheric NH_3 or NH_4^+. In Western Tasmania, Baseline (1981) shows very low concentrations of NH_4^+ (and NO_3^-) in rain. Ayers and Gras (1980) show similarly low NH_3 concentrations. Presumably the tropical areas have a high NH_3 output similar to that of tropical areas measured in the Northern Hemisphere (Lodge et al., 1974). Overall, one would expect a smaller inventory and smaller rainout rate with similar atmospheric lifetimes, although there is little experimental evidence.

4.7. Historical Data

Brimblecombe and Stedman (1982) show that from 1850 to 1980 the deposition of NH_4^+ has been relatively constant whereas that of NO_3^- has increased by an order of magnitude. The increase is discussed in Section 6.5. The constant NH_4^+ implies that anthropogenic sources are relatively small compared to soil sources (as suggested herein and by other authors) and that the soil sources have not changed in the last 130 yr. If, as suggested by Stuiver (1978) and Bolin (1977), biomass has decreased 7% during that period, apparently there is not a strong effect on NH_3 emissions. This is in accord with Dawson's (1977) simplified model in which NH_3 emission is related to "potential biomass" through a rainfall correlation. It is not in accord with our suggestions herein that pastureland is a much more effective NH_3 emitter than forests; however, the status of our data is such that a 7% or even 14% effect would be masked by random noise.

One would expect an effect on the gaseous NH_3 flux and inventory arising from human activities. The everpresent acidic SO_4^{2-} serves to immobilize NH_3 as NH_4^+ in atmospheric liquid water and prevent its return to NH_3 on droplet evaporation. Much as normal ambient ozone of 20–40 ppb is often removed to zero in urban environments by NO emission, so will ambient NH_3 be removed to NH_4^+ by the excess acidic SO_4^{2-} (and to some extent HNO_3). This implies that present-day European and eastern U.S. NH_3 values are much less (and NH_4^+ greater) now than they would have been 100 yr ago. If we are correct that prehistoric forests, tall grasses, and scrub are more conservative of NH_3 than present crop and range land, then prehistoric NH_3 and NH_4^+ values would both have been significantly less than at present.

5. NITROUS OXIDE

Nitrous oxide was discovered in the atmosphere spectroscopically by Adel (1938). He also realized that its photolysis at high altitude would be a source of atmospheric NO. The present status of N_2O has been discussed in detail recently by Weiss (1981). This section places Weiss' work in the context of the global picture developed herein and stresses only those areas that warrant further discussion.

Despite previous controversy, there is presently little doubt that the N_2O average mixing ratio is close to 302 ppb (1400 Tg N), slowly increasing at about 0.5 ppb yr^{-1} (2.3 Tg yr^{-1}), with the Northern Hemisphere having about 0.7 ppb (3.2 Tg) more than the Southern.

If, as is indicated in Figure 5, diffusion to the stratosphere and subsequent photodissociation [or reaction with $O(^1D)$] are the only sinks for N_2O, the atmosphere lifetime can be determined by diffusion calculations based on three-dimensional winds, or parameterized vertical diffusion (usually based, in turn, on other "tracers"). Best estimates of the lifetime by this process range between 100 and 175 yr. Weiss (1981) chooses 150 yr, giving a total loss rate of 9.3 Tg N yr^{-1}. On the basis of observations of N_2O in combustion sources, discussions of the observed rate of increase, and north–south difference, Weiss (1981) estimates a combustion source of 3 Tg yr^{-1} increasing at 3.5% per year. This leaves only 4 Tg yr^{-1} for all other sources including agricultural and other soils and from all water bodies. This seems too small. Kaplan et al. (1978) estimate that polluted estuaries alone produce as much

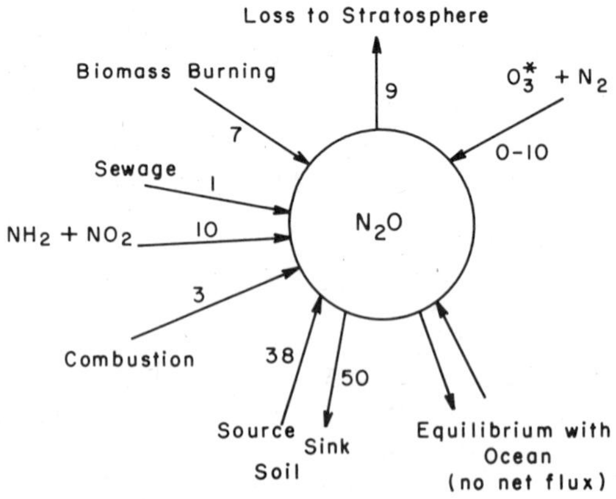

Figure 5. A possible balanced budget for present-day global tropospheric N_2O. Fluxes are Tg N $year^{-1}$.

Table 7. Estimated Sources and Sinks of N_2O

Source	Reference	Flux ($Tg\ N\ yr^{-1}$)
Soil source	Söderlund and Svensson (1976)	16–69
Biomass burning	Crutzen et al. (1979)	7
Sewage	Kaplan et al. (1978)	1
$NH_2 + NO_2$	This work	10[a]
$O_3^* + N_2$	This work[b]	10
Combustion for heat and power generation	Weiss (1981)	3
Oceanic source	Weiss (1981)[c]	0
Stratospheric sink	As discussed by Weiss (1981)	−9.3
Unbalanced sources		40–100

[a] This estimate should be regarded as an upper limit. This mechanism could be much less important than speculated here.

[b] By use of constants proposed by Prasad, again probably an upper limit.

[c] Weiss (1981) demonstrates that there is no *net* oceanic flux of N_2O.

as 1 Tg yr^{-1}, whereas Cohen and Gorden (1979) estimate 4–10 Tg yr^{-1}. We estimate in Section 4.3 that the sequence $NH_3 + OH \rightarrow NH_2 + H_2O$ followed by $NH_2 + NO_2 \rightarrow N_2O + H_2O$ produces a further 10 Tg yr^{-1} in the Northern Hemisphere alone, and Söderlund and Svensson (1975) estimate (albeit with considerable trepidation) that the soil source lies between 16 and 69 Tg yr^{-1}. Crutzen et al. (1979) estimate 7 Tg yr^{-1} from biomass burning, and there is a recent suggestion (Prasad, 1981) that the sequence

$$O + O_2 + M \rightarrow O_3^* + M$$

$$O_3^* + N_2 \rightarrow N_2O + O_2$$

is of some importance. Prasad shows its possible influence on the stratospheric N_2O budget but apparently neglects tropospheric oxygen atoms, which, with the use of his estimates of rate and quenching constants, would provide another 10 Tg N yr^{-1} of N_2O. Table 7 lists the known sources and sinks for N_2O. Seiler et al. (1978) have shown that, unlike the case for CO_2, growing plants are neither sources nor sinks.

McElroy et al. (1976) estimate an oceanic sink term of 50 Tg N yr^{-1}, a rate that neatly balances the budget in Table 7 and lowers the atmospheric lifetime of N_2O from 150 to about 24 yr. If Weiss is correct that a net oceanic sink of this magnitude does not exist, a soil sink of similar magnitude would be necessary to balance the source terms in Table 7, which do not seem to be discountable to a total as low as 9.3 Tg yr^{-1} as required by the Weiss (1981) budget. There is another argument against the oceanic sink. If the sink of 50 Tg yr^{-1} were assumed to be uniformly distributed between the Northern and Southern Hemispheres in proportion to their oceanic areas, the Southern Hemisphere sink would be 29 Tg yr^{-1} and the Northern Hemisphere sink would be 21 Tg yr^{-1}. Since the sources of N_2O from soil and from the NH_2 + NO_2 reaction are predominantly in the Northern Hemisphere, approximately 38 Tg yr^{-1} would be emitted in the Northern Hemisphere and 12 in the Southern Hemisphere, necessitating an interhemispheric flow of 17 Tg yr^{-1}. The observed difference in N_2O burden between the Northern and Southern Hemispheres is thought by Weiss (1981) to be 3.2 Tg N; Rasmussen and Peirotti (1978) estimate significantly less. Thus a 1-yr interhemispheric mixing time cannot support such a large flux, and the soil sink such as described by Delwiche (1978) is preferable since it occurs in the same hemisphere as the major source.

As the atmospheric lifetime is decreased by these considerations, it becomes proportionately more difficult to account for the difference in N_2O burden between the Northern and Southern Hemispheres and for the rate of increase in N_2O burden described by Weiss (1981). Fortunately, the NH_2 + NO_2 source, which we evaluated in Section 4.3, has certainly increased in the past courtesy of the increase in NO_x (Brimblecombe and Stedman, 1982). However, since the estimate depends on the triple correlation between NH_3, NO_2, and OH, more than the simple box modeling described here is required for estimation of an appropriate rate of increase. It is also possible that fertilizer use has increased the relatively large soil source. This has been discussed at length by others (CAST, 1976; Liu et al., 1977; McElroy et al., 1976) and is not treated further here.

Figure 5 shows a possible balanced budget for N_2O, based on a soil sink of 50 Tg N yr^{-1}. This gives a soil source of 38 Tg yr^{-1} if Prasad's (1981) novel path for N_2O formation is in fact negligible. If the constants are as large as he estimates, then the O_3^* + N_2 source becomes approximately 10 Tg yr^{-1}, and the soil source necessary to balance the budget must be decreased to only 28 Tg yr^{-1}.

These somewhat arbitrary (although reasonable) assumptions give a ratio of $N_2O:N_2$ ratio in global average denitrification of 38:155 = 0.25, similar to the value 0.23 estimated by Liu et al. (1977) for their parameterized 30-yr atmospheric lifetime of N_2O.

A final comment on N_2O is that if NO_x control is imposed on power plants, such control may well be achieved by an ammonia addition technique. It would be appropriate to determine carefully the yield of N_2O by this method, in case the control rids us of locally harmful tropospheric NO_x but contributes N_2O instead, ultimately a source of stratospheric NO_x.

6. HIGHER OXIDES OF NITROGEN

Figure 6 shows all the species and processes considered important to the tropospheric budget of the higher oxides of nitrogen. The sum of all the gas-phase species by nitrogen content (NO + NO_2 + NO_3 + HONO + $2N_2O_5$

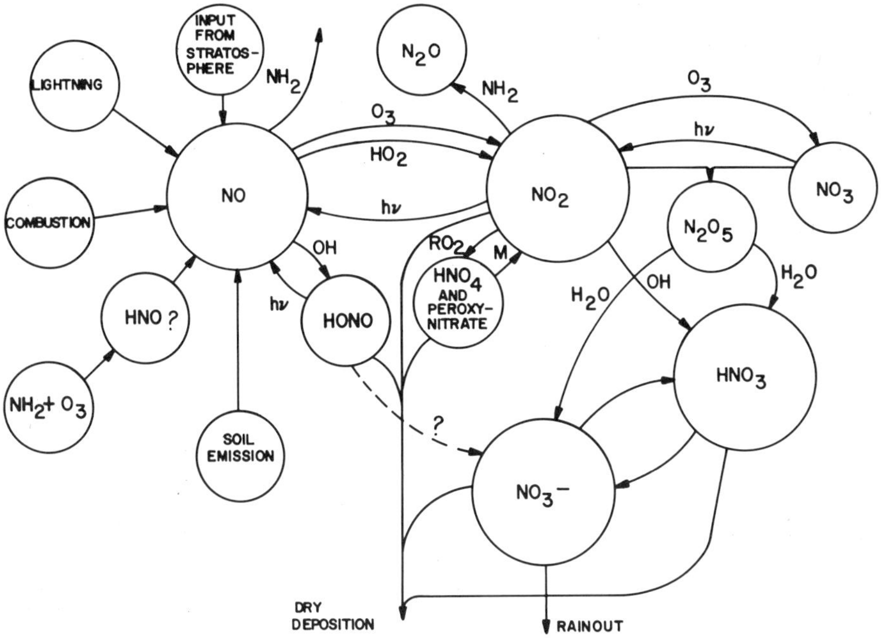

Figure 6. Schematic diagram of species and processes involved in the atmospheric cycles of the higher oxides of nitrogen.

+ HNO_3 + HNO_4) is often treated as a single variable NO_y (sometimes with NO_3^- added). During the day NO and NO_2 are coupled by the fast sequence of reactions

$$NO + O_3 \rightarrow NO_2 + O_2 \qquad [1]$$

$$NO_2 + h\nu \rightarrow NO + O \qquad [2]$$

$$O + O_2 + M \rightarrow O_3 + M, \qquad [3]$$

where the time constant for the coupling at noon is of the order of $j_{(NO_2)}^{-1}$, about 2 min (Harvey et al., 1977). The sum of NO + NO_2 (usually termed NO_x) is also frequently treated as a single variable. It is apparent from Figure 6 that most of the sources of NO_x are as NO (note, however, that EPA emission estimates are expressed by mass as NO_2).

Several recent discussions of tropospheric NO_x and of NO_x budgets have appeared (Bottenheim and Strausz, 1980; Lee and Schwartz, 1981; Liu et al., 1980; Logan et al., 1981), and new experimental data have been presented (Baseline, 1981; Cadle et al., 1982; Huebert, 1980; Huebert and Lazrus, 1980a,b; Kelly et al., 1980; Kley et al., 1981; Noxon, 1978, 1980, 1981; Platt and Perner, 1980; Schiff et al., 1979). Bauer (1982) has reviewed the literature on all sources of nitrogen oxides to the troposphere as a function of altitude and latitude. Bauer and Kowalczyk (1982) and Hameed et al. (1981) have discussed the latitude dependence of the lightning source. This flurry of activity has caused the review by National Research Council, (1978) to become dated, since that review accepts the erroneously large soil source originally proposed by Robinson and Robbins (1970), which has subsequently been shown to be incorrect by several authors (e.g., Dawson, 1977; Söderlund and Svensson, 1976).

The major pathway in Figure 6 starts with emissions of NO that are oxidized by ambient O_3 to NO_2. This NO_2, in turn, is oxidized diurnally by OH to HNO_3 and nocturnally to NO_3 (possibly terminating as HNO_3 or NO_3^-). The HNO_3 is then removed as NO_3^- in precipitation. A number of return paths and rapidly diurnally cycling loops (Figure 6) are the subject of active research. Some of these are discussed in the succeeding sections.

6.1. NO_2:NO Ratios

The three species NO, NO_2, and O_3 are involved in the rapid reactions [1]–[3]; of these, NO is generally present in the lowest concentration. Thus NO is the most appropriate species to test for local steady state. If only these three reactions are considered, then at steady state

$$\frac{d[\mathrm{NO}]}{dt} = j_{(\mathrm{NO_2})}[\mathrm{NO_2}] - k_3[\mathrm{O_3}][\mathrm{NO}] = 0.$$

One would then predict that the variable $p \equiv j_{(\mathrm{NO_2})}[\mathrm{NO_2}]/k_3[\mathrm{O_3}][\mathrm{NO}]$ would be constant and equal to unity. If other sources and sinks for NO are included, then

$$\frac{d[\mathrm{NO}]}{dt} = j_{(\mathrm{NO_2})}[\mathrm{NO_2}] + \text{sources} - k_3[\mathrm{O_3}][\mathrm{NO_2}] - \text{sinks}.$$

We know of no rapid sources of NO after the morning hours, and the only other possibly competitive process is the reaction

$$\mathrm{HO_2(RO_2)} + \mathrm{NO} \rightarrow \mathrm{HO(RO)} + \mathrm{NO_2}. \qquad [4]$$

Thus when [NO] is observed to be constant, the equation

$$\frac{d[\mathrm{NO}]}{dt} = j_{(\mathrm{NO_2})}[\mathrm{NO_2}] - k_3[\mathrm{O_2}][\mathrm{NO}] - k_4[\mathrm{HO_2}][\mathrm{NO}] = 0$$

should hold, provided the HO_2 term includes all possible RO_2 species. On the basis of this equation, if $p > 1$ is observed, a value of $[HO_2]$ may be predicted as $[HO_2] = [O_3](p - 1)k_3/k_4$. Since k_3/k_4 is approximately 2×10^{-3}, it follows that for $p = 2$, if $[O_3] = 40$ ppb, $[HO_2]$ would be 80 ppt. Logan et al. (1981) predict values closer to 10 ppt. Our group carried out the first measurements of the $[NO_2]$:[NO] ratio and other appropriate photochemical parameters ($j_{(NO_2)}$, $[O_3]$ and temperature) in 1973 in downtown Detroit (Stedman and Jackson, 1975). The observed and predicted ratios were apparently similar, without the need to include $HO_2(RO_2)$, although the data showed considerable scatter. We have significantly improved the instrumentation, and now in clean air (Kelly et al., 1980; Ritter et al., 1978) and in the urban environment (Stedman et al., 1981) find $[NO_2]:[NO]$ ratios considerably higher than predicted by the simple photochemical equilibrium described by reactions [1]–[3]. Other authors note similar discrepancies in clean air. Drummond (1977) shows $[NO_2]:[NO]$ ratios some three times larger than expected in clean air in Wyoming; Helas and Warneck (1981), find $[NO_2]:[NO] > 10$ in clean air over the northern Atlantic, and we have similar unpublished data from a first-class clean air site (Mt. John) in New Zealand.

There are three possible sources for this discrepancy: (1) there is a much larger than expected density of peroxy free radicals $HO_2(RO_2)$ that compete with ozone in the oxidation of NO to NO_2; (2) the NO_2 measurements are too high; or (3) the NO measurements are too low. Some models of urban air (Calvert, 1981) and clean air (Bottenheim and Strausz, 1980) predict large concentrations of $HO_2(RO_2)$. Bottenheim and Strausz predict a value of $p = 1.9$ in clean air at 55°N, closer to agreement with measurements. Calvert predicts p values from 1 up to >10 at low [NO] in urban air. This is comparable to our recently measured values (Stedman et al., 1981) of ~2.0. Nevertheless, few models of clean air predict the very large ratios of $[NO_2]:[NO]$ often observed in clean air [Helas and Warneck (1981); Kelly et al. (1980); and our unpublished work in New Zealand]. A graphic description of the discrepancies between measured and modeled ratios is given by Pommereau (1981), who compares the measured ratios given by Kelly et al. of $[NO_2]:[HNO_3]$, 5–8, and $[NO_2]:[NO]$, 4–10, with values given by Crutzen's (1979) model, which predicts 0.04 and 1, respectively.

All measurements of nitric oxide are made with chemiluminescent detectors. These devices and some of their problems are discussed in detail elsewhere (Kley and McFarland, 1980; Ridley and Hastie, 1981). The air sample is sucked through a flow-controlling device into a reactor where it mixes with a flow of excess ozone. The reaction $NO + O_3 \rightarrow NO_2 + O_2$ generates red light, which is measured photometrically. Chemiluminescent instruments suffer from three major problems. There is a variable "background" light intensity generated by the reaction of ozone presumably with dirt on the walls of the reactor; hence there is a need to constantly update the zero reading. The instrument necessarily has an inlet system, and surface reactions on that system have not been fully evaluated. Finally, to measure the more abundant species NO_2 or HNO_3 with the same instrument, one needs to convert these species to NO. A number of converters have been discussed in the literature. Surely the most unequivocal is the photochemical device described by Kley and McFarland (1980). It is, however, nontrivial to implement. Those who use other types of converter (Cox, 1977; Helas and Warneck, 1981; Kelly et al., 1980) are bound to be open to the criticism that some other molecule in the atmosphere (which has not been tested for in the screening process) is converted to either NO or another species that chemiluminesces with ozone. It is worth noting, however, that measurements of the $[NO_2]:[NO]$ ratio in the lower stratosphere (Ridley and Hastie, 1981), where NO_2 has been measured spectroscopically, again exceed expected values.

On the basis of these arguments, we do not believe that the many published NO_2 measurements in clean and dirty air are incorrect, nor that the models with which they disagree are wrong, but propose rather that low concentrations of NO may well be undermeasured by a chemiluminescent NO detector. This untested hypothesis is based on the observation that an extraordinary array of presumably highly surface active oxidizing agents are predicted to be present in clean air. Thus at ground level 45°N at noon, Logan et al. (1981) predict $[NO] = 3.6 \times 10^8$ molecule cm^{-3}, $[HO_2] = 3.2 \times 10^8\ cm^{-3}$ and $[H_2O_2] = 3.0 \times 10^{10}\ cm^{-3}$. Present knowledge of reactions of HO_2 or H_2O_2 does not rule out the suggestion that these compounds sit on the surface waiting to oxidize an unwary NO passing by. Similarly, at 6 km, 45°N, Logan et al. predict $HNO_4 > NO$ by a factor of 4. The effect of this material on the surface of the intake system is also unknown. If we are to continue to use chemiluminescent NO detectors for trace tropospheric measurements, apparently some very sophisticated laboratory tests are needed.

6.2. Measured and Modeled Concentrations

Despite the caveats given in the previous section, chemiluminescent measurements have helped our understanding of the nitrogen cycle by providing data that must be upper limits on $[NO_x]$ and that are much lower than estimates used in the early 1970s. There is the potential for undermeasurement of NO (as discussed in Section 6.3), which cannot be eliminated unless considerable testing is done. There is the possibility of overmeasuring NO_2 if an unknown and unexpected species interferes, but recent intercomparisons in downtown Los Angeles between our chemiluminescent technique and an unequivocal laser diode absorption system showed that most of the differences could be accounted for by the expected interference from PAN (Mackay and Schiff, 1982). Furthermore, the chemiluminescent systems also measured HNO_3 by difference after absorption on nylon wool. This measurement correlated very well with values obtained with the laser diode system, indicating that even in this extremely complex atmospheric soup, no unexpected interfering species were present.

Three other measurement techniques have contributed greatly to our understanding of the trophospheric concentrations and cycles of nitrogen oxides. These are HNO_3 filter collections, as discussed in detail by Huebert (1980) and Huebert and Lazrus (1980a,b); long path UV–visible absorption with the use of artificial light sources, as discussed in detail by Platt et al. (1979); and long path UV–visible absorption with the sun, moon, and sky as light sources, as discussed by Noxon (1978, 1980, 1981).

Noxon's most recent contribution shows NO_2 in the mid-Pacific troposphere averaging 30 ppt. Platt and Perner (1980) report typical sea-level values of 200 ppt, similar to the oceanic readings by Helas and Warneck (1981) with the use of chemiluminescence, and the mountaintop values of Kelly et al. (1980) also with chemiluminescence. The cleanest air we observed in New Zealand in late April 1981 [NO_x] averaged 100 ppt, with minimum excursion to 30 and the maximum excursion to 250 ppt. Baseline (1981) shows sea-level NO_x data from the Southern Hemisphere all between 100 and 200 ppt. If all these data are accepted at face value, a ground-level NO_x average concentration of 150 ppt is probably appropriate, with the caveat that strong source areas such as the United States and Europe should have much larger values. EPRI SURE (1981), averaging several stations in the northeastern United States, show a median NO_x of 6 ppb, 40 times the clean air New Zealand readings. One could attribute this factor all to anthropogenic emissions. This may not be correct since NO_x measured at the ground is a strong function of the nature of the site location. Even if the site is far from anthropogenic sources, the nature of the local terrain and its potential for soil emission and stagnant nocturnal inversions must be evaluated before mean data can be directly compared. At another apparently equally clean New Zealand site (except for a surrounding fertile valley) we observed a ground-level mean NO_2 concentration closer to 1 ppb with a nocturnal maximum. Since the EPRI sites are mostly valley sites, their 6-ppb median may be subject to similar caveats. Thus the anthropogenic New Zealand–United States difference apparently is a factor of 6–40. We expect a factor of 10–12 on the basis of historical deposition data (Brimblecombe and Stedman, 1982).

Huebert and Lazrus (1980a,b) and Huebert (1980) have made an extremely useful set of HNO_3 measurements. They find 40–80 ppt of HNO_3 in the clean lower troposphere and larger values in both the continental lower troposphere and the upper troposphere (100 ppt). This apparent contradiction with our prediction of the vertical profiles of soluble gases (Stedman et al., 1975) is discussed in Section 6.6.

Savoie and Prospero (1982), Huebert (1980) and Huebert and Lazrus (1978, 1980a,b) all report typical ground-level particulate NO_3^- concentrations of 80–160 ppt (1 ppb = 2.58 $\mu g\ NO_3^-\ m^{-3}$; 20°C, 1 atm). Clean air in Barrow, Alaska apparently contains 4 ppt NO_3^- (Barrie et al., 1981); however, the advent of long-range polluted air raises this value to 40–80 ppt. There has been considerable discussion concerning the measurement of particulate and vapor-phase NO_3^-, in view of the volatility of HNO_3 in acidic aerosols (negative NO_3^- artifact) and the tendency of HNO_3 to stick to surfaces (positivc NO_3^- artifact). This problem is discussed in more detail by Anderson in Chapter 9 of this volume. For this reason, it is necessary only to note that measurements of the type described by Huebert (1980), where both NO_3^- and HNO_3 are quantitatively collected, at least give a correct total $HNO_3 + NO_3^-$. If the $NO_3^- \rightleftharpoons HNO_3$ vaporization condensation equilibrium is as rapid as might be expected, perhaps this sum is the appropriate quantity to use for atmospheric chemical calculations. The conclusions by Savoie and Prospero (1982) that NO_3^- migrates from smaller to larger particle surface in the oceanic aerosol support this contention and imply that all measurement systems should attempt to measure at least the total $NO_3^- + HNO_3$ in case their attempt at speciation between gas and particle turns out to be incorrect.

Although estimation of average concentrations of species having such highly variable concentrations is of limited usefulness for budget considerations, we nonetheless present such estimates in Table 8. Because of the influence of the recent measurements, the total inventory is almost an order of magnitude less than the estimate by Söderlund and Svensson (1976), although the data used to generate Table 8 are too sparse and uncertain to be used in any detail. The implication that more than 20% of the inventory is in the atmosphere over only 4% of the surface area seems reasonable.

6.3. Other NO_y Species

Using Hampson's (1980) rate constants for formation and thermal decomposition of HNO_4, Logan et al. (1981), find this species to be an important NO_y species in the colder regions of the troposphere. Pommereau (1981) speculates that photodissociation may be fairly rapid, preventing a large concentration buildup. The role of HNO_4 as a NO_y species (and potential interferent in NO measurements) has yet to be fully evaluated, although it

Table 8. Assumed Average Concentrations of NO_y Species for Present Global Budget Calculations

Species	Assumptions	Concentration (ppt[a])	Inventory (Tg N)
NO_2	Uniform with altitude	50	0.1
NO	Diurnal average	5	0.01
HNO_3	Uniform with altitude	70	0.14
NO_3^-	Scale height 3 km	120	0.1
NO_x (anthropogenic)	Midlatitude populous continental land masses (2.1×10^{13} m^2), scale height 2 km	4000	0.1
Total			0.45

[a] Parts per trillion (10^{12}) by volume for NO_2, NO, and HNO_3, and the equivalent ppt (mole fraction) for NO_3^-.

certainly is highly soluble in water and yields H_2O_2 and HNO_3 on solution (Molina, 1982).

Nitrous acid (HONO) has been detected in urban atmospheres (Platt et al., 1980), although it has never been observed in clean air. This species is rapidly photolyzed by day to OH and NO, but it builds up by night by an unknown and presumably heterogeneous mechanism. Lee and Schwartz (1981) show that the rate of formation of HONO by the NO + NO_2 + $H_2O(l)$ mechanism is slow (see also Schwartz and White, this volume, Chapter 1); nevertheless, Thompson and Heikes (1982) suggest that nocturnal formation of HONO in clouds could be significant both to the NO_y and odd hydrogen free-radical budgets in clean air.

6.4. Sources of NO_y

The only source term about which there is little controversy is the anthropogenic input of 20 Tg N yr^{-1}, mostly in the midlatitudes of the Northern Hemisphere. Of this, about 50% derives from domestic and transportation sources and thus is emitted very close to ground level. The remaining 50% is industrial- and utility-generated. A large fraction of that emission is injected into the boundary layer with sufficient stack height plus plume rise that under stable conditions impact with the surface (potentially resulting in dry deposition) can be long delayed (Husar and Patterson, 1980).

The NO_y source arising from intentional biomass burning has been estimated to be as large as 19 Tg N yr^{-1} (Crutzen et al., 1979), but, as discussed in Section 3.3, this may be overestimated by as much as a factor of 4.

The other sources shown in Figure 6 are somewhat more controversial. The input from the stratosphere is estimated to be no larger than 1 Tg N yr^{-1}, mostly as HNO_3 (Liu et al., 1980). The lightning source has been discussed a great deal recently and, largely because of our present lack of knowledge of lightning statistics, is still very uncertain. The lowest recent estimates are 2–3 Tg yr^{-1} (Dawson, 1980; Levine, et al., 1981). This source is distributed with a maximum at the equator and at moderate altitudes (Hameed et al., 1981).

Bauer (1982) estimates soil NO_x emissions at 10 Tg N yr^{-1}, mostly from the more fertile and populous areas. Again, this emission is necessarily a ground-level source potentially subject to immediate dry deposition. Söderlund and Svensson (1975) estimate 1–14 Tg yr^{-1} as a soil source, but to balance their budget, show the total nonanthropogenic emission rate (otherwise unassigned) as 21–89 Tg yr^{-1}.

The question of oxidation of NH_3 as a source or sink for NO_x has been discussed earlier (Section 4.3). Since the source by way of NH_2 + O_3 is important only in the remote troposphere where $NO_x < 60$ ppt and NH_3 concentration is also very low, we estimate that this source cannot be larger than 1 Tg yr^{-1}. Figure 6 shows this source appearing as NO, but the reaction mechanism between NH_2 and O_3 is at present unknown (Kurasawa and Lesclaux, 1980). On the basis of analogy with our studies of PH_3, AsH_3, and SbH_3, a likely reaction would be

$$NH_2 + O_3 \rightarrow HO_2 + HNO$$

Since HNO does not react with O_2, it is possible that it could build up to spectroscopically detectable levels before reacting with any other molecule.

These estimates for the total of the above NO_y sources to the atmosphere are summarized in Table 9.

Aurela and Punkkinen (1981) have studied and continue to study the phenomenon of nitrogen oxide generation by means of electrical discharges at the points of pine needles. They believe that this process contributes nitrogen to the otherwise nitrogen-poor subarctic pine forests. If their observations are confirmed, this could be a significant source of NO_x and most importantly would imply that nitrogen oxide emissions (at least in some locations) could be ecologically beneficial.

The model described by Logan et al. (1981) requires a uniformly distributed source of NO_x to the clean troposphere of magnitude 10 Tg N yr^{-1}. Such a source strength is not incompatible with Table 9 provided the anthropogenic emissions are locally removed; however, Logan et al. use a rainout parameterization that gives an HNO_3 lifetime of 10 days, far too long for local removal.

We propose a soil source of NO of 10 Tg N yr^{-1} and of N_2O of 38 Tg N yr^{-1}. This is not in good agreement with the contention by Goreau et al. (1980) that equal quantities of NO and N_2O are emitted from soils, although our NO flux is in reasonable agreement with measurements by Galbally and Roy (1978). Since the necessary equipment is readily available, perhaps more studies of the N_2O:NO ratio in soil emissions would be valuable.

6.5. Sinks of NO_y

Söderlund and Svensson (1976) consider only two major sinks for tropospheric NO_y, dry deposition and rainout. They make use of Eriksson's (1952) data to estimate continental wet deposition and Söderlund's (1976) estimate of oceanic wet deposition (6–17 Tg N $year^{-1}$). If anything, this last value could be high because of continental influence. Unfortunately, Eriksson's estimates are based largely on rainfall analyses performed between 1880 and 1930. Analysis of recent rainfall NO_3^- data (Wilson et al., 1980) indicates that there has been a very significant increase in the NO_3^- component of rainfall. This leads us to suggest that the continental rainout rates derived from Eriksson's work (13–30 Tg yr^{-1}) may be somewhat underestimated.

The dry deposition rate estimated by Söderlund and Svensson (19–53 Tg N yr^{-1} over land and 6–17 over the oceans) is based on an assumed deposition velocity and much higher average concentrations than we now know

Table 9. Summary of Estimated of Sources of Atmospheric NO_y

Source	Flux (Tg N yr^{-1})
Input from the stratosphere	1[2][a]
Lightning	3[2]
$NH_2 + O_3$	1[2]
Combustion for energy	20[1.1]
Biomass burning	5[3]
Soil emission	10[2]
Total	40

[a] Estimated uncertainty factor as in Table 3.

to be present. Anderson (Chapter 9) recalculates continental and oceanic dry deposition to be 1–4 and 2–13 Tg yr^{-1} respectively, on the basis of modern values for [NO_x] in clean air. The studies of Lee and Schwartz (1981) suggest that dry deposition of NO_2 to water surfaces is much slower than had hitherto been expected. This understanding reduces our estimated oceanic dry deposition rate below Anderson's to 1–7 Tg yr^{-1}.

A third sink that we introduce into our calculations is the reaction $NH_2 + NO_2 \rightarrow N_2O + H_2O$. If our estimate (Section 4.3) of the correlation of OH, NH_3, and NO_2 in continental areas is correct, this is a further sink of 6 Tg N yr^{-1}.

The rates of the sink processes for atmospheric NO_x are summarized in Table 10.

The distribution of continental and oceanic sources and sinks shown in Tables 9 and 10 implies that 5 Tg N yr^{-1} is transported from continental sources to oceanic sinks. This amount of transport, which represents 15%

Table 10. Estimated Sinks for Atmospheric NO_y Species

Process		Flux (Tg N yr^{-1})[a]	Flux (Tg N yr^{-1})[b]
Rainout	Oceanic	6–17	7
	Continental	13–30	22
Dry deposition	Oceanic	1–7	3
	Continental	1–4	3
$NH_2 + NO_2$ reaction	Continental	6	6
Total		28–64	40

[a] Estimated values based on arguments given in the text.

[b] Estimated values based on the need to achieve a balance with the source terms. However, it must be emphasized that both source and sink rates have a considerable degree of uncertainty.

of the continental output, seems reasonable in view of the large populations of coastal areas. Approximate calculations of the relative contributions of sources and sinks in the Northern and Southern Hemispheres indicate sources and sinks of about 30 Tg yr^{-1} in the Northern Hemisphere and 10 in the Southern, further emphasizing the dominance of anthropogenic sources.

The observation of significantly increasing NO_3^- deposition over the last century (Brimblecombe and Stedman, 1982) clearly identifies humans as major contributors to atmospheric NO_x and sets an upper limit on the non-anthropogenic NO_x source strength close to the value given in Table 9.

6.6. Vertical Distribution and Lifetimes

On the basis of sources, sinks, and estimated concentrations in Tables 8–10, an average atmospheric lifetime of NO_y is obtained of 0.45 Tg/40 Tg yr^{-1} = 3 days, 75% by wet deposition (with a lifetime of about 4 days), and the rest by dry deposition and by reaction of NO_2 with NH_2. Such a value for a rainout lifetime would appear to be reasonable but close to the lower possible limit (Rodhe and Grandell, 1972). Unfortunately, if one considers the individual components HNO_3 and NO_3^-, which are regarded to have a negligibly slow return rate to the coupled pair NO–NO_2 (Logan et al., 1981), the required rainout lifetime becomes even shorter (<3 days). Similarly, the lifetime of NO_x is calculated to average only 1.8 days, with the major loss presumably the reaction with OH to generate HNO_3. We note that Logan et al. (1981) predict OH concentrations that support such a short lifetime.

The observed decrease in concentrations of all NO_x species (Huebert and Lazrus, 1980a,b; Kley et al., 1981) above the continental boundary layer is as expected for such a short-lived species with a major source at the ground. The vertical profiles of soluble gases discussed by Stedman et al. (1975) predict that all soluble materials should be transported downward by the same mechanism that vertically transports H_2O and that those that re-evaporate should match the observed H_2O distribution with a scale height of 2–3 km. this theory appears to explain adequately the observed vertical profiles of NH_3; however, it also predicts an increasing atmospheric lifetime $\tau(z)$ with altitude such that at 10 km the lifetime increases by approximately one hundredfold. For this reason the 1 Tg N yr^{-1} injected from the stratosphere has a disproportionately large influence on the global concentration patterns, relative to 39 Tg yr^{-1} from lower-altitude sources. Perhaps this is a reason why the apparently small estimated tropospheric input from high-flying aircraft (0.05 Tg yr^{-1}) (Bauer, 1982) cannot be neglected.

If the oceanic areas are only a sink for NO_x, one would expect the observed decrease in NO_y with decreasing altitude over remote oceanic areas (Huebert, 1980; McFarland et al., 1979).

There are, however, some problems with the lifetime estimations advanced here. Consideration of the atmospheric lifetimes (NO_2 roughly 2 days; $HNO_3 + NO_3^-$ roughly 3 days) leads one to expect that after more than 4 days of travel away from the source regions, HNO_3 and NO_3^- should markedly predominate over NO_2. This expectation is consistent with the averages estimated in Table 8 but has never been observed in any simultaneous determinations of HNO_3 and NO_2. In New Zealand we observed NO_2 frequently at approximately 100 ppt but HNO_3 rarely above 20 ppt. Kelly et al. (1980) report similar observations from western Colorado. Even if the NO_y were allowed to reach a steady-state flow through from NO_2 to HNO_3, the predicted ratio would be $[HNO_3]:[NO_2]$ roughly 1.5 : 1, again much larger than measured. Another problem is that the average lifetime of HNO_3 proposed here is 3.4 days, compared to the parameterized value of 10 days proposed by Logan et al. (1981). For resolution of this discrepancy, continental emissions would have to have an even shorter lifetime in order to allow for the longer lifetime estimated by Logan et al. (1981) for clean air. Some observations of free-radical NO_3 described in Section 6.7 may shed light on this apparent problem.

6.7. Diurnal Variations

The median diurnal variation observed in clean air at Mt. John, New Zealand (45°S, 500 m MSL) in March and April of 1981 is shown in Figure 7. The bars indicate the range of 70% of the data for 15 days where the NO_x maximum was less than 200 ppt. Measurable NO and HNO_3 were rarely

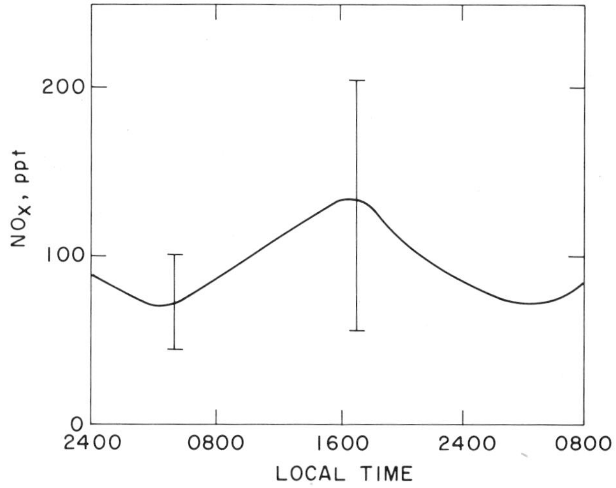

Figure 7. Average diurnal variation of NO_x at Mt. John, New Zealand on 15 days of uniformly clean air (NO_x maximum 200 ppt in March and April 1981). The line shows the median values; the bars enclose 70% of the observations.

observed during this period even though the instrumental detection limit was as low as 6 ppt for NO and 15 for HNO_3 (at times of slow fluctuations in NO_2). We observed a slow nocturnal NO_x decrease to a relatively noise-free predawn minimum, followed by a rise to a much noisier evening maximum. This is common in many published ground-level diurnal variations in clean air (Helas and Warneck, 1981; Kelly et al., 1980). This shape is very similar to simultaneously measured curves of ambient temperature (the average nocturnal temperature in the New Zealand study shown in Figure 7 was 7°C, peaking in the early evening at 10–15°C). The steady decrease at night would be expected on the basis of the reaction between NO_2 and O_3:

$$NO_2 + O_3 \rightarrow NO_3 + O_2;$$

the rate constant of this reaction is given (Hampson, 1980) as 1.9×10^{-17} $cm^3\ s^{-1}$ at 7°C. Ozone at Mt. John during this period was approximately constant and averaged 20 ppb; thus the expected decay lifetime of NO_2 is 29 h. This expected value is about 1.7 times longer than the observations; nevertheless, this is sufficiently similar to make the reaction with O_3 rather than dry deposition the probable fate of NO_2.

Model calculations by Bottenheim and Strausz (1980) and Graedel (1979) show a jump in NO_x at dawn in summer; this has not been observed by any investigators. In the winter Bottenheim and Strausz show a slow, evening-peaking NO_x variation in winter (although their peak is only 10% above their average). Logan et al. (1981) (at 45°N and 0 km) predict HNO_3 constant at 30 ppt and slightly larger than average NO_2 (about 20 ppt), with NO_x showing a dawn peak from N_2O_5 decomposition. We would certainly have observed HNO_3 with our system in New Zealand had it either averaged 30 ppt or been greater than $[NO_2]$.

Spectroscopic observations of free-radical NO_3 also show a discrepancy between modeling and measurements. Noxon et al. (1980) observed less NO_3 buildup than expected and was forced to postulate an unexplained tropospheric sink for NO_3 (or for N_2O_5, with which it reaches a rapid equilibrium). Platt et al. (1981) similarly observed less buildup of NO_3 than would be expected on the basis of the rate of loss of NO_2 and the known equilibrium constant for the reaction

$$NO_2 + NO_3 \rightleftharpoons N_2O_5.$$

They reevaluate this constant to $2.0 \times 10^{10}\ cm^{-3}$ at 10°C but still are left with an unexplained and rapid sink for NO_3 or N_2O_5. They suggest either rapid loss of N_2O_5 to aerosols as NO_3^- or possible recycling reactions of NO_3 such as

$$NO_3 + NO_2 \rightarrow NO + NO_2 + O_2$$

or

$$NO_3 + NO_3 \rightarrow NO + NO_2 + O_2$$

proceeding at rates much faster than measured by Graham and Johnston (1978). Unfortunately, neither solution is entirely satisfactory; the former would predict a rapid formation of NO_3^- aerosol, and thus NO_2 would behave like SO_2 and completely convert to aerosol. The fact that the N:S mole ratio in emissions in the northeastern United States is 0.5 and the observed $NO_3^-:SO_4^{2-}$ mole ratio in aerosols is about 0.08 (EPRI, 1981) makes this solution unlikely.

A potential solution to this problem would be the reaction

$$NO_3 + H_2O(l) \rightarrow H_2O_2 + NO_2. \quad [5]$$

This would not only provide the necessary sink for NO_3, but could also provide the mysterious source of H_2O_2 in air bubbled through water solutions (Heikes et al., 1982). According to present thermodynamic estimates reaction [5] apparently is at best 2 kcal mol^{-1} endothermic, but the thermodynamics of NO_3 (JANAF, 1971) are apparently subject to considerable uncertainty. A second possibility is HNO_4, which has already been shown to produce H_2O_2 in solution (Molina, 1982).

Another possibility is that the reactive hydrocarbons emitted from all anthropogenic sources except utilities in an approproximate mole ratio to NO_x of 2:1 (EPRI, 1981) are sufficiently reactive with NO_3 to remove it, possibly to organic nitrates; if so, however, these nitrates remain unidentified in the atmosphere. If this turns out to be the sink of NO_3 and if the sink is permanent and local, only the relatively hydrocarbon-free utility inputs would contribute to any long range impacts of NO_y in the atmosphere.

It is not clear what causes the extra variability and large signal associated with the late-afternoon NO_x peak. Helas and Warneck (1981) point out that their observed NO_x fluctuations can be interpreted as an air-mass phenomenon only if one can account for a time constant as short as 3 h. They observe an excess NO_x signal from a hot molybdenum converter over one obtained with a ferrous sulfate converter, which they tentatively interpret as HNO_3, but still cannot account for the afternoon peak and variability. The variability implies some sort of local source, but a local soil source would not be observed at a windy coastal station (Helas and Warneck, 1981) and would be expected not to peak in the evening, but rather later at night as a nocturnal inversion builds up. We observed a late-night peaking of NO_x at a fertile valley site south of Mt. John in New Zealand and interpret this as evidence for a local soil source.

This observation and similar observations at sites in the United States imply that site selection will be particularly crucial for NO_x measurements. The transport source can significantly pollute a thick boundary layer. Even if the site is located far from roads, NO_y from the soil source may dominate nocturnal measurements if fertile fields are close by. Thus if NO_x values are to be representative of a large airmass, probably only mountaintop sites can be used. This effect of site location can be seen most clearly in the SO_2 and NO_x data from EPRI (1981). The frequency distribution of NO_x from their clean air sites is almost lognormal, with a median of 6 ppb. Their SO_2 distribution shows a sharp downturn at about the sixtieth percentile with 25% of the data reading essentially zero. These SO_2 zeros arise when the nocturnal inversion cuts off the sensor from SO_2, most of which is emitted aloft. On the other hand, NO_x has several ground-level sources as discussed above, and thus zero readings are not observed even at night. However, since the nocturnal NO_x data are local in origin, it is then difficult to draw grand conclusions by comparing SO_2 and NO_x data sets since these data suffer from the fact that the truly regional nocturnal airmass SO_2 goes unmeasured while local NO_2, which may be a trivial contributor to the tropospheric column, can give rise to large NO_x readings.

6.8. Impact of Human Activities

The higher oxides of nitrogen, and hence nitrate in rain, are more affected by human activities than are any other components of the nitrogen cycle. Certainly energy release, a major fraction of biomass burning, and probably at least 50% of the soil emissions are anthropogenic in source. The primordial NO_x emission rate may thus have been as low as only 12 Tg N yr^{-1} compared to the present estimate of 40 Tg yr^{-1}. The effect of this increase in NO_x sources is difficult to estimate. In continental air masses NO_x input alone would depress the free radical concentration (Logan et al., 1981); however, the reactive hydrocarbons accompanying much of the NO_x may not only offer an extra direct NO_x sink, but may also serve to increase the free-radical concentrations and hence the rate of NO_2-to-HNO_3 conversion, as discussed in more detail by Anderson (Chapter 9). From the arguments developed here it is seen that the Southern Hemisphere may be almost unaffected by anthropogenic emissions whereas the Northern Hemisphere is dominated thereby.

7. CONCLUSIONS

In this chapter we have shown that atmospheric nitrogen content would be expected to be steady, or perhaps slowly increasing, over geologic time periods. By contrast, human activities in fixing nitrogen seem to put the biospheric cycles out of balance. This imbalance appears to be amplified by increased human mobilization of phosphorus; however, all these statements about the effect of human activities are hampered by lack of data, even over the last 100 y. The best information we can find shows a constant (soil sourced) NH_3 budget and a NO_x flux increasing by an order of magnitude in highly industrialized regions such as the United States. A consideration of the chemistry of OH, NH_3, and NO_x indicates that whereas ammonia can be a source of NO_x in the remote troposphere, in polluted continental regions it may be an important sink for NO_x and in the Southern Hemisphere, a source of N_2O. The only known effect of this flux is the increasing acidic nitrate contribution to precipitation. There may be others as yet undreamed of.

It could be considered self-serving to suggest that more research is needed, so we close by quoting Helas and Warneck (1981): "It is evident that the present data . . . pose more questions than they answer. We take this conclusion as a stimulus for further studies of NO_x in the troposphere.".

ACKNOWLEDGMENTS

We wish to thank B. Walunas, J. Walega, and G. Bocciardi for their help in preparing this manuscript and the Atmospheric Sciences Section and Division of International Programs of the National Science Foundation for their support of our studies on oxides of nitrogen.

REFERENCES

Adel, A. (1938). Further detail in the rock-salt prizmatic solar spectrum. *Astrophys. J.* **88,** 186–188.

Ångstrom, A. and Högberg, L. (1952). On the content of nitrogen (NH_4–N and NO_3–N) in atmospheric precipitation. *Tellus* **4,** 31–42.

Aurela, A. A. and Punkkinen, R. (1981). Electric nitrogen fixation by plants. *Rep. Kevo Subarctic Res. Stn.* **17,** 1–6.

Ayers, G. P. and Gras, J. L. (1980). Ammonia gas concentrations over the Southern Ocean. *Nature* **284,** 539–540.

Barrie, L. A., Hoff, R. M., and Daggupaty, S. M. (1981). The influence of mid-latitudinal pollution sources on haze in the Canadian mountains. *Atmosph. Environ.* **15,** 1407–1420.

Baseline (1981). Baseline Air Monitoring Report 1978, Australian Government Publishing Service, Canberra.

Bauer, E. (1982). Institute for Defense Analysis Report No. P-1619.

Bauer, E. and Kowalczyk, M. (1982). Institute for Defense Analysis Report No. P-1519.

Blanchard, D. D. and Syzdek, L. D. (1972). Concentrations of bacteria in jet drops from bursting bubbles. *J. Geophys. Res.* **77,** 5087–5099.

Bolin, B. (1977). Changes of land biota and their importance. *Science* **196,** 613–615.

Bottenheim, J. W. and Strausz, O. P. (1980), Gas-phase chemistry of clean air at 55°N latitude. *Environ. Sci. Technol.* **14,** 709–718.

Brimblecombe, P. and Pittman, J. (1980). Long term deposit at Rothamsted, Southern England. *Tellus* **32,** 261–267.

Brimblecombe, P. and Stedman, D. H. (1982). Evidence for a dramatic increase in the contribution of oxides of nitrogen to precipitation acidity. *Nature* **298,** 460–462.

Broecker, W. S. (1974). *Chemical Oceanography,* Harcourt Brace Jovanovich, New York.

Broecker, W. S., Takahashi, T., Simpson, H. J., and Peng, T. H. (1979). Fate of fossil fuel carbon dioxide and the global carbon budget. *Science* **206,** 409–418.

Burns, R. C. and Hardy, R. W. F. (1975). *Nitrogen Fixation in Bacteria and Higher Plants,* Springer-Verlag, Berlin.

Cadle, R. D. (1973). Particulate matter in the lower atmosphere, in *Chemistry of the Lower Atmosphere,* S. I. Rasool, Ed., Plenum Press, New York, pp. 69–120.

Cadle, S. H., Countess, R. J., and Kelly, N. A. (1982). Nitric acid and ammonia in urban and rural locations. *Atmosph. Environ.* (in press).

Calvert, J. G. (1981). The chemistry of the polluted troposphere. *Proceedings of the NATO Advanced Study Institute on Chemistry of the Polluted and Unpolluted Troposphere,* Greece.

CAST (1976). Council for Agricultural Science and Technology, *Effect of Increased Nitrogen Fixation on Stratospheric Ozone,* Report No. 53, Iowa State University, Ames.

Chappuis, J. (1880). Sur le spectre d'absorption de l'ozone, *C. R. Acad. Sci.,* (*Paris*) **97,** 985–986.

Cohen, Y. and Gordon, L. I. (1979). Nitrous oxide production in the ocean. *J. Geophys. Res.* **84,** 347–353.

Cowling, E. B. (1982). Acid precipitation in historical perspective. *Environ. Sci. Technol.* **16,** 110A–123A.

Cox, R. A. (1977). Some measurements of ground level NO, NO_2, and O_3 concentrations at an unpolluted maritime site. *Tellus* **29,** 356–362.

Crutzen, P. J. (1971). Ozone production rates in a oxygen–hydrogen–nitrogen atmosphere. *J. Geophys. Res.* **76,** 7311–7327.

Crutzen, P. J., Heidt, L. E., Krasnec, J. P., Pollack, W. H., and Seiler, W. (1979). Biomass burning as a source of atmospheric gases, CO_2 H_2, N_2O, NO, CH_3Cl, COS. *Nature* **282,** 253–256.

Dawson, G. A. (1977). Atmospheric ammonia from undisturbed land. *J. Geophys. Res.* **82,** 3125–3133.

Dawson, G. A. (1980). Nitrogen fixation by lightning. *J. Atmosph. Sci.* **37,** 174–178.

Delwiche, C. C. (1970). The nitrogen cycle. *Sci. Am.* **223,** 137.

Delwiche, C. C. (1978). Biological production and utilization of N_2O. *Pure Appl. Geophys.* **116,** 414–422.

Denmead, O. T., Freney, J. R., and Simpson, J. R. (1976). A closed ammonia cycle within a plant canopy. *Soil Biol. Biochem.* **8,** 161–164.

Drummond, J. (1977). Atmospheric measurements of nitric oxide using chemiluminescence. Ph.D. dissertation, University of Wyoming, Laramie.

EPRI (1981). Sulfate regional experiment: Results and implications. Electric Power Research Institute Report EA-2077-SY-LD.

Eriksson, E. (1952). Composition of atmospheric precipitation. *Tellus* **4,** 215–232.

Evdokimova, T. I., Grishina, L. H., Vasilyevskaya, V. D., Samoilova, E. M., and Bystriskaya, T. L. (1975). Biogeochemical cycles of elements in some natural zones of the European USSR, in *Nitrogen Phosphorus and Sulfur Global Cycles*, R. Söderlund and B. H. Svensson, Eds., SCOPE Report No. 7, Ecological Bulletin, 22, Stockholm, pp. 135–156.

Fahnestock, G. R. (1979). Carbon input to the atmosphere from forest fires. *Science* **204,** 209–210.

Fox, C. (1873). *Ozone and Antozone*, Churchill, London.

Galbally, I. E. and Roy, C. R. (1978). Loss of fixed nitrogen from soils by nitric oxide exhalation. *Nature* **275,** 734–735.

Georgii, H. W. and Gravenhorst, G. (1977). The ocean as a source or sink of reactive trace gases. *Pure Appl. Geophys.* **115,** 503–511.

Georgii, H. W. and Lenhard, U. (1978). Contributions to the atmospheric NH_3 budget. *Pure Appl. Geophys.* **116,** 385–392.

Goreau, T. J., Kaplan, W. A., Wofsy, S. C., McElroy, M. B., Valois, F. W., and Watson, S. W. (1980). Production of NO_2^- and N_2O by nitrifying bacteria at reduced concentrations of oxygen. *Appl. Environ. Microbiol.* **40,** 526–532.

Graedel, T. E. (1979). The kinetic photochemistry of the marine atmosphere. *J. Geophys. Res.* **84,** 273–286.

Graham, R. A. and Johnston, H. S. (1978). Photochemistry of NO_3 and the kinetics of the N_2O_5–O_3 system. *J. Phys. Chem.* **82,** 254–268.

Granat, L., Hallberg, R. O., and Rodhe, H. (1975). The global phosphorus cycle, in *Nitrogen Phosphorus and Sulfur Global Cycles*, R. Söderlund and B. H. Svensson, Eds., SCOPE Report No. 7, Ecological Bulletin, 22, Stockholm, pp. 89–134.

Hameed, S., Paidoussis, O. G., and Stewart, R. W. (1981). Implications of natural sources for the latitudinal gradients of NO_y in the unpolluted troposphere. *Geophys. Res. Lett.* **8,** 591–594.

Hampson, R. F. (1980). Chemical data sheets for atmospheric reactions. U.S. Department of Transportation, Report FAA-EE-80-17, Washington, D.C.

Hartley, W. N. (1881). On the absorption of solar rays by atmospheric ozone. *J. Chem. Soc.* **39,** 111–128.

Harvey, R. B., Stedman, D. H., and Chameides, W. L. (1977). Determination of the absolute rate of solar photolysis of NO_2. *J. Air Pollut. Control Assoc.* **27,** 663–666.

Heikes, B., Lazrus, A. L., Kok, G. L., Kunen, S. M., Gandrud, B. W., Gitlin, S. N., and Sperry, P. D. (1982). Evidence for aqueous phase hydrogen peroxide synthesis in the troposphere. *J. Geophys. Res.* **87,** 3045–3051.

Helas, G. and Warneck, P. (1981). Background NO_x mixing ratios in air masses over the North Atlantic Ocean. *J. Geophys. Res.* **86,** 7283–7290.

Holland, H. P. (1964). On the chemical evolution of the terrestrial and Cytherean atmospheres, in *The Origin and Evolution of Atmospheres and Oceans*, P. J. Brancazio and A. G. W. Cameron, Eds., Wiley, New York, pp. 86–101.

Huebert, B. J. (1980). Nitric acid and aerosol nitrate measurements in the equatorial Pacific region. *Geophys. Res. Lett.* **7,** 325–328.

Huebert, B. J. and Lazrus, A. L. (1978). Global tropospheric measurements of nitric acid vapor and particulate nitrate. *Geophys. Res. Lett.* **5,** 577–580.

Huebert, B. J. and Lazrus, A. L. (1980a). Tropospheric gas-phase and particulate nitrate measurements. *J. Geophys. Res.* **85,** 7322–7328.

Huebert, B. J. and Lazrus, A. L. (1980b). Bulk composition of aerosols in the remote troposphere. *J. Geophys. Res.* **85,** 7337–7344.

Husar, R. B. and Patterson, D. E. (1980). Regional scale air pollution sources and effects. *Ann. N. Y. Acad. Sci.* **338,** 399–417.

JANAF Thermochemical Tables, 2nd ed. (1971). D. R. Stull and H. Prophet, Eds., NSRDS-NBS-37, U.S. Government Printing Office, Washington, DC.

Johnston, H. S. (1971). Reduction of stratospheric ozone by nitrogen oxide catalyst from supersonic transport exhaust. *Science* **173,** 517.

Junge, C. E. (1963). *Air Chemistry and Radioactivity*, Academic Press, New York.

Kaplan, W. A., Elkins, J. W., Kolb, C. E., McElroy, M. B., Wofsy, S. C., and Duran, A. P. (1978). Nitrous oxide in fresh water systems: An estimate for the yield of atmospheric N_2O associated with disposal of human waste. *Pure Appl. Geophys.* **116,** 423–438.

Kelly, T. J., Stedman, D. H., Ritter, J. A., and Harvey, R. B. (1980). Measurement of oxides of nitrogen and nitric acid in clean air. *J. Geophys. Res.* **85,** 7417–7425.

Kley, D. and McFarland, M. (1980). Chemiluminescent detector for NO and NO_2, *Atmosph. Technol.* (National Center for Atmospheric Research) **12,** 63–68.

Kley, D., Drummond, J. W., McFarland, M., and Liu, S. C. (1981). Tropospheric profiles of NO_x. *J. Geophys. Res.* **86,** 3153–3163.

Kurasawa, H. and Lesclaux, R. (1979). Kinetics of the reaction NH_2 with NO. *Chem. Phys. Lett.* **66,** 602–607.

Kurasawa, H. and Lesclaux, R. (1980). Rate constant for the reaction of NH_3 with ozone in relation with atmospheric processes. *Chem. Phys. Lett.* **72,** 437–442.

Lazrus, A. L., Gandrud, B., and Cadle, R. D. (1971). Chemical composition of air filtration samples of the stratospheric sulfate layer. *J. Geophys. Res.* **76,** 8083.

Lee, Y. N. and Schwartz, S. E. (1981). Evaluation of the rate of uptake of nitrogen dioxide by atmospheric and surface liquid water. *J. Geophys. Res.* **86,** 11,971–11,984.

Leighton, P. A. (1961). *The Photochemistry of Air Pollution*, Academic Press, New York.

Lesclaux, R. and DeMissy, M. (1977). On the reaction of NH_2 radical with oxygen. *Nouv. J. Chim.* **1,** 443–444.

Lewis, G. N. and Randall, M. (1923). *Thermodynamics*, McGraw-Hill, New York, p. 568.

Levine, J. S., Rogowski, R. S., Gregory, G. L., Howell, W. E., and Fishman, J. (1981). Simultaneous measurements of NO_x, NO, and O_3 production in a laboratory discharge: Atmospheric implications. *Geophys. Res. Lett.* **8,** 357–360.

Levy, H. (1971). Normal atmosphere: Large radical and formaldyhyde concentration predictions. *Science* **173,** 141–143.

Liu, S. C., Cicerone, R. J., and Donahue, T. M. (1977). Sources and sinks of atmospheric N_2O and the possible ozone reduction due to industrial fixed nitrogen fertilizers. *Tellus* **29,** 251–263.

Liu, S. C., Kley, D., McFarland, M., Mahlman, J. D., and Levy, II, H. (1980). On the origin of tropospheric ozone. *J. Geophys. Res.* **85,** 7546–7552.

Lodge, J. P., Machado, P. A., Pate, J. B., Sheesley, D. C., and Wartburg, A. F. (1974). Atmospheric trace chemistry in the American humid tropics. *Tellus* **26,** 250–253.

Logan, J. A., Prather, M. J., Wofsy, S. C., and McElroy, M. B. (1981). Tropospheric chemistry: A global perspective. *J. Geophys. Res.* **86,** 7210–7254.

Lovelock, J. E. (1971). Air pollution and climatic change. *Atmosph. Environ.* **5,** 403–412.

Mackay, G. I. and Schiff, H. I. (1982). Private communication, Unisearch Co., Toronto, Ont.

Mackenzie, F. T. (1975). [Ref. in Broecker et al. (1979)] 169th American Chemical Society Meeting, Baltimore.

McConnell, J. C. and McElroy, M. B. (1973). Odd nitrogen in the atmosphere. *J. Atmosph. Sci.* **30,** 1465–1480.

McElroy, M. B., Elkins, J. W., Wofsy, S. C., and Yung, Y. L. (1976). Sources and sinks for atmospheric N_2O. *Geophys. Space Phys.* **14,** 143–150.

McFarland, M., Kley, D., Drummond, J. W., Schmeltekopf, A. L., and Winkler, R. J. (1979). Nitric oxide measurements in the equatorial Pacific region. *Geophys. Res. Lett.* **6,** 605–608.

Molina, M. S. (1982). Private communication, University of California, Irvine.

Munn, R. E. and Bolin, B. (1971). Global air pollution—Meteorological aspects. *Atmosph. Environ.* **5,** 363–402.

Miller, N. H. J. (1905). The amounts of nitrogen as ammonia and of chlorine in rainwater collected at Rothamsted. *J. Agric. Sci.* **1,** 280–303.

National Research Council (1977). *Nitrogen Oxides*, National Academy of Sciences, Washington, DC.

NBS (1968). *Selected Values of Chemical Thermodynamic Properties*, D. D. Wagman, W. H.

Evans, V. B. Parker, I. Halow, S. M. Bailey, and R. H. Schumm, Eds., Technical Note 270-3, U.S. National Bureau of Standards, Washington, DC.

Noxon, J. F. (1978). Tropospheric NO_2. *J. Geophys. Res.* **83,** 3051.

Noxon, J. F. (1981). NO_x in the Mid-Pacific troposphere. *Geophys. Res. Lett.* **8,** 1223–1226.

Noxon, J. F., Norton, R. B., and Marovich, E. (1980). NO_3 in the troposphere. *Geophys. Res. Lett.* **7,** 125.

Nriagu, J. O., Ed. (1976). *Nitrogen Cycling in Terrestrial Ecosystems*, Ann Arbor Science Publications, Ann Arbor, MI.

Perry, P. A., Atkinson, R., and Pitts, J. N. Jr. (1976). Rate constants for the reactions OH + $H_2S \rightarrow H_2O$ + SH and OH + $NH_3 \rightarrow H_2O$ + NH_2 over the temperature range 297-427°K. *J. Chem. Phys.* **64,** 3237–3239.

Pierrou, U. (1975). The global phosphorus cycle, in *Nitrogen Phosphorus and Sulfur Global Cycles*, R. Söderlund and B. H. Svensson, Eds., SCOPE Report No. 7, Ecological Bulletin, 22, Stockholm, pp. 75–88.

Pierson, W. R., Brachaczek, W. W., Truex, T. T., Butler, J. W., and Korniski, T. J. (1980). Ambient sulfate measurements on Allegheny Mountain and the question of atmospheric sulfate in the Northeastern United States. *Ann. N. Y. Acad. Sci.* **338,** 145–173.

Platt, U. and Perner, D. (1980). Direct measurement of atmospheric CH_2O, HNO_2, O_3, NO_2, and SO_2 by differential optical absorption in the near UV. *J. Geophys. Res.* **85,** 7453–7458.

Platt, U., Perner, D., and Pätz, H. W. (1979). Simultaneous measurement of atmospheric CH_2O, O_3, and NO_2 by differential optical absorption. *J. Geophys. Res.* **84,** 6329–6335.

Platt, U., Perner, D., Harris, G. W., Winer, A. M., and Pitts, Jr. J. N. (1980). Detection of NO_3 in the polluted troposphere by differential optical absorption. *Nature* **285,** 312–314.

Platt, U., Perner, D., Schröder, J., Kessler, C., and Toennissen, A. (1981). The diurnal variation of NO_3. *J. Geophys. Res.* **86,** 11965–11970.

Pommereau, J. B. (1981). Recherches sur le dioxyde d'azote NO_2 dans L'atmosphere de la terre. Ph.D. dissertation, Universite Pierre et Marie Curie, Paris.

Prasad, S. S. (1981). Excited ozone as a possible source of atmospheric N_2O. *Nature* **289,** 386–388.

Prospero, J. M., Mohnen, V., Jaenicke, R., Charlson, R. J., Delany, A. C., Moyers, J., and Zoller, W. (1982). Atmospheric aerosols: Cycles and measurements. *Rev. Geophys. Space Phys.* (in press).

Rahn, K. A. (1981). Relative importance of North America and Eurasia as sources of Arctic aerosol. *Atmosph. Environ.* **15,** 1447–1456.

Rasmussen, R. A. and Pierotti, D. (1978). Global and regional N_2O measurements. *Pure Appl. Geophys.* **116,** 405–413.

Ridley, B. A. and Hastie, D. (1981). Stratospheric odd nitrogen: NO measurements at 51°N in summer. *J. Geophys. Res.* **86,** 3162–3166.

Ritter, J., Stedman, D. H., and Kelly, T. J. (1978). Ground level measurements of NO, NO_2, and O_3 in rural air. in *Nitrogenous Air Pollutants–Chemical and Biological Implications*, D. Grosjean, Ed., Ann Arbor Science Publications, Ann Arbor, MI.

Robinson, E. and Robbins, R. C. (1970). Gaseous atmospheric pollutants from urban and natural sources, in *Global Effects of Environmental Pollution*, S. F. Singer, Ed., D. Reidel, Dordrecht, pp. 50–64.

Rodhe, H. and Grandell, J. (1972). On the removal time of aerosol particles from the atmosphere by precipitation scavenging. *Tellus* **24,** 442–454.

Rosswall, T. (1976). The internal nitrogen cycle between microorganisms, vegetation and soil, in *Nitrogen Phosphorus and Sulfur Global Cycles*, B. H. Svensson and R. Söderlund, Eds., SCOPE Report No. 7, Ecological Bulletin 22, Stockholm, pp. 157–168.

Savoie, D. L. and Prospero, J. M. (1982). Particle size distribution of nitrate and sulfate in the marine atmosphere. *Geophys. Res. Lett.* (in press).

Schiff, H. I., Pepper, D., and Ridley, B. (1979). Tropospheric NO measurements up to 7 kilometers. *J. Geophys. Res.* **84,** 7895–7897.

Seiler, W., Giehl, H., and Bunse, G. (1978). Influence of plants on atmospheric carbon monoxide and dinitrogen oxide. *Pure Appl. Geophys.* **116,** 439–451.

Sheih, C. M., Wesely, M. L., and Hicks, B. B. (1979). Estimated dry deposition velocities of sulfur over the Eastern U.S. and surrounding regions. *Atmosph. Environ.* **13,** 10, 1361–1368.

Smith, R. A. (1872). *Air and Rain*, Longmans, Green, London.

Söderlund, R. and Svensson, B. H. (1976). The global nitrogen cycle, in *Nitrogen, Phosphorus, and Sulfur Global Cycles*, B. H. Svensson and R. Söderlund, Eds., SCOPE Report No. 7, Ecological Bulletin, 22, Stockholm, pp. 23–74.

Stedman, D. H. and Jackson, J. O. (1975). The photostationary state in photochemical smog. *Internatl. J. Chem. Kinet. Symp.* **I,** 493–501.

Stedman, D. H., Chameides, W. L., and Cicerone, R. J. (1975). Vertical distribution of soluble gases in the troposphere. *Geophys. Res. Lett.* **2,** 333.

Stedman, D. H., Ritter, J., and Kelly, T. (1978). Budget of tropospheric nitrogen oxides. *Proceedings of AIChE*, Philadelphia, Pennsylvania.

Stedman, D. H., Shetter, R. E., and West, D. (1981). Measurements related to atmospheric oxides of nitrogen. Final report for Coordinating Research Council; Atmospheric & Oceanic Science Department, University of Michigan, Ann Arbor.

Stuhl, F. (1978). Absolute rate constant for the reaction $OH + NH_3 \rightarrow NH_2 + H_2O$. *J. Phys. Chem.* **59,** 535–537.

Stuiver, M. (1978). Atmospheric carbon dioxide and carbon reservoir changes. *Science* **199,** 253–258.

Stumm, W. (1973). The acceleration of hydrogeochemical cycling in phosphorus. *Water Resources* **7,** 131–144.

Söderlund, R. (1976), Wet deposition of inorganic nitrogen compounds over the sea. Report AC-34, International meteorological Institute, University of Stockholm, Stockholm, Sweden.

Thompson, A. M. and Heikes, B. G. (1982). Heterogeneous chemistry: Effect of NO_y in the unpolluted troposphere, presented at The Second Symposium on the Composition of the Non-Urban Troposphere, Williamsburg, VA.

Tsunogai, S. (1971). Ammonia in the oceanic atmosphere and the cycle of nitrogen compounds through the atmosphere and the hydrosphere. *J. Geochem.* **5,** 57–67.

Vijan, P. N. and Wood, G. R. (1981). Automated determination of ammonia by gas phase molecular absorption. *Anal. Chem.* **53,** 1447–1450.

Walker, J. C. G. (1977). *Evolution of the Atmosphere*, Macmillan, New York.

Weiss, R. F., (1981). The temporal and spatial distribution of tropospheric nitrous oxide. *J. Geophys. Res.* **86,** 7185–7196.

Wilson, J., Mohnen, V., and Kadlecek, J. (1980). Wet deposition in the Northeastern U.S., ARSC Publication 796, State University of New York (SUNY), Albany.

Wong, C. S. (1978). Atmospheric input of carbon dioxide from burning wood. *Science* **200,** 197.

11

CONSTRUCTION AND TESTING OF A SURROGATE CHEMICAL MECHANISM (SCHEME) FOR TROPOSPHERIC PHOTOCHEMICAL REACTIONS

Stuart Z. Levine and Stephen E. Schwartz

Environmental Chemistry Division
Brookhaven National Laboratory
Upton, New York 11973

---

1. INTRODUCTION

It is well known that primary air pollutants may undergo reactions in the ambient atmosphere, forming more noxious secondary pollutants such as ozone from NO_x–hydrocarbon reactions and sulfates from SO_2 oxidation reactions. It is also now apparent that because of atmospheric residence times ranging from hours to days (e.g., Rodhe, 1978), these pollutants will not only influence the environmental air quality immediately downwind of an emission source, but may also be transported and dispersed over hundreds to thousands of kilometers (Lyons et al., 1978; Wilson, 1978; Wolff et al., 1977). Therefore, to establish effective regional air-pollution controls as well as successfully assess the environmental impact of future energy strategies, it is necessary to acquire an understanding of the combined processes of atmospheric transport, dispersion, and chemical transformation. Such an understanding may, in turn, be gained largely from an examination of these processes by the use of a numerical transport model that incorporates a chemical model to describe the reaction system. Numerical models (Lagrangian and Eulerian) have been developed to simulate the meteorological factors involved in the movement of individual air parcels over long distances. Independently, models have also been developed that describe the reaction kinetics of the polluted troposphere under the conditions of spatial homogeneity (so-called box models). However, only limited progress has been made in merging these two types of models; i.e., in incorporating a chemical kinetic model as submodel in a combined chemical–meteorological model. In the case of Lagrangian transport models (Meyers et al., 1976; Sheih, 1977; Wendell et al., 1976) a sophisticated description of the chemical kinetics is precluded since these models are restricted to passive tracers or to materials whose removal can be described by first-order kinetics (exponential decay). In the case of Eulerian models [e.g., Duewer et al. (1978); see also Peters and Carmichael, this volume, Chapter 12], which in principle might incorporate more realistic (nonlinear) reaction kinetics, progress has been limited because the large number of species employed in available box models for the homogeneous gas-phase reaction system results in prohibitive requirements on computational time and space necessary to solve the partial differential equations describing the combined diffusion–advection–reaction problem for each species being modeled (Eschenroeder and Martinez, 1972; Seinfeld et al., 1972). Thus it is desirable to examine the extent to which existing models for the homogeneous gas-phase photochemistry and reaction kinetics of the troposphere may be simplified (i.e., decreasing the number of species in the model) while retaining an accurate description of the reaction kinetics.

As is now recognized, the chemical reaction system of the polluted troposphere is highly complex, involving hundreds of molecular and free-radical species. Perhaps the most complete treatment of this chemistry is the photochemical model due to Demerjian et al. (1974), which incorporates more than 200 species with the great majority of these arising from the explicit use of specific reactive hydrocarbons and their corresponding organic intermediates and sinks. Despite its comprehensiveness, the authors warn that even this model may be an oversimplification of the real atmosphere, which undoubtedly involves additional hundreds of organic compounds. Clearly, for modeling of the concentrations of each of these compounds, each organic species must be retained in the model. However, for many purposes, such as predicting ozone and sulfur chemistry, knowledge of the organic chemistry in such detail is neither necessary nor desirable; rather only enough detail to account for the influence of the organic system on species of interest is necessary. This has motivated work in reducing the number of organic species retained in the model while preserving within the model some measure of overall reactivity, e.g., ozone-forming potential. Considerable success in reducing such multispecies models has been achieved by Hecht et al. (1974), who greatly diminished the number of organic species required by grouping together compounds of similar properties. Such an approach has spawned a variety of reduced photochemical models that have been found useful in comparisons with ambient as well as laboratory data (e.g., Gelinas and Skewes-Cox, 1977; Muthukrishnan and Peters, 1977; Sander and Seinfeld, 1976; Yocke et al., 1975); however, even these parametrized models still incorporate tens of species and thus remain cumbersome for the purpose of serving as chemical submodels within transport–transformation models. Therefore, further reduction of the number of species in such chemical models continues to be of importance and is the subject of this chapter. We shall see that such further reduction may be accomplished with a minimum loss of accuracy or information related to the kinetics of the system.

The purpose of the present study is to examine the extent to which the number of species employed in parametrized models may be further diminished while adequately describing the currently known homogeneous chemistry of the polluted troposphere. Thus this work builds upon the advances already made for estimating yet unmeasured reaction rates (e.g., Calvert et al., 1978; Demerjian et al., 1974) as well as approximation techniques for grouping organic reactions according to "average" reactivities (Hecht et al., 1974) and uses these as a point of departure. To this end we have formulated a kinetic mechanism that is representative of present-day parametrized photochemical models and that, because of concern about long-range transport of atmospheric sulfur compounds, includes sulfur chem-

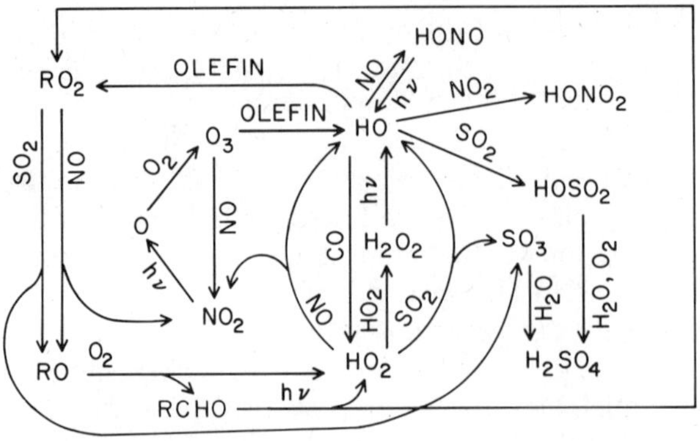

Figure 1. Reactions represented by the present models ATMOS and SCHEME.

istry as well. Some of the more important reactions are shown pictorially in Figure 1, where RCHO, RO, and RO_2 represent aldehyde and alkoxy and alkylperoxy radicals, respectively. This ATmospheric MOdel for Sulfur (ATMOS), consisting of reactions among the 30 species listed in Table 1, is used as the starting point in the present work, in which we address further means of reducing the required number of species and of developing methods for the construction of Surrogate CHEmical MEchanisms (SCHEMEs) to model pollutants such as NO, NO_2, SO_2, and O_3. Largely through the use of surrogate reactions and rate coefficients to retain the chemistry associated with omitted reactants, we have developed a 12-species SCHEME that is capable of maintaining its integrity and ability to replace the use of ATMOS for a broad range of chemical conditions. The species actively modeled by SCHEME are also given in Table 1.

In contrast to the present work, previous formulations by other research groups of chemical models incorporating fewer than 15 species have omitted potentially important and possibly interesting tropospheric chemistry by the exclusion of species such as H_2O_2, SO_2, and organics. In at least one recent case (Carmichael and Peters, 1979), the reduced model requires as input some estimate of NO and NO_2 concentration histories. Generally, these attempts at model reduction have made use of simplified mechanisms de-

Table 1. Chemical Species Used in ATMOS and SCHEME[a] Models

1. SO_2	7. HO_2	13. H_2O	19. $HONO_2$	25. CO
2. NO	8. RO_2	14. O_2	20. HO_2NO_2	26. CO_2
3. NO_2	9. $RCOO_2$	15. O	21. RONO	27. SO_3
4. HONO	10. H_2O_2	16. H_2	22. $RONO_2$	28. H_2SO_4
5. O_3	11. Olefin	17. NO_3	23. PAN	29. $ROSO_2$
6. HO	12. RCHO	18. N_2O_5	24. RO	30. H

[a] SCHEME employs only the species listed in the first two columns.

veloped against smog-chamber data for conditions similar to urban pollution so that they might be reasonably used in urban airshed models (Friedlander and Seinfeld, 1969; Hecht and Seinfeld, 1972; Reynolds et al., 1973). However, such mechanisms would not be expected to be applicable to more widely varying conditions of concentrations or concentration ratios that may be encountered during long-range transport, since the omission of reactive species and reactions involving these species severely diminishes the ability of these mechanisms to allow for potential shifts in the relative importance of competing atmospheric reactions. Such changes in chemistry become increasingly more likely as an air parcel ages and is transported away from the influence of the emission-source region.

A further difference between the methodology in the development of SCHEME and in previous attempts at model reduction is the manner in which steady-state approximations are used to limit the number of species and hence the number of differential equations necessary for description of the chemistry of a system. In previous studies steady-state treatment has been applied to species whose concentrations are coupled to one another (e.g., HO, HO_2, RO_2, HONO), thus requiring some form of iterative solution of the problem. Farrow and Edelson (1974) have shown that this type of calculation can actually increase, rather than decrease, the computational time involved in chemical modeling as well as introduce divergence from the concentration profiles obtained by the exact solution. In addition, the omission of potentially important reaction processes further restricts the range of conditions of model applicability by eliminating processes that would influence steady-state concentrations. However, by retaining the chemical influence of omitted species through the use of surrogates and by limiting the steady-state treatment to simple, uncoupled algebraic expressions, we have developed a 12-species surrogate mechanism that accurately reproduces the solutions obtained with ATMOS and does so with a 70–80% savings in execution time.

In addition to being more readily incorporable within atmospheric transport models, a surrogate mechanism such as SCHEME also permits, because of the diminished execution time, more extensive use of box-model calculations, including the modeling of smog-chamber data or testing the sensitivity of a computation to changes in species concentrations or rate constants. Although the development of SCHEME is carried out with the specific objective of describing the gas-phase homogeneous chemistry of the polluted troposphere, methods employed in its construction are general and would thus be applicable to other multispecies, multireaction systems (e.g., stratospheric or combustion chemistry). All simulation results presented in this chapter have been obtained by use of a modified version of a computer program (EPISODE) developed at the Lawrence Livermore Laboratory for integrating ordinary differential equations (Hindmarsh and Byrne, 1977).

2. ATMOSPHERIC MODEL FOR SULFUR (ATMOS)

2.1. Mechanism

In the years since Stephens (1966), Wayne (1962), and Leighton (1961) wrote their reviews on atmospheric photochemistry, it has become increasingly apparent and is now well recognized that the description of the homogeneous chemistry of the polluted troposphere—i.e., the formation of photochemical smog and the oxidation of SO_2—requires a complex series of elementary reactions involving several chain-carrying free radicals (HO, HO_2, CH_3O, CH_3O_2, $HCOO_2$, etc.) whose concentrations depend on the concentrations of trace atmospheric molecular constituents, including NO, NO_2, CO, SO_2, O_3, and organics as well as solar irradiation (e.g., Calvert et al., 1978; Demerjian et al., 1974; Finlayson-Pitts and Pitts, 1977; Graedel et al., 1976). A model of these processes that is representative of existing parametrized chemical models is given by the 44 reactions listed in Table 2. It is this ATmospheric MOdel for Sulfur (ATMOS) which is employed here to generate and test the validity of methods for the construction of reduced sur-

Table 2. ATmospheric MOdel for Sulfur (ATMOS)

Reaction	Rate Constant k or Photolytic Constant j	Units[a]
[A1] $NO_2 + h\nu \rightarrow NO + O$	4.8 (−1)[b,c]	min^{-1}
[A2] $O + O_2 \rightarrow O_3$	2.16(1)	
[A3] $NO + O_3 \rightarrow NO_2 + O_2$	2.37(1)	
[A4] $NO_2 + O_3 \rightarrow NO_3 + O_2$	4.77(−2)	
[A5] $NO + NO_3 \rightarrow 2NO_2$	1.4 (4)	
[A6] $NO_2 + NO_3 \rightarrow N_2O_5$	5.6 (3)	
[A7] $N_2O_5 \rightarrow NO_2 + NO_3$	1.5 (1)	min^{-1}
[A8] $N_2O_5 + H_2O \rightarrow 2HONO_2$	1.5 (−5)	
[A9] $NO + O \rightarrow NO_2$	3.0 (3)	
[A10] $NO_2 + O \rightarrow NO + O_2$	1.35(4)	
[A11] $NO_2 + O \rightarrow NO_3$	2.9 (3)	
[A12] $NO_2 + NO + H_2O \rightarrow 2HONO$	2.2 (−9)	$ppm^{-2}\,min^{-1}$
[A13] $2HONO \rightarrow NO_2 + NO + H_2O$	1.4 (−3)	
[A14] $HONO + h\nu \rightarrow HO + NO$	1.1 (−1)[c]	min^{-1}
[A15] $NO_2 + HO \rightarrow HONO_2$	2.1 (4)	
[A16] $NO + HO \rightarrow HONO$	1.7 (4)	
[A17] $NO + HO_2 \rightarrow NO_2 + HO$	3.0 (3)	
[A18] $H_2O_2 + h\nu \rightarrow 2HO$	1.5 (−3)[c]	min^{-1}
[A19] $OLEF + O \rightarrow \frac{1}{2}HO_2 + \frac{1}{2}RCOO_2 + RO_2$	3.0 (4)	
[A20] $OLEF + HO \rightarrow RCHO + RO_2$	7.5 (4)	
[A21] $OLEF + O_3 \rightarrow HO + RCOO_2 + RCHO$	5.0 (−2)	

Table 2. (*Continued*)

Reaction	Rate Constant k or Photolytic Constant j	Units[a]
[A22] $RCHO + h\nu \rightarrow \frac{3}{4}HO_2 + \frac{1}{4}RO_2 + \frac{1}{2}H_2 + CO$	5.0 (−3)[c]	
[A23] $RCHO + HO \rightarrow \frac{1}{2}HO_2 + \frac{1}{2}RCOO_2$	2.3 (4)	
[A24] $NO + RO_2 \rightarrow NO_2 + RO$	1.0 (4)	
[A25] $NO + RCOO_2 \rightarrow NO_2 + RO_2$	2.0 (3)	
[A26] $NO_2 + RCOO_2 \rightarrow PAN$	5.0 (2)	
[A27] $O_2 + RO \rightarrow HO_2 + RCHO$	2.4 (−2)	
[A28] $NO_2 + RO \rightarrow RONO_2$	4.9 (2)	
[A29] $NO + RO \rightarrow RONO$	2.5 (2)	
[A30] $2HO_2 \rightarrow H_2O_2 + O_2$	8.3 (3)	
[A31] $RO_2 + HO_2 \rightarrow O_2 + RO + HO$	1.0 (2)	
[A32] $2RO_2 \rightarrow O_2 + 2RO$	1.0 (2)	
[A33] $CO + HO \xrightarrow{O_2} HO_2 + CO_2$	4.5 (2)	
[A34] $NO_2 + HO_2 \rightarrow O_2 + HONO$	6.0	
[A35] $NO_2 + HO_2 \rightarrow HO_2NO_2$	3.0 (2)	
[A36] $HO_2NO_2 \rightarrow NO_2 + HO_2$	4.0	
[A37] $HONO + HO \rightarrow NO_2 + H_2O$	9.0 (3)	
[A38] $SO_2 + HO_2 \rightarrow HO + SO_3$	1.3	
[A39] $SO_2 + RO_2 \rightarrow RO + SO_3$	7.8	
[A40] $SO_2 + HO \xrightarrow{H_2O,\ O_2} HO_2 + H_2SO_4$	1.6 (3)	
[A41] $SO_2 + O \rightarrow SO_3$	8.4 (1)	
[A42] $SO_2 + RCOO_2 \rightarrow RO_2 + SO_3$	2.0 (−3)	
[A43] $SO_2 + RO \rightarrow ROSO_2$	8.9	
[A44] $SO_3 + H_2O \rightarrow H_2SO_4$	1.34(3)	

[a] All units are $ppm^{-1}\ min^{-1}$ unless otherwise noted.

[b] The notation 4.8(−1) represents 4.8×10^{-1}.

[c] Denotes primary photochemical rate constant j for solar zenith angle of 40°.

rogate models. The kinetics and mechanism described by his model have been extensively studied and reviewed with respect to pollution chemistry, and reasonably complete tabulations of rate constants are available (Hampson, 1980; Hampson and Garvin, 1978). The primary photolytic rate constants in Table 2 are given for a solar zenith angle of 40°, and all rate constants pertain to a temperature of 25°C; most rate constants here are thought to exhibit little or no temperature dependence, although the effect of temperature dependence on the rate of unimolecular decomposition processes has recently been noted by Carter et al. (1979). With perhaps a few exceptions, further review of the reactions and rate constants given in Table 2 is unnecessary here.

The reaction of HO_2 with nitric oxide,

$$NO + HO_2 \rightarrow NO_2 + HO, \qquad [A17]$$

has been identified as being important in the formation of photochemical smog (Calvert and McQuigg, 1975); however, the rate of this well-studied reaction has itself not been clearly established. For this study we have taken $k_{17} = 3 \times 10^3$ ppm^{-1} min^{-1}, which is an average of several recent determinations (Cox and Derwent, 1975; Hack et al., 1975; Howard and Evenson, 1977; Simonaitis and Heicklen, 1976). In addition, because this value is approximately an order of magnitude larger than that used by Demerjian et al. (1974) to approximate the relative rate for the oxidation of NO by the RO_2 and $RCOO_2$ radicals, we have correspondingly increased the values of k_{24} and k_{25} from the values estimated by those authors.

The formation of HO_2 by the HO-radical oxidation of CO proceeds through a two-step process involving the formation and loss of hydrogen atoms:

$$CO + HO \rightarrow H + CO_2$$

$$H + O_2\,(+ M) \rightarrow HO_2\,(+ M).$$

Because there does not appear to be any other important loss mechanism for hydrogen-atoms in the troposphere, it is reasonable to assume that the formation of HO_2 in the second step follows the first with a yield of approximately 100%. This allows for the elimination of hydrogen from ATMOS by summing the two steps to give reaction [A33]. The rate constant value of $k_{33} = 4.5 \times 10^2$ ppm^{-1} min^{-1} chosen for this reaction is taken from recent studies that show that the CO–HO reaction proceeds under atmospheric pressure at about twice the rate generally assumed from low-pressure measurements (Chan et al., 1977; Cox et al., 1976; Sie et al., 1976). In addition, new measurements or estimates of other HO reaction rates determined relative to the CO–HO reaction and based on the "high" value of k_{33} place k_{15}, k_{16}, and k_{40} at 2.1×10^4, 1.7×10^4, and 1.6×10^3 ppm^{-1} min^{-1}, respectively (Calvert et al., 1978; Cox et al., 1976).

The homogeneous gas-phase chemistry of SO_2 given by reactions [A38]–[A44] has been formulated from the recent extensive review by Calvert et al. (1978). Although it is generally accepted that the SO_2–HO reaction leads to the formation of $HOSO_2$ and that this species may react rapidly with oxygen to give $HOSO_2O_2$, the fate of the $HOSO_2O_2$ radical in the troposphere remains open to speculation and must await further laboratory studies. However, it is also generally accepted that the net result of the SO_2–HO reaction is the formation of H_2SO_4; therefore, for the present it appears reasonable to represent the overall process by the reaction

$$SO_2 + HO \xrightarrow{H_2O,O_2} HO_2 + H_2SO_4, \qquad \text{[A40]}$$

since this reaction leads to the expected sulfate formation while also producing HO_2 radicals, thus continuing any chain reactions that may in reality be carried by an $HOSO_2O_2$ radical. We note also that similar reactivities are expected for the HO_2 and $HOSO_2O_2$ radicals (Benson, 1978). In the case of the SO_2–HO_2 reaction

$$SO_2 + HO_2 \rightarrow HO + SO_3, \qquad [A38]$$

we have employed the value $k_{38} = 1.3$ ppm^{-1} min^{-1} (Payne et al., 1973). This value is markedly higher than has been indicated in recent studies of this reaction. Burrows et al. (1979) have indicated an upper limit for k_{38} some fiftyfold lower than the value given by Payne et al., if the reaction occurs as a two-body reaction as written, and Graham et al. (1979) have similarly indicated an upper limit to k_{38} approximately three orders of magnitude lower than that given by Payne et al. There is as yet no explanation for these discrepancies, and for the present we have decided to retain the earlier value due to Payne et al. since the later studies were carried out under conditions much less representative of the troposphere (i.e., low total pressure, rigorous exclusion of water vapor). We call attention, however, to the possible need for downward revision of k_{38} on the basis of future laboratory investigations.

The organic reactions shown in ATMOS have been obtained from the parametrized model of Hecht et al. (1974) or have been derived from the explicit model of Demerjian et al. (1974). For example, reactions of olefin (OLEF),

$$OLEF + O \rightarrow \tfrac{1}{2}HO_2 + \tfrac{1}{2}RCOO_2 + RO_2 \qquad [A19]$$

$$OLEF + HO \rightarrow RCHO + RO_2 \qquad [A20]$$

$$OLEF + O_3 \rightarrow HO + RCOO_2 + RCHO, \qquad [A21]$$

have been taken directly from the former study; however, instead of using rate constants for propylene, values for *trans*-2-butene have been selected as more representative of the abundance of alkenes in the polluted troposphere. The reactions

$$RCHO + h\nu \rightarrow \tfrac{3}{4}HO_2 + \tfrac{1}{4}RO_2 + \tfrac{1}{2}H_2 + CO \qquad [A22]$$

$$RCHO + HO \rightarrow \tfrac{1}{2}HO_2 + \tfrac{1}{2}RCOO_2, \qquad [A23]$$

have been derived from the aldehyde reaction sequence given by Demerjian et al. and assuming a typical polluted tropospheric mixture for the compounds of 50% CH_2O with the remainder being primarily CH_3CHO. For HO–aldehyde reactions ($k = 2.3 \times 10^4$ ppm^{-1} min^{-1}), such a mixture will give the yield of radicals shown in reaction [A23], where the HO_2 and $RCOO_2$ are produced from CH_2O and higher aldehydes, respectively. Also, in this mixture of aldehydes, CH_2O will photolyze to yield two HO_2 radicals at a rate ($j = 2.5 \times 10^{-3}$ min^{-1}) equal to that at which other aldehydes are yielding a single HO_2 and RO_2 radical. In addition, for each CH_2O photolyzing to give radical formation, another CH_2O photolyzes at twice the rate ($j = 5 \times 10^{-3}$ min^{-1}) to give unreactive H_2. The net result of these aldehyde photolysis reactions is summed in reaction [A22].

Obviously, any chemical model such as ATMOS is subject to change as the present knowledge of tropospheric chemistry is updated with new determinations of rate constants as well as new or revised reactions. While noting the inevitability of such changes in ATMOS, it is also expected that the methods to be presented for the construction of Surrogate CHEmical MEchanisms are so generally applicable to any model that SCHEME should be readily adjustable to accommodate the change. As becomes apparent later in this chapter, adjustments of the rate constants listed in Table 2 are easily and directly handled by SCHEME because all surrogate rate coefficients are presented in terms of functional relationships that are simply evaluated regardless of what k value may become more appropriate for use in ATMOS. In addition, even the inclusion into ATMOS of reactions involving additional species may be readily handled without increasing the species content of SCHEME if these additional species are stable reactants or products or if they are reactive intermediates whose steady-state concentrations may be approximated by simple uncoupled relationships.

2.2. Smog-Chamber Data

A customary and useful test of any model is to apply it to simulations of smog-chamber experiments; however, the interpretation of chamber data and the validation of models against such data is complicated by the highly ill-defined and elusive nature of what are commonly known as "chamber effects" (Bufalini et al., 1977). These effects include the influence of the chamber wall itself and differences between chambers and the real atmosphere, e.g., higher surface: volume ratios and different spectral distributions of irradiation sources in chambers. Also, unsuspected chemical perturbations to a system under study may occur as a result of unknown materials present within the initial reaction mixture. These effects lead to discrepancies between experimental and model predictions that can be attributable to incomplete descriptions of the experimental conditions, and not to inadequacies in the model. Therefore, computer simulation runs should not be carried out to force a model to fit chamber observations, but rather to determine whether enough detail have been included in the model to at least predict the bulk trends observed.

An up-to-date and careful set of chamber experiments that are useful for model comparison has been carried out at the Statewide Air Pollution Research Center at the University of California, Riverside (UCR). Figure 2 shows concentration data obtained by the UCR group (Pitts et al., 1977) for their EC-11 chamber run, which we have chosen for simulation comparisons because the initial NO_x–olefin concentrations employed in the study are similar to that of a polluted urban environment. However, before trying to model the observed concentration profiles, it was first necessary to modify ATMOS for some specific experimental conditions of the chamber run. Thus

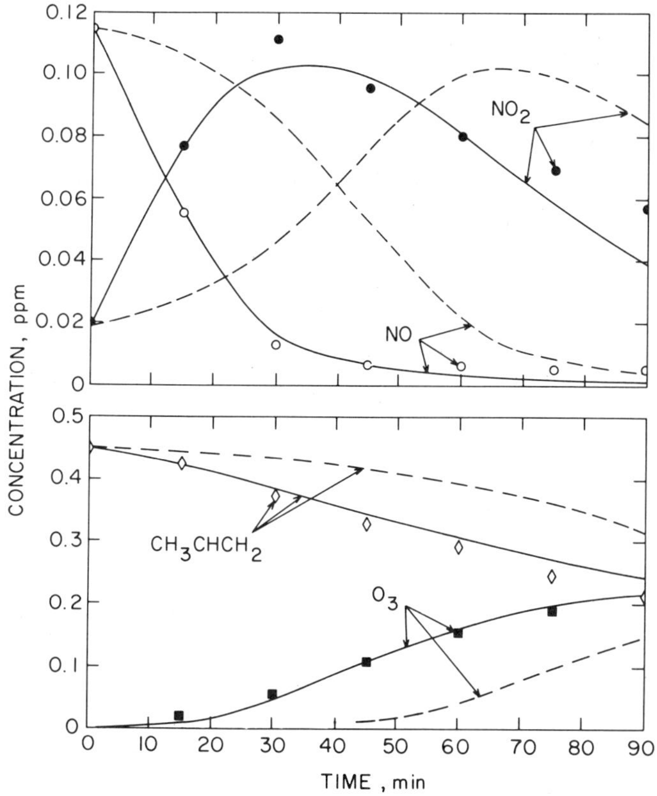

Figure 2. Simulation of smog-chamber data [Pitts et al. (1977), run EC-11], using ATMOS as modified for experimental conditions (see text), employing initial HONO concentrations of zero (---) or 8 ppb (——).

all photolytic rate constants were adjusted relative to the experimentally observed j_1 value of 0.22 min^{-1}. In addition, the rate constants for olefin reactions (k_{19}, k_{20}, k_{21}) were adjusted (6.8×10^3, 2.5×10^4, 1.5×10^{-2} ppm^{-1} min^{-1}) to reflect the use of propylene as the only alkene reported present during the course of the experiment. Without further modification of the model, we have carried out a simulation using ATMOS with the initial concentrations reported for NO, NO_2, O_3, CO, and CH_3CHCH_2. This simulation has also been repeated, assuming the presence of some initial but unreported HONO concentration, since heterogeneous formation of HONO during gas handling may be significant and the need for HONO as an additional radical source has been postulated in previous modeling studies of the UCR data (Duewer et al., 1978; Whitten and Hogo, 1977). The results of the present simulations are shown in Figure 2, where it is apparent that ATMOS is not only capable of predicting the general trends observed in the experiment, but also gives particularly good fits to the data when HONO is assumed to be initially present at its NO_x–H_2O equilibrium concentration. The agreement possible between ATMOS predictions and chamber observations gives the confidence desired in this model.

2.3. Simulation Results

Using ATMOS as listed in Table 2, we have simulated the 6-hour irradiation of a tropospheric mixture of several important reactants at initial concentrations typical of urban air-pollution levels so that the resulting computer predictions might be used as a "standard" for the development of a surrogate model. Thus, to a mixture of NO, NO_2, and SO_2 at 0.075, 0.025, and 0.01 ppm, we have included olefinic hydrocarbon and CO at 0.1 and 10 ppm, respectively, as well as background H_2O and O_2 at 2×10^4 and 2×10^5 ppm, respectively. The concentration-time profiles obtained in this simulation run, which are in comparatively good agreement with the predictions of previous modeling studies employing similar initial conditions (Calvert and McQuigg, 1975), are shown in part by the solid curves given in Figure 3. The points shown in this and all succeeding figures may be ignored for the present since they are the results of simulations with a reduced surrogate model to be discussed later.

The simulation curves given in Figure 3 are the theoretical predictions of what may be expected for the simplest model of atmospheric transport

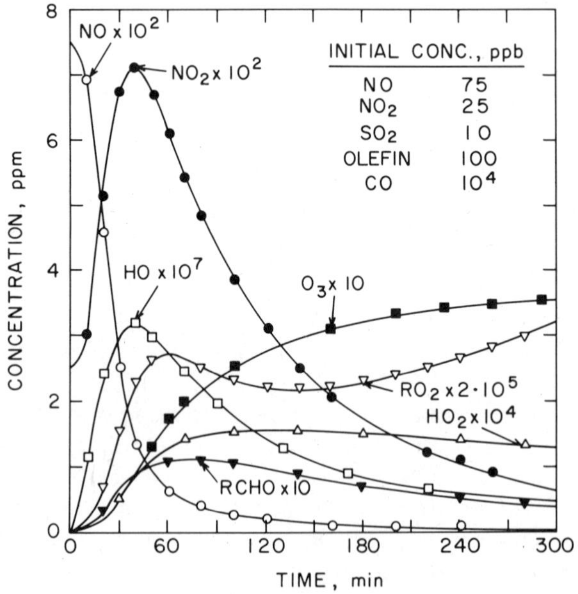

Figure 3. Comparison of simulation results predicted by ATMOS (curves) and SCHEME (points) for the "standard" initial pollutant concentration mixture shown.

Table 3. HO and HO_2 Formation and Loss Rates[a] Predicted by ATMOS at 10 and 300 min into Simulation Run with "Standard" Initial Conditions[b]

	$d[HO]/dt$		$d[HO_2]/dt$	
Reaction Number	10 min	300 min	10 min	300 min
HO Formation and HO_2 Loss Reactions				
[A14]	0.577	0.058		
[A17]	13.1	1.33	−13.1	−1.33
[A18]	—[c]	1.11		
[A21]	0.386	0.018		
[A30]			−0.007	−2.92
[A31]	—[c]	0.002	—[c]	−0.002
[A34]			−0.011	−0.051
[A35]			−0.547	−2.55
[A38]	—[c]	0.015	—[c]	−0.015
Overall rate	14.1	2.53	−13.7	−6.87
HO Loss and HO_2 Formation Reactions				
[A15]	−0.630	−0.062		
[A16]	−1.22	−0.003		
[A19]			0.045	—[c]
[A20]	−7.31	−0.004		
[A22]			0.354	1.50
[A23]	−0.222	−0.425	0.111	0.213
[A27]			7.95	0.544
[A33]	−4.62	−2.04	4.62	2.04
[A36]			0.526	2.56
[A37]	−0.005	—[c]		
[A40]	−0.016	−0.006	0.016	0.006
Overall rate	−14.0	−2.54	13.6	6.86

[a] Units = ppm $min^{-1} \times 10^4$.

[b] Initial NO, NO_2, SO_2, olefin, CO, H_2O, and O_2 concentrations are 0.075, 0.025, 0.01, 0.1, 10, 2×10^4, and 2×10^5 ppm, respectively.

[c] Formation or loss rate less than 10^{-7} ppm min^{-1}.

processes, namely, that in which an air parcel moves out of the source emission region into a "clean" environment without experiencing any dilution. However, even for such a simple box-model picture of the atmosphere, the chemical changes occurring within the system are quite complex, as exemplified by the rates of radical-species reactions given in Tables 3 and 4. For example, it is clear from these tables that at early times (10 min) the radical oxidation of NO to NO_2 in reaction [A17] dominates the chain-reaction rates for the loss of HO_2 and the formation of HO. Although the HO thus formed will then react to some extent with both NO_2 (5%) and NO (9%), it is removed principally by olefin (52%) in reaction [A20] to produce RO_2 and aldehyde and by CO (33%) in reaction [A33] to reform the HO_2

Table 4. RO and RO_2 Formation and Loss Rates[a] Predicted by ATMOS at 10 and 300 min into Simulation Run with "Standard" Initial Conditions[b]

	$d[RO]/dt$		$d[RO_2]/dt$	
Reaction Number	10 min	300 min	10 min	300 min
RO Formation and RO_2 Loss Reactions				
[A24]	8.01	0.531	−8.01	−0.531
[A31]	—[c]	0.002	—[c]	−0.002
[A32]	—[c]	—[c]	—[c]	—[c]
[A39]	—[c]	0.011	—[c]	−0.011
Overall rate	8.01	0.544	−8.01	−0.544
RO Loss and RO_2 Formation Reactions				
[A19]			0.09	—[c]
[A20]			7.31	0.004
[A22]			0.118	0.50
[A25]			0.491	0.04
[A27]	−7.95	−0.544		
[A28]	−0.023	—[c]		
[A29]	−0.028	—[c]		
[A42]			—[c]	—[c]
[A43]	—[c]	—[c]		
Overall rate	−8.00	−0.544	8.01	0.544

[a] Units = ppm $min^{-1} \times 10^4$.

[b] Initial conditions as in Table 3.

[c] Formation or loss rate less than 10^{-7} ppm min^{-1}.

radical. In turn, the further oxidation of NO to NO_2 in reaction [A24] reduces the RO_2 radical to RO, which then reacts with molecular oxygen to complete the cycle by forming HO_2 as well as aldehyde. This sequence of reactions results in an early buildup of NO_2 and aldehyde at the expense of NO and olefin, respectively.

As the NO and NO_2 become chemically depleted from the system through the formation of sinks such as $HONO_2$ and PAN, the previously unimportant HO_2–radical recombination reaction becomes increasingly more important so that after 5 hours of simulation time the formation of H_2O_2 in reaction [A30] represents 43% of the HO_2 loss mechanism while reaction [A17] now accounts for only 19%. The remaining 38% of HO_2 loss is due to the formation of peroxynitric acid in reaction [A35]; however, this reaction does not significantly influence the HO_2 concentration level since the HO_2NO_2 rapidly dissociates back to HO_2 and NO_2 in reaction [A36]. Also, at this later time the photolysis of H_2O_2 in reaction [A18] competes efficiently with reaction [A17] as a source of the HO radical, which now reacts predominantly with an essentially unaltered CO concentration to complete the cycle by reforming HO_2. In addition, aldehydes have now increased in concentration to a level at which the photolytic process in reaction [A22] becomes an important source of both HO_2 and RO_2, with the latter again leading to the formation of HO_2 and aldehyde through reactions [A24] and [A27].

As also indicated in Tables 3 and 4, SO_2 is oxidized during the course of the simulation run principally by reactions with HO_2, RO_2, and HO and to a much lesser extent (<1%) by reactions with $O(^3P)$, $RCOO_2$, and RO (reactions [A38]–[A43], respectively); the rate of oxidation by $O(^3P)$, not given in Tables 3 and 4, is generally less than 10^{-8} ppm min^{-1}. Thus it is clear that HO_2, RO_2, and HO account for essentially all the SO_2 conversion. As shown in Figure 4*a*, at 10 min into the simulation the HO radical is responsible for nearly 90% of the SO_2 oxidation rate, whereas HO_2 and RO_2 each contribute about 5%. With the passage of time, a decrease in NO_x and olefin concentrations results in a corresponding decrease in the importance

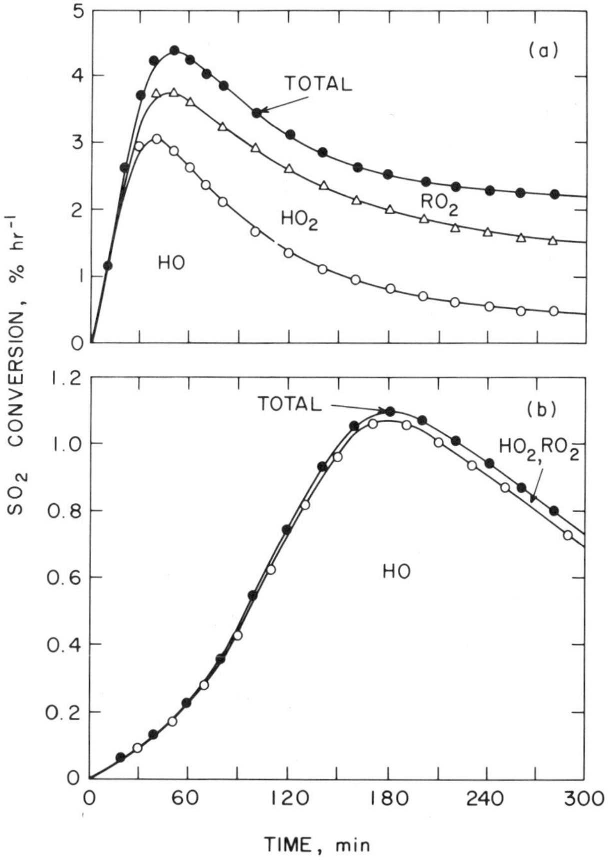

Figure 4. Contributions to the total SO_2 oxidation rates predicted by ATMOS (curves) and SCHEME (points) for (a) the "standard" initial pollutant concentration mixture and (b) a tenfold increase in the initial NO_2 and SO_2 concentrations.

of primary HO radical sources (e.g., reactions [A1], [A2], and [A21]), whereas aldehyde photolysis becomes increasingly important for initiation of HO_2 and RO_2 radical chains; therefore, after 5 h of simulation time the HO is responsible for only 20% of the SO_2 oxidation, with HO_2 and RO_2 now contributing 47 and 33%, respectively.

Although the free radicals in the NO_x–hydrocarbon system are responsible for SO_2 oxidation, a further important point to note from Tables 3 and 4 is that the fraction of HO_2, RO_2, and HO radicals that react with SO_2 is quite small under typical pollution conditions (although we note that these fractions may become high for stack plume conditions). Consequently, it is seen that sulfur reactions exert only a minor influence on the NO_x–O_3–hydrocarbon system; thus reactions [A38]–[A44] may be eliminated from ATMOS by those not interested or concerned with the SO_2 conversion processes.

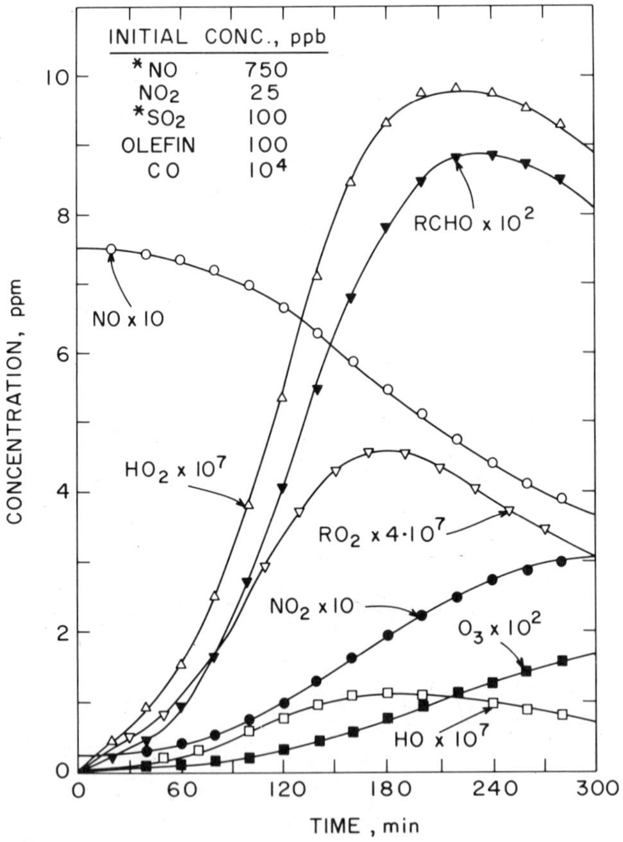

Figure 5. Simulation results obtained with ATMOS (curves) and SCHEME (points) for the initial pollutant mixture shown, representing a tenfold increase in initial concentrations of NO and SO_2 above the "standard" concentration.

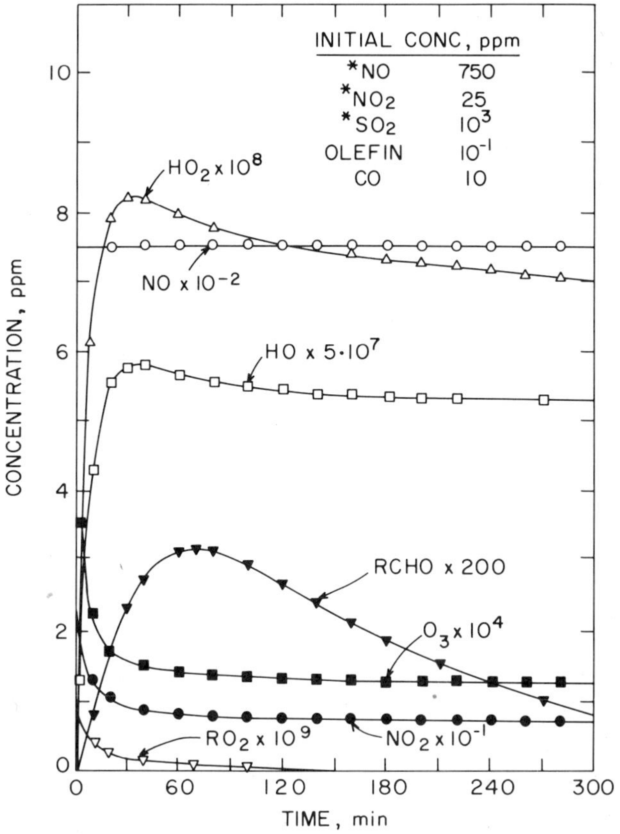

Figure 6. Simulation results obtained with ATMOS (curves) and SCHEME (points) for the initital pollutant concentration mixture shown, representing 10^3-, 10^4-, and 10^5-fold increases above the "standard" pollutant concentrations of NO_2, NO, and SO_2, respectively.

Although the above simulation results have been used as the primary source of information in developing a reduced model, additional simulation runs have been carried out with ATMOS in which the initial concentrations of NO, NO_2, SO_2, or olefin have been increased and decreased by an order of magnitude as well as being set equal to the more extreme values that might be representative of a power plant plume close to the emission stack. In addition, a simulation has been carried out employing the "standard" initial concentration mixture but with a tenfold decrease in all photolytic rate constants contained in ATMOS. Some of the results of these simulation runs are given in Figures 4*b*–10 (solid curves). It should be emphasized that the conditions of these simulations were chosen not to represent realistic atmospheric situations, but rather to exercise ATMOS over a broad range of kinetic conditions to permit the validation of any reduced model acting as a surrogate for the more complete model. For example, in comparison to the curves given in Figure 3, Figures 5 and 6, obtained for high initial NO concentrations, show suppressions in the predicted O_3 and HO concentrations, which correspond to increases in the relative importance of NO reactions [A3] and [A16]. This, in turn, diminishes the importance of the RO_2 and HO_2 formation reactions [A20] and [A33], which, coupled with the increased importance of RO_2 and HO_2 removal by reactions [A24] and [A17], reduce by several orders of magnitude the maximum predicted RO_2 and HO_2 concentrations. As illustrated in Figure 4*b*, such changes in radical concentration levels reduce the HO_2 and RO_2 contributions to the SO_2 conversion processes while also decreasing the overall oxidation rate. Although there is a greater net suppression in the maximum radical concentration levels achieved in Figure 6 than in Figure 5, it is apparent from these figures that at early times the net rate of increase of radical concentrations is greater for the conditions represented by Figure 6. Similar trends are also seen in a comparison of Figures 3 and 7 as a result of a boost in the early NO_2 photolysis rate due to the higher NO_2 concentrations employed in the starting conditions represented by Figures 6 and 7.

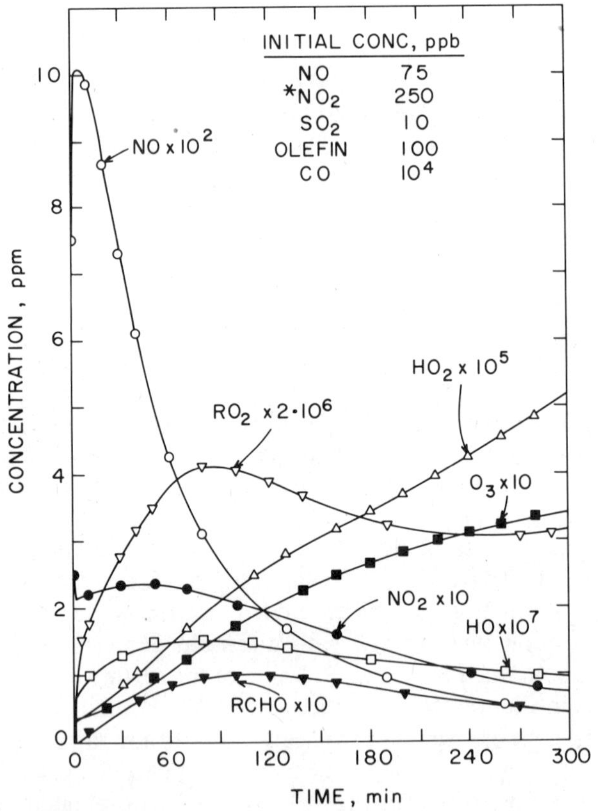

Figure 7. Simulation results obtained with ATMOS (curves) and SCHEME (points) for the initial pollutant concentration mixture shown, representing a tenfold increase in initial concentration of NO_2 above the "standard" concentration.

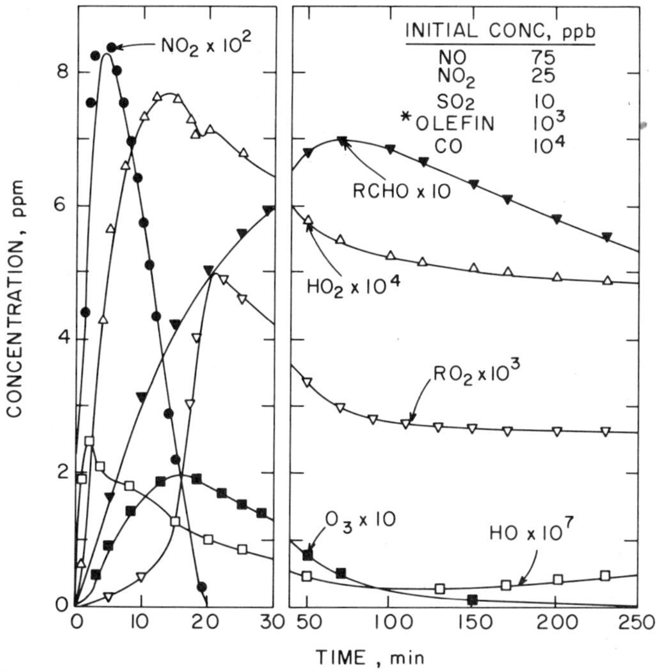

Figure 8. Simulation results obtained with ATMOS (curves) and SCHEME (points) for the initial pollutant concentration mixture shown, representing a tenfold increase in initial concentration of olefin above the "standard" concentration.

As shown by a comparison of Figures 3 and 8, increases in the net formation rates of radicals at early times also occur in the presence of excess olefin concentrations. In this case increased rates of reactions [A20] and [A21] relative to [A3], [A15], [A16], [A33], and [A40] lead to enhanced production of both RO_2 and RCHO with a corresponding increase in the importance of aldehyde photolysis. Thus, as shown in Figure 8, whereas the O_3 goes through a maximum and is subsequently depleted by reaction [A21], the increasing importance of reaction [A22] helps to maintain the elevated HO_2 and RO_2 concentrations and contributes as well to the appearance of a second maximum on the $[HO_2]$ profile (i.e., at 20 min). This increase in the HO_2 and RO_2 radical concentration levels results in the rapid and complete conversion of SO_2 to sulfate as shown in Figure 9. For instance, at 10 min into the run the SO_2 oxidation rate is 28% h^{-1} with the HO, HO_2, and RO_2 being responsible, respectively, for 6, 20, and 74% of the overall rate.

As expected and as shown in Figure 10 for the "standard" initial concentration condition, a reduction in photolytic rate constants inhibits the chemical activity of the system so that, in contrast to the predictions obtained with high initial $[NO_2]$ or [OLEF], there is now a decrease in the net formation rates of radicals that results in concentration maxima occurring at times much later than those represented in Figure 3. Clearly, the requirement

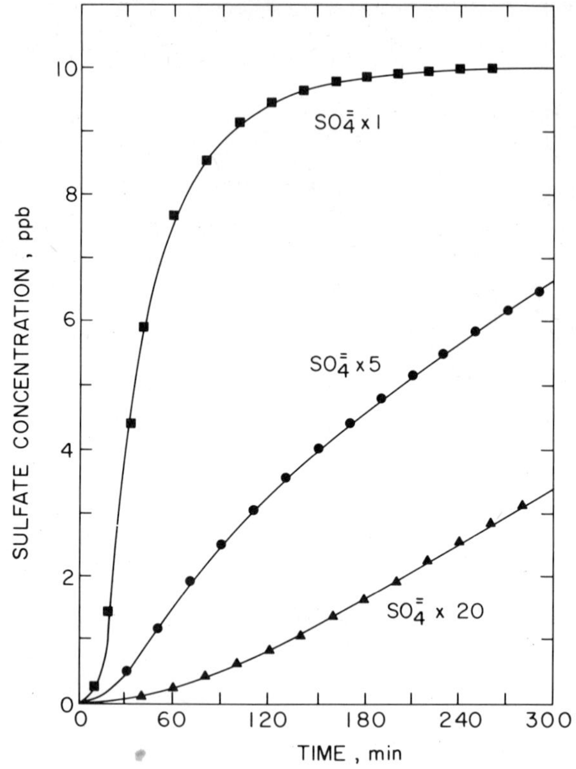

Figure 9. Sulfate concentration profiles obtained with ATMOS (curves) and SCHEME (points) for standard initial pollutant concentrations (●) and a tenfold increase (■) or decrease (▲) above or below the initial olefin concentration.

that a reduced model accurately reproduce the simulation results of the parent mechanism for the wide range of conditions represented in Figures 3–10 presents a severe test for the validity of such a surrogate model.

3. METHODS FOR MODEL REDUCTION

Several methods have been employed in eliminating species from ATMOS and developing a surrogate model. As shown by the examples given below, these methods of model reduction are such that the surrogate no longer consists of a set of elementary reactions. Rather, the reduced model consists in large part of composite reactions and rate coefficients (γ values) that at first glance may appear strange and even unacceptable to the chemist; i.e., some surrogate reactions may contain a species as both reactant and product whereas others may not even appear to conserve matter. However, for all species explicitly modeled by the surrogate mechanism, these methods of model reduction retain both the reaction stoichiometry and rate information necessary to give a set of ordinary differential equations representing the kinetics associated with the parent model (i.e., ATMOS).

3.1. Sink Species

Several of the chemical constituents modeled in ATMOS are end products or sinks of reaction processes and do not themselves undergo any appreciable subsequent reaction in the polluted troposphere that could affect the chemistry of the other species being modeled. Therefore, these reaction sinks, which include $HONO_2$, PAN, RONO, $RONO_2$, H_2SO_4, $ROSO_2$, H_2, and CO_2, may be eliminated from ATMOS even if they are produced in relatively fast and important reactions. Thus reaction [A15], for example, is replaced by surrogate reaction

$$NO_2 + HO \rightarrow (HONO_2), \qquad [S12]$$

where the parentheses indicate that the $HONO_2$ has not been explicitly

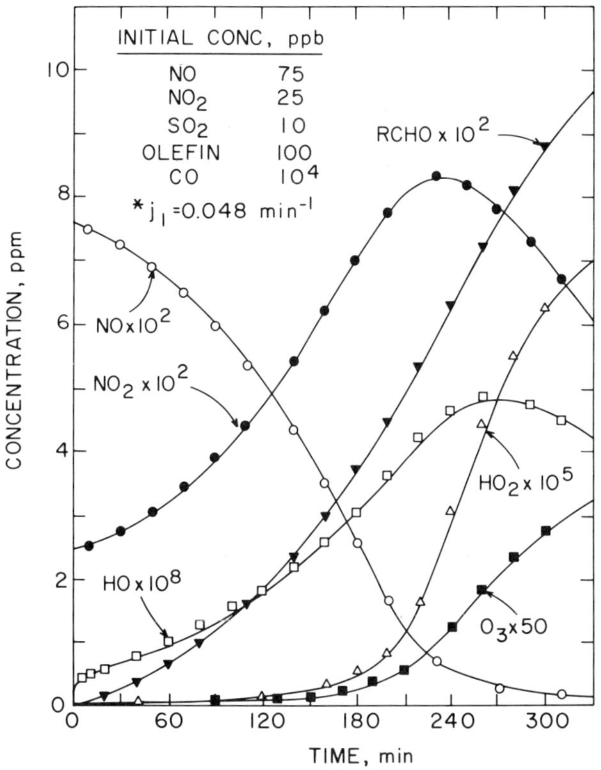

Figure 10. Simulation results obtained with ATMOS (curves) and SCHEME (points) for the "standard" initial pollutant concentrations but with a tenfold decrease in photolytic rate constants, j_1, j_{14}, j_{18}, and j_{22}.

included in the model; i.e., no differential equation is required for $HONO_2$ for execution of the model. Thus the incorporation of $HONO_2$ within the model becomes optional to the user. Similarly, reaction [A40] becomes

$$SO_2 + HO \rightarrow HO_2 + (H_2SO_4). \qquad [S43]$$

Note that we have introduced reaction numbers here preceded by the letter S to indicate reactions that are included in the final surrogate mechanism.

This method of handling sink species must by nature result in the loss of information concerning the concentration history of the species being eliminated, with the possible exception of total sulfate ($H_2SO_4 + ROSO_2$), which might be computed by mass balance assuming the absence of other sink processes (e.g., deposition). In general, if one is interested in the time-concentration profile of a particular sink species, e.g., $HONO_2$ or H_2SO_4, this species should be explicitly retained in the mechanism.

3.2. Stable Species

In addition to stable–unreactive sinks, some reactive species modeled by ATMOS are present in such excess that they also may be considered stable because their concentrations are essentially unaffected by the chemical reactions taking place. These stable species include the CO, H_2O, and O_2 molecules. Of these, CO by far shows the greatest change, and its concentration decreases by only ~1% over the course of a 6-h simulation run. As in the case of the sinks, these stable species may be simply eliminated from reactions in which they are products, since their formation in these reactions will not enhance the rates of any process in which they act as reactants. In contrast, such a stable species cannot be simply eliminated from a reaction in which it participates as a reactant, because its presence is necessary in determining the rate of that reaction; however, in this case the species may still be eliminated in a straightforward and even trivial manner by incorporating its concentration in the corresponding rate coefficient γ. Reaction [A33], for example, can be replaced by

$$HO \rightarrow HO_2, \qquad [S39]$$

having rate coefficient $\gamma_{39} = k_{33}[CO]$; note that the product sink species CO_2 is also eliminated.

3.3. Unimportant Reactions

An analysis of the time dependence of species concentrations and reaction rates obtained with ATMOS for all reasonable pollution conditions shows that some reactions may be considered unimportant in influencing, directly or indirectly, the rates of formation or loss of pollutants such as NO, NO_2,

SO_2, and O_3. For example, HO_2NO_2 has been eliminated from ATMOS by deleting reactions [A35] and [A36], since the first-order decay of this species in the latter reaction quickly reforms the HO_2 and NO_2 consumed in the former reaction (see Table 3). In addition, it is apparent from Table 4 that for "standard" pollution conditions, essentially all the RO radical formed in the system reacts with O_2 in reaction [A27]; thus it might have been possible to eliminate the $RONO_2$, RONO, and $ROSO_2$ species by deleting reactions [A28], [A29], and [A43], respectively, rather than having treated these species as sinks (Section 3.1). In fact, this method of model reduction was applied to the RO reactions with some success in a first attempt at formulating a surrogate model (Levine and Schwartz, 1978); however, differences were found between the predictions of the surrogate model and ATMOS for some simulation runs involving initial concentration mixtures other than the "standard" condition. For example, the substantial levels of NO_2 and NO predicted by ATMOS in the simulation run with high initial NO (see Figure 5) are responsible for the loss of some RO by reactions [A28] and [A29], respectively. Although these reactions together account for less than 4% of the RO loss at any given instant during this run, a surrogate model that assumes all the RO to react with O_2 will after 6 hours of simulation time give errors of 10–40% in the concentrations predicted for NO_2, NO, and O_3. Thus care must be taken in reducing a chemical model by this method of deleting apparently unimportant reactions. In the present study this method has been limited to the elimination of HO_2NO_2 as described above.

3.4. Rate-Determining Reactions

The rapid hydrolysis of SO_3 in the troposphere by reaction [A44] will quickly establish a low-level steady-state concentration of this species (10^{-13} ppm predicted for "standard" conditions). Consequently, in the apparent absence of any other competing reaction path, the SO_3-to-H_2SO_4 conversion will be determined entirely by the rate of SO_3 formation. The SO_3 molecule has thus been eliminated by deleting reaction [A44] and replacing SO_3 with H_2SO_4 in reactions [A38], [A39], [A41], and [A42]; e.g., reaction [A38] is replaced by

$$SO_2 + HO_2 \rightarrow HO + (H_2SO_4). \qquad [S42]$$

Similarly, as indicated above, for the "standard" pollution condition the important loss mechanism in establishing the RO steady state is reaction [A27] so that the rate of this reaction is determined by the rate of RO production in reactions [A24], [A31], [A32], and [A39]. Under these conditions RO could be eliminated by deleting reaction [A27] (in addition to reactions [A28], [A29], and [A43]) and replacing RO in all its formation steps by the products of this loss step. However, for the same reason given above, this method of model reduction must also be treated with caution and for the present study has not been applied to the elimination of RO or any species other than SO_3 (as well as H, as discussed in Section 2.1.

3.5. Steady-State Approximations

The concentration of an intermediate species X being formed at a rate $F(t)$ and removed with a characteristic lifetime $\tau_X(t)$ is well approximated by the relationship

$$[X](t) = F(t)\tau_X(t),$$

at times substantially exceeding $\tau_X(t)$, provided neither $F(t)$ nor $\tau_X(t)$ varies greatly over such a characteristic lifetime (Leighton, 1961, pp. 109–113). For a species meeting this criterion, the steady-state condition applies (i.e., $d[X]/dt = 0$), and its concentration may be related to and thus approximated from the concentrations of those reactants responsible for the formation and loss rates. In the present study such relationships are used to construct a set of surrogate reactions and rate coefficients. This treatment obviates the necessity of explicitly modeling the steady-state species N_2O_5, NO_3, $O(^3P)$, and RO while retaining the chemical influence of these species on the remainder of the system. As is shown below, this method of model reduction is cumulative in that a surrogate reaction replacing any one of these species while still incorporating another may be further surrogated during the removal of the latter.

We proceed with this method by first applying it to the elimination of N_2O_5. The steady-state concentration of N_2O_5 is established by reactions [A6]–[A8] and is given by the relationship

$$[N_2O_5]_{ss} = k_6\tau_{N_2O_5}[NO_2][NO_3],$$

where $\tau_{N_2O_5}$ is defined by

$$\tau_{N_2O_5}^{-1} = k_7 + k_8[H_2O].$$

This relationship is readily derived by making the steady-state assumption that $d[N_2O_5]/dt = 0$ and hence equating the N_2O_5 formation and loss rates; i.e., any N_2O_5 produced by the NO_2–NO_3 reaction

$$NO_2 + NO_3 \rightarrow N_2O_5, \qquad [A6]$$

being immediately lost in forming the products of reactions [A7] and [A8],

$$N_2O_5 \rightarrow NO_2 + NO_3 \qquad [A7]$$

$$N_2O_5 + H_2O \rightarrow 2HONO_2. \qquad [A8]$$

By employing this steady-state relationship, the rate of N_2O_5 loss in reaction

[A7] ($R_{A7} = k_7[N_2O_5]$) may be written as

$$R_{A7} = k_6 k_7 \tau_{N_2O_5}[NO_2][NO_3],$$

which is the rate of N_2O_5 formation in reaction [A6],

$$R_{A6} = k_6[NO_2][NO_3],$$

times the fraction of N_2O_5 reacting in reaction [A7],

$$\frac{R_{A7}}{R_{A7} + R_{A8}} = \frac{k_7[N_2O_5]_{ss}}{k_7[N_2O_5]_{ss} + k_8[N_2O_5]_{ss}[H_2O]} = k_7\tau_{N_2O_5}.$$

Thus R_{A7} is the rate of the process in which the NO_2–NO_3 reaction leads directly to the products of reaction [A7], and the surrogate reaction describing this process is obtained simply by the addition of reactions [A6] + [A7]:

$$NO_2 + NO_3 \rightarrow NO_2 + NO_3, \qquad \text{[N1]}$$

which according to our expression for R_{A7} must have a rate coefficient $\gamma_{N1} = k_6 k_7 \tau_{N_2O_5}$. This reaction is a null process in that it does not represent any net chemical change and hence is not included in the surrogate mechanism (we introduce reaction numbers preceded by the letter N here to indicate surrogate reactions that are not included in the final reduced model).

In a manner similar to that just described for treating the process by which reaction [A6] leads to the products of reaction [A7], the process whereby an NO_2–NO_3 reaction yields the formation of two nitric acid molecules by reaction [A8] is given by the reaction sum [A6] + [A8],

$$NO_2 + NO_3 \rightarrow 2(HONO_2), \qquad \text{[N2]}$$

and proceeds at a rate equal to that of reaction [A8],

$$R_{N2} = R_{A8} = k_8[N_2O_5][H_2O].$$

By again employing the $[N_2O_5]_{ss}$ approximation, the rate of reaction [N2] is expressed as

$$R_{N2} = k_6 k_8 \tau_{N_2O_5}[NO_2][NO_3][H_2O],$$

so that

$$\gamma_{N2} = k_6 k_8 \tau_{N_2O_5}[H_2O].$$

Note that the H_2O has been included in γ_{N2} as a stable species achieving the elimination of this species as an explicit reagent. Although reaction [N2] retains the kinetic information contained in reactions [A6]–[A8] and alone suffices to account for N_2O_5 formation and loss, this reaction also contains kinetic information related to the NO_3 steady state and hence is not included in the reduced model but is subjected to further surrogation along with the remaining reactions involving NO_3.

The reactions leading to the formation of NO_3 are

$$NO_2 + O_3 \rightarrow NO_3 + O_2 \qquad [A4]$$

$$NO_2 + O \rightarrow NO_3, \qquad [A11]$$

and those leading to removal are

$$NO + NO_3 \rightarrow 2NO_2 \qquad [A5]$$

$$NO_2 + NO_3 \rightarrow 2(HONO_2), \qquad [N2]$$

where [N2] was obtained above in the surrogation for N_2O_5 reactions. The resulting expression for $[NO_3]_{ss}$ is

$$[NO_3]_{ss} = k_4\tau_{NO_3}[NO_2][O_3] + k_{11}\tau_{NO_3}[NO_2][O]$$

where

$$\tau_{NO_3}^{-1} = k_5[NO] + \gamma_{N2}[NO_2].$$

It should be noted that an evaluation of τ_{NO_3} will not give the characteristic NO_3 lifetime since we have excluded the null reaction [N1] as part of the NO_3 loss mechanism; however, this does not invalidate the above expression for $[NO_3]_{ss}$ since the elimination of reaction [N1] equally affects the formation and loss rates responsible for establishing the steady-state concentration. We now address the rate of the surrogate reaction obtained by the sum [A4] + [A5],

$$NO_2 + O_3 + NO \rightarrow 2NO_2 + (O_2), \qquad [A4+A5]$$

which may be evaluated as R_{A4} times the fraction of the NO_3 loss rate that occurs by [A5],

$$R_{A4+A5} = R_{A4}\frac{R_{A5}}{R_{A5} + R_{N2}} = k_4k_5\tau_{NO_3}[NO_2][O_3][NO].$$

Denoting the sum [A4] + [A5] as surrogate reaction [S1] and deleting the stable species O_2, we obtain

$$NO_2 + O_3 + NO \rightarrow 2NO_2 \qquad [S1]$$

with rate coefficient

$$\gamma_1 = k_4k_5\tau_{NO_3}.$$

We proceed similarly to obtain the three other surrogate reactions resulting from the elimination of NO_3,

$$2NO_2 + O_3 \rightarrow 2(HONO_2) \qquad [S2]$$

$$NO_2 + NO + O \rightarrow 2NO_2 \qquad [N3]$$

$$2NO_2 + O \rightarrow 2(HONO_2), \qquad [N4]$$

with rate coefficients

$$\gamma_2 = k_4 \gamma_{N2} \tau_{NO_3}$$

$$\gamma_{N3} = k_5 k_{11} \tau_{NO_3}$$

and

$$\gamma_{N4} = k_{11} \gamma_{N2} \tau_{NO_3}.$$

Reactions [S1] and [S2] are included in our surrogate mechanism (Table 5), whereas [N3] and [N4] are subjected to further surrogation to eliminate the $O(^3P)$ atom.

The $O(^3P)$ steady-state concentration is determined by the formation reaction [A1] and the loss reactions [A2], [A9], [A10], [A19], and [A41] as well as the temporary surrogate reactions [N3] and [N4]. The resulting steady-state relationship obtained with these reactions is

$$[O]_{ss} = j_1 \tau_O [NO_2]$$

where

$$\tau_O^{-1} = k_2[O_2] + k_9[NO] + k_{19}[OLEF] + k_{41}[SO_2] + (k_{10} + \gamma_{N3}[NO] + \gamma_{N4}[NO_2])[NO_2]$$

By employing this $[O]_{ss}$ relationship in the several rate expressions for oxygen-atom loss, we obtain rate coefficients that correspond to the surrogate reactions resulting from the addition of reaction [A1] with reactions [A2], [A9], [A10], [A19], [A41], [N3], and [N4]. Of the seven surrogate reactions thus obtained, only five are incorporated within the reduced model and are shown by reactions [S3]–[S7] in Table 5; the remaining two, obtained by the additions of reactions [A1] + [A9] and [A1] + [N3], are null processes.

In the case of RO the construction of a set of surrogate reactions and rate coefficients is with one modification identical to the procedure for NO_3 elimination. The reaction stoichiometry of RO in the formation reaction [A32] is 2, compared to unity in all other reactions involving this species. Thus the addition of reaction [A32] with any of the RO loss steps does not directly delete this species from the reaction mechanism. To correct this situation, reaction [A32] is rewritten as

$$2RO_2 \rightarrow RO + RO_2 + \tfrac{1}{2}(O_2) \qquad [N5]$$

which retains all the kinetic information contained in reaction [A32] if we set $\gamma_{N5} = 2k_{32}$. With reaction [N5] now replacing [A32], the RO steady-state relationship may be written as

$$[RO]_{ss} = \tau_{RO}[RO_2](k_{24}[NO] + k_{31}[HO_2] + k_{39}[SO_2] + \gamma_{N5}[RO_2])$$

where

$$\tau_{RO}^{-1} = k_{27}[O_2] + k_{28}[NO_2] + k_{29}[NO] + k_{43}[SO_2].$$

By employing this $[RO]_{ss}$ expression in a manner analogous to that used in the elimination of NO_3, one readily derives a set of rate expressions and thus rate coefficients corresponding to the surrogate reactions obtained by the addition of each formation reaction ([A24], [A31], [A39], [N5]) to each loss reaction ([A27], [A28], [A29], [A43]). The resulting total of 16 surrogate reactions and rate coefficients are given by reactions [S20]–[S35] in Table 5.

Although the steady-state concentrations of HO and $RCOO_2$ might also be represented by fairly simple and uncoupled algebraic expressions, the steady-state approximation method is not presently applied to the elimination of these species. In the case of HO this procedure becomes highly tedious, requiring a minimum of 42 surrogate reactions for removal of this radical as an explicitly modeled species. In contrast to the case for HO, only nine reactions would be necessary for replacement of the $RCOO_2$ radical; however, we have found that a reduced model involving the elimination of $RCOO_2$ by the steady-state approximation method failed to accurately reproduce the simulation results of ATMOS for the condition of high initial olefin because the condition for validity of the steady-state approximation on $[RCOO_2]$ was no longer satisfied during the course of that simulation run. This failure to satisfy steady-state conditions for $RCOO_2$ in the presence of high olefin concentrations appears to reflect a potential inadequacy in the present version of the parent model ATMOS. As indicated in Figures 8 and 9, the presence of high olefin concentrations results in the rapid loss of NO, NO_2, and SO_2 from the reaction mixture, thus effectively eliminating all sinks for the $RCOO_2$ radical. In actuality, other sink paths would become important under such conditions, e.g., radical–radical reactions. Therefore, in addition to reactions [A25], [A26], and [A42], future versions of ATMOS should contain such additional $RCOO_2$ sink reactions. In the meantime, however, we have retained $RCOO_2$ as an actively modeled species for preservation of the match between SCHEME and ATMOS.

4. SURROGATE CHEMICAL MECHANISM (SCHEME)

4.1. Mechanism

Using the methods of model reduction outlined in Section 3, we have developed the surrogate mechanism SCHEME that simulates the chemistry of ATMOS but employs only the first 12 species listed in Table 1. This SCHEME, consisting of 44 surrogate reactions, is presented in Table 5. Also given are expressions for the rate coefficients (γ values) corresponding to these reactions as obtained by the methods described above. As mentioned earlier, these methods of surrogation retain both reaction stoichiometry and rate information from ATMOS so that despite the strange appearance of SCHEME, a set of ordinary differential equations may be derived in the usual manner to describe the kinetics of the 12 species being modeled. These differential equations are given in Table 6.

With respect to utilizing this SCHEME, it is convenient to note that each γ value may be expressed as the product of up to three factors, a constant factor K comprised of elementary rate constants for reactions of ATMOS in Table 2, a second constant factor C representing the concentrations of stable species, and a variable τ term. In practice, the K factors may be input during computer compile time, and C factors may be set at the beginning of execution, whereas τ values would have to be calculated during execution and continuously updated for use at each time step, using the appropriate reactant concentrations determined in the previous time step. This last procedure is analogous to using variable concentrations in the integration of species rate expressions. The separation of γ values into these three components is of considerable user convenience since it permits any γ to be readily evaluated for modifications in the selection of elementary rate constants, concentrations of stable species, or reactions determining the τ relationships.

4.2. Comparison with ATMOS

To determine the capability of the reduced model to accurately reproduce the modeling results obtained with ATMOS, simulations have been carried out with SCHEME for the same widely varying initial concentration conditions employed in the ATMOS runs presented above. Comparisons of the simulation results obtained with SCHEME (points) and with ATMOS (curves) are shown in Figures 3–10, where the choice of curves and points for the representation of the predictions of ATMOS and SCHEME, respectively, is purely arbitrary and used only for the sake of clarity. As shown by these Figures 3–10, the concentration profiles obtained with SCHEME for the various sets of chemical conditions are in excellent agreement with those obtained with ATMOS, in general differing by less than 1% over the 6-h course of simulated irradiation time. Thus it is clear that our methods of model reduction have retained all the kinetic information necessary for SCHEME to accurately replace ATMOS even over the wide range of conditions chosen to put SCHEME to the test.

In addition to reducing the number of species required to model a reaction system, a further advantage of using SCHEME is a substantial savings in the computer time (CPU time) required to carry out a simulation run. For example, the actual CPU times (seconds on a CDC 7600) required in executing both ATMOS and SCHEME to obtain the simulation results given in Figure 3 are shown in Figure 11 as functions of simulation time (minutes). The slope of the linear portion of the curve shown for the ATMOS run is 0.27 CPU seconds per simulation minute, which is 3.3 times larger than the slope obtained with the use of SCHEME; thus if the simulation is run with the reduced model, a 70% savings in execution time results. Although the present computer code has not been optimized for execution speed, this savings in exccution time is probably representative of that achievable with such an optimized program. Also, note that an extrapolation of these lines back to the start of the simulated irradiation period gives an intercept for ATMOS that is approximately 30 CPU seconds larger than that for SCHEME. This difference in CPU intercepts reflects the ability of the reduced model to more readily solve the initial problem associated with suddenly exposing the reaction mixture to an irradiation source. This additional advantage of the use of SCHEME will be of maximum importance in simulating conditions where such sudden perturbations to a system are possible, as they are in smog-chamber experiments.

4.3. Model Alterations

In this section we address the ability of the methods developed for model reduction to be readily adapted to potential changes in the parent mechanism. As an example, we consider the introduction into ATMOS of a reaction sequence for the formation of hydroxyl radical by the reaction of $O(^1D)$

Table 5. SCHEME

	Surrogate Reaction[a]	Rate Coefficient[b]
[S1]	$NO_2 + O_3 + NO \rightarrow 2NO_2$	$k_4k_5\tau_{NO_3}$
[S2]	$2NO_2 + O_3 \rightarrow 2(HONO_2)$	$k_4k_6k_8[H_2O]\tau_{N_2O_5}\tau_{NO_3}$
[S3]	$NO_2 \rightarrow NO + O_3$	$j_1k_2[O_2]\tau_O$
[S4]	$2NO_2 \rightarrow 2NO$	$j_1k_{10}\tau_O$
[S5]	$NO_2 + OLEF \rightarrow NO + \frac{1}{2}HO_2 + \frac{1}{2}RCOO_2 + RO_2$	$j_1k_{19}\tau_O$
[S6]	$NO_2 + SO_2 \rightarrow NO + (H_2SO_4)$	$j_1k_{41}\tau_O$
[S7]	$3NO_2 \rightarrow NO + 2(HONO_2)$	$j_1k_6k_8k_{11}[H_2O]\tau_{N_2O_5}\tau_{NO_3}\tau_O$
[S8]	$NO + O_3 \rightarrow NO_2$	k_3
[S9]	$NO_2 + NO \rightarrow 2HONO$	$k_{12}[H_2O]$
[S10]	$2HONO \rightarrow NO_2 + NO$	k_{13}
[S11]	$HONO \rightarrow HO + NO$	j_{14}
[S12]	$NO_2 + HO \rightarrow (HONO_2)$	k_{15}
[S13]	$NO + HO \rightarrow HONO$	k_{16}
[S14]	$NO + HO_2 \rightarrow NO_2 + HO$	k_{17}
[S15]	$H_2O_2 \rightarrow 2HO$	j_{18}
[S16]	$OLEF + HO \rightarrow RCHO + RO_2$	k_{20}
[S17]	$OLEF + O_3 \rightarrow HO + RCOO_2 + RCHO$	k_{21}
[S18]	$RCHO \rightarrow \frac{3}{4}HO_2 + \frac{1}{4}RO_2$	j_{22}
[S19]	$RCHO + HO \rightarrow \frac{1}{2}HO_2 + \frac{1}{2}RCOO_2$	k_{23}
[S20]	$NO + RO_2 \rightarrow NO_2 + HO_2 + RCHO$	$k_{24}k_{27}[O_2]\tau_{RO}$
[S21]	$NO + RO_2 + NO_2 \rightarrow NO_2 + (RONO_2)$	$k_{24}k_{29}\tau_{RO}$
[S22]	$2NO + RO_2 \rightarrow NO_2 + (RONO)$	$k_{24}k_{29}\tau_{RO}$
[S23]	$RO_2 + NO + SO_2 \rightarrow NO_2 + (ROSO_2)$	$k_{24}k_{43}\tau_{RO}$

Table 5. (*Continued*)

	Surrogate Reaction[a]	Rate Coefficient[b]
[S24]	$RO_2 + HO_2 \rightarrow RCHO + HO + HO_2$	$k_{31}k_{27}[O_2]\tau_{RO}$
[S25]	$RO_2 + HO_2 + NO_2 \rightarrow HO + (RONO_2)$	$k_{31}k_{28}\tau_{RO}$
[S26]	$RO_2 + HO_2 + NO \rightarrow HO + (RONO)$	$k_{31}k_{29}\tau_{RO}$
[S27]	$SO_2 + RO_2 + HO_2 \rightarrow HO + (ROSO_2)$	$k_{31}k_{43}\tau_{RO}$
[S28]	$2RO_2 \rightarrow HO_2 + RCHO + RO_2$	$2k_{32}k_{27}[O_2]\tau_{RO}$
[S29]	$NO_2 + 2RO_2 \rightarrow RO_2 + (RONO_2)$	$2k_{32}k_{28}\tau_{RO}$
[S30]	$NO + 2RO_2 \rightarrow RO_2 + (RONO)$	$2k_{32}k_{29}\tau_{RO}$
[S31]	$SO_2 + 2RO_2 \rightarrow RO_2 + (ROSO_2)$	$2k_{32}k_{43}\tau_{RO}$
[S32]	$SO_2 + RO_2 \rightarrow HO_2 + RCHO + (H_2SO_4)$	$k_{39}k_{27}[O_2]\tau_{RO}$
[S33]	$SO_2 + RO_2 + NO_2 \rightarrow (H_2SO_4) + (RONO_2)$	$k_{39}k_{28}\tau_{RO}$
[S34]	$NO + SO_2 + RO_2 \rightarrow (H_2SO_4) + (RONO)$	$k_{39}k_{29}\tau_{RO}$
[S35]	$2SO_2 + RO_2 \rightarrow (H_2SO_4) + (ROSO_2)$	$k_{39}k_{43}\tau_{RO}$
[S36]	$NO + RCOO_2 \rightarrow NO_2 + RO_2$	k_{25}
[S37]	$NO_2 + RCOO_2 \rightarrow (PAN)$	k_{26}
[S38]	$2HO_2 \rightarrow H_2O_2$	k_{30}
[S39]	$HO \rightarrow HO_2$	$k_{33}[CO]$
[S40]	$NO_2 + HO_2 \rightarrow HONO$	k_{34}
[S41]	$HONO + HO \rightarrow NO_2$	k_{37}
[S42]	$SO_2 + HO_2 \rightarrow HO + (H_2SO_4)$	k_{38}
[S43]	$SO_2 + HO \rightarrow HO_2 + (H_2SO_4)$	k_{40}
[S44]	$SO_2 + RCOO_2 \rightarrow RO_2 + (H_2SO_4)$	k_{42}

[a] Compounds shown in parentheses are sink species not otherwise included in SCHEME.

[b] Indices of rate constants k and photolytic constants j refer to reactions of ATMOS (Table 2).

$$\tau_{N_2O_5}^{-1} = k_7 + k_8[H_2O].$$

$$\tau_{NO_3}^{-1} = k_5[NO] + k_6k_8\tau_{N_2O_5}[H_2O].$$

$$\tau_O^{-1} = k_2[O_2] + k_9[NO] + k_{19}[OLEF] + k_{41}[SO_2]$$

$$+ [NO_2](k_{10} + k_5k_{11}\tau_{NO_3}[NO] + k_6k_8k_{11}\tau_{NO_3}\tau_{N_2O_5}[H_2O]).$$

$$\tau_{RO}^{-1} = k_{27}[O_2] + k_{28}[NO_2] + k_{29}[NO] + k_{43}[SO_2].$$

formed by ozone photolysis. Although this source of HO is generally not important in a polluted troposphere, it can become increasingly more important in clean air. This sequence of reactions is given by reactions [A45]–[A48] in Table 7, where, as before, O is used to represent the ground-state $O(^3P)$. This reaction sequence introduces into ATMOS a new reactive intermediate $O(^1D)$ and employs as well the O, O_2, and H_2O species that have previously been surrogated in the formulation of SCHEME.

The surrogation of reactions [A45]–[A48] is readily accomplished by first employing the method described in Section 3.2 to eliminate the stable O_2 and H_2O species while applying the steady-state approximation method for

Table 6. Differential Equations Derived from SCHEME

Species, Y	$d[Y]/dt$ = Sources − (Sinks)
1. SO_2	$-(R_6^a + R_{27} + R_{35} + R_{42} + R_{43} + R_{44} + R_A{}^b)$
2. NO	$R_B + R_3 + R_{10} + R_{11} - (R_C + R_D + R_E + R_9 + R_{13} + R_{22} + R_{26} + R_{34})$
3. NO_2	$R_C + R_D + R_{10} + R_{41} - (R_B + 2R_2 + R_3 + 2R_7 + R_9 + R_{12} + R_{25} + R_{29} + R_{33} + R_{37} + R_{40})$
4. HONO	$2R_9 + R_{13} + R_{40} - (2R_{10} + R_{11} + R_{41})$
5. O_3	$R_3 - (R_D + R_2 + R_{17})$
6. HO	$R_F + R_H + R_{11} + 2R_{15} + R_{17} + R_{24} - (R_G + R_{12} + R_{13} + R_{16} + R_{19} + R_{41})$
7. HO_2	$R_G + \frac{1}{2}R_5 + \frac{3}{4}R_{18} + \frac{1}{2}R_{19} + R_{20} + R_{28} + R_{32} - (R_F + R_H + 2R_{38} + R_{40})$
8. RO_2	$R_5 + R_{16} + \frac{1}{4}R_{18} + R_{36} + R_{44} - (R_A + R_E + R_H + R_{24} + R_{28} + R_{29})$
9. $RCOO_2$	$\frac{1}{2}R_5 + R_{17} + \frac{1}{2}R_{19} - (R_{36} + R_{37} + R_{44})$
10. H_2O_2	$R_{38} - (R_{15})$
11. OLEF	$-(R_5 + R_{16} + R_{17})$
12. RCHO	$R_{16} + R_{17} + R_{20} + R_{24} + R_{28} + R_{32} - (R_{18} + R_{19})$

[a] The notation R_6 represents the rate of surrogate reaction [S6] listed in Table 5.

[b] The following reactions are grouped for convenience because of repeated occurrence in the rate expressions:

$$R_A = R_{23} + R_{31} + R_{32} + R_{33} + R_{34} + R_{35}$$

$$R_B = 2R_4 + R_5 + R_6 + R_7$$

$$R_C = R_{14} + R_{22} + R_{23} + R_{36}$$

$$R_D = R_1 + R_8$$

$$R_E = R_{20} + R_{21} + R_{30}$$

$$R_F = R_{14} + R_{42}$$

$$R_G = R_{39} + R_{43}$$

$$R_H = R_{25} + R_{26} + R_{27}$$

the elimination of $O(^1D)$. In this manner reactions [A46]–[A48] are replaced by sums of reactions [A46] + [A47],

$$O_3 \rightarrow O, \qquad \text{[N6]}$$

and reactions [A46] + [A48],

$$O_3 \rightarrow 2HO, \qquad \text{[S45]}$$

having rate coefficients

$$\gamma_{N6} = j_{46}k_{47}\tau_{O(^1D)}$$

$$\gamma_{45} = k_{46}k_{48}\tau_{O(^1D)}[H_2O]$$

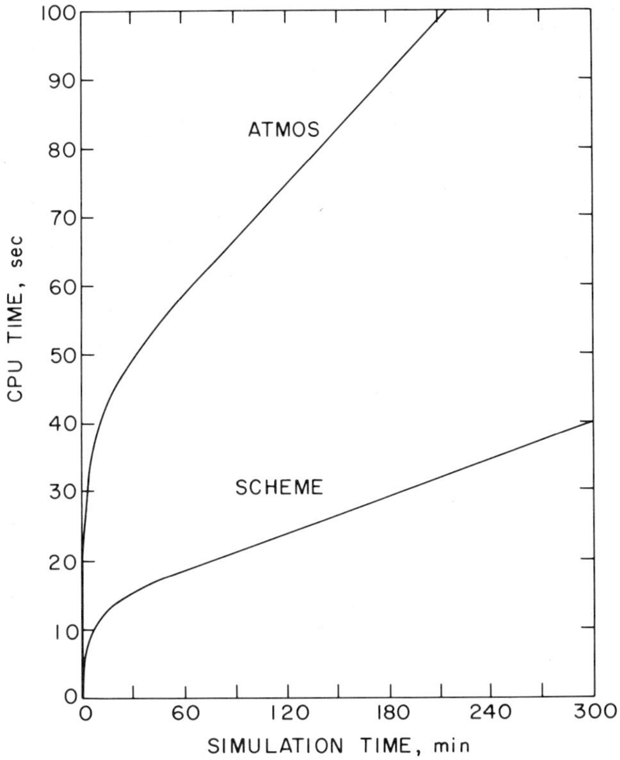

Figure 11. Execution times (CDC 7600) as a function of simulation time with ATMOS and SCHEME models for the "standard" initial pollutant concentration mixture shown in Figure 3.

Table 7. Reaction Sequence for O_3 Photolysis

Reaction	Rate Constant k or Photolytic Constant j	Units
[A45] $O_3 + h\nu \rightarrow O + O_2$	$2.1(-2)^{a,b}$	min^{-1}
[A46] $O_3 + h\nu \rightarrow O(^1D) + O_2$	$3.4(-3)^{b}$	min^{-1}
[A47] $O(^1D) \rightarrow O$	7.4(10)	min^{-1}
[A48] $O(^1D) + H_2O \rightarrow 2HO$	2.4(5)	$ppm^{-1}\,min^{-1}$

[a] The notation $2.1(-2)$ represents 2.1×10^{-2}.

[b] Denotes primary photochemical rate constant j for solar zenith angle of 40°.

where

$$\tau_{O(^1D)}^{-1} = k_{47} + k_{48}[H_2O].$$

With O_2 having already been eliminated from reaction [A45], it is seen that reactions [N6] and [A45] represent the same net process; thus both may now be combined into a single reaction,

$$O_3 \rightarrow O, \qquad [N7]$$

having a rate coefficient

$$\gamma_{N7} = \gamma_{N6} + j_{45}.$$

Reaction [N7] is simply an oxygen-atom formation process not previously included in ATMOS so that its contribution to the $O(^3P)$ steady state will require the addition of a second source term (i.e., $\gamma_{N7}\tau_O[O_3]$) to the $[O]_{ss}$ relationship given in Section 3.5. Because reactions [N7] and [A1] are independent and parallel sources of oxygen-atoms, our previous surrogation of $O(^3P)$ remains unaltered; however, the introduction of the $\gamma_{N7}\tau_O[O_3]$ source term into the $[O]_{ss}$ expression generates a second and parallel set of surrogate reactions that, with the exclusion of one null reaction ($O_3 \rightarrow O_3$), is given by reactions [S46]–[S51] in Table 8. As illustrated by this example, it is possible to directly treat additions and other alterations to ATMOS by one or more methods of model reduction and thereby construct the appropriate set of surrogate reactions corresponding to the change in the parent model.

Because of the relatively low O_3 photolytic rate ($k_{45} + k_{46} = 2.4 \times 10^{-2}$ min^{-1}), the inclusion of reactions [A46]–[S48] in ATMOS and of [S45]–[S51] in SCHEME does not significantly alter the simulation results shown in

Table 8. A SCHEME to Replace Reactions [A45]–[A48] of Table 7

Surrogate Reaction[a]		Rate Coefficient[b]
[S45]	$O_3 \rightarrow 2HO$	$j_{46}k_{48}\tau_{O(^1D)}[H_2O]$
[S46]	$NO + O_3 \rightarrow NO_2$	$k_9\gamma_{N7}\tau_O$
[S47]	$NO_2 + O_3 \rightarrow NO$	$k_{10}\gamma_{N7}\tau_O$
[S48]	$OLEF + O_3 \rightarrow \frac{1}{2}HO_2 + \frac{1}{2}RCOO_2 + RO_2$	$k_{19}\gamma_{N7}\tau_O$
[S49]	$SO_2 + O_3 \rightarrow (H_2SO_4)$	$K_{41}\gamma_{N7}\tau_O$
[S50]	$NO_2 + NO + O_3 \rightarrow 2NO_2$	$k_5k_{11}\gamma_{N7}\tau_{NO_3}\tau_O$
[S51]	$2NO_2 + O_3 \rightarrow 2(HONO_2)$	$k_6k_7k_{11}\gamma_{N7}[H_2O]\tau_{N_2O_5}\tau_{NO_3}\tau_O$

[a] Compounds shown in parentheses are sink species not otherwise included in SCHEME.

[b] For k and j values, refer to Tables 2 and 7. For $\tau_{N_2O_5}$, τ_{NO_3}, and τ_O, refer to Table 5.

$\gamma_{N7} = j_{45} + j_{46}k_{47}\tau_{O(^1D)}$.

$\tau_{O(^1D)}^{-1} = k_{47} + k_{48}[H_2O]$.

Figures 3–10. However, we have found that the extremely high value of the effective rate coefficient for removal of $O(^1D)$, $k_{47} + k_{48}[H_2O] = 7.9 \times 10^{10}\ min^{-1}$, introduces additional stiffness and instability (Gelinas, 1972) into the problem requiring an order of magnitude decrease in the initial time-step size employed in solutions with ATMOS. This requirement is not imposed on solutions with SCHEME, thus enabling the user of SCHEME to realize an additional advantage of obtaining stable solutions of otherwise stiff systems.

5. SUMMARY

In this chapter we have presented methods for constructing SCHEMEs by which modelers of tropospheric chemistry and other reactive systems may significantly reduce the number of actively modeled species necessary for description of the kinetics of the system. These methods of model reduction have been applied to the surrogation of a particular set of reactions and rate constants as given by the parent mechanism ATMOS. However, these methods are sufficiently general in nature to readily allow for adjustments in the surrogate mechanism SCHEME whenever changes appear necessary in the parent mechanism. The ability of SCHEME to replace ATMOS over a wide range of kinetic conditions was demonstrated (Figures 3–10). Because the rate coefficients for surrogate reactions are expressed as functions of elementary rate constants in the parent mechanisms (Tables 5 and 8), the user of SCHEME may readily reevaluate any of these rate coefficients as necessary for differing conditions of insolation or temperature or to reflect changes that may become necessary in the selection of rate constant values in the parent mechanism. Also, as shown in Section 4.3, the surrogate mechanism may be readily adjusted to reflect alterations in the reactions of the parent mechanism even if the change involves the addition of new reactive intermediates.

In concluding this chapter, we note that versions of SCHEME have already been successfully tested and employed with models describing reaction kinetics under dynamic atmospheric conditions. In a comparison study of simulation results obtained with a parent mechanism and its surrogate, Levine (1980) has demonstrated both the applicability and cost-efficiency of employing SCHEMEs with models involving rapid dilution such as would be experienced in expanding stack plumes. Similar comparisons and conclusions were made by Easter et al. (1982), who also report, as do Bottenheim and Strausz (1982), the use of SCHEMEs to model chemically reactive power plant plumes. Finally, we would again point out that the methodology employed in the development of a surrogate mechanism from its parent mechanism is general and would thus be applicable to multispecies, multireaction processes other than those of atmospheric systems.

ACKNOWLEDGMENTS

This work was performed under the auspices of the United States Department of Energy under contract No. DE-AC02-76CH00016. The authors thank Dr. Robert Adamowicz for his contributions to the development of the computer code employed in this study.

REFERENCES

Benson, S. W. (1978). Thermochemistry and kinetics of sulfur-containing molecules and radicals. *Chem. Rev.* **78,** 23–35.

Bufalini, J. J., Walter, T. A., and Bufalini, M. M. (1977). Contamination effects on ozone formation in smog chambers. *Environ. Sci. Technol.* **11,** 1181–1185.

Bottenheim, J. W. and Strausz, O. P. (1982). Modelling study of a chemically reactive power plant plume. *Atmosph. Environ.* **16,** 85–97.

Burrows, J. P., Cliff, D. I., Harris, G. W., Thrush, B. A., and Wilkinson, J. P. T. (1979). Atmospheric reactions of the HO_2 radical studied by laser magnetic resonance spectroscopy. *Proc. R. Soc. (Lond.)* **A368,** 463–481.

Calvert, J. G. and McQuigg, R. D. (1975). The computer simulation of rates and mechanisms of photochemical smog formation. *Internatl. J. Chem. Kinet. Symp.* **1,** 113–153.

Calvert, J. G., Su, F., Bottenheim, J. W., and Strausz, O. P. (1978). Mechanism of the homogeneous oxidation of sulfur dioxide in the troposphere. *Atmosph. Environ.* **12,** 197–226.

Carmichael, G. R. and Peters, L. K. (1979). Numerical simulation of the regional transport of SO_2 and sulfate in the eastern United States. *Fourth Symposium on Turbulence, Diffusion, and Air Pollution, American Meteorological Society,* Reno, NV.

Carter, W. P. L., Winer, A. M., Darnall, K. R., and Pitts, J. N., Jr. (1979). Smog chamber studies of temperature effects in photochemical smog. *Environ. Sci. Technol.* **13,** 1094–1100.

Chan, W. H., Uselman, W. M., Calvert, J. G., and Shaw, J. H. (1977). The pressure dependence of the rate constant for the reaction: $HO + CO \rightarrow H + CO_2$. *Chem. Phys. Lett.* **45,** 240–244.

Cox, R. A. and Derwent, R. G. (1975). Kinetics of the reaction of HO_2 with NO and NO_2. *J. Photochem.* **4,** 139–153.

Cox, R. A., Derwent, R. G., and Holt, P. M. (1976). Relative rate constants for reactions of OH radicals with H_2, CH_4, CO, NO, and HONO at atmospheric pressure and 296°K. *J. Chem. Soc. Faraday I* **72,** 2031–2043.

Demerjian, K. L., Kerr, J. A., and Calvert, J. G. (1974). The mechanism of photochemical smog formation, in *Advances in Environmental Science Technology,* Vol. 4, J. N. Pitts, Jr. and R. L. Metcalf, Eds., Wiley-Interscience, New York, pp. 1–266.

Duewer, W. H., Walton, J. J., Grant, K. E., and Walker, H. (1978). *Livermore Regional Air Quality (LIRAQ) Application to St. Louis,* Lawrence Livermore Laboratory, Report UCRL-52432.

Easter, R. C., Hales, J. M., Sverdrup, G. M., and Spicer, C. W. (1982). Plume conversion rates in the SURE region, Vol. 3. Electric Power Research Institute, Palo Alto, CA, Report EA-2793.

Eschenroeder, A. Q. and Martinez, J. R. (1972). Concepts and applications of photochemical smog models, in *Photochemical Smog and Ozone Reactions, Advances in Chemistry*

Series, No. 113, R. F. Gould, Ed., American Chemical Society, Washington, DC, pp. 101–168.

Farrow, L. A. and Edelson, D. (1974). The steady-state approximation: Fact or fiction? *Internatl. J. Chem. Kinet.* **6,** 787–800.

Finlayson-Pitts, B. J. and Pitts, J. N., Jr. (1977). The chemical basis of air quality: Kinetics and mechanisms of photochemical air pollution and application to control strategies, in *Advances Environmental Science Technology,* Vol. 7, J. N. Pitts, Jr. and R. L. Metcalf, Eds., Wiley-Interscience, New York, pp. 75–162.

Friedlander, S. K. and Seinfeld, J. H. (1969). A dynamic model of photochemical smog. *Environ. Sci. Technol.* **3,** 1175–1181.

Gelinas, R. J. (1972). Stiff systems of kinetic equations—a practitioner's view. *J. Computat. Phys.* **9,** 222–236.

Gelinas, R. J. and Skewes-Cox, P. D. (1977). Tropospheric photochemical mechanisms. *J. Phys. Chem.* **81,** 2468–2479.

Graedel, T. E., Farrow, L. A., and Weber, T. A. (1976). Kinetic studies of the photochemistry of the urban troposphere. *Atmosph. Environ.* **10,** 1095–1116.

Graham, R. A., Winer, A. M., Atkinson, R., and Pitts, J. N., Jr. (1979). Rate constants for the reaction of HO_2 with HO_2, SO_2, CO, N_2O, *trans*-2-butene, and 2,3-dimethyl-2-butene at 300 K. *J. Phys. Chem.* **83,** 1563–1566.

Hack, W., Hoyerman, K., and Wagner, H. G. (1975). The reaction $NO + HO_2 \rightarrow NO_2 + OH$ with $OH + H_2O_2 \rightarrow HO_2 + H_2O$ as an HO_2-source. *Internatl. J. Chem. Kinet. Symp.* **1,** 329–339.

Hampson, R. F. (1980). Chemical kinetic and photochemical data sheets for atmospheric reactions, U.S. Federal Aviation Agency Report FAA-EE-80-17.

Hampson, R. F., Jr. and Garvin, D. (1978). *Reaction Rate and Photochemical Data for Atmospheric Chemistry—1977.* U.S. National Bureau of Standards Special Publication 513.

Hecht, T. A. and Seinfeld, J. H. (1972). Development and validation of a generalized mechanism for photochemical smog. *Environ. Sci. Technol.* **6,** 47–57.

Hecht, T. A., Seinfeld, J. H., and Dodge, M. C. (1974). Further development of generalized kinetic mechanism for photochemical smog. *Environ. Sci. Technol.* **8,** 327–339.

Hindmarsh, A. C. and Byrne, G. D. (1977). *EPISODE: An Effective Package for the Integration of Systems of Ordinary Differential Equations, Rev. 1,* Lawrence Livermore Laboratory, Report UCID-30112.

Howard, C. J. and Evenson, K. M. (1977). Kinetics of the reaction of HO_2 with NO. *Geophys. Res. Lett.* **4,** 437–440.

Leighton, P. A. (1961). *Photochemistry of Air Pollution,* Academic Press, New York.

Levine, S. Z. (1980). A model for Stack Plume REactions with Atmospheric Dilution (SPREAD). *Symposium on Plumes and Visibility,* Grand Canyon, AZ, November 14–18; Brookhaven National Laboratory Report BNL-28332.

Levine, S. Z. and Schwartz, S. E. (1978). Construction of Surrogate CHEmical MEchanisms (SCHEMEs) for atmospheric photochemical systems. *J. Photochem.* **9,** 104–106.

Lyons, W. A., Dooley, J. C., and Whitby, K. T. (1978). Satellite detection of long-range pollution transport and sulfate aerosol hazes. *Atmos. Environ.* **12,** 621–631.

Meyers, R. E., Cederwall, R. T., Ohmstede, W. D., and Kampe, W. (1976). Transport and diffusion using a diagnostic mesoscale model employing mass and total energy conservation constraints, *Third Symposium on Atmospheric Turbulence, Diffusion, and Air Quality, American Meteorological Society,* Boston.

Muthukrishnan, S. and Peters, L. K. (1977). A mechanistic model for photochemical smog incorporating sulfur oxide reactions. *AIChE Symp. Ser.* **73,** 43–49.

Payne, W. A., Stief, L. J., and David, D. D. (1973). A kinetics study of the reaction HO_2 with SO_2 and NO. *J. Am. Chem. Soc.* **95,** 7614–7619.

Pitts, J. N., Jr., Darnall, K. R., Winer, A. M., and McAfee, J. M. (1977). *Mechanisms of Photochemical Reactions in Urban Air. Vol. II. Chamber Studies,* U.S. Environ. Protect. Agency, Report EPA-600/3-77-014h.

Reynolds, S. D., Roth, P. M., and Seinfeld, J. H. (1973). *Atmosph. Environ.* **7,** 1033.

Rodhe, H. (1978). Budgets and turn-over times of atmospheric sulfur compounds. *Atmosph. Environ.* **12,** 671–680.

Sander, S. P. and Seinfeld, J. H. (1976). Chemical kinetics of homogeneous atmospheric oxidation of sulfur dioxide. *Environ. Sci. Technol.* **10,** 1114–1123.

Seinfeld, J. H., Reynolds, S. D., and Roth, P. M. (1972). Simulation of urban air, in *Photochemical Smog and Ozone Reactions, Advances in Chemistry Series* 113, R. F. Gould, Ed., American Chemical Society, Washington, DC, pp. 58–100.

Sheih, C. M. (1977). Application of a statistical trajectory model to the simulation of sulfur pollution over northeastern United States. *Atmosph. Environ.* **11,** 173–179.

Sie, B. K. T., Simonaitis, R., and Heicklen, J. (1976). The reaction of OH with CO. *Internatl. J. Chem. Kinet.* **8,** 85–98.

Simonaitis, R. and Heicklen, J. (1976). Reactions of HO_2 with NO and NO_2 and of OH with NO. *J. Phys. Chem.* **80,** 1–7.

Stephens, E. R. (1966). Reactions of oxygen atoms and ozone in air pollution. *Internatl. J. Air Water Pollut.* **10,** 649.

Wayne, L. G. (1962). *The Chemistry of Urban Atmospheres. Technical Progress Report III.* Los Angeles County Air Pollution Control District.

Wendell, L. L., Powell, D. C., and Drake, R. L. (1976). A regional scale model for computing deposition and ground level air concentration of SO_2 and sulfates from elevated and ground sources, *Third Symposium on Atmospheric Turbulence, Diffusion and Air Quality, American Meteorological Society,* Boston.

Whitten, G. Z. and Hogo, H. (1977). *Mathematical Modeling of Simulated Photochemical Smog,* Environmental Protection Agency, Report EPA-600/3-77-011.

Wilson, W. E. (1978). Sulfates in the atmosphere: A progress report on project MISTT. *Atmosph. Environ.* **12,** 537–547.

Wolff, G. T., Lioy, P. J., Wight, G. P., Meyers, R. E., and Cederwall, R. T. (1977). *An Investigation of Long-Range Transport of Ozone Across the Midwestern and Eastern United States,* Environmental Protection Agency, Report EPA-600/3-77-001a.

Yocke, M. A., Mundkur, P., and Liu, M. (1975). Numerical simulation of reactive plumes, *68th Meeting of the American Institute of Chemical Engineering,* Los Angeles.

12

MODELING OF TRANSPORT AND CHEMICAL PROCESSES THAT AFFECT REGIONAL AND GLOBAL DISTRIBUTIONS OF TRACE SPECIES IN THE TROPOSPHERE

Leonard K. Peters

Department of Chemical Engineering
University of Kentucky
Lexington, Kentucky 40506

Gregory R. Carmichael

Chemical and Materials Engineering Program
University of Iowa
Iowa City, Iowa 52242

1. INTRODUCTION

The establishment of the budgets and the realization of an understanding of cause and effect of changes in cycles of the major biogeochemicals of the troposphere are major challenges facing atmospheric scientists today. This knowledge is important because the chemical composition of the troposphere and, particularly, the presence of certain trace species can greatly influence health, weather, climate, and agriculture. For example, trace species may influence climate through their effects on the radiation balance, visibility by conversion to light-scattering aerosols, or the ecosystem through effects such as acid rain formation.

To better understand the troposphere, it is necessary to know: (1) which species are present and at what concentrations and with what distribution (spatial, temporal, geographic); (2) which processes control the cycle; (3) how these species interact with climate factors; and (4) how humans impact on the cycle (both now and in the future). The answers to these questions must be obtained through a combination of laboratory, field, and modeling studies.

Modeling plays an absolutely essential role in unraveling the answers to these questions because it provides a means of closing the loop between field experiment results and our understanding of the chemical and physical processes of trace species in the troposphere. The necessity of modeling analysis is due to the fact that field measurement programs, even with extensive spatial coverage, identify a subset of the troposphere for only limited periods of time, whereas laboratory studies elucidate only a limited number of processes under highly idealized conditions. Verified models must be used to project the state of the troposphere at different times and under different conditions.

Modeling analysis can also provide guidance for experimental design. For example, models can (1) provide information on which species should be present in the troposphere (e.g., OH, H_2O_2, PAN) and thus suggest what sensors need to be developed; (2) predict the magnitude of trace species concentrations and thus indicate what instrument detection limits are needed to verify the presence of the species; (3) estimate the temporal and spatial distribution of trace species, and thereby indicating what sampling strategies should be used; and (4) estimate the variability and lifetime, thus suggesting optimum sampling frequency.

The analysis of Crutzen and Fishman (1977) provides a good example of how important findings evolve from modeling efforts. The main point realized from their study was that additional CO sources are needed to balance the budget of CO. This effort has led to additional studies by Crutzen and co-workers suggesting that the oxidation of naturally emitted hydrocarbons may produce considerable quantities of CO (Zimmerman et al., 1978) and that tropical biomass burning may also emit large quantities of CO to the atmosphere (Crutzen et al., 1979). Field studies designed to quantify these sources have subsequently been initiated.

Since modeling analysis plays such a critical role in developing an understanding of tropospheric chemistry cycles, it is important that model design and development be given appropriate consideration. The structure of tropospheric models is inherently process oriented instead of cycle oriented. This is because the transport, chemistry, dry deposition (surface removal), wet removal, and primary source processes occurring in the troposphere determine the cycles of trace species and the models must reflect that. Furthermore, cycles attempt to describe the long-term average mass balance of a species, whereas the dynamic processes can show extreme excursions from the mean behavior. Thus, to develop meaningful models, it is necessary to characterize (mathematically) these chemical and physical processes affecting the cycle. In addition, each process included in the model must be treated in appropriate detail. For example, if in a particular application the transport and chemistry are equally important, the model must treat the transport and chemistry in the same detail. A very detailed chemistry model with little or no transport would be of marginal value, as would a transport model with no chemistry.

Finally, each individual atmospheric process is itself a very complex phenomenon. However, it is not necessary to wait until the details of the phenomena are completely understood to simulate the process. Processes can be incorporated into a model framework by utilizing chemical, dynamic, and thermodynamic parameterizations. Thus it is possible to develop realistic models that include the governing processes and consider a large number of chemical species. The formulation of such models is the topic of this chapter.

2. MATHEMATICAL BASIS

2.1. Lagrangian versus Eulerian Analysis

The turbulent transport of material in the atmosphere can be analyzed from either an Eulerian or a Lagrangian approach. The difference between these approaches lies in the way the position in the field is identified. In the Eulerian approach the concentration of material is described at a given point and time with the coordinates fixed and independent of time. Therefore, the coordinate position $\mathbf{x}$ (e.g., in Cartesian coordinate system $\mathbf{x} = x, y, z$) and time are independent variables.

In the Lagrangian approach the concentration of material is described for a particular fluid element as it travels with the flow (Zak, this volume, Chapter 7). The coordinate position $\mathbf{x}$ is that of the fluid element and in the general case depends on time. Therefore, in the Lagrangian approach the coordinates are dependent variables, and the fluid element is identified by its position in the field relative to that at some arbitrary time, usually $t = 0$.

The mathematical equation describing the time-averaged concentration of a nonreactive species using the Eulerian approach is

$$\frac{\partial \bar{c}}{\partial t} + \boldsymbol{\nabla} \cdot (\bar{\mathbf{v}}\bar{c}) = -\boldsymbol{\nabla} \cdot (\overline{\mathbf{v}'c'}) + \bar{S}, \tag{1}$$

where $\mathbf{v}$ is the velocity vector, c is the concentration, and $\bar{S}$ is the source term. The first term on the right-hand side of Eq. (1) is introduced by the time-averaging procedure and represents the turbulent fluxes. This term introduces two new variables, $\mathbf{v}'$ and c', the turbulent fluctuations of the velocity and concentration, respectively. Therefore, to solve Eq. (1), it is also necessary to have equations for $\overline{\mathbf{v}'c'}$. Equations for $\overline{\mathbf{v}'c'}$ can be derived, but they ultimately depend on terms such as $\overline{v_i'v_j'c'}$. Each attempt to derive an expression for the terms introduced by the time-averaging process introduces new, higher-order terms. Thus there is no rigorous way of "closing" the system.

The most common method of closure for Eq. (1) is to model the turbulent fluxes analogously to Fickian diffusion; i.e., as a product of a diffusion coefficient and the mean concentration gradient,

$$-\overline{\mathbf{v}'c'} = \mathsf{K} \cdot \nabla \bar{c}, \tag{2}$$

where K, the eddy diffusivity tensor, is a function of the fluid flow but not of the material property (Seinfeld, 1975). This closure method is referred to as *K-theory*. Higher-order closure models are being developed, but at present they are computationally expensive (Donaldson, et al., 1972; Lewellen and Teske, 1975, 1976), and some results indicate that their effect is not substantial (Kewley, 1978).

The mathematical basis of Lagrangian transport of a nonreactive material is

$$\bar{c}(\mathbf{x}, t) = \int_0^t \int_{-\infty}^{\infty} P(\mathbf{x}, t \mid \mathbf{x}', t')\, S(\mathbf{x}', t')\, d\mathbf{x}'\, d\mathbf{t}', \tag{3}$$

where $P(\mathbf{x}, t \mid \mathbf{x}', t')$ is the probability density that a particle released at $\mathbf{x}'$ at time t' will be found at point $\mathbf{x}$ at time t and S is the function describing the source distribution. The Lagrangian approach is free from the closure problem associated with the Eulerian approach. However, the solution of Eq. (3) requires that the probability density function be specified. Techniques available for determining the probability density function include evaluation by use of Monte Carlo techniques, numerical turbulence models, and some assumption of the functional form (Lamb et al., 1979). It is interesting to note that if the process is assumed to be Markovian (i.e., the process is stochastic and the future state depends only on the present state and on the transition probabilities from the present to the future state), Eq. (3) reduces to

$$\frac{\partial \bar{c}}{\partial t} + \nabla \cdot (\bar{\mathbf{v}}\bar{c}) = \nabla \cdot \mathsf{K} \cdot \nabla \bar{c} + \bar{S}, \tag{4}$$

which is the same equation arising from the Eulerian approach with use of K-theory closure. A schematic diagram illustrating the relationship between Eulerian and Lagrangian approaches to turbulent diffusion modeling is presented in Figure 1.

Most of the regional-scale transport models currently in use are a form of Lagrangian model frequently referred to as *trajectory models* (Shannon, 1981). These models are generally formulated under assumptions of no horizontal turbulent diffusion, no convergent or divergent flows, and no wind shear. In these models parcels of material are emitted from each source and are advected with the mean wind with the parcel's location computed at equal time intervals. By following a fixed mass of air, trajectory models avoid lengthy integration of Eq. (4) and permit the general verification of pollutant distribution by tracing parcels from its source area.

On the other hand, Eulerian models employ numerical integration of Eq. (4) and provide solutions over the entire modeling region, rather than along

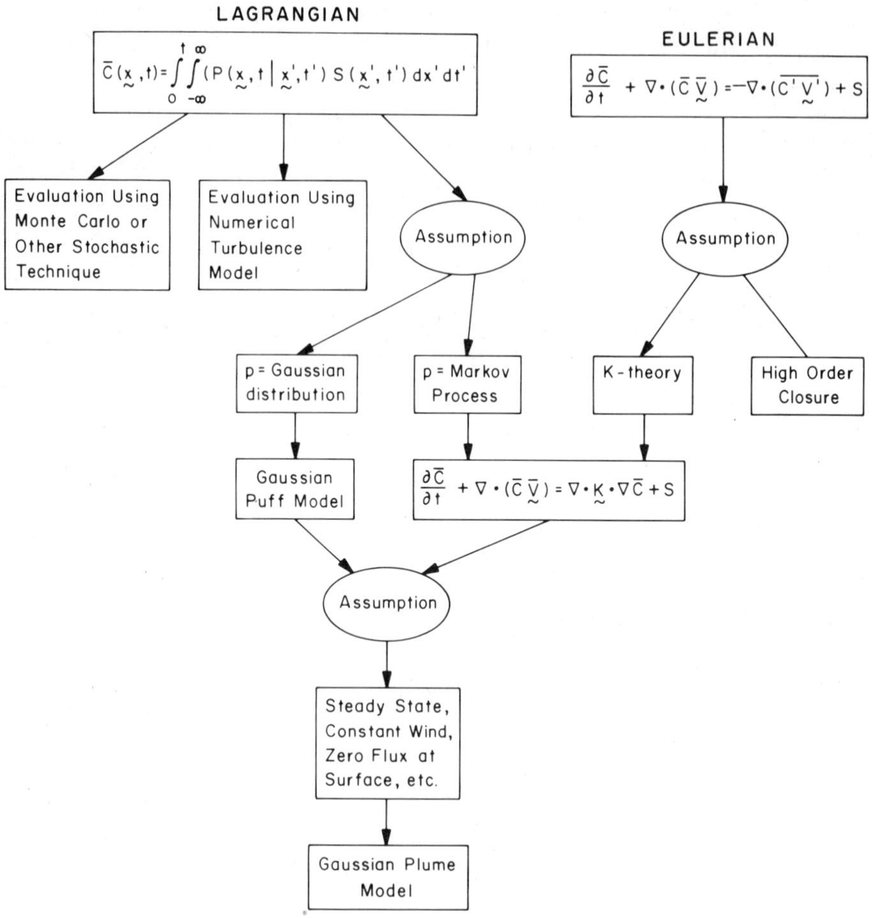

Figure 1. Schematic diagram of relationship between Eulerian and Lagrangian approaches to turbulent transport modeling. After Lamb et al. (1979).

trajectories. These grid models are also able to include the effects of wind shear and vertical air motions.

When the material of interest is chemically reactive, the modeling of the turbulent transport is further complicated. The basic equation for reactive flows with use of the Eulerian approach is

$$\frac{\partial \bar{c}_i}{\partial t} + \nabla \cdot (\bar{\mathbf{v}}\bar{c}_i) = \nabla \cdot \mathsf{K} \cdot \nabla \bar{c}_i + \bar{S}_i + \bar{R}_i, \tag{5}$$

where $\bar{R}_i$ is the rate of formation by chemical reaction and subscript i designates the ith species. In most applications $\bar{R}$ is evaluated by using the mean concentrations $\bar{c}$. However, it must be emphasized that in obtaining Eq. (5), the time averaging of R may introduce additional terms; e.g., if R

$= -kc^2$, then $\bar{R} = -k(\bar{c}^2 + \overline{c'^2})$. So again, a closure problem arises when turbulent flow is considered. The influence of turbulence on the reaction rate is most significant in near-source areas. In addition to these considerations, the reaction term can greatly complicate the system because $\bar{R}$ can be highly nonlinear and can, therefore, introduce complicated coupling of the equations.

Only first-order chemical reactions, where each molecule decays individually and does not affect the decay of others, can be rigorously (and easily) modeled with the Lagrangian approach. For higher-order reactions, the superposition principle inherent in Eq. (3) is no longer valid. On the basis of these considerations, along with the fact that the type of information usually desired is the value of the species concentration at a fixed point, the Eulerian approach and specifically Eq. (5) is the most appropriate basis for the modeling of regional- and global-scale transport and chemical processes.

2.2. One-, Two-, and Three-Dimensional Models

The transport of reactive species in the troposphere can be described by Eq. (5). The coordinate system chosen in the application of this equation is based on the problem of interest. If the problem is regional, a rectangular coordinate system is generally used. In this case the gradient operator is that for rectangular coordinates; i.e.,

$$\nabla = \frac{\partial}{\partial x}\mathbf{i} + \frac{\partial}{\partial y}\mathbf{j} + \frac{\partial}{\partial z}\mathbf{k}.$$

For applications to global problems, the gradient operator would be that for spherical coordinates (Bird et al., 1960).

Regardless of the coordinate system used, Eq. (5) in full form represents a complicated equation. However, it can be simplified in many applications. For example, if all transport processes are ignored and no sources are considered, Eq. (5) reduces to

$$\frac{d\bar{c}_i}{dt} = \bar{R}_i. \tag{6}$$

This equation describes the change of concentration in a batch or closed container. In atmospheric applications this equation is used to investigate chemical reaction mechanisms and to model laboratory chamber studies (e.g., Falls and Seinfeld, 1978). By use of $u = dx/dt$, Eq. (6) can also be used to study the chemical changes that occur in a parcel of air as it travels with mean velocity u. These studies can provide estimates of the amounts of material (e.g., SO_x and NO_x compounds) emitted locally that make it to the background troposphere (e.g., Bazzell and Peters, 1981).

Equation (6) represents a simple time-dependent model. A one-dimensional model often used is

$$\frac{\partial \bar{c}_i(z, t)}{\partial t} = \frac{\partial}{\partial z}\left(K_z(z, t)\frac{\partial \bar{c}_i(z, t)}{\partial z}\right) + \bar{S}_i(z, t) + \bar{R}_i(z, t). \tag{7}$$

[To obtain Eq. (7) from Eq. (5), it is necessary to define $\bar{c}_i(z) = \int_{-\infty}^{+\infty}\int_{-\infty}^{+\infty} \bar{c}_i(x, y, z)dx\,dy$]. Models based on Eq. (7) have been quite helpful in identifying basic processes affecting a species balance (e.g., Chameides et al., 1977; Wofsy et al., 1972). However, the reduction of atmospheric problems to the vertical dimension alone can dangerously oversimplify the problem. This approach neglects the horizontal variability in emission rates, transport processes, and photochemical reactions. Two-dimensional models, which include limited horizontal as well as vertical transport, can treat the transport processes in more detail and, in so doing, can provide an estimate of spatial distributions. In global applications, two-dimensional models are usually height–latitude models (Fishman et al., 1979; Rodhe and Isaksen, 1980). These models incorporate important inhomogeneities, such as differences between various latitudinal belts of emissions (natural and anthropogenic) and of photon fluxes.

In general, tropospheric trace distributions exhibit large variations in all three dimensions. Therefore, simplification of models to one or two dimensions severely restricts the possibility of making meaningful comparisons between model calculations and measured data. This is particularly true for those compounds that have a residence time in the atmosphere shorter than the characteristic transport time. Thus modeling for such species requires three-dimensional models. Both regional (for SO_x compounds) and global (CH_4–CO) three-dimensional combined transport–chemistry models have been developed (Carmichael and Peters, 1980; Peters and Jouvanis, 1979). In addition, detailed three-dimensional global transport models with limited chemistry have also been developed (e.g., Mahlman et al., 1980).

2.3. Boundary Conditions

In addition to the governing equations [Eq. (5)], the boundary conditions must be specified. At the surface the boundary condition typically takes the form

$$K_z \frac{\partial c}{\partial z} = v_d c - Q, \tag{8}$$

where v_d is the deposition velocity and Q is the surface emission flux. (Deposition is discussed in more detail in Section 3.2.)

At the other boundaries the condition is that for the prescribed total flux; i.e.,

$$\mathbf{v}\, c - \mathsf{K} \cdot \nabla c = \mathbf{F} \tag{9}$$

The form of the prescribed flux $\mathbf{F}$ will depend on whether there is inflow or outflow at that boundary. The most typically used conditions are a zero diffusive flux (i.e., $-\mathsf{K} \cdot \nabla c = 0$) under outflow, and the total flux equal to the advection flux under inflow conditions (i.e., $\mathbf{F} = \mathbf{v}c_b$, where c_b represents a background concentration). However, these conditions are restricted to applications where sources and sinks are located far from boundaries and reaction rates are sufficiently fast that much of the conversion has already occurred by the time the air mass reaches the boundary. Alternative conditions have been proposed (Carmichael and Peters, 1980).

2.4. Numerics

In general, closed-form analytic solutions to Eq. (5) do not exist for the initial and boundary conditions encountered in complex situations. Therefore, the solutions frequently must be obtained numerically. However, numerical simulation of this equation is difficult for several reasons: (1) the problem is multidimensional; (2) the horizontal transport is usually convection dominated; (3) the boundary conditions are mixed; and (4) both slow and fast chemical reactions can be important.

There are many methods available for solving atmospheric transport problems. The methods available can be broadly classified into the categories of finite-difference schemes, variational methods, and particle-in-cell techniques. The methods can be further classified as explicit or implicit, depending on whether each succeeding integration step is direct (explicit) or requires the simultaneous solution of the difference equations (implicit). In general, implicit methods are computationally more difficult than explicit methods, but they frequently are numerically stable over larger ranges of spatial and temporal step sizes.

The most troublesome numerical problems involve dissipation, dispersion, and aliasing. Numerical dissipation errors, which can be introduced when first-order numerical approximations are used, tend to smooth initially peaked distributions. Dispersion is common to all methods and arises from the approximation of the advection terms. In the numerical solution, all Fourier components lag the true solution, and each wave number component travels with a different speed. This is unlike the real solution where all components travel with the same speed. This lagging causes the different Fourier components to spread apart, or disperse, as the numerical solution proceeds. This error is larger for the shorter wavelengths. Aliasing is associated with energy exchange between Fourier components and will occur in problems involving spatially varying wind fields. The Fourier components interact in such a way that energy cascades from long wavelengths to shorter ones; if no dissipative mechanism is present to remove the energy from the shortest wavelength, the energy will flip over and reappear in the long wavelengths. The many numerical techniques available reflect different attempts to decrease or compensate for these errors.

The selection of a numerical technique for a specific application must be based on a number of considerations, including numerical stability, accuracy, nonnegativity of computed concentrations, computer storage requirements, computing time requirements, and adaptability of the method to the solution of the governing equations. Fortunately, a number of performance tests have been conducted that compare directly many of the numerical techniques available (Carmichael et al., 1980a; Crowley, 1968; Molenkamp, 1968; Pepper et al., 1980). The results of Carmichael et al. (1980a) and Pepper et al. (1980) indicate that the pseudospectral method and the finite-element methods exhibit high accuracy on practical discretizations when applied to distributed source problems. In addition, direct comparisons between pseudospectral and finite-element methods in one- and two-dimensional problems have found the pseudospectral method to be slightly more accurate but at the same time to require substantially more computer time. In addition, the pseudospectral method requires a delicate handling of the boundary conditions. This is in contrast to the flexibility of the finite-element method.

It is our opinion that the finite-element method offers the best combination of flexibility and accuracy. In addition, the finite-element approach has several other advantages: (1) ability to accurately represent complex surface geometries; (2) natural treatment of boundary conditions; (3) ability to easily employ nonuniform, nonrectangular meshes; and (4) preservation of certain conservative properties. The finite-element method has been used by the authors in a regional scale SO_2–sulfate model (Carmichael and Peters, 1980) and in a global CH_4–CO model (Kitada and Peters, 1982).

2.5. Model Inputs

Inputs necessary for solution of Eq. (5) can be classified broadly into emissions, meteorological, reaction, and surface removal data. The emission input data include the specification of the emission rates of each transported species over the entire spatial and temporal range of interest (i.e., $\mathbf{x}$ and t). The necessary meteorological inputs consist of the transport variables, $\mathbf{v}$ and K, and the reaction rate variables of temperature, water vapor content, solar actinic flux, and cloud cover. The meteorological variables must also be specified over the entire domain. Inputs classified as reaction and surface removal include the chemical reaction mechanism, rate expressions and rate constants, and dry deposition velocities (which necessitate input of surface type and land-use data).

The required input data can be obtained from observational data and/or can be generated from primitive equation models (e.g., general circulation models). If observational data are used, the problem arises that observations are made only at discrete locations and times. Therefore, temporal and spatial interpolation are required for generation of the input fields. In addition, many of the important input data are seldom measured, for example, the vertical velocity component and K. These quantities must then be estimated from those variables that are measured. Thus there are large uncertainties in input fields generated from observational data.

The alternative to using the observational data directly is to generate the input fields by use of a primitive equation model. For global-scale inputs, a general circulation model can be used (Mahlman et al., 1980), whereas for regional applications a mesoscale model is more applicable (Tarbell et al., 1981). These models, which involve solving the Navier–Stokes equations as well as equations for the energy balance and the water vapor balance,

Table 1. Summary of Inputs Needed to Solve the Atmospheric Diffusion Equation [Eq. (5)]

Input	Level of Detail	Sources of Error
Wind velocities	Variation of $\mathbf{v}$ with $\mathbf{x}$ and t	Measurements available only at a few locations, generally at ground level, at discrete times; error in (1) variation of winds horizontally, (2) variation of winds vertically, and (3) determination of vertical velocity
Eddy diffusivities	Variation of K with $\mathbf{x}$ and t	Few direct measurements available; K must be inferred from theory; both magnitude and vertical variation uncertain
Chemical reaction mechanism	Rate equation for each c_i	Inaccurate rates because of (1) inability to simulate atmosphere in the laboratory, (2) unknown reactions, and (3) unknown rate constants
Source emissions	Emission rate as a function $\mathbf{x}$ and t	Inaccurate knowledge of (1) variability of source activity and (2) emission factors.
Boundary conditions	Location of upper vertical boundary as a function of position and t	Lack of data or adequate model of temperature structure of atmosphere
	At surface, need mass-transfer coefficient and concentration of absorbing gas	Lack of data on the concentration in the absorbing phase and incomplete knowledge of deposition process

provide as output temporally varying three-dimensional meteorological fields. Furthermore, these models can in principle be constrained with the observational fields.

The generation of the meteorological fields from primitive equation models constrained with observational data may eliminate many of the uncertainties in the meteorological fields. However, there remain large uncertainties in the other input fields. Some of the uncertainties in the input parameters are summarized in Table 1.

3. TRANSPORT AND REMOVAL PROCESSES

The distribution of trace gases in the troposphere is determined by the distribution of source emissions and by the transport, removal, and reaction processes that occur. For example, trace gases are transported by the mean winds both horizontally and vertically, mixed vertically by turbulent diffusion, are removed by interactions with surfaces (e.g., soil, vegetation, particulate and dispersed water droplet surfaces, etc.), and transformed by chemical reaction.

However, these processes are complicated and are not fully understood. Therefore, to incorporate them into the model framework [i.e., Eq. (5)], one must characterize them mathematically. This often necessitates the use of chemical, dynamic, and thermodynamic parameterization. Furthermore, even processes that are quite well understood may require parameterization to maintain some balance of the details among the different processes that are treated in the model.

The characterization of the transport and removal processes are discussed in this section, and the chemical processes are discussed in Section 4.

3.1. Dynamic Processes

Atmospheric dynamical processes generally refer to those processes that are responsible for moving momentum, heat, and mass from one point to another through atmospheric motions. This transport can occur through movement of large air masses, eddy transport, and a vast range of intermediate scales. Detailed treatment of such processes can be found in numerous references (Csanady, 1973; Monin and Yaglom, 1971; Tennekes and Lumley, 1972). In this section, the processes of convective transport, vertical mixing, and stratospheric intrusion are briefly considered.

3.1.1. Convective Transport

The uneven distribution of energy resulting from latitudinal variations in insolation and from differences in the absorptivity of the Earth's surface leads to the large-scale air motions of the Earth. The temperature difference between the poles and the equator and those between the continents and the oceans supply the driving forces for the global motions, e.g., the polar easterlies, the midlatitude westerlies, and the tropical trade winds. Local winds result from local temperature differences arising from the irregularities of land masses and their surface temperature. The magnitude and fluctuations of these temperature differences are determined by the incoming solar radiation, the composition of the atmosphere, and the characteristics of the Earth's surface. These smaller-scale motions are often classified into cyclone and anticyclone systems.

The motions of the Earth's atmosphere are also influenced by the relative motion of the Earth's surface; and near the surface, air circulation is retarded as a result of the frictional effects. This layer of frictional influence is called the *planetary boundary layer*, and its thickness, which depends on the stability of the air, varies typically from a few hundred to a few thousand meters. Within the boundary layer there exists a layer extending from the Earth's surface to 50–100 m, known as the *surface layer*. Motions above this layer of approximately constant shearing stress are further influenced by the rotation of the Earth (i.e., the Coriolis force).

Above the planetary boundary layer is a layer that is essentially decoupled from the Earth's surface. This region is called the *geostrophic layer*. The wind direction in the geostrophic layer commonly changes direction with altitude; this shear is referred to as the *Ekman spiral*.

It is apparent that the wind field is three dimensional and time dependent. Therefore, the observed trace gas distribution at any time will be highly dependent on the wind field prior to and during the period of observation. In addition, large- and small-scale circulation patterns are important in both regional- and global-scale problems. These factors necessitate the use of three-dimensional, time-dependent wind fields for characterization of the transport by the mean winds. The problems associated with obtaining a three-dimensional wind field were outlined in the previous section.

3.1.2. Vertical Mixing

Mixing is one of the principal processes by which material is distributed vertically in the atmosphere. The intensity of vertical mixing is frequently classified in terms of stability, which, in turn, is measured by the vertical temperature gradient or lapse rate.

It is often useful to classify the regions of the atmosphere on the basis of the temperature profile. The regions of importance to regional- and global-scale transport are the troposphere and the stratosphere, separated by the tropopause. The troposphere is further classified into the boundary layer and the free troposphere. Mixing within the boundary layer can be intense, whereas mixing between the boundary layer and the free troposphere and between the free troposphere and the stratosphere are restricted. Since virtually all emissions of trace gases occur within the lowest kilometer of the atmosphere, it is essential to understand the mixing of this layer.

The vertical temperature structure of the boundary layer is the result of heating and cooling of the Earth's surface. When the surface acts as a heat source, heat is transferred from the surface to the air above, affecting the lowest layers first while having little effect on the upper layers. As this heating continues, an unstable or superadiabatic lapse rate develops to permit the warm air to rise as convection currents. The height to which this region of intense mixing grows depends on the intensity and duration of the heating, mechanical turbulence, and the initial lapse rate, which acts as a resistance to the rising convective currents. When the surface acts as a heat sink, a stable lapse rate results since the air nearest the surface cools while the air above remains relatively warm. Mixing within stable layers is restricted. As the cooling continues, an inversion develops.

The diurnal variation of the mixing behavior within the planetary boundary layer is shown schematically in Figure 2 (Carson, 1973). Under typical conditions over a homogeneous surface, convection begins shortly after sunrise as the Earth's surface warms. These convective currents then begin breaking down the surface-based nocturnal inversion H_I and permit the mixing depth H_M to rise into the overlying slightly stable air. This process continues until midafternoon, when the surface heat source has weakened to the point that it can no longer support penetrative convection. Beyond that time, vigorous and deep mixing diminishes as stable conditions begin to develop in the boundary layer. The mixing depth is somewhat ambiguous during this transition until net cooling at the surface reestablishes the ground-based nocturnal inversion. The air above the nocturnal inversion gradually becomes increasingly stable through radiation heat loss.

The wind field adjusts to the reduction in frictional stress in the air above the nocturnal inversion by accelerating forming so-called nocturnal jets.

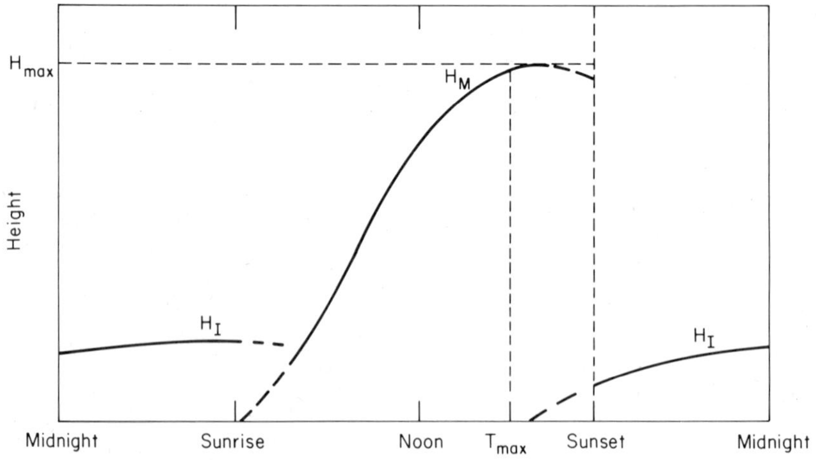

Figure 2. Schematic diagram of diurnal variation of the mixing-layer height. After Carson (1973).

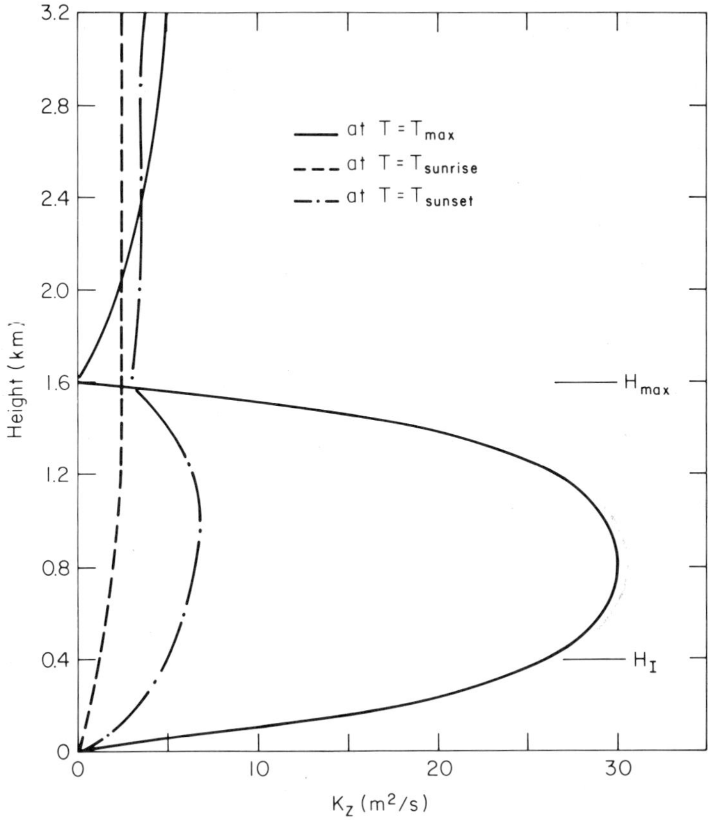

Figure 3. Eddy-diffusivity profiles corresponding to different times in the diurnal cycle of the mixing-layer height shown in Figure 2.

These nocturnal jets occur frequently and attain maximum velocity just above the top of the inversion layers. The average increase of layer-averaged winds at night over daytime may be 20%. Material in these nocturnal jets can be transported over large downwind distances with little vertical spread (Wolff et al., 1979).

Vertical mixing can be conveniently incorporated into the mathematical model [i.e., Eq. (5)] through the use of the vertical eddy diffusivity, K_z. In this treatment the mixing-layer dynamics and the accompanying vertical mixing are characterized by the eddy diffusivity profiles. Unfortunately, eddy diffusivities have rarely been measured; consequently, eddy coefficients must be estimated from data that are available.

A parameterization procedure consistent with theory and observation has been developed (Carmichael and Peters, 1980) on the basis of the work by Myrup and Ranzieri (1976). Expressions for K_z are parameterized in terms of the friction velocity u_* and the Monin–Obukhov length scale L. This parameterization technique has been tested and found to mimic the mixing-layer dynamics (Carmichael et al., 1980). Typical boundary-layer eddy-diffusivity profiles are presented in Figure 3. Within the mixing layer the eddy-diffusivity profiles have a broad maximum that falls to near zero at the base of the inversion. This "roll-off" restricts the vertical mixing of material contained within the mixing layer with that outside. The eddy-diffusivity profile also "rolls off" at the tropopause for similar reasons. The maximum mixing-layer height H_{max} can be estimated by the method described by Holzworth (1964), and the mixing-layer height varies with time qualitatively as presented in Figure 2.

3.1.3. Stratospheric Intrusion

The tropopause represents a nonmaterial boundary that separates relatively well mixed tropospheric air from the more stable air of the stratosphere. From a modeling standpoint, this function of the tropopause can be modeled by use of a roll-off in eddy-diffusivity values as just discussed. However, on occasion stratospheric air can penetrate through the tropopause. This process can serve as a significant source of tropospheric material. For example, ozone produced in the stratosphere can be transported directly into the troposphere during stratospheric intrusion events (Johnson et al., 1979). This may be a significant source of tropospheric ozone.

Stratospheric penetration can occur by a number of processes, including vertical motions associated with jets and waves, tropopause lifting, severe cumulus convection, and mean meridional circulation (Falconer et al., 1978). However, the most widely accepted mechanism is through cyclogenetic events where stratospheric air extrudes beneath the frontal jet streams that arise as upper tropospheric cyclones deepen. This process typically occurs at the midlatitudes.

These events can be included in global models by using the boundary conditions discussed in Section 2.3, provided that the vertical velocity and the location of the events can be characterized from observed data or predicted from the derived meteorological fields. Work in this area has been described by Danielson (1968).

3.2. Air–Surface Exchange

3.2.1. Gases

The interaction between trace gases and the surface over which they are transported is a process that occurs continuously. The surfaces can be semi-continuous, such as lakes, cornfields, or coniferous forests, or can be dispersed as in hydrometeors. In addition, these surfaces can act as a source or sink or be passive to the transport processes. The general condition describing the interaction with surfaces is

$$F_z = -\mathsf{K}_z \frac{\partial c_g}{\partial z}\bigg|_{\text{surface}} = k_l(c_{l_b} - c_{l_i}), \tag{10}$$

$$F_z = \mathsf{K}_z \frac{\partial c_g}{\partial z}\bigg|_{\text{surface}} = k_g\,(c_{g_b} - c_{g_i}), \tag{11}$$

or

$$F_z = \mathsf{K}_z \frac{\partial c_g}{\partial z}\bigg|_{\text{surface}} = K_g\,(c_{g_b} - c_g^*), \tag{12}$$

where F_z is the flux of material, c_g is the gas-phase concentration, c_l is the liquid-phase concentration (the subscripts b and i indicate the bulk and interfacial locations, respectively), and k_l, k_g, and K_g are the individual liquid-phase, individual gas-phase, and overall gas-phase mass-transfer coefficients, respectively.

The flux of material is toward the surface (i.e., the surface is a sink) if $c_{g_b} > c_{g_i}$ and away from the surface (i.e., a source) if $c_{g_b} < c_{g_i}$. When material is removed at the ground surface, the processes are those commonly referred to as dry deposition. Wet deposition processes are those where dispersed hydrometeor surfaces act as sinks.

For sufficiently dilute systems, Henry's law can be used to relate gas-phase and solution-phase concentrations at the interface. If $\mathcal{H}$ is the Henry's law constant, then for equilibrium at the interface

$$c_{g_i} = \mathcal{H}\, c_{l_i}, \tag{13}$$

and c_g^* is the gas-phase concentration that would be in equilibrium with the bulk liquid composition, i.e.,

$$c_g^* = \mathcal{H}\, c_{l_b}. \tag{14}$$

Furthermore, K_g is related to k_g and k_l by way of

$$\frac{1}{K_g} = \frac{1}{k_g} + \frac{\mathcal{H}}{k_l}. \tag{15}$$

The value for k_g can be estimated from mass-transfer analogies. For mass transfer to dispersed liquid water droplets, the Frössling equation can be used (Skelland, 1974), and for mass transfer to rough surfaces, an expression derived by Owen and Thomson (1963) can be used. The value of k_l depends on many factors, but the degree of turbulence in the fluids on both sides of the interface and the chemical reactivity of gas in the liquid are perhaps the most important.

In practice, dry deposition to the Earth's surface is frequently characterized and modeled in terms of the deposition velocity v_d,

$$v_d(z) = \frac{F_z(z)}{c_g(z)}, \tag{16}$$

where v_d is an experimentally determined parameter. Unfortunately, however, most studies of dry deposition have been conducted in wind tunnels. Whereas these studies provide useful information on the physical, chemical, and biological processes involved in the mass-transfer processes, they are not easily extrapolated to characterization of the processes occurring naturally.

The number of gases that have been investigated in the field is small (e.g., O_3, SO_2, NO_2, NO_x, CO_2). Most of this available field information has been obtained by use of micrometeorological flux-gradient relationships, but these techniques require a high degree of experimental accuracy. Current trends are away from these gradient methods and toward use of eddy correlation and radioactive tracer techniques. Regardless of the experimental technique, measurements of deposition at the same site vary appreciably, as do measurements at different sites over the same surface type and those over different surface types. The deposition velocities are found to vary with surface roughness, atmospheric stability, surface type, and surface wind speed [Hicks and Wesely (1978); see also Wesely, this volume, Chapter 8].

To apply these deposition data to estimate deposition under different conditions, it is necessary to isolate the meteorological factors and the factors characteristic of the surface. This can be done by interpreting v_d as the overall mass-transfer coefficient and using an equation analogous to Eq. (15); i.e.,

$$\frac{1}{v_d} = r_a + r_s, \tag{17}$$

where r_a is the aerodynamic resistance dependent on the surface roughness and meteorological conditions and r_s is the surface resistance. Under the special case where $c_g^* = 0$, then $v_d = K_g$, $r_a = 1/k_g$ and $r_s = \mathcal{H}/k_l$. Thus it is important to note that the deposition velocity is a more restrictive case of the more general definition contained in Eq. (12).

The surface resistance can be regarded as a property of the surface and will vary for surface type (e.g., cornfield, oat field, open sea) and from gas to gas. This resistance can be established from deposition experiments. By determining v_d and estimating the aerodynamic resistance from correlations for k_g (which depends on surface wind speed, which, in turn, depends on the atmospheric stability and surface roughness), r_s for a specific gas can be estimated from Eq. (17).

Knowledge of r_s can then be used to develop generalized parameterization techniques for estimating v_d under different meteorological conditions and surface types. Such procedures have been developed by Sheih et al. (1979) and Carmichael (1979).

3.2.2. Particles

Particles also are removed from the atmosphere by interaction with surfaces, but the mechanisms for removal differ from those for gases. The process of dry deposition of particles can be described by the general equation (Slinn, 1976)

$$v_d = v_s + \frac{\epsilon}{1 + \epsilon} \alpha u_* + \frac{\delta(D)}{1 + \epsilon} \frac{u_*}{u_z} \frac{E_j}{B}, \tag{18}$$

where v_s is the settling velocity, E_j is a collection efficiency dependent on particle size, ϵ is related to the canopy filtration efficiency, $\delta(D)$ is a resuspension parameter dependent on particle size, and α and B are constants near unity. For grass and grasslike surfaces, the combination of large canopy efficiency and a small resuspension parameter allows the last term of Eq. (18) to be neglected. Measurements by Hicks and Wesely (1978) have shown that the deposition rates of submicron particles are comparable to those of soluble trace gases.

The dry deposition process is incorporated into the mathematical model by use of the boundary condition [Eq. (8)]. This condition accounts for both deposition and for surface sources. The removal of material by dispersed surfaces is not as easily incorporated into the mathematical model.

3.3. Clouds

Important processes affecting the regional and global distributions of trace gases occur in clouds and during precipitation events. Clouds act as giant vacuum cleaners and process large quantities of boundary-layer air (see Figure 4). This is regarded as one of the major processes by which boundary-layer air is transported to the free troposphere. (Indeed, it is the boundary-layer source of moisture that allows the clouds to intensify and persist.) Material can even be injected directly into the stratosphere by penetrating cumulus clouds (this is believed to be the primary process for stratospheric injection in the tropics) (Scott, 1982).

Clouds do more than simply transport material aloft. They also act as an active filter by removing soluble gases and particulate material. These removal processes tend to concentrate gaseous material inside the dispersed cloud hydrometeors. Additionally, chemical reactions may occur within the cloud droplets, often at rates much faster than in the gas phase. For example, the formation of sulfate can occur at rates greater than 20% per hour in clouds compared to the typical rate of 1% per hour in the gas phase (Hegg and Hobbs, 1981). In addition, clouds can greatly affect the vertical distribution of material by releasing absorbed material and reaction products during cloud evaporation.

Even though the importance of cloud processes is widely recognized, the incorporation of these processes into tropospheric trace gas models remains

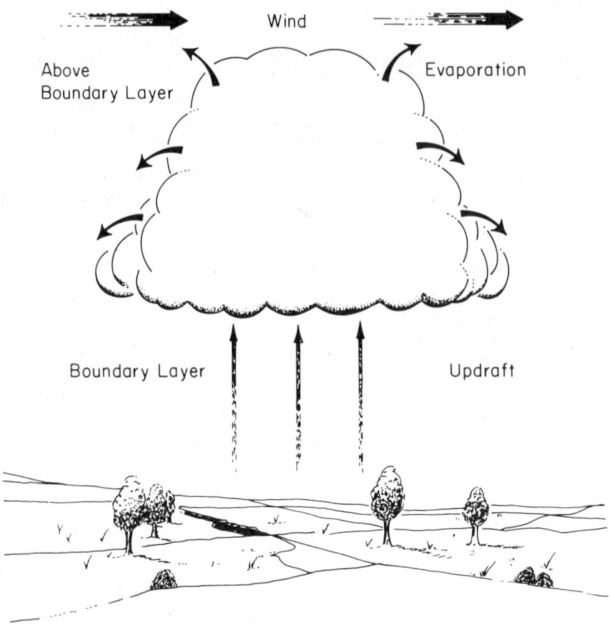

Figure 4. Schematic diagram of the important transport processes involving clouds.

difficult because the individual mechanisms (e.g., aerosol capture, liquid-phase oxidation, cloud droplet accretion) are complex and occur on spatial scales that are much smaller than the size of most grid systems. Further complications arise because the chemical composition of the hydrometeors determines the extent that the dispersed liquid phase can interact with the trace gases. It thus becomes necessary to model both the gas and liquid phases.

Simple chemical models combined with kinematic storm models are being developed (Scott, 1982) and are beginning to provide valuable information on cloud processes. These studies, along with studies of the wet removal of trace gases from the atmosphere (e.g., Barrie, 1978; Carmichael and Peters, 1979a; Reda and Carmichael, 1981) and the reactions occurring in solution (Schwartz and White, Chapter 1; Lee and Schwartz, 1981; Overton et al., 1979) will provide information necessary for development of meaningful parameterization of cloud processes.

4. CHEMICAL PROCESSES

The time scale of chemical processes that effect change on regional and global scales can be considerably longer than that required to effect change on the local scale. A notable example is CO, which can be considered inert for urban modeling. On the global scale, however, reaction with OH radicals leading to oxidation to CO_2 becomes important. This time scale for chemical change is an important consideration when formulating the loss and/or generation terms for inclusion in the partial differential equation describing the three-dimensional transport and chemistry of a particular species [Eq. (5)].

Identification of chemical time scales that are shorter than that of the transport processes can lead to substantial simplifications through judicious and careful application of the pseudo-steady-state approximation. In this section we discuss complex chemistry mechanisms that are normally nonlinear but that can be conveniently linearized on the time scale of the transport. This leads to more efficient solution methods for the partial-differential equations. Detailed treatment of representative homogeneous chemistry schemes is presented; additionally, introductory discussions of a method for including heterogeneous processes into these models are noted.

4.1. Homogeneous Chemistry

The CH_4 oxidation sequence is an important component in the global balances of CH_4 and CO, representing a major sink for CH_4 and a principal source of CO. The structure of this reaction scheme is illustrated in Figure 5. The oxidation of CH_4 by OH radicals leads to CH_3 radical formation followed by a free-radical chain reaction sequence, formation of CH_2O, and ultimately, formation of CO. The longest-lived species in this chain between CH_4 and CO is CH_2O, which has a lifetime during daylight on the order of

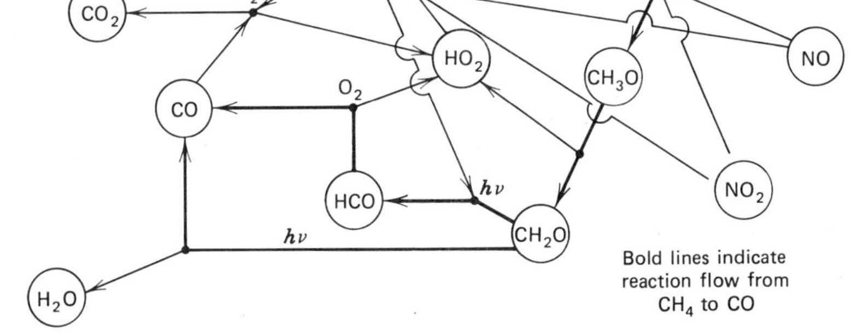

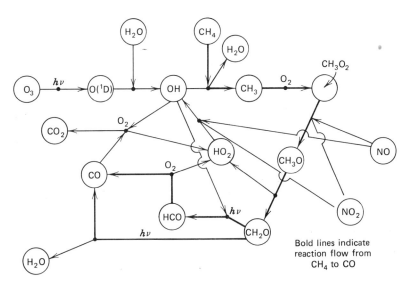

Figure 5. The principal pathways for oxidation of CH_4 to CO and CO_2 following the chemical reaction scheme of Kitada and Peters (1980).

minutes. At nighttime, there appear to be no homogeneous gas-phase reactions either producing or consuming CH_2O, so that chemical source and sink terms are nonexistent under these conditions. The lifetimes of the other intermediate species can be measured in fractions of a second.

The detailed reaction sequence is shown in Table 2. If the concentrations of NO, NO_2, O_3, H_2O, O_2, and M are presumed to be fixed, and if the steady-state approximation is used for CH_3, CH_3O_2, and CH_3O radicals, the reaction rates for CH_4, CO, and CH_2O can be written as follows:

$$R_{CH_4} = -k_{19}[CH_4][OH] \tag{19}$$

$$R_{CO} = (k_{23} + k_{24} + k_{25}[OH])[CH_2O] - k_{28}[CO][OH] \tag{20}$$

$$R_{CH_2O} = k_{19}[CH_4][OH] - (k_{23} + k_{24} + k_{25}[OH])[CH_2O]. \tag{21}$$

Here the rate constants refer to the numbered reactions in Table 2. Note that k_{23} and k_{24} are photolytic rate constants, which become zero at night. For an OH concentration that is established by means of photostationary state considerations, Eqs. (19)–(21) represent coupled but linear reaction generation terms. Recognition of this enables one to utilize linear equation solvers for the numerical integration. Before applying Eqs. (19)–(21), it must be determined (1) whether the photostationary state is valid for OH and (2) under what constraints the steady-state approximation can be extended to CH_2O.

We have performed a number of numerical tests in static simulations with a mechanism very similar to that in Table 2 to test the validity of the steady-state hypothesis for OH (and other radical species) (Peters and Jouvanis, 1979). The concentration of OH was found to satisfy two necessary prerequisites—it was quite low with respect to CO, CH_4, and CH_2O, and its response time to changes in these species concentrations was on the order of 10^{-6} s. Thus point 1 (in the preceding paragraph) apparently can be answered in the affirmative. Relative to point 2, it seems sufficient to conduct comparative simulations. In the first such simulation typical diurnally varying OH concentrations and photolytic rate constants were used to solve all three equations, whereas in the second simulation $[CH_2O]$ was maintained at steady state. Results for these simulations demonstrated that no significant differences in the time-concentration profiles could be detected. This simple calculation illustrated that even for a species such as CH_2O with relatively long residence time (on the order of minutes or so), the steady-state approximation can be exploited to linearize the chemistry provided abrupt changes in the concentrations of longer-lived species such as CH_4 and CO do not occur. This is generally true for long-lived species.

The foregoing example illustrates a chemistry scheme that can be linearized as a consequence of the various species possessing widely varying chemical time constants, with the longer-lived species showing relatively gradual changes in concentration. A second important case is that for SO_2, whose chemistry can be linearized because it has little influence on the

Table 2. Reaction Mechanism for CH_4–CO Chemistry[a]

Reaction	Rate Constant[b]
[1] $O_3 + h\nu \rightarrow O(^1D) + O_2$	$1.89 \times 10^{-4} \exp(-1.93/\cos\theta)$[c]
[2] $O(^1D) + H_2O \rightarrow 2OH$	2.3×10^{-10}
[3] $O(^1D) + M \rightarrow O + M$	$2.0 \times 10^{-11} \exp(107/T)$[d]
[4] $O + O_2 + M \rightarrow O_3 + M$	$1.1 \times 10^{-34} \exp(510/T)$
[5] $NO_2 + h\nu \rightarrow NO + O$	$1.55 \times 10^{-2} \exp(-0.48/\cos\theta)$
[6] $NO + O_3 \rightarrow NO_2 + O_2$	$9.0 \times 10^{-13} \exp(-1200/T)$
[7] $NO + HO_2 \rightarrow NO_2 + OH$	8.1×10^{-12}
[8] $NO + CH_3O_2 \rightarrow NO_2 + CH_3O$	$3.3 \times 10^{-12} \exp(-500/T)$
[9] $NO + OH \rightarrow HNO_2$	2.0×10^{-12}
[10] $NO + NO_2 + H_2O \rightarrow 2HNO_2$	6.0×10^{-38}
[11] $HNO_2 + h\nu \rightarrow NO + OH$	$2.75 \times 10^{-3} (\cos\theta)$
[12] $NO_2 + OH \rightarrow HNO_3$	2.0×10^{-12}
[13] $NO_2 + O_3 \rightarrow NO_3 + O_2$	$1.2 \times 10^{-13} \exp(-2450/T)$
[14] $NO_2 + NO_3 \rightarrow N_2O_5$	$1.48 \times 10^{-13} \exp(861/T)$
[15] $N_2O_5 \rightarrow NO_2 + NO_3$	$1.24 \times 10^{14} \exp(-10317/T)$
[16] $N_2O_5 + H_2O \rightarrow 2HNO_3$	1.0×10^{-20}
[17] $HNO_3 + h\nu \rightarrow OH + NO_2$	$9.88 \times 10^{-7} (\cos\theta)$
[18] $NO + NO_3 \rightarrow 2NO_2$	1.9×10^{-11}
[19] $CH_4 + OH \rightarrow CH_3 + H_2O$	$2.36 \times 10^{-12} \exp(-1710/T)$
[20] $CH_3 + O_2 \rightarrow CH_3O_2$	2.2×10^{-12}
[21] $2CH_3O_2 \rightarrow 2CH_3O + O_2$	1.6×10^{-13}
[22] $CH_3O + O_2 \rightarrow CH_2O + HO_2$	$1.6 \times 10^{-13} \exp(-3300/T)$
[23] $CH_2O + h\nu \rightarrow HCO + H$	$3.92 \times 10^{-5} \exp(-0.825/\cos\theta)$
[24] $CH_2O + h\nu \rightarrow H_2 + CO$	$1.23 \times 10^{-4} \exp(-0.652/\cos\theta)$
[25] $CH_2O + OH \rightarrow HCO + H_2O$	$3.0 \times 10^{-11} \exp(-250/T)$
[26] $HCO + O_2 \rightarrow HO_2 + CO$	5.7×10^{-12}
[27] $H + O_2 + M \rightarrow HO_2 + M$	$2.08 \times 10^{-32} \exp(290/T)$
[28] $CO + OH \rightarrow CO_2 + H$	$2.1 \times 10^{-13} \exp(-115/T) + 7.3 \times 10^{-33} n_m$[e]
[29] $HO_2 + OH \rightarrow H_2O + O_2$	3.0×10^{-11}
[30] $2HO_2 \rightarrow H_2O_2 + O_2$	2.5×10^{-12}
[31] $2OH + M \rightarrow H_2O_2 + M$	$1.25 \times 10^{-32} \exp(900/T)$
[32] $HO_2 + O_3 \rightarrow OH + 2O_2$	$1.4 \times 10^{-14} \exp(-580/T)$
[33] $H_2O_2 + h\nu \rightarrow 2OH$	$1.81 \times 10^{-5} \exp(-0.3297/\cos\theta)$
[34] $H_2O_2 + OH \rightarrow HO_2 + H_2O$	$1.0 \times 10^{-11} \exp(-750/T)$
[35] $HO_2 + NO_2 \rightarrow HO_2NO_2$	1.78×10^{-12}
[36] $HO_2NO_2 \rightarrow HO_2 + NO_2$	$1.26 \times 10^{16} \exp(-11700/T)$
[37] $HO_2 + O_3 \rightarrow OH + 2O_2$	$1.5 \times 10^{-12} \exp(-1000/T)$

[a] After Kitada and Peters (1980).

[b] Units are cubic centimeters, molecules, and seconds.

[c] θ represents the solar zenith angle.

[d] T represents the absolute temperature, K.

[e] n_m is the density of air expressed as molecules per cubic centimeter.

concentrations of the important radicals OH and HO_2 that oxidize SO_2 itself. Figure 6 is a schematic representation of this chemistry. Current knowledge indicates that the homogeneous gas-phase oxidation of SO_2 results principally from reaction with HO_2 and OH radicals by way of (Carmichael, 1979; Graedel et al., 1976)

$$SO_2 + OH \rightarrow HSO_3 \tag{38}$$

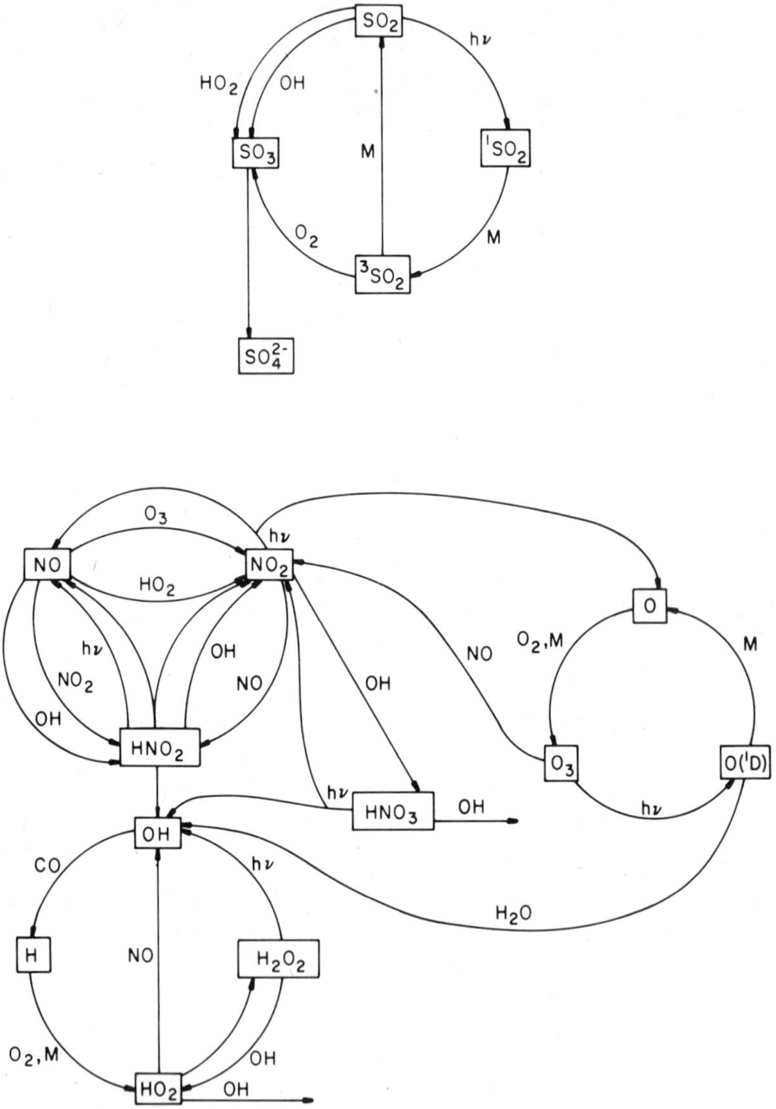

Figure 6. Schematic diagram of the homogeneous gas-phase chemistry of SO_2 and generation of OH and HO_2 radicals.

and

$$SO_2 + HO_2 \rightarrow SO_3 + OH. \qquad [39]$$

These reactions are obviously linear in the SO_2 concentration if the radical concentrations are essentially independent of SO_2. Such is, in fact, the situation since the OH and HO_2 concentrations depend primarily on HNO_x and H_xO_y pathways, reaction with CO, and formation by reaction of $O(^1D)$ with H_2O. Thus

$$\frac{d[SO_2]}{dt} = -(k_{38}\,[OH] + k_{39}\,[HO_2])\,[SO_2] \qquad (22)$$

This linearization scheme has been extensively tested against completely coupled gas-phase chemistry (Carmichael and Peters, 1979b, 1981a). The evaluation included three phases: (1) comparison of the SO_2 oxidation rate with data from Kocmond and Yang (1976); (2) comparison of the SO_2 oxidation rate with the data given by Bradstreet (1973); and (3) comparison of the predicted concentrations of several species with those predicted by detailed models. The important parameters all agreed quite closely. But equally important, the computer savings was significant. The simplified scheme ran in 2–3% of the time of the detailed mechanism when the computation was done on an IBM 370. Thus linearization is very important, and efforts toward that objective can be well worthwhile.

Not all atmospheric homogeneous chemistry mechanisms, of course, can be linearized in this manner to yield accurate results. Probably, the best-known case is that where reactive hydrocarbons are present in substantial quantities—the photochemical smog system. This system has been studied in nontransporting cases by numerous researchers, including Seinfeld and co-workers (Falls and Seinfeld; 1978; Falls et al., 1979; Hecht et al., 1974) in a Lagrangian framework (Bazzell and Peters, 1981; Rodhe et al., 1981), and in an Eulerian configuration (Reynolds et al., 1974).

A generalized linearization scheme for nonlinear generation terms uses a Taylor series expansion from one time step to the next time step at each grid point. Consider that the chemical reaction scheme has p total species, of which m are being advected. (It seems desirable, if not absolutely necessary, to advect as few species as possible in multidimensional grid formulations to conserve computing time.) The generation terms can then be written for each of the m advected species and $p - m$ algebraic relationships on the basis of steady-state or other considerations; i.e.,

$$\frac{\partial c_i}{\partial t} + \nabla \cdot (\bar{\mathbf{v}}\bar{c}_i) = \nabla \cdot \mathsf{K} \cdot \nabla\, \bar{c}_i + f_i\,(c_i, \ldots, c_p), \qquad i = 1, m, \qquad (23)$$

$$g_j(c_i, \ldots, c_p) = 0, \qquad j = 1, p - m. \qquad (24)$$

It is, of course, possible that $m = p$. The generation terms can be linearized by use of a Taylor series expansion from time t^n to t^{n+1}; i.e.,

$$f_i^{n+1} \approx f_i^n + \sum_{k=1}^{p} (c_k^{n+1} - c_k^n) \left.\frac{\partial f_i}{\partial c_k}\right|^n, \qquad i = 1, m, \tag{25}$$

and

$$g_j^{n+1} \approx g_j^n + \sum_{k=1}^{p} (c_k^{n+1} - c_k^n) \left.\frac{\partial g_j}{\partial c_k}\right|^n = 0, \qquad j = 1, p - m. \tag{26}$$

In principle, $p - m$ of the variables that are at steady state can be eliminated from the generation terms [Eq. (25)] by application of Eq. (26) for the $p - m$ species. It is generally desirable to perform this elimination by hand, as far as is practicable, to reduce the number of equations that must be solved simultaneously. One can note that the discretized form of Eq. (25) by use of a Crank–Nicholson scheme (i.e., $f_i = \frac{1}{2}(f_i^n + f_i^{n+1})$) also results in a linear system of equations.

4.2. Heterogeneous Aerosol Processes

There are two aspects of gas-particle interactions that merit attention in the context of the current discussions. First, these processes obviously affect the composition of the particulate phase. But equally important, the gas-phase composition can also be substantially affected, leading to an apparent irreversible loss of the gas-phase species.

For the present purposes, the aerosol particles are assumed to be spherical and to have the physicochemical properties of water. This does not seem too unrealistic since particles in the troposphere can quite regularly be covered with a layer of water. Obviously, some species would be more likely than others to interact with aerosol particles. There are several limitations relative to this representation of the heterogeneous processes. First, in-cloud and below-cloud scavenging by rain are not considered. Second, the model aerosol particle does not consider absorption by nonaqueous surfaces. Although that phenomenon could be included, knowledge of the fraction of particles with surfaces of that nature would be necessary. Finally, adsorption–desorption phenomena, which would lead to the appearance of reversible reaction phenomena, are not included. Thus processes such as reactant adsorption and product desorption or product incorporation into the particle are not accounted for. The end result of this last assumption is that the interaction is irreversible.

Many species present in the troposphere can potentially interact with the aerosol population. The extent to which this interaction takes place depends primarily on the solubility of the species in the particle and the concentration of the species surrounding the particle. The species concentration is important from two aspects. First, the heterogeneous removal rate varies directly as the species concentration. There is, however, a more subtle effect associated with the capacity of the particles to absorb or hold the species. For a species that is not very soluble in the particle, the particle phase can rapidly become saturated with respect to that species. As a result, the particle activity toward that species will be greatly reduced, and the particle would be active for only part of its residence time in the atmosphere.

Luther and Peters (1982) have described the heterogeneous removal rate of a gaseous species by an aerosol, using an effective first-order loss rate constant. Two mass conservation balances were used to develop the loss rate constant expression, which incorporates K_g, the overall mass-transfer coefficient based on the gas-phase concentration. A linear relationship for the species distribution between the gas phase and particle phase was also assumed,

$$[\mathrm{A}^*(D)] = \mathcal{H}\,[\mathrm{A}_a(D)], \tag{27}$$

where $[\mathrm{A}_a(D)]$ is the concentration of A in aerosol particles of diameter D and $[\mathrm{A}^*(D)]$ is the concentration of the gas-phase concentration of A if the gas and particle phases are in equilibrium.

To proceed further, specific assumptions on the particle age and particle size distribution functions must be made that describe the removal of species A from the gas phase by this mechanism. There is considerable latitude in these assumptions. Luther and Peters (1982) assumed that an individual particle's residence time in the atmosphere depends only on the average residence time for that size (Jaenicke, 1978) and that particles of that size are not present longer than τ_D. Their result was

$$\frac{d[\mathrm{A}]}{dt} = -\left\{\int_{D_0}^{D_\infty} K_g \pi D^2 n(D) \frac{D}{6\mathcal{H}K_g\tau_D} \times \left[1 - \exp\left(-\frac{6\mathcal{H}K_g\tau_D}{D}\right)\right] dD\right\} [\mathrm{A}] \tag{28}$$

The quantity within the braces can thus be identified as the heterogeneous loss constant; i.e.,

$$k = \int_{D_0}^{D_\infty} \frac{\pi D^3 n(D)}{6\mathcal{H}\tau_D}\left[1 - \exp\left(-\frac{6\mathcal{H}K_g\tau_D}{D}\right)\right] dD. \tag{29}$$

The variable $n(D)$ designates the number of particles of size D present per unit volume of air.

It is instructive to investigate the limiting cases for very insoluble gases ($\mathcal{H} \to \infty$), very soluble gases ($\mathcal{H} \to 0$), and very short and long particle residence times. These limits can be summarized as follows:

$$k = \int_{D_0}^{D_\infty} \pi D^2 n(D) k_g(D) dD, \qquad \mathcal{H} \to 0 \quad \text{or} \quad \tau_D \to 0$$

$$k = 0, \qquad \mathcal{H} \to \infty \quad \text{or} \quad \tau_D \to \infty.$$

Neither limit is surprising. The heterogeneous removal rate constant is controlled by the gas-phase exchange resistance for infinitely soluble gases or for particles with a very short residence time; the case of $\tau_D \rightarrow 0$ corresponds to an instantaneous renewal of particle surface. On the other hand, the heterogeneous loss constant approaches zero for the insoluble species or for very long residence time particles.

Equation (29) shows that information on the equilibrium distribution of the species between the gas and particle phases ($\mathcal{H}$) and the rate of transfer to the surface (K_g) must be known. These data are not always available. In addition, the residence times and size distribution of the particles must be known.

The overall mass-transfer coefficient used in Eq. (29) can be written for a linear equilibrium relationship as

$$K_g = \frac{1}{1/k_g + \mathcal{H}/k_l}. \tag{30}$$

It is conventional to divide the aerosol particles into size regimes according to the Knudsen number ($\mathrm{Kn} \equiv \lambda/D$), where λ is the gaseous mean free path. The first is the noncontinuum regime ($\mathrm{Kn} \gtrsim 10$), in which the rate of absorption of a species into the particle can be analyzed by use of gas–kinetic theory. The gas-phase mass-transfer coefficient for this regime can be written as

$$k_g^{(n)} = \frac{\alpha u}{4} \tag{31}$$

where α is an accommodation coefficient dependent on the fraction of collisions of molecules with the particles that are effective and u is the gas–kinetic velocity. The upper limit on α is unity; this can be used to determine the maximum effect that the heterogeneous processes might have. Recently, Baldwin and Golden (1980) suggested that heterogeneous radical reactions would need approximately unit collisional reaction probability to have any significance in tropospheric chemistry. They measured the collisional reaction probability with sulfuric acid for various species present in the troposphere and found relatively low values (e.g., $\alpha \sim 5 \times 10^{-4}$ for the OH radical). However, their results were probably affected strongly by the chemical nature of the two species involved.

The second regime for relatively large particles is the continuum regime ($\mathrm{Kn} \lesssim 0.1$), for which the rate of absorption of a species into the particle can be calculated by classic diffusional mass-transfer theory,

$$k_g^{(c)} = \frac{2\mathcal{D}_{\mathrm{A,air}}}{D}, \tag{32}$$

where $\mathcal{D}_{\mathrm{A,air}}$ is the diffusion coefficient of species A in air. In the transition regime ($0.1 \lesssim \mathrm{Kn} \lesssim 10$) the suggestion by Fuchs and Sutugin (1971) can be used to obtain

$$k_g^{(t)} = \left(\frac{1}{1 + l\mathrm{Kn}}\right) k_g^{(c)}, \tag{33}$$

where

$$l = \frac{4/3 + 0.71\ \mathrm{Kn}^{-1}}{1 + \mathrm{Kn}^{-1}}. \tag{34}$$

Equation (33) matches the noncontinuum and continuum regimes in a smooth manner. Finally, classic diffusion theory can be used for liquid-phase mass transfer, providing the following approximation for the liquid-phase mass-transfer coefficient,

$$k_l = \frac{2\mathcal{D}_{\mathrm{A,H_2O}}}{D}. \tag{35}$$

Several physicochemical properties and parameters are important to the overall analysis. The distribution of the gas species between the air and particle phases (described by $\mathcal{H}$) is crucial. Luther and Peters assumed that the distribution can be described by use of solubility data in water, but additional data are needed for better calculations. It also seems apparent that the entire particle population is not active toward the gas species. Current data, however, do not permit better estimates.

It can be observed that the quantity D/τ_D ranges from about 3×10^{-12} to 2×10^{-9} cm s^{-1} (Jaenicke, 1978). If $\mathcal{H}K_g$ is greater than this range, the

Table 3. Heterogeneous Loss Constants

Species	Henry's Law Constant $\left(\frac{\text{Mole Fraction}}{\text{Mole Fraction}}\right)$	Heterogeneous Loss Constant (s^{-1})[a]
NO	17.5	1.41×10^{-8}
CO	39.4	6.25×10^{-9}
O_3	3.7	6.70×10^{-8}
CH_4	33.2	7.42×10^{-9}
H_2O_2	—[b]	6.20×10^{-4}
CH_2O	—[b]	7.23×10^{-4}
HNO_2	—[b]	7.31×10^{-4}
HNO_3	—[b]	6.53×10^{-4}
NO_2	—[b]	7.56×10^{-4}

[a] Calculated according to Eq. (29) for an aerosol having a particle surface area of 5.2 μm^2 cm^{-3} and a particle residence-time distribution given by Jaenicke (1978).

[b] Assumed to be very soluble so that the heterogeneous loss constant is gas-phase controlled.

exponential term is unimportant in the evaluation of Eq. (29), and $\mathcal{H}$ alone becomes the critical unknown parameter. Finally, the accommodation coefficient α directly affects the gas-particle exchange rate, and only limited data are available for accurate assessment of its role.

Table 3 lists several heterogeneous loss constants calculated according to Eq. (29) with α taken to be unity. Many of these rate constants are quite large because of the values used for $\mathcal{H}$. Some indicate that the heterogeneous rates could dominate the known homogeneous processes. The most striking example is for HNO_3, where the homogeneous loss processes are on the order of 10^{-6}–10^{-7} s^{-1}, compared to 6.53×10^{-4} s^{-1} for heterogeneous processes. Other differences, although less substantial, are also noted for CH_2O and H_2O_2.

5. SUB-GRID-SCALE PROCESSES

When analyzing the physicochemical phenomena that establish the concentrations of trace species on a regional or global scale it is not practical to evaluate all space and time scales. For example, Eulerian analysis requires grid scales of $10 \sim 10^3$ km, depending on the species and overall scale of the process(es) being considered. As a consequence, careful consideration must be given to the sub-grid-scale phenomena occurring. After all, it is, to a large extent, these sub-grid-scale phenomena that are forcing the system. Specifically, problems of trace pollutant species tend to be source dominated, and individual sources occupy regions considerably smaller than a single grid. Additionally, the species are chemically reacting at concentrations substantially different from the grid- or volume-averaged concentration. Linear chemical processes are not affected by these considerations, but nonlinear chemistry can impose severe constraints and restrictions on results generated without such considerations. Although this has generally been recognized and some formulations have been suggested, there appears to have been no meaningful incorporation of such procedures into large-scale grid models.

Examples of two important sub-grid-scale processes are discussed in this section. These include the transport of reactive plumes on the subgrid scale and mixing and chemical reaccion.

5.1. Transport of Reactive Plumes from Urban Regions

5.1.1. General Considerations

Urban areas obviously represent regions of large anthropogenic pollutant production. Such areas occupy a small fraction of the total land area, but the major amount of anthropogenic sources is contained therein. These small geographic regions thus impact greatly on the background troposphere.

Background air has been variously defined as air that is as clean as one would have expected to find prior to the advent of technological civilization, the least modified air that presently exists globally, or air that is adjacent to but outside of a given area, e.g., rural versus urban (Stampfer and Anderson, 1975). Many studies have been undertaken to determine concentration levels of pollutants in clean air and have shown wide ranges of background levels for species such as NO, NO_2, and O_3, in part due to varying effects from anthropogenic sources.

In a previous paper Bazzell and Peters (1981) studied the phenomena occurring as a chemically reacting plume, with specific initial concentrations of NO, NO_2, O_3, and hydrocarbons, emerged from an urban area and was advected to the background troposphere. They used a 56-step, lumped kinetic mechanism for photochemical smog developed by Falls and Seinfeld (1978) and Falls et al. (1979). Fifteen time-dependent differential equations and 11 algebraic equations that describe the mechanism were solved simultaneously.

Since a plume emerging from an urban area is also being diluted while the chemical reactions are occurring, the kinetic equations were modified to simulate both phenomena. A dilution parameter was used to represent the phenomena occurring as the urban plume disperses in the atmosphere and as the individual chemical species in the plume are diluted. Let $f_i(k_1, k_2, \ldots, k_n, c_1, c_2, \ldots, c_m)$ represent the chemical kinetics of species i. The plume behaves as a reactor of continuously increasing volume V. Ignoring surface deposition and entrainment of background air, we obtain

$$\frac{d(Vc_i)}{dt} = Vf_i(k_1, k_2, \ldots, k_n, c_1, c_2, \ldots, c_m). \tag{36}$$

Differentiation of the left-hand side of Eq. (36) and rearrangement yields

$$\frac{dc_i}{dt} = f_i(k_1, k_2, \ldots, k_n, c_1, c_2, \ldots, c_m) - c_i \frac{d \ln V}{dt}. \tag{37}$$

The second term on the right-hand side represents dilution due to the spreading of the plume, which can be added to the kinetic rate expressions represented by the functions f_i.

If dispersion of the plume in the direction of advection is neglected, the cross section of the plume at any time can be related to the standard deviations of the dispersion. Thus if B is a proportionality constant, the area is $B\sigma_y\sigma_z$, where σ_z and σ_z are the horizontal and vertical standard deviations from the centerline of the plume. Bazzell and Peters used the Gifford–Pasquill atmospheric stability categories to estimate these standard deviations as functions of downwind distance (Gifford, 1961; Pasquill, 1962).

The concentration profiles for NO, NO_2, O_3, and hydrocarbons exhibited characteristics of actual urban plume activity. The NO, NO_2, and hydrocarbon levels decreased slowly with downwind distance, whereas the O_3 concentration increased to significant levels ($\sim$ 0.095 ppm) before peaking late in the afternoon. For a plume emergence time of 11:00 a.m. only 15–20% of the initial NO_x and 25–30% of the initial olefins remained in the plume at 10:00 p.m. The implications of that result is that the greatest portion of the species generated in an urban area could be transformed by the time the material has traveled a few hundred kilometers or less. The fraction transformed was found to increase as the HC:NO_x ratio increased.

Although diurnal variations were detected for the concentration of NO_x (= NO + NO_2), the magnitude of the peak on the second day was quite small in comparison to the primary peak; this result implies that relatively little NO_x was stored overnight as species that could reproduce NO and NO_2 by photolytic processes (e.g., HNO_2) on the following day. Atmospheric stability was found to be relatively unimportant in affecting the fractions of NO_x and hydrocarbons remaining in the plume, although the actual concentration profiles were quite different.

5.1.2. Effect of HC:NO_x Ratio

Simulations examining the effect of the initial HC:NO_x ratio on the rate of NO_x depletion were reported by Bazell and Peters (1981). For the case where the initial NO_x concentration was large compared to the initial reactive hydrocarbon concentration (HC:NO_x ratio = 7.2 $\mu g\ \mu g^{-1}$), the reactive hydrocarbon was expended before all of the NO was converted to NO_2. Significant O_3 was not formed, since an appreciable amount of NO remained after the hydrocarbons had been depleted and consumed the O_3. For a high HC:NO_x ratio (HC:NO_x = 28.8 $\mu g\ \mu g^{-1}$), NO was rapidly converted to NO_2 with little consumption of the hydrocarbons, and O_3 accumulated as the NO concentration decreased.

Spicer et al. (1979) reported that very high O_3 concentrations (~0.2 ppm) in plumes downwind of large northeastern U.S. cities can be attributed primarily to the character of the urban emissions. The study by Bazzell and Peters has shown similar results but indicates that O_3 levels downwind of cities such as Houston (HC:NO_x ~ 13–18 $\mu g\ \mu g^{-1}$) might not be as high.

Few data have been reported on the study of such phenomena. Spicer (1980) reported NO_x transformation rates in transported urban plumes. To study NO_x reactions in an urban plume, it is important that the plume be isolated and that subsequent emissions into the plume be negligible after the air mass leaves the city. Therefore, the studies by Spicer were performed in plumes transported over the desert (Phoenix, Arizona) and over the ocean (Boston, Massachusetts). Specific air parcels, e.g., the 8:00 a.m. air mass, were sampled just downwind of the city at midmorning and then periodically throughout the day.

Spicer reported NO_x transformation–removal rates for transported Boston plumes in the range of 0.14–0.24 h^{-1}. The results were plotted in the integrated form of a first-order rate equation, $\ln([NO_x]_e/[NO_x]_t)$ versus time, where $[NO_x]_e$ is the expected NO_x concentration calculated from the measured initial $[NO_x]$ and tracer species data if no consumption is occurring, and $[NO_x]_t$ is the concentration of NO_x actually measured at time t. The slope of the straight line represents the pseudo-first-order rate constant for the removal of NO_x from the plume.

Spicer reported that during the early stages of NO_x transformation, NO is converted to NO_2 with little net loss of NO_x. Since such behavior can cause curvature in the first-order plot, the initial plume transverse was always taken 1–2 h after the parcel left the urban area. Bazzell and Peters determined the pseudo-first-order rate constant for simulated results with a high initial HC:NO_x ratio (28.8 μg μg^{-1}), most representative of industrialized cities in the northeastern United States, to be 0.18 h^{-1}, which is in the range reported for Boston. The first-order behavior of this study is shown in Figure 7, although it can be seen that first-order kinetics do not precisely describe the Boston simulation.

Spicer reported NO_x transformation–removal rates for Phoenix of less than 0.05 h^{-1}. Simulated results with a lower HC:NO_x ratio (14.4 μg μg^{-1}) were used to calculate a pseudo-first-order rate constant for NO_x transformation–removal that might be representative of such U.S. cities. The rate constant was 0.046 h^{-1}, and this simulation is also shown in Figure 7. This study indicates that the low rate of NO_x depletion in Phoenix was due largely to the low HC:NO_x ratio.

5.1.3. Formation of Sulfuric and Nitric Acid

A similar approach to study the formation of sulfuric and nitric acid during long-range transport in the atmosphere was taken by Rodhe et al. (1981). They employed a simple photochemical model and compared the results

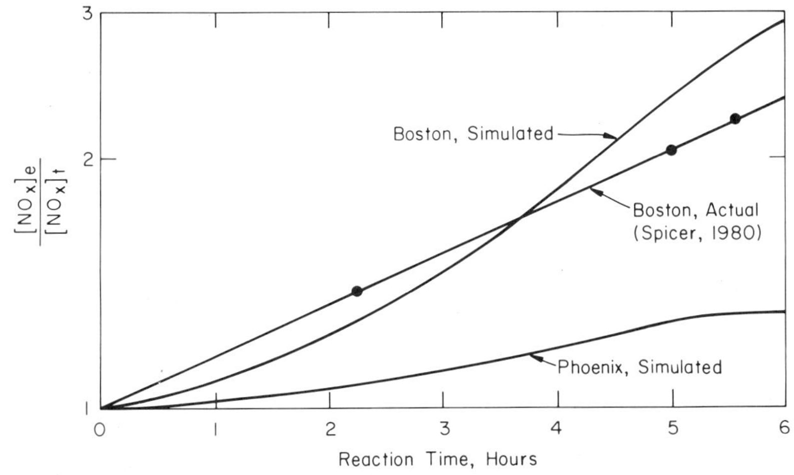

Figure 7. Semilogarithmic plot of simulation results for the transport of NO_x in urban plumes and comparison with the data of Spicer (1980) for the Boston and Phoenix plumes. After Bazzell and Peters (1981).

with observations of sulfate and nitrate in precipitation at various distances from source areas in northern Europe.

The observations and model simulations indicated that HNO_3 is formed more rapidly than H_2SO_4. To a large extent, this phenomenon is associated with the common dependence of the oxidation of both SO_2 and NO_x on the OH radical concentration. Since the rate of oxidation of NO_2 by OH is about an order of magnitude faster than that of SO_2, HNO_3 preferentially forms. As NO_x is consumed, SO_2 can compete more effectively for the OH. Thus H_2SO_4 formation can dominate only at longer distances downwind, giving rise to the observation of higher H_2SO_4 concentrations than HNO_3 concentrations at these greater distances.

5.2. Mixing and Chemical Reaction

On the subgrid scale, plumes from different sources can mix and react. It is obvious that large-scale transport–chemistry models cannot explicitly calculate all these interactions and must parameterize them. The rates at which chemical transformations occur in the atmosphere are dependent on complex interactions between turbulent mixing and the chemical kinetics. A useful parameter for describing the role of turbulence on chemical reaction rates is the ratio of the time scales of chemical kinetics τ_k and turbulent mixing τ_m.

Atmospheric chemical transformations can be classified into three regimes (Donaldson and Hilst, 1972): (1) the slow reaction ($\tau_k/\tau_m \gg 1$); (2) the very rapid reaction ($\tau_k/\tau_m \ll 1$); and (3) the moderate reaction rate ($\tau_k/\tau_m \sim 1$). In the first case atmospheric turbulence can induce chemical homogeneity before any significant reaction can occur, and the reaction rate can be described in terms of the mean concentrations. In the second case the reaction rate is controlled by the rate of mixing, but it also depends on the manner in which the reactants are introduced (e.g., premixed or segregated). In the third case ($\tau_k/\tau_m \sim 1$) a complex coupling between the reaction rate and the turbulence is expected.

There have been a number of methods developed to model turbulent mixing with chemical reactions. These methods vary considerably in conceptual approach, computational complexity, number of free parameters, and conditions of applicability. Advances have been made in this field during the past decade, but additional study can improve our understanding and add to the development of modeling techniques that will accurately describe the phenomena. As apparent as the need is for the further development of detailed models, there is also a distinct need for simple models.

Carmichael and Peters (1981b) have described a simple one-parameter model developed by Ghodsizadeh (1978) and applied it to nitric oxide oxidation in the near-source portion of the plume. The model, which is based on the notion of mixing and chemical reaction occurring as a series of kinetic steps, can handle a variety of chemical reaction types: irreversible or reversible; fast or slow; or multiple chemical reactions. This model has been compared to more complex models (e.g., Shu et al., 1978).

An atmospheric problem of considerable significance that may be limited by mixing is the injection of a plume containing NO_x into a background atmosphere containing O_3. This problem is used here to illustrate the procedures. The dominant chemical reactions in the plume after its release into the atmosphere are

$$NO + O_3 \longrightarrow NO_2 + O_2 \qquad [6]$$

and

$$NO_2 + h\nu \xrightarrow[O_2]{} NO + O_3. \qquad [5+4]$$

The mixing and reaction can be envisioned as two consecutive but separate steps; i.e.,

$$\text{Unmixed reactants} \xrightarrow{k_m} \text{mixed reactants} \underset{k_5, O_2}{\overset{k_6}{\rightleftharpoons}} \text{products}$$

$$(A = NO, B = O_3) \qquad (M = NO, N = O_3) \qquad (R = NO_2),$$

where M and N represent A and B, respectively, in the mixed state. In this example it is assumed that NO_2 is formed only by reaction [6] and not present initially. Under these conditions NO_2 exists only in the mixed state. The mixing step is viewed as a pseudokinetic step governed by the parameter k_m, which describes the mixing intensity, and the concentration of unmixed reactants. The chemical reaction step is determined by the rate constants k_5 and k_6 and the concentrations of mixed reactants.

The mixing step can be described by any one of a number of rate laws, depending on the system. One assumption is that the concentrations of A and B in the various plume regions are uniform (i.e., concentration gradients are not considered directly) and the rate of production of mixed-state reactants is proportional to the amount of reactants in the unmixed state. This, of course, is analogous to the mass-action law conventionally used in describing the kinetics of a chemical reaction process. [Ghodsizadeh (1978) has discussed other forms for the mixing expression.] On this basis, unmixed A can pass to the mixed state M by mixing either with unmixed B or some some already mixed volume; i.e.,

$$\frac{df_A}{dt} = -k_m f_A (1 - f_A) \qquad (38)$$

and

$$\frac{df_B}{dt} = -k_m f_B (1 - f_B), \qquad (39)$$

where f_A and f_B are the fractions of the volume that are unmixed A and B, respectively. The total volume consists of unmixed A, unmixed B, and mixed A and B; i.e.,

$$f_{\mathrm{A}} + f_{\mathrm{B}} + f_{\mathrm{M}} = 1, \tag{40}$$

where f_{M} is the fraction of the total volume that is mixed.

The rates of change of the concentrations of the species are (Carmichael and Peters, 1981b)

$$\frac{dc_{\mathrm{M}}}{dt} = -\frac{c_{\mathrm{A}_0}}{f_{\mathrm{A}_0} f_{\mathrm{M}}} \frac{df_{\mathrm{A}}}{dt} - k_6 c_{\mathrm{M}} c_{\mathrm{N}} + \frac{k_5 F_{\mathrm{A}} c_{\mathrm{A}_0}}{f_{\mathrm{M}}} - c_{\mathrm{M}} \frac{d(\ln f_{\mathrm{M}})}{dt}, \tag{41}$$

$$\frac{dc_{\mathrm{N}}}{dt} = -\frac{c_{\mathrm{B}_0}}{f_{\mathrm{B}_0} f_{\mathrm{M}}} \frac{df_{\mathrm{B}}}{dt} - k_6 c_{\mathrm{M}} c_{\mathrm{N}} + \frac{k_5 F_{\mathrm{A}} c_{\mathrm{A}_0}}{f_{\mathrm{M}}} - c_{\mathrm{N}} \frac{d(\ln f_{\mathrm{M}})}{dt}, \tag{42}$$

and

$$\frac{dF_{\mathrm{A}}}{dt} = \frac{f_{\mathrm{M}}}{c_{\mathrm{A}_0}} k_6 c_{\mathrm{M}} c_{\mathrm{N}} - k_5 F_{\mathrm{A}}, \tag{43}$$

where c_{A_0} and c_{B_0} are the initial concentrations of A and B averaged over the total volume; f_{A_0} and f_{B_0} are the initial unmixed fractions; c_{M}, c_{N}, and c_{R} are the concentrations of M, N, and R in the mixed volume; and F_{A} is the fractional conversion of A,

$$F_{\mathrm{A}} = \frac{c_{\mathrm{R}}}{c_{\mathrm{A}_0}} f_{\mathrm{M}}. \tag{44}$$

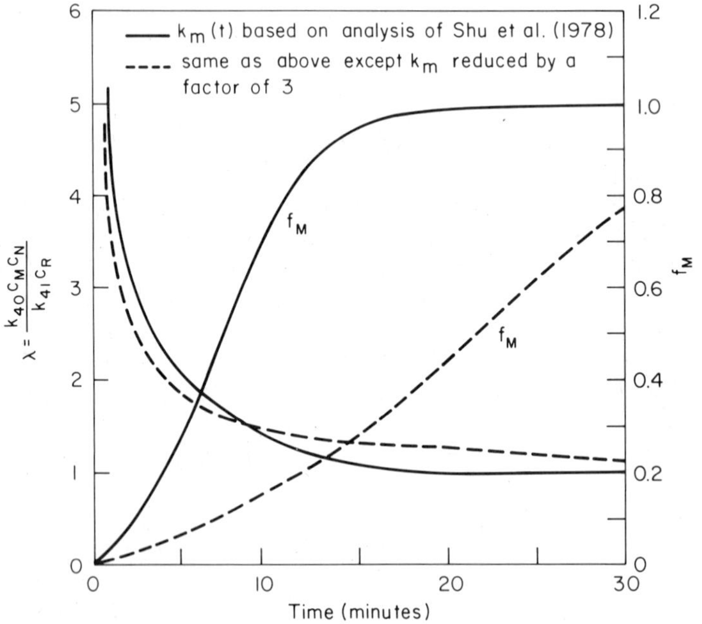

Figure 8. The approach to the photostationary state for an NO_x plume mixing with background O_3. After Carmichael and Peters (1981b).

The first terms on the right-hand side of Eqs. (41) and (42) represent the injection of material into the mixed volume by entrainment of unmixed A and B. The last terms on the right-hand side describe the change in mixed region concentration due to the growth of the mixed volume. The middle two terms represent the usual concentration changes due to chemical reactions occurring within the mixed volume.

In the absence of other pollutants such as hydrocarbons, the NO_x–O_3 system approaches the pseudosteady state, which requires that $\lambda \equiv (k_6\,[O_3]\,[NO])/(k_5\,[NO_2]) = 1$. Observed values of λ are generally greater than 1 (Calvert, 1976; Hegg et al., 1976; Kewley and Post, 1978), indicative of the mixing of segregated packets of O_3-rich and NO-rich air. The results of the "mixing-reaction in series" model support the view. Predicted values of λ versus time are shown in Figure 8. When f_M is small, λ is large, but $\lambda \rightarrow 1$ as $f_M \rightarrow 1$. In addition, λ decreases more rapidly as k_5 increases. These results are consistent with other studies of NO_x plumes (Hegg et al., 1976; White, 1977) and the modeling work of Bilger (1978).

These results, along with the simplicity and flexibility of the method, suggest that this approach may prove to be a useful tool for parameterizing sub-grid-scale phenomena.

6. CONCLUDING REMARKS

In this chapter we have discussed several important aspects of regional and global transport model development. The presentation has not included all the significant phenomena occurring but has tried to emphasize some of the processes that are common to many species.

Regional- and global-scale transport–chemistry models of trace species will be an important area of atmospheric sciences over the next decade as we try to increase our knowledge and understanding of the impact of anthropogenic sources of pollution on the environment. We conclude this presentation by citing several important problems associated with our understanding of regional and global processes by model analysis. This area of the atmospheric sciences is still quite new, and all the important problems cannot be cited or even envisioned at this time. Nevertheless it is hoped that these citations will illustrate the breadth of scientific disciplines needed to address these complex problems.

6.1. Mass-Conservative Wind Field

It is generally recognized that mass-conservative wind fields are required as input to Eulerian models to avoid introducing systematic numerical errors. Several methods have been suggested to generate mass-consistent wind fields, ranging from direct differencing procedures, to variational methods, to iterative interpolation methods (Goodin et al., 1980; Liu and Goodin, 1976; Sherman, 1978). These schemes all recognize that there are observation errors in the horizontal winds and attempt to adjust the wind field to generate mass conservation while maintaining the general sense of the large-scale flow field. Unfortunately, this procedure generally results in a vertical velocity that may be more representative of errors in the observed horizontal winds than of the actual vertical motion.

Some success has been achieved in developing procedures whereby the divergence at each grid point is reduced. On the regional and global scales, the simulation time period of interest is quite long (days to weeks to months), and the residual divergence in the flow must be correspondingly decreased to minimize inaccuracies from this source. For example, if one is content with a 10% error introduced by this process, residual divergences at each grid point must be less than 10^{-6} s^{-1} for a 1-day simulation, 10^{-7} s^{-1} for a 10-day simultation, and 10^{-8} s^{-1} for a 100-day simulation. This places great strains on the wind-adjustment scheme, especially when it should also not significantly distort the flow patterns. Current schemes probably reduce the residual divergence to a level appropriate for the 5–10-day simulation time.

6.2. Sub-Grid-Scale Phenomena

As previously noted, sub-grid-scale processes are important in driving regional- and global-scale transport and chemical processes. This area has received relatively little attention, although one can readily identify a number of important problem areas. The discussions in Section 5 introduced two areas that deserve attention. Since considerable chemical conversion may occur on a time scale typical of the transport over one grid length, the effective source strength that should be used in the Eulerian analysis can be significantly reduced. Second, large-scale plumes that originate from different emission sources can interact on the subgrid scale and substantially alter the effective source distribution. Finally, many sources of anthropogenic pollution are emitted from so-called point sources. It has been conventional to sum these point source strengths over the grid volume and replace the sum by the volume-averaged value. Although this is convenient, it introduces errors into the analysis, and there appears to have been little consideration given to this source of error. In our opinion, these three problem areas merit careful analysis.

6.3. Comparison of Model Results with Field Experiments

It is ultimately desirable to extend the results of field experiments to other atmospheric conditions through modeling analysis. Such extrapolations require carefully verified models. At a minimum, the verification procedures must consider that the field data are generally quite limited and the model input data are usually incomplete. These considerations imply that point-by-point comparisons and their associated statistics, although generally desirable, are much too restrictive.

Associated with verification is the general area of model sensitivity analysis. Regional and global Eulerian models require numerous input parameters for estimation of the output concentration fields, and the variation of output with variation of the input parameters (commonly referred to as *sensitivity coefficients*) can elucidate the crucial variables in the analysis. The transport–chemistry models are complex, computationally expensive, and time consuming to run. As a result, efficient sensitivity analysis schemes are needed as part of the model verification process. In addition, efficient and novel sensitivity analysis schemes may prove useful in (1) analyzing the models to determine the dominant physical and chemical processes affecting the output variables and (2) establishing acceptable bounds of the input variables (e.g., emissions source strengths) when constrained by the field observations.

6.4. Space and Time Interpolation

The observed input data are very greatly limited in both space and time. Generally, polynomial weighting functions have been used to fill in the grid points, intermediate in both space and time to the field observation data. Although these procedures are simple to use and frequently yield smooth input data fields, the physical basis for such approximations is very limited. This is especially true for observation densities (in both space and time) that are less than those of the important physical and meteorological phenomena occurring.

There is a critical need for space and time interpolation procedures that include consideration of the physical phenomena. Whereas the mesoscale and general circulation models are an important step in that direction, their overall complexity indicates that simpler schemes may be more appropriate for routine use. Use of the mesoscale and general circulation models is probably restricted to very specific situations where relatively high density data networks are available, e.g., during comprehensive field studies.

ACKNOWLEDGMENTS

The authors gratefully acknowledge assistance from the National Aeronautics and Space Administration and from the MAP3S/RAINE Program for support of these research efforts.

NOMENCLATURE

All the equations presented are dimensionally consistent. The units included with each variable are only typical.

$[A_a(D)]$ Concentration of species A in aerosol particles of diameter D (μg m^{-3}) (of solution)

$[A^*(D)]$ Concentration of species A in the gas phase if the gas and particles are in equilibrium (μg m^{-3})

B Constant in Eq. (18), dimensionless; or proportionality constant for calculating plume cross-section, dimensionless

c Concentration of species (μg m^{-3})

c_g Concentration of species in gas phase (μg m^{-3})

c_g^* Concentration of species in gas phase if the gas and bulk phase are in equilibrium (μg m^{-3})

c_ℓ Concentration of species in the bulk absorbing liquid phase (μg m^{-3})

D Particle diameter (μm)

$\mathcal{D}_{A,B}$ Diffusivity of species A in B (m^2 s^{-1})

E_j Collection efficiency of particles at the surface, dimensionless

f_A, f_B Fractions of the total volume that are unmixed species A or B, dimensionless

f_i Generation term by reaction of species i that is being advected (μg m^{-3} s^{-1})

f_M Fraction of the total volume that is mixed, dimensionless

$\mathbf{F}$ Flux vector (μg m^{-2} s^{-1})

F_A Fractional conversion of species A defined by Equation (44), dimensionless

F_z Flux in the vertical direction (μg m^{-2} s^{-1})

g_j Algebraic expression describing the generation of species j that is at steady state, dimensionless

H_I Height of surface-based nocturnal inversion (m)

H_M Height of mixing layer (m)

$\mathcal{H}$ Henry's law constant, dimensionless

k Reaction rate constant for first-order reaction (s^{-1}), for second-order reaction (cm^3 molecule^{-1} s^{-1}), for third-order reaction (cm^6 molecule^{-2} s^{-1})

k_g Gas-phase mass-transfer coefficient (m s^{-1})

k_l Liquid-phase mass-transfer coefficient (m s^{-1})

k_m Mixing-intensity parameter (s^{-1})

K Eddy-diffusivity tensor (m^2 s^{-1})

K_g Overall gas-phase mass-transfer coefficient (m s^{-1})

K_z Eddy diffusivity in vertical direction (m^2 s^{-1})

Kn Knudsen number (λ/D), dimensionless

l Parameter defined by Eq. (34), dimensionless

L Monin–Obukhov length scale (m)

n_m	Density of air (molecules cm^{-3})
$[NO_x]_e$	Expected NO_x concentration if reaction is not occurring ($\mu g\ m^{-3}$)
$[NO_x]_t$	NO_x concentration at time t ($\mu g\ m^{-3}$)
$P(\ )$	Probability density function (m^{-1})
r_a	Aerodynamic resistance ($s\ m^{-1}$)
r_s	Surface resistance ($s\ m^{-1}$)
Q	Surface emissions source strength ($\mu g\ m^{-2}\ s^{-1}$)
R	Generation term accounting for transformation by chemical reaction ($\mu g\ m^{-3}\ s^{-1}$)
S	Source distribution function ($\mu g\ m^{-3}\ s^{-1}$)
t	Time (s)
T	Temperature (K)
u	Gas kinetic velocity ($m\ s^{-1}$)
u_z	Horizontal wind velocity of height z ($m\ s^{-1}$)
u_*	Friction velocity ($m\ s^{-1}$)
$\mathbf{v}$	Velocity vector ($m\ s^{-1}$)
v_d	Deposition velocity ($m\ s^{-1}$)
v_s	Particle settling velocity ($m\ s^{-1}$)
V	Plume volume (m^3)
$\mathbf{x}$	Position vector (m)
x	Horizontal coordinate (m)
y	Horizontal coordinate (m)
z	Vertical coordinate (m)

Greek Letters

α	Constant in Eq. (18), dimensionless
$\delta(D)$	Resuspension parameter dependent on particle diameter ($m\ s^{-1}$)
ϵ	Canopy filtration efficiency, dimensionless
λ	Gas mean free path (m); or parameter describing photostationary state of NO, NO_2, and O_3, dimensionless
θ	Solar zenith angle (°)
τ_D	Residence time in the atmosphere of particles of diameter D (s)
τ_k	Time scale of chemical reaction (s)
τ_m	Time scale of turbulent mixing (s)
σ_y, σ_z	Cross-wind and vertical standard deviations describing the plume cross section (m)

Subscripts

b	Bulk phase; or background
i	Species i; or interface

j	Species j
0	Initial

Superscripts

$\bar{\xi}$	Time-averaged value of the variable ξ
ξ'	Fluctuating component of the variable ξ
n	Time level

REFERENCES

Baldwin, A. C. and Golden, D. M. (1980). Heterogeneous atmospheric reactions: Atom and radical reactions with sulfuric acid. *J. Geophys. Res.* **85,** 2888–2889.

Barrie, L. A. (1978). An improved model of reversible SO_2 washout by rain. *Atmosph. Environ.* **12,** 407–412.

Bazzell, C. C. and Peters, L. K. (1981). The transport of photochemical pollutants to the background troposphere. *Atmosph. Environ.* **15,** 957–968.

Bilger, R. W. (1978). The effect of admixing fresh emissions on the photostationary state relationship in photochemical smog. *Atmosph. Environ.* **12,** 1109–1118.

Bird, R. B., Stewart, W. E., and Lightfoot, E. N. (1960). *Transport Phenomena*, Wiley, New York.

Bradstreet, J. W. (1973). Effects of nitric oxide on the photochemical oxidation of sulphur dioxide in dilute gas–air mixtures. 66th Annual Meeting of the Air Pollution Control Association, Chicago, Illinois.

Calvert, J. G. (1976). Test of the theory of ozone generation in Los Angeles atmosphere. *Environ. Sci. Technol.* **10,** 248–256.

Carmichael, G. R. (1979). Development of a regional transport/transformation/removal model for SO_2 and sulfate in the Eastern United States. Ph.D. dissertation, Department of Chemical Engineering, University of Kentucky.

Carmichael, G. R. and Peters, L. K. (1979a). Some aspects of SO_2 absorption by water—generalized treatment. *Atmosph. Environ.* **13,** 1505–1513.

Carmichael, G. R. and Peters, L. K. (1979b). Numerical simulation of the regional transport of SO_2 and sulfate in the Eastern United States. *AMS Fourth Symposium on Turbulence, Diffusion, and Air Pollution: Preprints*, January, Reno, NV, pp. 337–343.

Carmichael, G. R. and Peters, L. K. (1980). The transport, chemical transformation, and removal of SO_2 and sulfate in the Eastern United States, in *Atmospheric Pollution 1980, Studies in Environmental Science*, Vol. 8, M. M. Benarie, Ed., Elsevier, Amsterdam, pp. 31–36.

Carmichael, G. R. and Peters, L. K. (1981a). Regional transport and deposition of SO_x in the Eastern United States. Presented at AIChE Meeting, November, New Orleans, LA.

Carmichael, G. R. and Peters, L. K. (1981b). Application of the mixing-reaction in series model to NO_x–O_3 plume chemistry. *Atmosph. Environ.* **15,** 1069–1074.

Carmichael, G. R., Kitada, T., and Peters, L. K. (1980a). Application of a Galerkin finite element method to atmospheric transport problems. *Comput. Fluids* **8,** 155–176.

Carmichael, G. R., Yang, D.-Y., and Lin, C. (1980b). A numerical technique for the investigation of the transport and dry deposition of chemically reacting pollutants. *Atmosph. Environ.* **14,** 1433–1438.

Carson, D. J. (1973). The development of a dry inversion capped convectively unstable boundary layer. *Quart. J. R. Meteorol. Soc.* **99,** 450–467.

Chameides, W. L., Liu, S. C., and Cicerone, R. J. (1977). Possible variations in atmospheric methane. *J. Geophys. Res.* **82,** 1795–1798.

Crowley, W. P. (1968). Numerical advection experiments. *Mon. Weather Rev.* **96,** 1–11.

Crutzen, P. J. and Fishman, J. (1977). Average concentrations of OH in the Northern Hemisphere troposphere and the budgets of CH_4, CO, and H_2. *Geophys. Res. Lett.* **4,** 321–324.

Crutzen, P. J., Heidt, L. C., Krasnec, J. D., Pollock, W. H., and Seiler, W. (1979). Biomass burning as a source of the atmospheric gases CO, H, N_2O, NO, CH_3Cl, and COS. *Nature* **282,** 253–256.

Csanady, G. T. (1973). *Turbulent Diffusion in the Environment*, Reidel, Dordrecht, The Netherlands.

Danielson, E. F. (1968). Stratospheric–tropospheric exchange based on radioactivity, ozone, and potential vorticity. *J. Atmosph. Sci.* **25,** 502–518.

Donaldson, C. D. and Hilst, G. R. (1972). Effect of inhomogeneous mixing on atmospheric photochemical reactions. *Environ. Sci. Technol.* **6,** 812–816.

Donaldson, C. D., Sullivan, R. D., and Rosenbaum, H. (1972). A theoretical study of the generation of atmospheric–clear air turbulence. *AIAA J.* **10,** 162–170.

Falconer, P. D., Pratt, R., and Mohnen, V. A. (1978). The transport cycle of atmospheric ozone and its measurement from aircraft and at the earth's surface, in *Man's Impact on the Troposphere (Lectures in Tropospheric Chemistry)*, J. S. Levine and D. R. Schryer, Eds., NASA Reference Publication 1022, pp. 109–147.

Falls, A. H. and Seinfeld, J. H. (1978). Continued development of a kinetic mechanism for photochemical smog. *Environ. Sci. Technol.* **12,** 1398–1406.

Falls, A. H., McRae, G. J., and Seinfeld, J. H. (1979). Sensitivity and uncertainty of reaction mechanisms for photochemical air pollution. *Internatl. J. Chem. Kinet.* **11,** 1137–1162.

Fishman, J., Soloman, S., and Crutzen, P. J. (1979). Observational and theoretical evidence in support of a significant *in situ* photochemical source of tropospheric ozone. *Tellus* **31,** 432–446.

Fuchs, N. A. and Sutugin, A. G. (1971). Highly dispersed aerosols, in *Topics in Current Aerosol Research*, G. M. Hidy and J. R. Brock, Eds. Pergamon Press, New York.

Ghodsizadeh, Y. (1978). Simultaneous mixing and chemical reaction. Ph.D. dissertation, Department of Chemical Engineering, Case Western Reserve University, Cleveland, OH.

Gifford, F. A. (1961). Uses of routine meteorological observations for estimating atmospheric dispersion. *Nucl. Safety* **2,** 47–51.

Goodin, W. R., McRae, G. J., and Seinfeld, J. H. (1980). An objective analysis technique for constructing three-dimensional urban-scale wind fields. *J. Appl. Meteorol.* **19,** 98–108.

Graedel, T. E., Farrow, L. A., and Weber, T. A. (1976). Kinetic studies of the photochemistry of the urban troposphere. *Atmosph. Environ.* **10,** 1095–1116.

Hecht, T. A., Seinfeld, J. H., and Dodge, M. C. (1974). Further development of generalized mechanism for photochemical smog. *Environ. Sci. Technol.* **8,** 327–339.

Hegg, D. A. and Hobbs, P. V. (1981). Cloud water chemistry and production of sulfates in clouds. *Atmosph. Environ.* **15,** 1597–1604.

Hegg, D. A., Hobbs, P. V., and Radke, L. F. (1976). Reactions of nitrogen oxides, ozone, and sulfur in power plant plumes. Electric Power Research Institute Report EA-270, 126 pp.

Hicks, B. B. and Wesely, M. L. (1978). An examination of some micrometeorological methods for measuring dry deposition. U.S. EPA Report EPA-600/7-78-116, July.

Holzworth, G. C. (1964). Estimates of mean maximum mixing depths in the contiguous United States. *Monthly Weather Rev.* **92,** 235–242.

Jaenicke, R. (1978). Über die dynamik atmosphärischer aitkenteilchen. *Ber. Bunsenges. Phys. Chem.* **82,** 1198–1202.

Johnson, W. B., Viezee, W., Cavanagh, L. A., Ludwig, F. L., Singh, H. B., and Danielson, E. F. (1979). Measurements of stratospheric ozone penetrations into the lower troposphere. *AMS Fourth Symposium on Turbulence, Diffusion and Air Pollution: Preprints*, January, Reno, NV, pp. 355–362.

Kewley, D. J. (1978). Atmospheric dispersion of a chemically reacting plume. *Atmosph. Environ.* **12,** 1895–1900.

Kewley, D. J. and Post, K. (1978). Photochemical ozone formation in the Sydney airshed. *Atmosph. Environ.* **12,** 2179–2184.

Kitada, T. and Peters, L. K. (1980). A model of CO–CH_4 global transport/chemistry: I. Chemistry model. *J. Japan. Soc. Air Pollut.* **15,** 91–108.

Kitada, T. and Peters, L. K. (1982). A three-dimensional transport–chemistry analysis of CO and CH_4 in the troposphere. *Second Symposium on the Composition of the Non-Urban Troposphere*, May, Williamsburg, VA, pp. 96–101.

Kocmond, W. C. and Yang, J. Y. (1976). Sulfur dioxide photooxidation rates and aerosol formation mechanisms: A smog chamber study. Calspan Corporation, U.S. EPA Report EPA-600/3-76-090; National Technical Information Service PB-260910.

Lamb, R. G., Hogo, H., and Reid, L. E. (1979). A Lagrangian Monte Carlo model of air pollution transport, diffusion, and removal processes. *AMS Fourth Symposium on Turbulence, Diffusion, and Air Pollution: Preprints*, January, Reno, NV, pp. 381–388.

Lee, Y.-N. and Schwartz, S. E. (1981). Evaluation of the rate of uptake of nitrogen dioxide by atmospheric and surface liquid water. *J. Geophys. Res.* **86,** 11971–11983.

Lewellen, W. S. and Teske, M. (1975). Turbulence modeling and its application to atmospheric diffusion. Part I: Recent program development, verification, and application. U.S. EPA Report EPA-600/4-75-0169; National Technical Information Service PB-253450.

Lewellen, W. S. and Teske, M. (1976). Second-order closure modeling of diffusion in the atmospheric boundary layer. *Boundary-Layer Meteorol.* **10,** 69–90.

Liu, C. Y. and Goodin, W. R. (1976). An iterative algorithm for objective wind field analysis. *Monthly Weather Rev.,* **104,** 784–792.

Luther, C. J. and Peters, L. K. (1982). The possible role of heterogeneous aerosol processes in the chemistry of CH_4 and CO in the troposphere, in *Heterogeneous Atmospheric Chemistry*, D. R. Schryer, Ed., Geophysical Monograph 26, American Geophysical Union, Washington, pp. 264–273.

Mahlman, J. D., Levy, H., and Moxim, W. J. (1980). Three-dimensional tracer structure and behavior as simulated in two ozone precursor experiments. *J. Atmosph. Sci.* **37,** 655–685.

Molenkamp, C. R. (1968). Accuracy of finite-difference methods applied to the advection equation. *J. Appl. Meteorol.* **7,** 160–167.

Monin, A. S. and Yaglom, A. M. (1971). *Statistical Fluid Mechanics*, MIT Press, Cambridge, MA.

Myrup, L. O. and Ranzieri, A. J. (1976). A consistent scheme for estimating diffusivities to be used in air quality models. National Technical Information Service. PB-272 484.

Overton, J. H., Jr., Aneja, V. P., and Durham, J. L. (1979). Production of sulfate in rain and raindrops in polluted atmospheres. *Atmosph. Environ.* **13,** 355–367.

Owen, P. R. and Thomson, W. R. (1963). Heat transfer across rough surfaces. *J. Fluid Mech.* **15,** 321–334.

Pasquill, F. (1962). *Atmospheric Diffusion*, Van Nostrand, London, pp. 366–380.

Pepper, D. W., Cooper, K. E., and Baker, A. J. (1980). An investigation of multidimensional computational models for calculating pollutant transport, in *Developments in Theoretical and Applied Mechanics*, Vol. 10, J. E. Stoneking, Ed., University of Tennessee, pp. 397–412.

Peters, L. K. and Jouvanis, A. A. (1979). Numerical simulation of the transport and chemistry of CH_4 and CO in the troposphere. *Atmosph. Environ.* **13,** 1443–1462.

Reda, M. and Carmichael, G. R. (1981). Non-isothermal effects on SO_2 absorption by water droplets—I. Model development. *Atmosph. Environ.* **15,** 145–150.

Reynolds, S. D., Liu, M. K., Hecht, T. A., Roth, P. M., and Seinfeld, J. H. (1974). Modeling of photochemical air pollution—III. Evaluation of the model. *Atmosph. Environ.* **8,** 563–596.

Rodhe, H. and Isaksen, I. (1980). Global distribution of sulfur compounds in the troposphere estimated in a height/latitude transport model. *J. Geophys. Res.* **85,** 7401–7409.

Rodhe, H., Crutzen, P., and Vanderpol, A. (1981). Formation of sulfuric and nitric acid in the atmosphere during long-range transport. *Tellus* **33,** 132–141.

Scott, B. C. (1982). Predictions on in-cloud conversion rates of SO_2 to SO_4 based upon a simple chemical and kinematic storm model. *Atmosph. Environ.* **16,** 1735–1752.

Seinfeld, J. H. (1975). *Air Pollution: Physical and Chemical Fundamentals*, McGraw-Hill, New York, p. 268.

Shannon, J. V. (1981). A model of regional long-term average sulfur atmospheric pollution, surface removal, and net horizontal flux. *Atmosph. Environ.* **15,** 689–702.

Sheih, C. M., Wesely, M. L., and Hicks, B. B. (1979). Estimated dry deposition velocities over the eastern United States and surrounding dry regions. *Atmosph. Environ.* **13,** 1361–1368.

Sherman, C. A. (1978). A mass-consistent model for wind fields over complex terrain. *J. Appl. Meteorol.* **17,** 312–319.

Shu, W. R., Lamb, R. G., and Seinfeld, J. H. (1978). A model of second-order chemical reactions in turbulent fluid—Part II. Application to atmospheric plumes. *Atmosph. Environ.* **12,** 1695–1704.

Skelland, A. H. P. (1974). *Diffusional Mass Transfer*, Wiley, New York, p. 276.

Slinn, W. G. N. (1976). Dry deposition and resuspension of aerosol particles—a new look at some old problems. Presented at the Atmospheric-Surface Exchange of Particulate and Gaseous Pollutants (1974), ERDA Symposium Series Conf-740921, pp. 1–40.

Spicer, C. W. (1980). The rate of NO_x reaction in transported urban air, in *Atmospheric Pollution 1980, Studies in Environmental Science*, Vol. 8, M. M. Benarie, Ed., Elsevier, Amsterdam, pp. 181–186.

Spicer, C. W., Joseph, D. S., Sticksel, P. R., and Ward, G. F. (1979). Ozone sources and transport in the northeastern United States. *Environ. Sci. Technol.* **13,** 975–985.

Stampfer, J. F. and Anderson, J. A. (1975). Locating the St. Louis urban plume at 80 and 120 km and some of its characteristics. *Atmosph. Environ.* **9,** 301–313.

Tarbell, T. C., Warner, T. T., and Anthes, R. A. (1981). An example of the initialization of the divergent wind component in a mesoscale numerical weather prediction model. *Mon. Weather Rev.* **109,** 77–95.

Tennekes, H. and Lumley, J. L. (1972). *A First Course in Turbulence*, MIT Press, Cambridge, MA.

Wesely, M. L., Hicks, B. B., Dannevik, W. P., Frisella, S., and Husar, R. B. (1977). An eddy-correlation measurement of particulate deposition from the atmosphere. *Atmosph. Environ.* **11,** 561–563.

White, W. H. (1977). NO_x–O_3 photochemistry in power plant plumes: Comparison of theory with observation. *Environ. Sci. Technol.* **11,** 995–1000.

Wofsy, S. C., McConnell, J. C., and McElroy, M. B. (1972). Atmospheric CH_4, CO, and CO_2. *J. Geophys. Res.* **77,** 4477–4492.

Wolff, G. T., Monson, P. R., and Ferman, M. A. (1979). On the nature of the diurnal variation of sulfates at rural sites in the eastern United States. *Environ. Sci. Technol.* **13,** 1271–1276.

Zimmerman, P. R., Chatfield, R. B., Fishman, J., Crutzen, P. J., and Hanst, P. L. (1978). Estimates on the production of CO and H_2 from the oxidation of hydrocarbon emissions from vegetation. *Geophys. Res. Lett.* **5,** 679–682.

INDEX

* The index was prepared by Eloise Gmur.